Springer Series in Optical Sciences Volume 15
Edited by David L. MacAdam

Springer Series in Optical Sciences

Edited by David L. MacAdam

Volume 1 **Solid-State Laser Engineering**
By W. Koechner

Volume 2 **Table of Laser Lines in Gases and Vapors** 2nd Edition
By R. Beck, W. Englisch, and K. Gürs

Volume 3 **Tunable Lasers and Applications**
Editors: A. Mooradian, T. Jaeger, and P. Stokseth

Volume 4 **Nonlinear Laser Spectroscopy**
By V. S. Letokhov and V. P. Chebotayev

Volume 5 **Optics and Lasers** An Engineering Physics Approach
By M. Young

Volume 6 **Photoelectron Statistics** With Applications to Spectroscopy and Optical Communication By B. Saleh

Volume 7 **Laser Spectroscopy III**
Editors: J. L. Hall and J. L. Carlsten

Volume 8 **Frontiers in Visual Science**
Editors: S. J. Cool and E. L. Smith III

Volume 9 **High-Power Lasers and Applications**
Editors: K.-L. Kompa and H. Walther

Volume 10 **Detection of Optical and Infrared Radiation**
By R. H. Kingston

Volume 11 **Matrix Theory of Photoelasticity**
By P. S. Theocaris and E. E. Gdoutos

Volume 12 **The Monte Carlo Method in Atmospheric Optics**
By G. I. Marchuk, G. A. Mikhailov, M. A. Nazaraliev, R. A. Darbinjan, B. A. Kargin, and B. S. Elepov

Volume 13 **Physiological Optics**
By Y. Le Grand and S. G. El Hage

Volume 14 **Laser Crystals** Physics and Properties
By A. A. Kaminskii

Volume 15 **X-Ray Spectroscopy**
By B. K. Agarwal

Volume 16 **Holographic Interferometry** From the Scope of Deformation Analysis of Opaque Bodies By W. Schumann and M. Dubas

Volume 17 **Nonlinear Optics of Free Atoms and Molecules**
By D. C. Hanna, M. A. Yuratich, D. Cotter

Volume 18 **Holography in Biology and Medicine**
By G. von Bally

B. K. Agarwal

X-Ray Spectroscopy

An Introduction

With 188 Figures

Springer-Verlag Berlin Heidelberg New York 1979

Professor BIPIN K. AGARWAL, Ph. D.

Department of Physics, X-Ray Laboratory, University of Allahabad
Allahabad 211002, India

ISBN 3-540-09268-4 Springer-Verlag Berlin Heidelberg New York
ISBN 0-387-09268-4 Springer-Verlag New York Heidelberg Berlin

Library of Congress Cataloging in Publication Data. Agarwal, Bipin K. 1931—. X-ray spectroscopy. (Springer series in optical sciences; v. 15) Bibliography: p. Includes indexes. 1. X-ray spectroscopy. I. Title. QC482.S6A34 537.5′352 79-415

Printed in Germany

Offset printing: Beltz Offsetdruck, Hemsbach/Bergstr. Bookbinding: J. Schäffer oHG, Grünstadt.
2153/3130-543210

To my Mami Ji
in memory

Preface

Röntgen's discovery of X-rays in 1895 launched a subject which became central to the development of modern physics. The verification of many of the predictions of quantum theory by X-ray spectroscopy in the early part of the twentieth century stimulated great interest in this area, which has subsequently influenced fields as diverse as chemical physics, nuclear physics, and the study of the electronic properties of solids, and led to the development of techniques such as Auger, Raman, and X-ray photoelectron spectroscopy.

The improvement of the theoretical understanding of the physics underlying X-ray spectroscopy has been accompanied by advances in experimental techniques, and the subject provides an instructive example of how progress on both these fronts can be mutually beneficial. This book strikes a balance between historical description, which illustrates this symbiosis, and the discussion of new developments. The application of X-ray spectroscopic methods to the investigation of chemical bonding receives special attention, and an up-to-date account is given of the use of extended X-ray absorption fine structure (EXAFS) in determining interatomic distances, which has attracted much attention during the last decade.

This monograph is intended to be used as a basic text for a one-year course at postgraduate level, and aims to provide the general background that is essential to enable the reader to participate fruitfully in the growing research activity in this field.

Sri Arvind Agarwal and Dr. Chitra Dar have helped me in many ways in the preparation of the manuscript. My whole family joined me in preparing and checking the index.

Allahabad, April 1979 *B.K. Agarwal*

Contents

1. Continuous X-Rays

X-rays are electromagnetic radiations in the wavelength region around 1Å. An X-ray tube is operated by accelerating electrons to high energy and then allowing them to strike a metallic target. Within the target, the motions of electrons are affected by the strong fields of the atomic nuclei, and the electrons are slowed or even stopped within small distances. In other words, the electrons are strongly decelerated. As a result *bremsstrahlung* (*deceleration radiation*) or *continuous radiation* (*white radiation*) is produced. It would therefore be useful to study the nature of the field of a charged particle that is in uniform motion and then decelerated.

1.1 Invariance of Charge

The magnitude of a charge Q at rest can be determined by measuring the force F on a test charge q, a certain distance r away and at rest. This is based on Coulomb's law

$$F = qQ/r^2 \quad . \tag{1.1}$$

The electric field E at that point is defined as the force on q per unit charge

$$E = F/q = Q/r^2 \quad , \text{ or } \quad Q = E4\pi r^2/4\pi \quad . \tag{1.2}$$

This suggests that the amount of charge in a region can be defined in terms of the surface integral of the electric field $\underline{E}$ over a surface S that encloses the region, $\int \underline{E} \cdot \underline{n}da$,

$$Q = \frac{1}{4\pi} \int_{\text{Surface S}} \underline{E} \cdot \underline{n}da \quad . \tag{1.3}$$

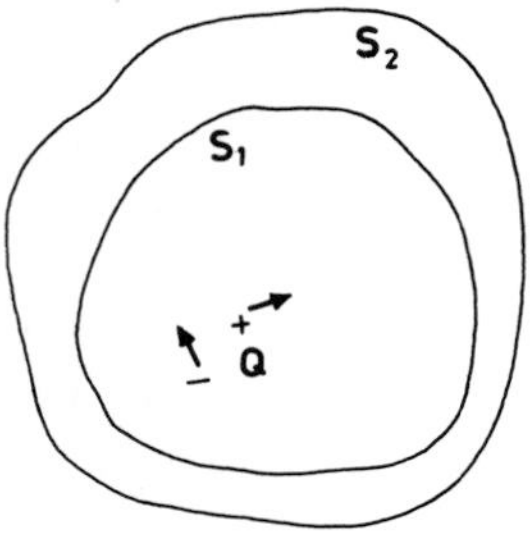

Fig. 1.1. Gauss' law remains true for the field of moving charges. The flux of E through S_2 is the same as that through S_1, evaluated at the same instant of time

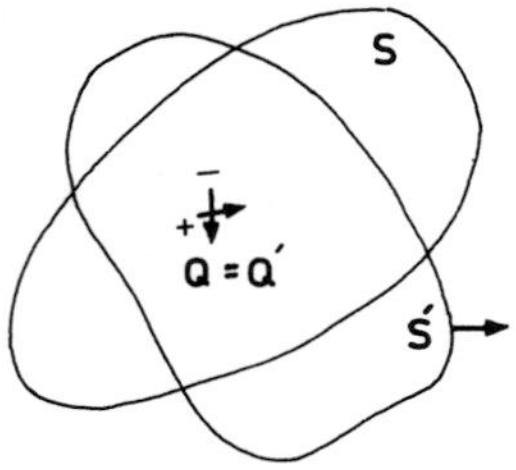

Fig. 1.2. The charge is the same in all frames of reference

Here $\underline{n}$ is the outwardly directed unit normal to the surface at that point, and da is an element of surface area. For a spherical surface (1.3) obviously gives (1.2). The value of Q so determined does not depend upon the size or shape of the surface (Fig.1.1). This is Gauss' law for stationary charges. It is also true when the charges are *moving* (Fig.1.1); this is an experimental fact. For example, atoms and molecules remain neutral in spite of the different internal motions of the protons and the electrons.

Another experimental fact is that measurements of charge made from all reference frames give the same value of the charge. Suppose the inertial frame F' is moving with respect to the inertial frame F. If S' is a closed surface in F' that encloses the same charged particles that were enclosed by S in F at time t, then we must have (Fig.1.2)

$$\int_{S(t)} \underline{E} \cdot d\underline{a} = \int_{S'(t')} \underline{E}' \cdot d\underline{a}' \quad . \tag{1.4}$$

This is the statement of the relativistic *invariance of charge*. The charge is a scalar, an invariant number, with respect to the Lorentz transformation. This fact completely determines the nature of the field of a moving charge.

1.2 Electric Field Measured in a Moving Frame of Reference

Consider the stationary parallel-plate condenser of Fig.1.3a, in the frame F. The surface-charge densities on the positive and negative plates are $+\sigma$ and $-\sigma$, respectively. By Gauss' law $E = 4\pi\sigma$ for an observer in F. This is the only field component inside the condenser. The electric field is zero outside the condenser. There being no moving charges, the magnetic field is zero, $H = 0$.

Fig.1.3b shows the same condenser as seen by an observer in the frame F', which is moving parallel to E toward the left, with respect to F, with speed v. To the observer in F', the plate spacing has changed, because of Lorentz contraction, from d to $d\sqrt{1-\beta^2}$, where $\beta = v/c$, c being the velocity of light. However, the surface charge density has not changed, $\sigma = \sigma'$, because the plates themselves have not contracted. Because the distance d does not enter into the expression for the field, we must have, by Gauss' law

$$E_{\parallel} = 4\pi\sigma = 4\pi\sigma' = E'_{\parallel} \quad , \quad (H=0) \quad . \tag{1.5}$$

Thus E is unchanged by translation parallel to itself. No H field arises in this case, as can be easily seen from the following. By symmetry considerations, there can be no H field transverse to the direction of motion. If it existed, which way could it point? There is no preferred direction transverse to the motion. Moreover, the charges do not produce a magnetic field parallel to their motion. Therefore, we must have $H = 0$ in this case.

Figure 1.4b shows the same condenser as seen by an observer in F' that is moving, perpendicular to the E field, with respect to F. Now the plate spacing

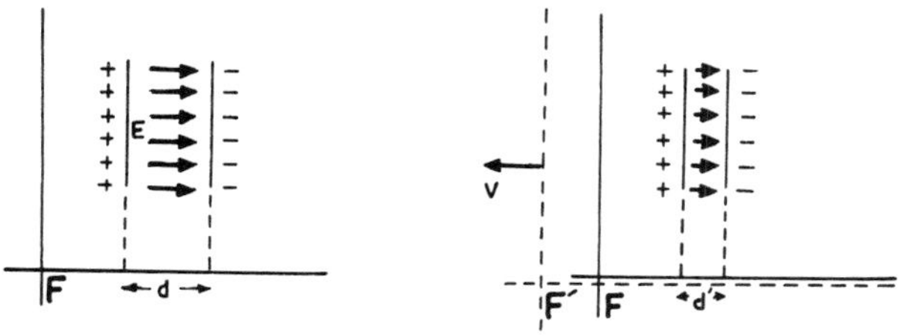

(a) (b)

Fig. 1.3. Parallel-plate condenser (a) at rest in the frame F, and (b) as seen from the frame F' that moves parallel to E. The plate spacing, as measured in F, is d and as measured in F' is $d' = d\sqrt{1-\beta^2}$, $\beta = v/c$

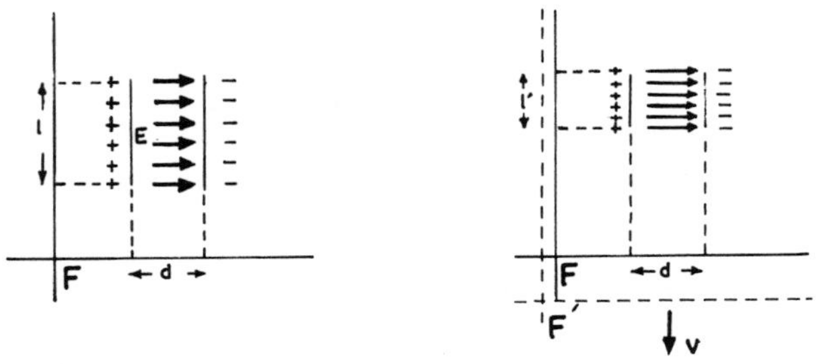

(a) (b)

Fig. 1.4. Parallel-plate condenser (a) at rest in the frame F, and (b) as seen from the frame F' moving perpendicular to E. The plate spacing is unchanged, but the height l is contracted to $l' = l\sqrt{1-\beta^2}$, resulting in an increase of the charge density by a factor $1/\sqrt{1-\beta^2}$. This is indicated by a closer vertical spacing of charges in (b)

is unchanged, but the plate height is contracted from l to $l' = l\sqrt{1-\beta^2}$. But total charge is invariant. So the charge density measured in F' must be greater than σ, in the ratio $1/\sqrt{1-\beta^2}$. Thus, the field perpendicular to a translation is

$$E'_\perp = 4\pi \frac{\sigma}{\sqrt{1-\beta^2}} = \gamma E_\perp \quad , \quad \gamma = \frac{1}{\sqrt{1-\beta^2}} \geqq 1 \quad , \quad (\text{if } H=0) \quad . \tag{1.6}$$

(In this case, there is a magnetic field due to the motion of the charge density). Note that $\gamma > 1$.

1.3 Field of a Point Charge Moving with Uniform Velocity

A point charge Q is at rest at the origin in the frame F (Fig.1.5a). In this frame, the electric field $\underline{E}$ is radial, spherically symmetrical and has the magnitude Q/r^2. In the xz plane its components are

$$E_x = \frac{Q}{r^2}\cos\theta = \frac{Qx}{(x^2+z^2)^{3/2}} \quad ,$$

$$E_z = \frac{Q}{r^2}\sin\theta = \frac{Qz}{(x^2+z^2)^{3/2}} \quad . \tag{1.7}$$

The field is radial as $E_z/E_x = z/x$.

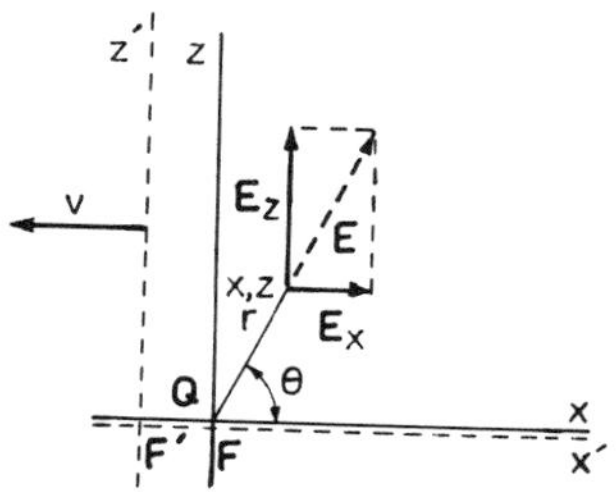

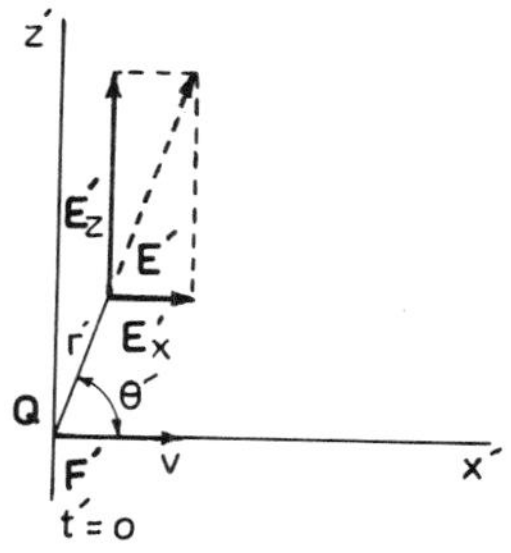

Fig. 1.5. The electric field of a point charge Q (a) in the rest frame F, and (b) in the frame F' in which Q moves with a given constant velocity v

Let the frame F' move with speed v toward the left (Fig.1.5b). The Lorentz transformations are

$$x = \gamma(x'-vt') \quad , \quad y = y' \quad , \quad z = z' \quad , \quad t = \gamma\left(t' - \frac{v}{c^2}x'\right) \quad . \tag{1.8}$$

The clocks read zero when $x = 0$ and $x' = 0$ coincide.

At the instant $t' = 0$, when Q passes the origin in F', we have

$$E'_x = E_x = \frac{Q\gamma x'}{[(\gamma x')^2 + z'^2]^{3/2}} \quad ,$$

$$E'_z = \gamma E_z = \frac{\gamma Q z'}{[(\gamma x')^2 + z'^2]^{3/2}} \quad . \tag{1.9}$$

The field remains radial as $E'_z/E'_x = z'/x'$. Clearly

$$E'^2 = E'^2_x + E'^2_z = \frac{\gamma^2 Q^2(x'^2+z'^2)}{[(\gamma x')^2+z'^2]^3} = \frac{Q^2(x'^2+z'^2)}{\gamma^4(x'^2+z'^2-\beta^2 z'^2)^3}$$

$$= \frac{Q^2(1-\beta^2)^2}{(x'^2+z'^2)\left(1-\frac{\beta^2 z'^2}{x'^2+z'^2}\right)^3} \quad . \tag{1.10}$$

Using $r' = (x'^2+z'^2)^{\frac{1}{2}}$ and $z' = r' \sin\theta'$ (Fig.1.5b), we have

$$E' = \frac{Q}{r'^2}\,\frac{1-\beta^2}{(1-\beta^2 \sin^2\theta')^{3/2}} = \frac{Q}{r'^2} \quad \text{for} \quad \beta \to 0 \quad . \tag{1.11}$$

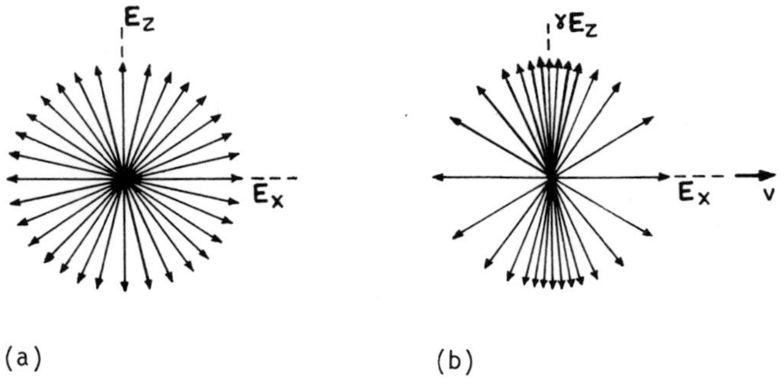

Fig. 1.6. Lines of force of a positive charge (a) at rest, (b) moving, $\beta \sim 4/5$. The lines of force tend to concentrate in a slab perpendicular to the direction of motion, (b), in a three-dimensional picture

Thus, if β^2 is not negligible, the field is stronger at right angles to the motion than in the direction of motion, at the same distance from Q. If we indicate the intensity of the field by the density of lines of force, a two-dimensional picture would be as shown in Fig.1.6. In a three-dimensional picture, the lines of force tend to concentrate in a slab perpendicular to the direction of motion.

It may be noted that the field strength due to a stationary or uniformly moving charge falls off as $1/r^2$ and is *nonradiating* in the sense that it does not contribute to the energy flux over a large sphere. The reason is as follows. The energy flux is given by a quadratic expression in field strengths. Therefore, in this case it will vary as $1/r^4$. The integral over a spherical surface, $\int(\text{energy flux})r^2 d\Omega$, will then involve $1/r^2$ in the integrand. As a result, for large r $(r \to \infty)$, the contribution will become negligible.

1.4 Radiation from an Accelerated or Decelerated Charged Particle

Consider an electron that is initially at rest at $x = 0$ (Fig.1.7). It is suddenly (in negligible time interval Δt) accelerated up to a high speed $\Delta v, \Delta\beta = 4/5$, at $t = 0$. The electron moves away along the x axis with this speed. We observe the situation after a time $t \gg \Delta t$ (Fig.1.7). Because no signal can travel with a speed greater than c, any observer at a point outside the sphere or radius $c(t+\Delta t)$ around the origin must see the field of a charge at rest at the origin (static field) as he could not have received the information

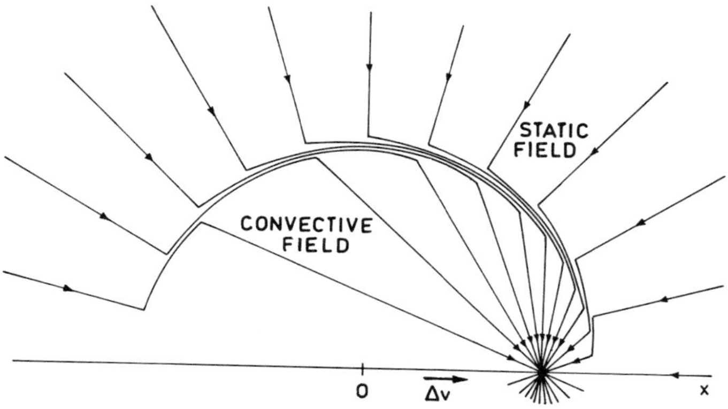

Fig. 1.7. Field at time t due to an electron, initially at rest at $x = 0$, that is suddenly accelerated at $t = 0$, and moves away with uniform velocity Δv, $\beta = 4/5$

that the electron had started to move. Similarly, for all points inside the sphere of radius ct, the field at time t must be due to a uniformly moving electron (convective field) with the lines having the appropriate Lorentz-contracted configuration of Fig.1.6b, converging to the electron at a distance Δvt from the origin.

There exists a thin spherical shell of width $c\Delta t$ within which the transition from the static field to the convective field occurs. The shell of adjustment expands with the velocity c around $x = 0$. Because lines of force cannot break, within this shell the lines of force will run so as to connect the outer (static) and inner (convective) field lines (Fig.1.7).

For bremsstrahlung (braking radiation) we are interested in the case of an electron that has been moving with uniform velocity Δv until it reaches the origin ($x=0$) at $t = 0$, where it is suddenly stopped (decelerated) within a short interval of time Δt. It remains at rest thereafter. The resulting field after a time t is shown in Fig.1.8. The static field is within a sphere of radius ct. The convective Lorentz-contracted field is outside a sphere of radius $c(t+\Delta t)$. The transition spherical shell of small width $c\Delta t$ moves out with the velocity c. The news that the electron has stopped cannot reach, by time t, any point farther than ct from the origin. The convective field therefore will continue to follow the charge as if it has continued to move at its original speed and will converge to the position where the electron would be if it had not stopped.

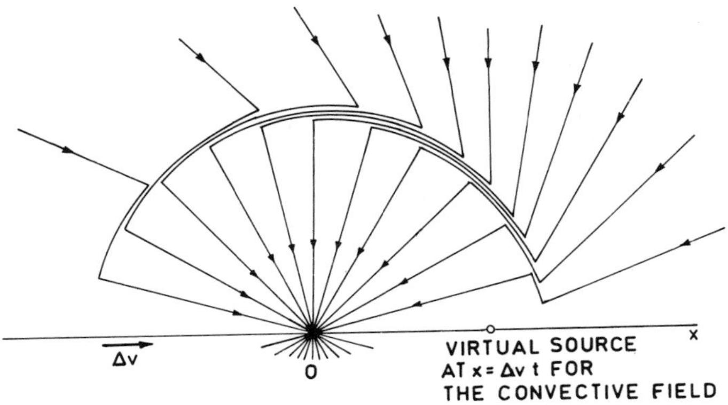

Fig. 1.8. Field due to a decelerated electron which stopped at $x = 0$, at $t = 0$

From (1.7,8) we have

$$\frac{E_z}{E_x} = \tan\theta \quad , \quad \frac{E_z'}{E_x'} = \tan\theta' = \frac{\gamma E_z}{E_x} = \gamma\tan\theta \quad , \quad \gamma = \frac{1}{\sqrt{1-\beta^2}} \quad . \tag{1.12}$$

It is clear from this and the resulting construction (Fig.1.8) that the field lines, connecting the inner field to the outer field, must have a transverse or nonradial component. For $\beta = 4/5$, Fig.1.8 shows that there exists an intense field within the transition shell, with field lines that run almost perpendicular to the radius vector from the origin. We thus have *an outgoing wave of transverse electric field radiated outward with velocity c by the decelerated electron.*

1.5 Transverse Radiation Field due to the Acceleration of an Electron to Low Velocity ($\beta < 1/3$)

For an electron at rest, accelerated suddenly within a short interval of time Δt to a velocity $\Delta v, \Delta\beta < 1/3$, we have $\theta' \simeq \theta$. Thus, after a time t, the field lines look like that shown in Fig.1.9 (compare with Fig.1.7). For low velocity, the lines of force will not be Lorentz-contracted in the inside region and will simply move parallel to themselves a distance $t\Delta v \sin\theta$.

During the time Δt of the acceleration, the wave advances a distance $c\Delta t$. The part AB of the field line inside the shell is straight when the accelera-

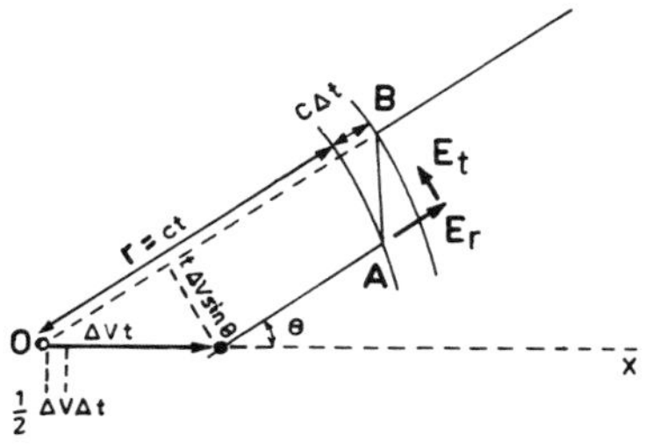

Fig. 1.9. Electron at rest is accelerated within a time Δt to a slow velocity Δv. The acceleration is $a = \Delta v/\Delta t$

tion $a = \Delta v/\Delta t$ is uniform. Let us resolve the electric field $\underline{E}$ at any point on AB into a radial component E_r (parallel to OB) and a perpendicular component E_t. Then E_t/E_r is approximately (ratio of corresponding sides of similar triangles)

$$\frac{E_t}{E_r} = \frac{t\Delta v \sin\theta}{c\Delta t} = \frac{at}{c}\sin\theta \quad . \tag{1.13}$$

Using $E_r = e/r^2$ and $t = r/c$, we get

$$E_t = \frac{ea}{c^2 r}\sin\theta \quad . \tag{1.14}$$

Thus, within the transition shell, we have two components

$$E_r \propto 1/r^2 \quad \text{and} \quad E_t \propto 1/r \quad . \tag{1.15}$$

For large r, E_r becomes negligible and only E_t remains. In other words, at great distances, we shall be left with a pulse of a *transverse field* travelling outward with velocity c. Its energy flux will vary as $1/r^2$, making the integral over a spherical surface independent of r. A finite contribution will be obtained even for large r. Therefore, it is a *radiating field*.

The above simple calculation shows that an accelerated (or decelerated) electron generates a *transverse radiation field*. This result is of interest in connection with the theory of continuous X-rays. The calculation of the associated transverse magnetic field is more involved and requires a general treatment. We now prepare the ground for such a calculation.

1.6 Maxwell's Equations

For a charge that moves with velocity $\underline{v}$ at a given point and a given time, the current density $\underline{j}$ is

$$\underline{j} = \rho \underline{v} \quad , \tag{1.16}$$

where ρ is the *charge density*. If the electric and magnetic field strengths are $\underline{E}(x,y,z,t)$ and $\underline{H}(x,y,z,t)$, the *Maxwell equations* for a given distribution of charge and current are

$$\nabla \times \underline{E} + \frac{1}{c}\frac{\partial \underline{H}}{\partial t} = 0 \quad \text{(Faraday's law)} \quad , \tag{1.17a}$$

$$\nabla \cdot \underline{H} = 0 \quad \text{(absence of free magnetic poles)} \quad , \tag{1.17b}$$

$$\nabla \times \underline{H} - \frac{1}{c}\frac{\partial \underline{E}}{\partial t} = \frac{4\pi}{c}\underline{j} \quad \text{(generalized Ampere's law)} \quad , \tag{1.17c}$$

$$\nabla \cdot \underline{E} = 4\pi\rho \quad \text{(Coulomb's law)} \quad . \tag{1.17d}$$

In view of (1.17b), $\underline{H}$ can always be expressed as the curl of a vector potential $\underline{A}$

$$\underline{H} = \nabla \times \underline{A} \quad . \tag{1.18a}$$

Then the homogeneous equation (1.17a) becomes

$$\nabla \times \left(\underline{E} + \frac{1}{c}\frac{\partial \underline{A}}{\partial t}\right) = 0 \quad .$$

This shows that $\underline{E} + \frac{1}{c}\,\partial\underline{A}/\partial t$ can be expressed as the gradient of a *scalar potential* ϕ

$$\underline{E} + \frac{1}{c}\frac{\partial \underline{A}}{\partial t} = -\nabla\phi \quad . \tag{1.18b}$$

These definitions of $\underline{H}$ and $\underline{E}$ in terms of $\underline{A}$ and ϕ can now be substituted in the last two inhomogeneous Maxwell equations. Using curl curl = grad div $-\nabla^2$, we finally get

$$\frac{1}{c^2}\frac{\partial^2 \underline{A}}{\partial t^2} - \nabla^2\underline{A} + \nabla\left(\nabla\cdot\underline{A} + \frac{1}{c}\frac{\partial\phi}{\partial t}\right) = \frac{4\pi}{c}\underline{j} \quad , \tag{1.19a}$$

$$-\nabla^2\phi - \frac{1}{c}\nabla\cdot\frac{\partial \underline{A}}{\partial t} = 4\pi\rho \tag{1.19b}$$

A vector is uniquely defined only when both its curl and divergence are specified. Although the curl of $\underline{A}$ is given in (1.18a), the div of $\underline{A}$ remains to be specified. For this, we introduce the *Lorentz condition*

$$\nabla\cdot\underline{A} = -\frac{1}{c}\frac{\partial\phi}{\partial t} \quad ; \tag{1.20}$$

it uncouples (1.19a,b). We are left with two inhomogeneous equations, one for ϕ and one for $\underline{A}$,

$$\nabla^2\phi - \frac{1}{c^2}\frac{\partial^2\phi}{\partial t^2} = -4\pi\rho(\underline{x},t) \quad , \tag{1.21a}$$

$$\nabla^2\underline{A} - \frac{1}{c^2}\frac{\partial^2\underline{A}}{\partial t^2} = -\frac{4\pi}{c}\,\underline{j}(\underline{x},t) \quad . \tag{1.21b}$$

These two equations along with (1.20), form a set of equations that are equivalent in all respects to Maxwell's equations. Therefore, it is enough to solve the inhomogeneous equations (1.21), which have the familiar form of wave equations with source terms on the right.

1.7 Coulomb Potential

For a static field (1.21a) reduces to the *Poisson equation*

$$\nabla^2\phi(\underline{x}) = -4\pi\rho(\underline{x}) \quad , \tag{1.22}$$

which has a solution

$$\phi(\underline{x}) = \int \frac{\rho(\underline{x}')}{|\underline{x}-\underline{x}'|}\,d^3x' \quad . \tag{1.23}$$

To verify (1.23), operate with ∇^2 on both sides.

$$\nabla^2\phi = \nabla^2\int\frac{\rho(\underline{x}')}{|\underline{x}-\underline{x}'|}\,d^3x' = \int \rho(\underline{x}')\nabla^2\left(\frac{1}{|\underline{x}-\underline{x}'|}\right)d^3x' \quad . \tag{1.24}$$

To evaluate $\nabla^2(1/|\underline{x}-\underline{x}'|)$, translate the origin to $\underline{x}'$. Then we consider $\nabla^2(1/R)$, where $R = |\underline{x}|$,

$$\nabla^2\left(\frac{1}{R}\right) = \frac{1}{R}\frac{d^2}{dR^2}\left(R\frac{1}{R}\right) = \frac{1}{R}\frac{d^2}{dR^2}(1) = 0 \quad \text{for } R \neq 0 \quad . \tag{1.25}$$

At $R = 0$, the expression is undefined, so we must use a limiting process. We integrate $\nabla^2(1/R)$ over a small volume dV that contains the origin, and use the divergence theorem to obtain a surface integral

$$\int_{dV} \nabla^2\left(\frac{1}{R}\right)d^3x = \int_{dV} \nabla \cdot \nabla\left(\frac{1}{R}\right)d^3x = \int_S \underline{n} \cdot \nabla\left(\frac{1}{R}\right)dS$$
$$= \int_S \frac{\partial}{\partial R}\left(\frac{1}{R}\right)R^2 d\Omega = -4\pi \quad . \tag{1.26}$$

Results (1.25,26) can be combined as

$$\nabla^2(1/R) = -4\pi\delta^3(\underline{x}) \quad , \tag{1.27}$$

or, more generally, as

$$\nabla^2\left(\frac{1}{|\underline{x}-\underline{x}'|}\right) = -4\pi\delta^3(\underline{x}-\underline{x}') \quad , \tag{1.28}$$

where $\delta^3(\underline{x})$ is the *Dirac delta function* defined by the properties

$$\delta^3(\underline{x}) = 0 \quad \text{if } \underline{x} \neq 0 \quad ,$$
$$\int_V \delta^3(\underline{x})d^3x = \begin{cases} 1 & \text{if } \underline{x} = 0 \text{ is in volume V} \\ 0 & \text{if } \underline{x} = 0 \text{ is not in volume V} \end{cases} \quad ,$$
$$\int_V \delta^3(\underline{x})f(\underline{x})d^3x = f(0) \quad \text{if } \underline{x} = 0 \text{ is in volume V} \quad . \tag{1.29}$$

It now immediately follows that (1.23) is a solution of (1.22)

$$\nabla^2\phi = \int \rho(\underline{x}')\left[-4\pi\delta^3(\underline{x}-\underline{x}')\right]d^3x' = -4\pi\rho(\underline{x}) \quad , \tag{1.30}$$

where we have used the last relation of (1.29). For a point charge, $\rho = e\delta^3(\underline{x})$ and from (1.27)

$$\phi = e/R \quad (\textit{Coulomb potential}) \quad , \tag{1.31}$$

as expected.

1.8 Retarded Potentials

The solution of the inhomogeneous linear equation (1.21a) consists of a particular solution with the right-hand side plus a general solution of the homogeneous equation (without the right-hand side).

To find the particular solution, we divide the whole space into infinitely small regions (cells) and determine the field produced by the *point* charge located in one of these cells. That is, we first find a solution for a point charge de, and then later sum over all the charge elements $de = \rho dV$, where dV is the cell volume, in the appropriate charge distribution $\rho(\underline{x},t)$. Equation (1.21a) being linear in ϕ, the actual field will be the sum of the fields produced by all such elements.

We wish to solve (1.21a) for a moving point charge, $\partial^2\phi/\partial t^2 \neq 0$. The charge de in a given cell is, in general, a function of time, We choose the origin of coordinates in the cell under consideration, where the point charge de(t) is located. Then (1.21a) becomes

$$\nabla^2\phi - \frac{1}{c^2}\frac{\partial^2\phi}{\partial t^2} = -4\pi\, de(t)\delta^3(\underline{R}) \quad , \tag{1.32}$$

where $\underline{R}$ is the distance from the origin. Everywhere, except at the origin, $\delta^3(\underline{R}) = 0$. Therefore, the equation

$$\nabla^2\phi - \frac{1}{c^2}\frac{\partial^2\phi}{\partial t^2} = 0 \qquad (|\underline{R}| \neq 0) \quad , \tag{1.33}$$

is satisfied everywhere except at the origin, whereas in a small volume element dV that surrounds the origin

$$\int_{dV}\left[\nabla^2\phi - \frac{1}{c^2}\frac{\partial^2\phi}{\partial t^2}\right]dV = -4\pi\, de(t) \quad , \tag{1.34}$$

must be satisfied.

In our case, ϕ has central symmetry, $\phi(\underline{R},t) = \phi(R,t)$. Therefore, in spherical coordinates (1.33) becomes

$$\frac{1}{R^2}\frac{\partial}{\partial R}\left(R^2\frac{\partial\phi}{\partial R}\right) - \frac{1}{c^2}\frac{\partial^2\phi}{\partial t^2} = 0 \quad . \tag{1.35}$$

Putting $\phi(R,t) = \chi(R,t)/R$, we get

$$\frac{\partial^2\chi}{\partial R^2} - \frac{1}{c^2}\frac{\partial^2\chi}{\partial t^2} = 0 \quad . \tag{1.36}$$

This is just a one-dimensional wave equation, which is solved by any function of $t-(R/c)$ or $t+(R/c)$. To verify this, let

$$T = t - \frac{R}{c} \quad , \tag{1.37}$$

and let $f(T)$ be any function of T that can be twice differentiated. Clearly

$$\frac{\partial f}{\partial R} = \frac{\partial f}{\partial T}\frac{\partial T}{\partial R} = -\frac{1}{c}\frac{\partial f}{\partial T} \quad , \quad \frac{\partial^2 f}{\partial R^2} = \frac{\partial^2 f}{\partial T^2}\frac{\partial T}{\partial R} = \frac{1}{c^2}\frac{\partial^2 f}{\partial T^2}$$

$$\frac{\partial f}{\partial t} = \frac{\partial f}{\partial T}\frac{\partial T}{\partial t} = \frac{\partial f}{\partial T} \quad , \quad \frac{\partial^2 f}{\partial t^2} = \frac{\partial^2 f}{\partial T^2} \quad ,$$

showing that $f(T)$ is a solution of (1.36).

The solution $f(t-R/c)$ represents a plane wave that is propagated *outward* along the polar axis R with a velocity c. The reason is that the field has the same value when $t-R/c=\text{constant}$, that is, $R=\text{constant}+ct$.

Similarly, $f(t+R/c)$ represents a solution for a wave moving *inward*. Because we are interested in the outward-propagated wave, we can take the solution to be

$$\chi = \chi\left(t-\frac{R}{c}\right) \quad , \text{ or } \quad \phi = \frac{\chi\left(t-\frac{R}{c}\right)}{R} \quad . \tag{1.38}$$

So far, the form of χ is arbitrary. It can be chosen so that (1.34), or (1.32), is satisfied for the potential at the origin. To this end, note that as $R\to 0$, the potential increases to infinity. Consequently, its derivatives with respect to the coordinates increase more rapidly than its time derivative. Then, as $R\to 0$ in (1.34), or (1.32), we can ignore the term $\partial^2\phi/\partial t^2$, and we get a static field equation that leads to the Coulomb law (Sect.1.7). Thus, near the origin, (1.38) must go over to the Coulomb law and we should put $\chi(t)=de(t)$; that is

$$\phi = \frac{de\left(t-\frac{R}{c}\right)}{R} \quad . \tag{1.39}$$

It is now a simple matter to find the solution of (1.21a) for an arbitrary distribution of charges $\rho(\underline{x},t)$. In (1.39), we write $de=\rho dV$ and sum over the whole space

$$\phi(\underline{x},t) = \int \frac{\rho\left(\underline{x}',t-\frac{R}{c}\right)dV'}{R} \quad , \tag{1.40}$$

$$R^2 = (x-x')^2 + (y-y')^2 + (z-z')^2 \quad , \quad dV' = dx'\,dy'\,dz' \quad ,$$

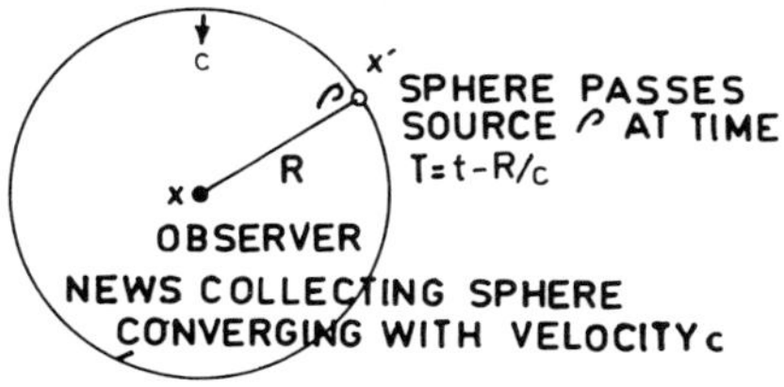

Fig. 1.10. Sphere collecting news for retarded potential at $\underline{x}$. The time of observation is t

where $R = |\underline{x}-\underline{x}'|$ is the distance from the volume element dV' to the point of observation. We can write the result briefly as

$$\phi(\underline{x},t) = \int \frac{[\rho(\underline{x}',T)]_{ret}}{R}\, dV' \quad . \tag{1.41a}$$

It is called the *retarded scalar potential*. $[\rho]_{ret}$ means that ρ is to be taken at the *retarded time* $T = t - R/c$.

This is one of the most interesting results in electromagnetic theory. It says that an observer at the field point $\underline{x}$ is affected at a given time t only by what a source at $\underline{x}'$ was doing at an earlier time T. Because $t - T = R/c$, the earlier time T precedes the present time t by just the amount of time it takes the information to travel the distance R from the source point $\underline{x}'$ to the field point $\underline{x}$, namely R/c. In general, an effect at one space-time point is due to all disturbances throughout space that preceded the effect by exactly the time interval necessary for the disturbances to propagate from their points of origin to the point of observation.

A retarded potential can be realized in the following way. Let an observer be situated at the point $\underline{x}$ (Fig.1.10). Around this point, consider a sphere contracting with a radial velocity c. Arrange the event in such a way that this news-collecting sphere converges on the point $\underline{x}$ at the time of observation t. Then this sphere must have passed the source of the electric field $\rho(\underline{x}')$ at $\underline{x}'$ at the earlier time $T = t - R/c$. Because $R = |\underline{x}-\underline{x}'|$, it is the retarded distance measured at the time of passage of the sphere at the source; $R = R(T) = c(T-t)$, where t is the time of observation.

A similar analysis of (1.21b) gives the *retarded vector potential*

$$\underline{A}(\underline{x},t) = \frac{1}{c}\int_V \frac{\underline{j}(\underline{x}',t-|\underline{x}-\underline{x}'|/c)}{|\underline{x}-\underline{x}'|}\, dV' = \frac{1}{c}\int_V \frac{[\underline{j}(\underline{x}',T)]_{ret}}{R}\, dV' \quad . \tag{1.41b}$$

To the solution (1.41) of the inhomogeneous equations (1.21) we can still add the solution $\phi_0(\underline{x},t)$ and $\underline{A}_0(\underline{x},t)$, respectively, of the same equations without the right-hand side or source terms. The quantities ϕ_0 and $\underline{A}_0$ are determined by the initial conditions. As we shall not be dealing with such initial conditions, we can neglect ϕ_0 and $\underline{A}_0$. Usually, we will be given conditions at large distances.

1.9 Lienard-Wiechert Potentials

The retarded potentials due to a moving charge distribution ρ are given by (1.41). Let us now study the implications of (1.41) in detail.

Consider a situation in which all parts of a given charge distribution in any small volume element are moving systematically with a velocity $\underline{v}$ (Fig. 1.11). As the news-collecting sphere shrinks with velocity c, the charge distribution also moves with velocity $\underline{v}$. Therefore, the amount of charge crossed at the earlier time T by the sphere in time dT is (see Fig.1.11)

$$de = [\rho]_{ret}\, dSdR - [\rho]_{ret} \frac{\underline{v}\cdot\underline{R}}{R}\, dTdS \quad . \tag{1.42}$$

Because $dSdR = dV'$ and $dT = dR/c$, we have

$$[\rho]_{ret}\, dV' = \frac{de}{\left[1 - \dfrac{\underline{v}\cdot\underline{R}}{cR}\right]_{ret}} \quad , \tag{1.43}$$

or

$$\frac{[\rho]_{ret}\, dV'}{R} = \frac{de}{s} \quad , \tag{1.44}$$

where

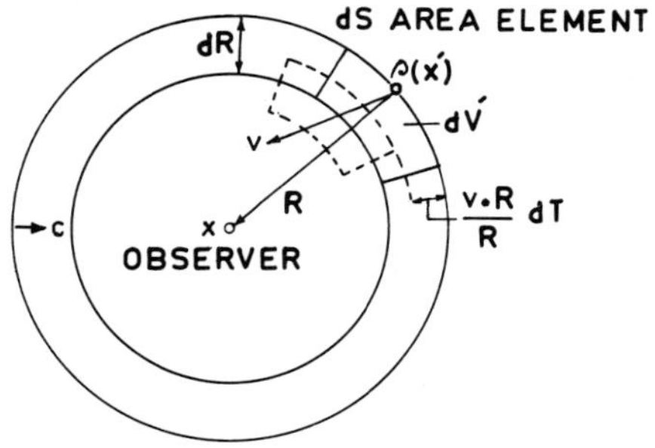

Fig. 1.11. Element dS of sphere collecting news from the charge distribution in motion with velocity $\underline{v}$

$$s \equiv \left[R - \frac{\underline{v}\cdot\underline{R}}{c}\right]_{t-R/c} = \left[R\left(1 - \frac{\underline{v}\cdot\underline{n}}{c}\right)\right]_T \quad . \tag{1.45}$$

The retarded potentials thus become

$$\phi = \int \frac{de}{s} \quad , \tag{1.46a}$$

$$\underline{A} = \frac{1}{c}\int \frac{\underline{v}\ de}{s} \quad . \tag{1.46b}$$

The field due to a moving point charge is the properly taken limit of the field due to the moving charge distribution in a small volume element. We can use (1.46) to calculate the potentials due to a moving *point* charge. When viewed from a large distance, the moving charge distribution in a small volume element will appear as a point charge. Clearly, in such a limit of a point charge the distance-dependent terms are slowly varying. In (1.46a), we can take 1/s outside the integral sign and write $\int de = e$ as the known charge of a point electron. Thus,

$$\phi(\underline{x},t) = \left[\frac{e}{s}\right]_{ret} \quad , \tag{1.47a}$$

$$\underline{A}(\underline{x},t) = \frac{1}{c}\left[\frac{e\underline{v}}{s}\right]_{ret} \quad . \tag{1.47b}$$

These are called *Lienard-Wiechert potentials* of a single electron. All quantities in (1.47) have to be taken at the retarded time $T = t - R/c$. For $\underline{v} = 0$, (1.47) reduces to

$$\phi = \frac{e}{R} \quad , \quad \underline{A} = 0 \quad , \tag{1.48}$$

as expected.

1.10 Radiation from an Accelerated Charge

The field strengths $\underline{E}$, $\underline{H}$ can be found from (1.47) by differentiation

$$\underline{E} = -\nabla\phi - \frac{1}{c}\frac{\partial \underline{A}}{\partial t} \quad , \tag{1.49a}$$

$$\underline{H} = \nabla \times \underline{A} \quad . \tag{1.49b}$$

The derivatives of $\phi(\underline{x},t)$ and $\underline{A}(\underline{x},t)$ have to be taken with respect to the time t of observation and coordinates $\underline{x}$ of the point of observation. However, (1.47) expresses the potentials as functions of $T = t - R/c$. The quantities $\underline{R}$ and $\underline{v}$ that occur in (1.47) are also functions of the retarded time T. The motion of the particle at the time T is determined by $\underline{R}(T)$ and $\partial\underline{R}/\partial T = -\underline{v}(T)$ (the minus sign appears because $\underline{R}$ is the radius vector from the charge e to the point of observation $\underline{x}$, and not the reverse). Therefore, it is necessary to express the derivatives of ϕ, $\underline{A}$ with respect to t in terms of the derivatives with respect to T.

The retarded time T and the retarded distance R are related by the retardation condition

$$R[\underline{x},\underline{x}'(T)] = f(\underline{x},T) = \left[(x-x')^2+(y-y')^2+(z-z')^2\right]^{\frac{1}{2}} = c(t-T) \quad , \tag{1.50}$$

which connects the field- and source-point variables. We can use (1.50) to relate the partial derivative $\partial/\partial t$ to $\partial/\partial T$, by keeping point of observation x,y,z fixed.

Let us first differentiate the identity $R^2 = \underline{R} \cdot \underline{R}$. Using $\partial\underline{R}(T)/\partial T = -\underline{v}(T)$, we get

$$\left(\frac{\partial R}{\partial T}\right)_{\underline{x}} = - \frac{\underline{R}\cdot\underline{v}}{R} \quad , \tag{1.51}$$

where the suffix $\underline{x}$ indicates that $\underline{x}$ has been kept fixed. Now differentiate the relation $R = c(t-T)$ with respect to t

$$\frac{\partial R}{\partial t} = \frac{\partial R}{\partial T}\frac{\partial T}{\partial t} = - \frac{\underline{R}\cdot\underline{v}}{R}\frac{\partial T}{\partial t} = c\left(1 - \frac{\partial T}{\partial t}\right) \quad , \tag{1.52}$$

or

$$\frac{\partial T}{\partial t} = \frac{1}{1 - \dfrac{\underline{R}\cdot\underline{v}}{Rc}} = \frac{R}{s} \quad . \tag{1.53}$$

In other words,

$$\frac{\partial}{\partial t} = \frac{R}{s}\frac{\partial}{\partial T} \quad . \tag{1.54}$$

Similarly, differentiating the same relation with respect to the coordinates and keeping t fixed, we have

$$\nabla R = (\nabla)_T R + \frac{\partial R}{\partial T}\nabla T = \frac{\underline{R}}{R} - \frac{\underline{R}\cdot\underline{v}}{R}\nabla T = -c\nabla T \quad . \tag{1.55}$$

Here $(\nabla)_T$ represents differentiation with respect to the first argument of the function $f(\underline{x},T)$ in (1.50) with T kept fixed, and we have used the rules

$$\left(\frac{\partial f}{\partial x}\right)_t = \left(\frac{\partial f}{\partial x}\right)_{t,T} + \left(\frac{\partial f}{\partial T}\right)_{x,t}\left(\frac{\partial T}{\partial x}\right)_t \quad , \tag{1.56a}$$

$$\nabla R = \left(\underline{i}_x \frac{\partial}{\partial x} + \underline{i}_y \frac{\partial}{\partial y} + \underline{i}_z \frac{\partial}{\partial z}\right)\left[(x-x')^2+(y-y')^2+(z-z')^2\right]^{\frac{1}{2}} = \frac{\underline{R}}{R} \quad . \tag{1.56b}$$

From (1.55)

$$\nabla T = -\frac{\underline{R}}{cs} \quad , \tag{1.57}$$

$$\nabla = (\nabla)_T - \frac{\underline{R}}{cs}\frac{\partial}{\partial T} \quad . \tag{1.58}$$

These relations transform the differential operators from the coordinates of the field point to those of the moving electron.

For the derivatives of $s = R - \underline{v}\cdot\underline{R}/c$, we get

$$\frac{\partial s}{\partial t} = \frac{\partial s}{\partial T}\frac{\partial T}{\partial t} = \left(\frac{\partial R}{\partial T} - \frac{1}{c}\frac{\partial \underline{R}}{\partial T}\cdot\underline{v} - \frac{1}{c}\underline{R}\cdot\frac{\partial \underline{v}}{\partial T}\right)\frac{R}{s} = \left(-\frac{\underline{R}\cdot\underline{v}}{R} + \frac{v^2}{c} - \frac{\underline{R}\cdot\dot{\underline{v}}}{c}\right)\frac{R}{s} \quad , \tag{1.59}$$

$$\nabla s = (\nabla)_T\left(R - \frac{\underline{v}\cdot\underline{R}}{c}\right) - \frac{\underline{R}}{cs}\frac{\partial}{\partial T}\left(R - \frac{\underline{v}\cdot\underline{R}}{c}\right) = \left(\frac{\underline{R}}{R} - \frac{\underline{v}}{c}\right) - \frac{\underline{R}}{cs}\left(-\frac{\underline{R}\cdot\underline{v}}{R} + \frac{v^2}{c} - \frac{\underline{R}\cdot\dot{\underline{v}}}{c}\right) \quad , \tag{1.60}$$

where $\underline{v}$ depends only on T and $\dot{\underline{v}} = \partial\underline{v}/\partial T$. Clearly

$$\frac{\partial \underline{v}}{\partial t} = \frac{\partial \underline{v}}{\partial T}\frac{\partial T}{\partial t} = \dot{\underline{v}}\,\frac{R}{s} \quad , \tag{1.61}$$

$$\nabla \times \underline{v} = \left[(\nabla)_T - \frac{\underline{R}}{cs}\frac{\partial}{\partial T}\right] \times \underline{v} = -\frac{\underline{R}\times\dot{\underline{v}}}{cs} \quad . \tag{1.62}$$

From (1.47,49)

$$\frac{1}{e}\underline{E} = -\nabla\frac{1}{s} - \frac{1}{c^2}\frac{\partial}{\partial t}\frac{\underline{v}}{s} = \frac{1}{s^2}\nabla s - \frac{1}{c^2 s}\frac{\partial \underline{v}}{\partial t} + \frac{\underline{v}}{c^2 s^2}\frac{\partial s}{\partial t} \quad , \tag{1.62a}$$

$$\frac{1}{e}\underline{H} = \frac{1}{c}\nabla \times \frac{\underline{v}}{s} = \frac{1}{cs}\nabla \times \underline{v} - \frac{1}{cs^2}(\nabla s\times\underline{v}) \quad , \tag{1.62b}$$

where we have used $\nabla \times (\psi\underline{B}) = \psi(\nabla\times\underline{B}) + (\nabla\psi)\times\underline{B}$. Making proper substitutions, we get

$$\frac{1}{e}\,\underline{E}(\underline{x},t) = \frac{1}{s^2}\,\frac{1}{R}\left(\underline{R} - \frac{\underline{v}R}{c}\right) - \frac{R}{cs^3}\left(-\frac{\underline{R}\cdot\underline{v}}{R} + \frac{v^2}{c} - \frac{\underline{R}\cdot\dot{\underline{v}}}{c}\right)$$

$$- \frac{1}{c^2 s}\,\frac{R}{s}\,\dot{\underline{v}} + \frac{\underline{v}}{c^2 s^2}\,\frac{R}{s}\left(-\frac{\underline{R}\cdot\underline{v}}{R} + \frac{v^2}{c} - \frac{\underline{R}\cdot\dot{\underline{v}}}{c}\right)$$

$$= \frac{R - \frac{\underline{v}\cdot\underline{R}}{c}}{s^3}\,\frac{1}{R}\left(\underline{R} - \frac{\underline{v}R}{c}\right) - \frac{1}{cs^3}\left(-\frac{\underline{R}\cdot\underline{v}}{R} + \frac{v^2}{c}\right)\left(\underline{R} - \frac{\underline{v}R}{c}\right)$$

$$- \frac{1}{c^2 s}\,\frac{R}{s}\,\dot{\underline{v}} + \frac{\underline{R}\cdot\dot{\underline{v}}}{c}\left(\frac{\underline{R}}{cs^3} - \frac{\underline{v}}{c^2 s^2}\,\frac{R}{s}\right)$$

$$= \frac{1}{s^3}\left(\underline{R} - \frac{\underline{v}R}{c}\right)\left(1 - \frac{v^2}{c^2}\right) - \frac{R - \frac{\underline{v}\cdot\underline{R}}{c}}{c^2 s^3}\,R\dot{\underline{v}} + \frac{1}{c^2 s^3}\left(\underline{R} - \frac{\underline{v}R}{c}\right)\underline{R}\cdot\dot{\underline{v}}$$

$$= \frac{1}{s^3}\left(\underline{R} - \frac{\underline{v}R}{c}\right)\left(1 - \frac{v^2}{c^2}\right) + \frac{1}{c^2 s^3}\left\{\underline{R}\times\left[\left(\underline{R} - \frac{\underline{v}R}{c}\right)\times\dot{\underline{v}}\right]\right\} \quad , \qquad (1.63a)$$

where in the last step we have combined the two terms that contain $\dot{\underline{v}}$ by using $\underline{A}\times(\underline{B}\times\underline{C}) = \underline{B}(\underline{A}\cdot\underline{C}) - \underline{C}(\underline{A}\cdot\underline{B})$. Similarly, from (1.62b) we get

$$\frac{1}{e}\,\underline{H}(\underline{x},t) = \nabla\times\frac{\underline{v}}{cs} = \frac{1}{c}\left(\nabla\frac{1}{s}\right)\times\underline{v} + \frac{1}{cs}(\nabla\times\underline{v})$$

$$= -\frac{1}{cs^2}\left\{\left[\left(\frac{\underline{R}}{R} - \frac{\underline{v}}{c}\right) - \frac{\underline{R}}{sc}\left(-\frac{\underline{R}\cdot\underline{v}}{R} + \frac{v^2}{c} - \frac{\underline{R}\cdot\dot{\underline{v}}}{c}\right)\right]\times\underline{v}\right\} + \frac{1}{cs}\left(-\frac{\underline{R}\times\dot{\underline{v}}}{cs}\right)$$

$$= -\frac{\underline{R}\times\underline{v}}{cs^3}\left(R - \frac{\underline{R}\cdot\underline{v}}{c}\right)\frac{1}{R} + \frac{\underline{R}\times\underline{v}}{c^2 s^3}\left(-\frac{\underline{R}\cdot\underline{v}}{R} + \frac{v^2}{c} - \frac{\underline{R}\cdot\dot{\underline{v}}}{c}\right) - \frac{\underline{R}\times\dot{\underline{v}}}{c^2 s^2}$$

$$= \frac{\underline{v}\times\underline{R}}{cs^3}\left(1 - \frac{v^2}{c^2}\right) - \frac{1}{c^2 s^3}\left[(\underline{R}\times\underline{v})\frac{\underline{R}\cdot\dot{\underline{v}}}{c} + \left(R - \frac{\underline{R}\cdot\underline{v}}{c}\right)(\underline{R}\times\dot{\underline{v}})\right]$$

$$= \frac{\underline{v}\times\underline{R}}{cs^3}\left(1 - \frac{v^2}{c^2}\right) + \frac{1}{c^2 s^3}\,\frac{\underline{R}}{R}\times\left\{\underline{R}\times\left[\left(\underline{R} - \frac{R\underline{v}}{c}\right)\times\dot{\underline{v}}\right]\right\} = \frac{\underline{R}\times\underline{E}}{R} \quad . \qquad (1.63b)$$

In (1.63), all quantities on the right side are understood to refer to the retarded time $T = t - R/c$. Clearly, the magnetic field $\underline{H}$ is perpendicular to $\underline{E}$ and to the retarded radius vector $\underline{R}$.

The fields $\underline{E}$ and $\underline{H}$ given by (1.63) are composed of two separate parts. The *velocity fields*, independent of acceleration, vary as $1/R^2$ for large distances and so are nonradiating. The *acceleration fields* depend linearly upon the acceleration $\dot{v}$, vary as $1/R$ for large distances and so give rise to a finite energy flow through any large sphere, i.e., to an *emission of radiation* (bremsstrahlung).

1.11 Radiation at Low Velocities

If the charge is accelerated, but its velocity is small, $v/c \ll 1$, we can write

$$\underline{R} - \frac{\underline{v}}{c} R \approx \underline{R} \quad , \quad s = R - \frac{\underline{R}\cdot\underline{v}}{c} \approx R \quad . \tag{1.64}$$

If we consider only the radiation fields ($\underline{E}_{rad}$ and $\underline{H}_{rad}$), which vary as $1/R$, we get

$$\frac{1}{e}\,\underline{E}_{rad} = \frac{1}{c^2 s^3}\left\{\underline{R}\times\left[\left(\underline{R} - \frac{\underline{v}R}{c}\right)\times\dot{\underline{v}}\right]\right\} \approx \frac{1}{c^2R^3}[\underline{R}\times(\underline{R}\times\dot{\underline{v}})] \quad , \tag{1.65a}$$

$$\frac{1}{e}\,\underline{H}_{rad} = \frac{1}{c^2 s^3}\,\frac{\underline{R}}{R}\times\left\{\underline{R}\times\left[\left(\underline{R} - \frac{\underline{v}R}{c}\right)\times\dot{\underline{v}}\right]\right\} \approx \frac{1}{c^2R^4}\left\{\underline{R}\times[\underline{R}\times(\underline{R}\times\dot{\underline{v}})]\right\}$$

$$= \frac{1}{c^2R^4}\left\{\underline{R}\times[(\underline{R}\cdot\dot{\underline{v}})\underline{R}-(\underline{R}\cdot\underline{R})\dot{\underline{v}}]\right\} = \frac{1}{c^2R^4}(-\underline{R}\times\dot{\underline{v}}R^2) = \frac{1}{c^2R^2}(\dot{\underline{v}}\times\underline{R}) \quad . \tag{1.65b}$$

Clearly, as in (1.14)

$$|\underline{E}_{rad}| = \frac{e}{c^2R^3}|\underline{R}\times(\underline{R}\times\dot{\underline{v}})| = \frac{e}{c^2R^3}|\underline{R}|\,|\underline{R}\times\dot{\underline{v}}|\ \sin\frac{\pi}{2}$$

$$= \frac{e}{c^2R}\,\dot{v}\,\sin\Theta = |\underline{H}_{rad}| \qquad (v/c \ll 1) \quad , \tag{1.66}$$

where Θ is the angle between $\underline{R}$ and $\dot{\underline{v}}$. Here E is expressed in electrostatic and H in electromagnetic units to preserve the symmetry of our expression. We shall also drop the suffix "rad" in the future. The situation is shown in Fig.1.12.

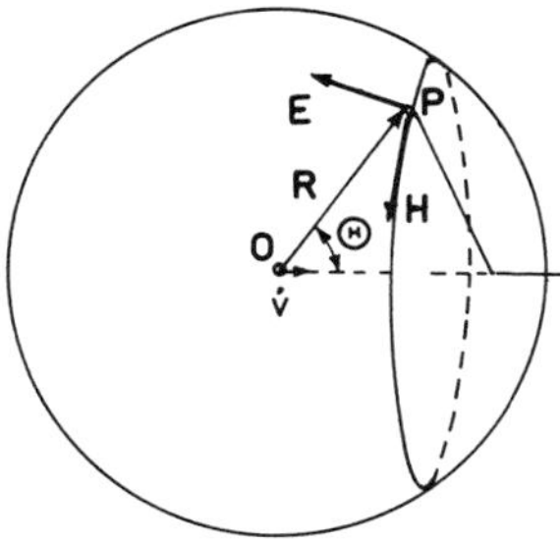

Fig. 1.12. Radiation field at P due to an accelerated positive charge at O. $\underline{E}$ is perpendicular to $\underline{R}$ and lies in the plane of $\dot{\underline{v}}$ and $\underline{R}$. $\underline{H}$ is normal to the plane of $\dot{\underline{v}}$ and $\underline{R}$

Vectors $\underline{E}$ and $\underline{H}$ are perpendicular to each other and both are perpendicular to $\underline{R}$. The vector $\underline{E}$ lies in the plane of $\dot{\underline{v}}$ and $\underline{R}$, (1.65a). In other words, the radiation is *polarized* in the plane that contains $\dot{\underline{v}}$ and $\underline{R}$. Equation (1.66) shows that E vanishes for $\theta = 0$ and has its maximum value for $\theta = \pi/2$, when $v/c \lesssim 1/10$ (low velocity).

The intensity $I(\theta)$ of this radiation is given by the magnitude of the Poynting vector[1] $\underline{S}$

$$I(\theta) = |\underline{S}| = \frac{c}{4\pi}|\underline{E}\times\underline{H}| = \frac{c}{4\pi} E^2 = \frac{e^2}{4\pi c^3 R^2} \dot{v}^2 \sin^2\theta \ , \ v/c << 1 \quad . \tag{1.67}$$

Equation (1.67) gives the well-known $\sin^2\theta$ angular distribution shown in Fig.1.13. A three-dimensional model of the distribution would look like a fat tire with its axis parallel to $\dot{\underline{v}}$. The rate at which an electron loses energy by radiation is obtained by integrating over a spherical surface around the electron. This is the *power*, or energy per unit time

$$-\frac{dU}{dT} = P = \int_0^\pi I(\theta) \cdot 2\pi R \sin\theta \cdot R d\theta$$

$$= \frac{e^2 \dot{v}^2}{2c^3} \int_0^\pi \sin^3\theta d\theta = \frac{2}{3}\frac{e^2 \dot{v}^2}{c^3} \quad \text{(in Gaussian units)} \quad . \tag{1.68}$$

This is called *Larmor's formula* for a nonrelativistic accelerated charge.

It is often useful to express (1.68) in terms of *electric dipole moment* instead of the electric charge of an electron. If an atom is neutral when the accelerated electron is at $x = 0$, a dipole moment $D = ex$ is created when the electron moves to x. We can write $e\dot{v} = ed^2x/dt^2 = \ddot{D}$. Then

---

[1]Multiply (1.17c) by $\underline{E}$ and (1.17a) by $\underline{H}$ and combine the results to get

$$\frac{1}{c}\underline{E} \cdot \frac{\partial \underline{E}}{\partial t} + \frac{1}{c}\underline{H} \cdot \frac{\partial \underline{H}}{\partial t} = -\frac{4\pi}{c}\underline{j} \cdot \underline{E} - [\underline{H}\cdot(\nabla\times\underline{E}) - \underline{E}\cdot(\nabla\times\underline{H})] \quad .$$

Using $\nabla\cdot(\underline{A}\times\underline{B}) = \underline{B}\cdot(\nabla\times\underline{A}) - \underline{A}\cdot(\nabla\times\underline{B})$, we have

$$\frac{\partial}{\partial t}\left(\frac{E^2+H^2}{8\pi}\right) = -\underline{j} \cdot \underline{E} - \frac{c}{4\pi}\nabla \cdot (\underline{E}\times\underline{H}) \quad .$$

The vector $\underline{S} = (c/4\pi)\underline{E}\times\underline{H}$ is called the *Poynting vector*. Comparison with the equation of continuity suggests that $(E^2+H^2)/8\pi = E^2/4\pi$ is the energy per unit volume or *energy density*. Consequently, $(c/4\pi)E^2$ represents the energy that passes normally through unit area in unit time. Because $|\underline{S}| = (c/4\pi)E^2$, it represents the *intensity* of the radiation at the instant under consideration.

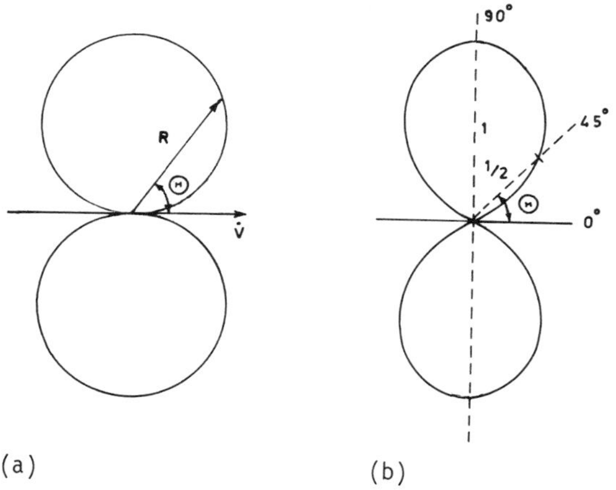

Fig. 1.13. Angular distribution of radiation of an electron, $v \ll c$: (a) $E \propto \sin\theta$, (b) $I(\theta) \propto \sin^2\theta$

$$E = H = \frac{\ddot{D}}{rc^2}\sin\theta \quad , \tag{1.69}$$

$$P = \frac{2}{3c^3}\ddot{D}^2 \quad . \tag{1.70}$$

1.12 Polarization of Continuous X-Rays

According to classical electromagnetic theory, if X-rays are electromagnetic waves, they should be completely polarized with the electric vector in the same plane as the deceleration of cathode electrons. This was first demonstrated by BARKLA [1.1], using the experimental arrangement shown in Fig.1.14. The primary X-ray beam from the X-ray tube was scattered by a scattering block of graphite. The scattered intensity of the X-ray beam was measured in two mutually perpendicular directions. The X-ray tube was first placed so that the flow of electrons in the tube was parallel to the z axis. Denoting the intensity of the X-ray beam in a parallel direction to the tube axis by $I_\parallel$ (ionization in detector 1) and that at right angles (detector 2) by $I_\perp$, we can define the magnitude of the polarization by

$$P_{pol} = \frac{I_\perp - I_\parallel}{I_\perp + I_\parallel} \quad . \tag{1.71}$$

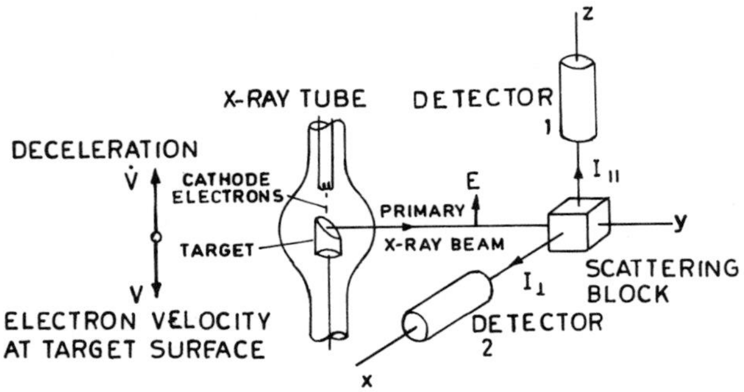

Fig. 1.14. Barkla's experiment for detecting polarization of primary X-rays

BARKLA found that $I_\perp$ was greater than $I_\parallel$ by about 20 percent. Same result was obtained with the tube axis parallel to x. Thus the primary beam was partly polarized. This was a most important discovery as it proved that X-rays were transverse electromagnetic waves.

If the deceleration $\dot{\underline{v}}$ of the electron inside the target is parallel to the z axis then the resulting primary X-ray beam will be completely polarized with electric vector $\underline{E}$ parallel to the z axis. The electron bound to an atom in the scattering block will therefore receive an acceleration $\underline{a}$ in the same direction

$$\underline{a} = \frac{\text{Force}}{\text{mass}} = \frac{e\underline{E}}{m} \quad . \tag{1.72}$$

This accelerated electron will in turn radiate a transverse electromagnetic wave (X-rays) according to (1.69), which at the point of observation P will be

$$E_P = \frac{e|\underline{a}|}{c^2R}\sin\theta = \frac{e^2|\underline{E}|}{mc^2R}\sin\theta = \begin{cases} 0 & \text{for } \theta = 0 \quad , \\ e^2E/mc^2R & \text{for } \theta = \pi/2 \quad . \end{cases} \tag{1.73}$$

Therefore, BARKLA should have recorded $I_\parallel = 0$, giving $P_{pol} = 100$ percent. The fact that he got $I_\parallel \neq 0$ and only $I_\parallel < I_\perp$ shows that some of the primary radiation is due to electrons that are not being decelerated parallel to the z axis in the thick target. After one collision with the target atom, the direction of $\dot{\underline{v}}$ for the cathode electron changes. We can expect complete polarization for a very thin target in which the cathode electrons do not have their direction of motion altered before they suffer a collision, resulting in deceleration and accompanying radiation. Thin aluminium foil targets in the X-ray tube have been used by KULENKAMPFF [1.2]. They have shown a strong polarization of primary X-rays in the predicted plane. DASANNACHARYA [1.3] extended this work for several thicknesses of Al foils and different tube voltages.

1.13 The Case of $\dot{\underline{v}}$ Parallel to $\underline{v}$

This case is of special interest for the theory of the production of continuous X-rays. If $\dot{\underline{v}}$ and $\underline{v}$ are along the *same* direction, $\underline{v}\times\dot{\underline{v}}=0$, then for any v, small or large, the radiation fields are

$$\underline{E} = \frac{e}{c^2s^3}[\underline{R}\times(\underline{R}\times\dot{\underline{v}})] \quad , \tag{1.74a}$$

$$\underline{H} = \frac{eR}{c^2s^3}(\dot{\underline{v}}\times\underline{R}) \quad . \tag{1.74b}$$

These results differ from (1.65) only by the factor

$$\frac{R^3}{s^3} = \frac{R^3}{\left(R-\frac{\underline{R}\cdot\underline{v}}{c}\right)^3} = \frac{1}{\left(1-\frac{v}{c}\cos\theta\right)^3} \quad , \tag{1.75}$$

where θ is the angle between $\underline{R}$ and the common direction of $\dot{\underline{v}}$ and $\underline{v}$. The qualitative effect of this relativistic weighting factor is to increase the radiation in the forward direction (Fig.1.15). The lobes of Fig.1.13 are tilted in the direction of $\underline{v}$ owing to this weighting factor.

From (1.66,75), we can write

$$E = H = \frac{R^3}{s^3}\frac{e\dot{v}}{c^2R}\sin\theta = \frac{e\dot{v}}{c^2R}\frac{\sin\theta}{\left(1-\frac{v}{c}\cos\theta\right)^3} \quad . \tag{1.76}$$

The corresponding instantaneous intensity for one decelerated electron is

$$I(\theta) = \frac{c}{4\pi}E^2 = \frac{e^2\dot{v}^2}{4\pi c^3R^2}\frac{\sin^2\theta}{\left(1-\frac{v}{c}\cos\theta\right)^6} \quad . \tag{1.77}$$

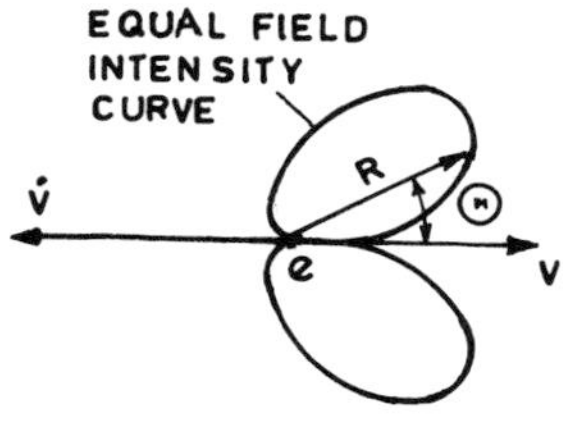

Fig. 1.15. Radiation pattern for a fast electron decelerated along its line of motion

For $v/c \ll 1$, it reduces to (1.67). The direction θ_{max} of the propagation of the greatest intensity is determined by the condition $dI/d\theta = 0$. Solving for $\cos\theta$, we get

$$\cos\theta_{max} = \frac{1}{4\frac{v}{c}}\left[\left(1+24\frac{v^2}{c^2}\right)^{\frac{1}{2}} - 1\right] \quad . \tag{1.78}$$

1.14 Sommerfeld's Theory for the Spatial Distribution of Continuous X-Rays

The incident electrons will be stopped by a succession of collisions with the target atoms in an X-ray tube. Each collision results in deceleration of the electron and emission of radiation. If the target is thick, such multiple collisions deflect the electrons irregularly. STOKES [1.4] assumed that continuous X-rays are produced in irregular electromagnetic pulses due to the irregular decelerations of the electrons in the target. THOMSON [1.5] for purpose of calculation simply assumed that the cathode electron is brought to *rest* within a very short distance inside the target by a *uniform deceleration along its original line of motion.*

SOMMERFELD [1.6] used this assumption to calculate the total radiated energy traversing unit area at the point of observation P, due to stopping of the electron

$$I_{tot}(\theta) = \int I(\theta)dt \quad . \tag{1.79}$$

The integration is over the entire time of the radiation pulse, that is, until the electron comes to rest. I is given by (1.77). If t is the time at which the radiation reaches P, which left the electron at the instant T, then $t = T + R/c$ and $dt = dT + dR/c$. From Fig.1.16 we see $dR = -vdT \cos\theta$ and so $dt = \left(1-\frac{v}{c}\cos\theta\right)dT$, in agreement with (1.53).

The energy emitted by the electron in time dT is located in the volume between two spheres, one of radius $R + cdT$ about O and the other of radius R around O', where $OO' = vdT$. The energy that crosses unit area at P in observer's time dt, during which the entire pulse is received, is $I(\theta)cdt = I(\theta)(c-v\cos\theta)dT$. Because $\dot{v} = dv/dT$, we can write $cdt = (c-v\cos\theta)dv/\dot{v}$ and

$$\begin{aligned} I_{tot}(\theta) &= \int I(\theta)dt = \int I(\theta)\left(1-\frac{v}{c}\cos\theta\right)\frac{dv}{\dot{v}} \\ &= \int \frac{e^2\dot{v}^2}{4\pi c^3R^2}\,\frac{\sin^2\theta}{\left(1-\frac{v}{c}\cos\theta\right)^6}\left(1-\frac{v}{c}\cos\theta\right)\frac{dv}{\dot{v}} \quad . \end{aligned} \tag{1.80}$$

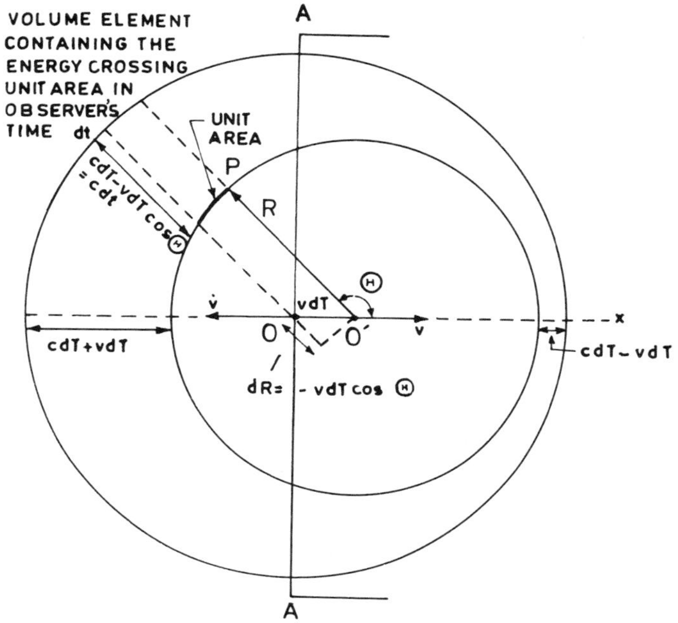

Fig. 1.16. Location of energy radiated by an electron incident on a target AA, as the electron moves from O to O' in time dT

The integral over dv extends from the initial velocity v_0 of the electron to the final zero velocity. If we assume that $\dot{v}$ is constant while the velocity decreases from v_0 to 0, we get

$$I_{tot}(\Theta) = \frac{e^2\dot{v}\,\sin^2\Theta}{4\pi c^3R^2}\int\limits_{v_0}^{0}\frac{dv}{\left(1-\frac{v}{c}\cos\Theta\right)^5}$$

$$= \frac{e^2\dot{v}\,\sin^2\Theta}{16\pi c^2R^2\,\cos\Theta}\left[\frac{1}{\left(1-\frac{v_0}{c}\cos\Theta\right)^4}-1\right] \quad . \tag{1.81}$$

For small v/c

$$I_{tot}(\Theta) = \frac{e^2\dot{v}\,\sin^2\Theta}{4\pi c^3R^2}\int\limits_{v_0}^{0}dv = \frac{e^2\dot{v}v_0}{4\pi c^3R^2}\sin^2\Theta \quad . \tag{1.82}$$

Thus, the classical theory predicts the energy-distribution polar diagram for radiation. In Fig.1.17 we plot the intensity as a function of angle Θ, according to (1.81), for three values of v_0/c. The movement of the direction of max-

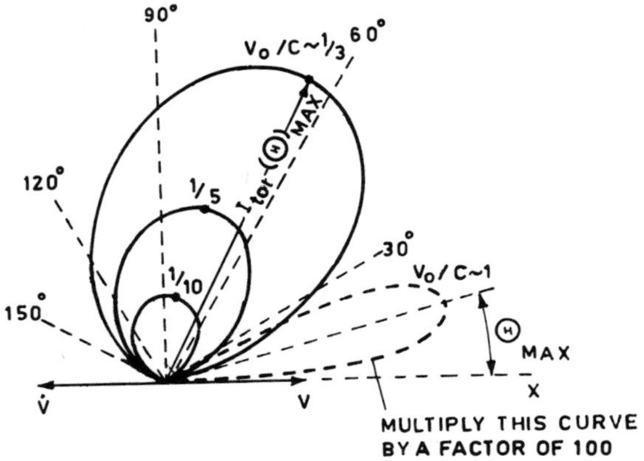

Fig. 1.17. Theoretical radiation pattern for an electron decelerated in its direction of motion, when the detector is placed at a given distance R. The solid dot represents the direction of maximum intensity $I_{tot}(\Theta)(\max)$. The dotted curve is for $v_0/c \sim 1$ and has been reduced by a factor of about 100 for convenience. $v_0/c = 1/3$ corresponds to about 30 keV kinetic energy of the electron

imum intensity toward the direction of incident-electron velocity (forward direction) with the increase of the incident velocity is easily seen. An X-ray tube is like a low-velocity linear accelerator ($v_0/c \lesssim 1/3$). For a high-velocity accelerator ($v_0/c \sim 1$), the radiation is confined to a very narrow cone in the direction of the motion (dotted curve in Fig.1.17).

Sommerfeld's condition, that the bombarding electrons are decelerated along the direction of incident motion, is likely to be satisfied in two cases: 1) extremely thin target, and 2) thick target of very low atomic number. In both of these cases, we expect no significant deflection of the electrons from their original direction. The radiation loss from the electron while it is being accelerated between the cathode and the target of the X-ray tube is negligible, because this acceleration over a large distance is very slow in comparison with the rapid deceleration inside the solid target. In fact, the X-ray tube acts like a linear accelerator in which the motion of the electron is essentially one-dimensional. From (1.68) we have for an *accelerated* electron

$$P = \frac{2}{3}\frac{e^2}{m^2c^3}\left(\frac{dp}{dt}\right)^2 = \frac{2}{3}\frac{e^2}{m^2c^3}\left(\frac{dE}{dx}\right)^2 \quad ,$$

where $p = mv$ and the rate of change of momentum is equal to the change of energy E of the particle per unit distance, dE/dx. This shows that, for linear motion, the power radiated depends only on the external forces ($E = eV$, V being tube voltage) that determine the rate of change of the electron energy with distance and not on the actual energy or momentum of the electron. The ratio of power radiated to power supplied by the tube voltage is

$$\frac{P}{dE/dt} = \frac{2}{3}\frac{e^2}{m^2c^3}\frac{1}{v}\frac{dE}{dx} \xrightarrow[\beta\to 1]{} \frac{2}{3}\frac{e^2/mc^2}{mc^2}\frac{dE}{dx} \quad .$$

Thus, the radiation loss will be unimportant unless the gain of energy is of the order of $mc^2 = 0.51$ MeV in a distance of $e^2/mc^2 = 2.8\times 10^{-13}$ cm, or $\sim 2\times 10^{15}$ keV cm^{-1}. In an X-ray tube, the distance between the filament and the target is of the order of a cm and $V \sim 20$ kV. Therefore, radiation loss is negligible compared to the energy gain of the electron accelerated in the direction of and by the electric field eV. In the target, owing to rapid deceleration in collisions, dE/dx has a large value, which increases with Z. For circular accelerators, like the synchrotron, the radiation loss is greater. The Sommerfeld theory has been put in a useful form by SCHEER and ZEITLER [1.7].

1.15 Frequency Spectrum of Continuous X-Rays

The radiated energy is spread over a range of frequencies. This frequency spectrum can be obtained by Fourier analysis of the radiation field.

The radiation takes place when the incident electron suddenly changes its direction during a collision with one of the atoms in the target. If we assume zero collision time and $v \ll c$ then we can write the resulting acceleration and field as

$$\dot{v} = \delta(t_0 - t)\Delta v \quad , \quad \int \dot{v}\, dt = \Delta v \quad , \tag{1.83}$$

$$E(t) = \frac{e \sin\theta}{c^2 R}\Delta v\; \delta(t_0 - t) \quad , \tag{1.84}$$

where t_0 is the instant at which the radiation takes place. In (1.84), the instantaneous field is expressed in the observer's time because we wish to consider a frequency spectrum in terms of the observer's frequencies ω.

We introduce the Fourier transform

$$E(t) = \int_0^{+\infty} E(\omega)\, e^{-i\omega t}\, d\omega \quad , \tag{1.85}$$

where ω is the angular frequency, $\omega = 2\pi\nu$. It is customary to integrate only over positive frequencies, because the sign of the frequency has no physical meaning. The inverse is

$$E(\omega) = \frac{1}{\pi}\int_{-\infty}^{+\infty} E(t)\, e^{i\omega t}\, dt \quad . \tag{1.86}$$

We can write, from (1.84,86),

$$E(\omega) = \frac{e \sin\Theta}{\pi c^2 R}\, \Delta v\, e^{i\omega t_0} \quad . \tag{1.87}$$

The total energy radiated is given by the integral of the Poynting vector over the surface of a sphere and over the time during which the change of velocity takes place. For $v \ll c$

$$-U = \int \left(-\frac{dU}{dT}\right) dT = \frac{c}{4\pi}\iint E^2\, dt d\sigma \quad . \tag{1.88}$$

By Parseval's formula[2]

$$\int_{-\infty}^{+\infty} |E|^2\, dt = \pi\int_0^{\infty} |E(\omega)|^2\, d\omega \quad . \tag{1.89}$$

Therefore, for the energy radiated in a frequency interval $d\omega$, we have

$$-U(\omega) d\omega = \pi\, \frac{c}{4\pi}\int_{\sigma} |E(\omega)|^2\, d\sigma d\omega \quad . \tag{1.90}$$

For the spectrum (1.87)

$$-U(\omega) d\omega = \frac{e^2 (\Delta v)^2}{4\pi^2 c^3 R^2}\, d\omega \int_0^{2\pi} d\Phi \int_0^{\pi} \sin^2\Theta\, R^2 \sin\Theta d\Theta = \frac{4e^2}{3c}\left(\frac{\Delta v}{c}\right)^2 \frac{d\omega}{2\pi} \quad . \tag{1.91}$$

---

[2]Parseval's formula is

$$\pi\int_0^{\infty} |g(\omega)|^2\, d\omega = \int_{-\infty}^{+\infty} |f(t)|^2\, dt \quad , \text{ where}$$

$$f(t) = \int_0^{+\infty} g(\omega)\, e^{-i\omega t}\, d\omega \quad , \quad g(\omega) = \frac{1}{\pi}\int_{-\infty}^{+\infty} f(t)\, e^{i\omega t}\, dt \quad .$$

Proof:

$$\int_0^{+\infty} |g(\omega)|^2\, d\omega = \int_0^{\infty} g^*(\omega) g(\omega)\, d\omega = \int_0^{+\infty} g^*(\omega)\, \frac{1}{\pi}\int_{-\infty}^{+\infty} f(t)\, e^{i\omega t}\, d\omega dt \quad ,$$

$$\int_{-\infty}^{+\infty} |f(t)|^2\, dt = \int_{-\infty}^{+\infty} f(t) \int_0^{\infty} g^*(\omega)\, e^{i\omega t}\, d\omega dt = \pi\int_0^{\infty} |g(\omega)|^2\, d\omega \quad . \qquad \text{QED}$$

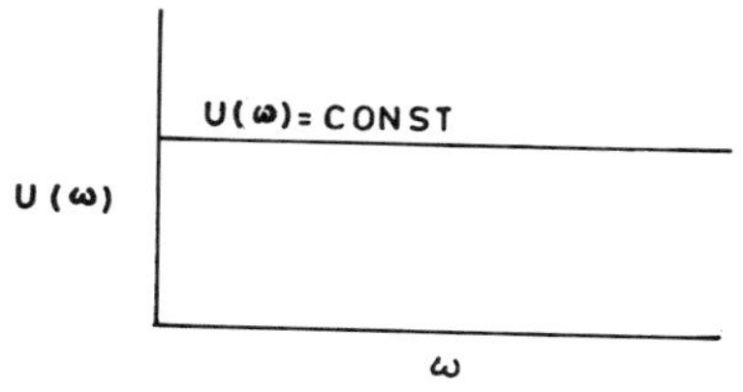

Fig. 1.18. Frequency spectrum of continuous X-rays, according to the classical theory

Equation (1.91) shows that the spectrum of radiation is independent of frequency ω (Fig.1.18). The spectrum extends to high ω values, owing to the simplifying assumption of zero collision time. A Fourier analysis of a finite-collision-time process will damp the very-high-frequency components but *not* abruptly, as observed (see below).

1.16 Experimental Spectral and Spatial Distributions

NICHOLAS [1.8] studied the spectral distribution of continuous X-rays from thin-foil targets, using a crystal to analyze the frequency (or wavelength) spectrum and a photographic-comparison method to measure the intensity. Foils were ^{13}Al 0.7 μm thick and ^{79}Au 0.99 μm thick. The voltage on the X-ray tube was 45 kV and angles were 40°, 90°, and 140° to the forward direction of the incident electrons. His results are shown in Fig.1.19. The data are in agreement with the classical theory (Fig.1.18). A new feature is the appearance of a *cutoff* at a frequency ν_{max} for a given voltage on the X-ray tube, which is not predicted by the classical theory.

The spectral and spatial distributions from thin foils have also been studied by KULENKAMPFF [1.2], BOHM [1.9], HONERJÄGER [1.10], and HARWORTH and KIRKPATRICK [1.11,12]. KULENKAMPFF and BOHM used 0.6 μm (6000 Å) thick foil of ^{13}Al as target. HONERJÄGER used thinner ^{13}Al foils (100, 350 and 1000 Å)

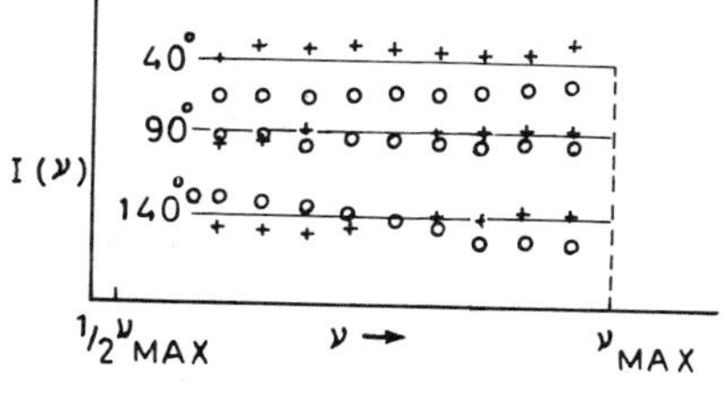

Fig. 1.19. Thin-target spectra at different angles. o Al and + Au data are not on the same scale of intensity

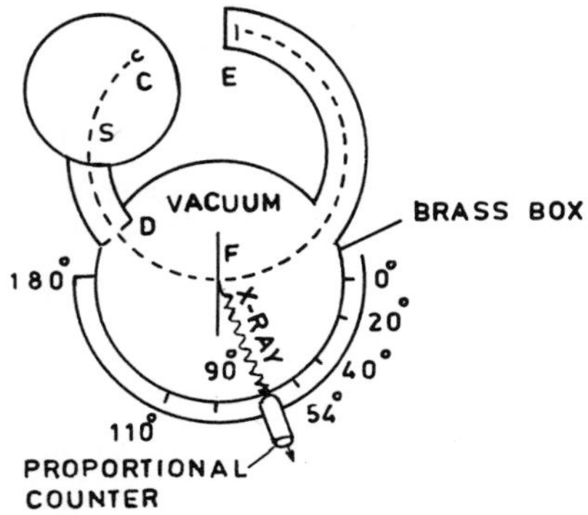

Fig. 1.20. Apparatus for the study of continuous X-ray spectrum of thin foils

and ^{28}Ni foil of 150 Å thickness (1 Å = 10^{-10} m). HARWORTH and KIRKPATRICK used 199 Å thick Ni foil. As the radiation from such thin foils is very weak, it is not possible to obtain reliable analysis of the spectrum with a crystal. These workers used selective absorbers to select out portions of the spectrum, and used ionization chambers as detectors. The Ross balanced-filter technique can be used to isolate portions of the spectrum. Ag-Pd metal sheets pass a band of 485 to 508 XU and Se-As pair of powdered metals pass the band 979 to 1044 XU. Note that 1 kXU = 1000 XU = 1.00202 Å.

Recently, KERSCHER and KULENKAMPFF [1.13] studied the spectral and spatial distribution of the intensity of the continuous spectrum from an ^{13}Al foil of thickness 250 Å for 34 keV electrons. Their apparatus is shown in Fig.1.20. It consists of a flat, brass vacuum chamber placed between the poles of an electromagnet (not shown). The cathode C emits an electron beam that is accelerated by a voltage of 34 kV, exits through the slit S and moves along a circle that passes through the diaphragm D and the foil F under study. Finally, the electron beam is collected by the electrode E placed at the end of a curved leg so as to isolate the X-rays that may be emitted by it. The detector for the X-rays, emitted by the foil F, is a proportional counter that moves along an arc that extends from -5° to $+185^\circ$. X-rays from F come out of the box from a long slot on one side of the chamber to fall on the counter. The slot is covered by an Al foil to maintain the vacuum. The proportional counter, used with a single-channel pulse-height analyzer, measures the intensity (number of photons) as well as the energy (frequency) of the X-rays received by it.

The measured distribution of intensity along the spectrum is given in Fig.1.21. These straight lines were obtained at various angles θ to the forward direction of the bombarding electrons. The slopes of the lines nearly

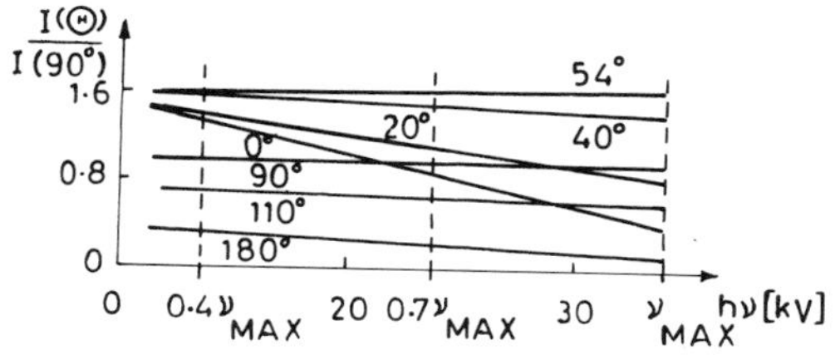

Fig. 1.21. The distribution of spectral intensity for 250 Å Al foil at V = 34 kV for various angles θ

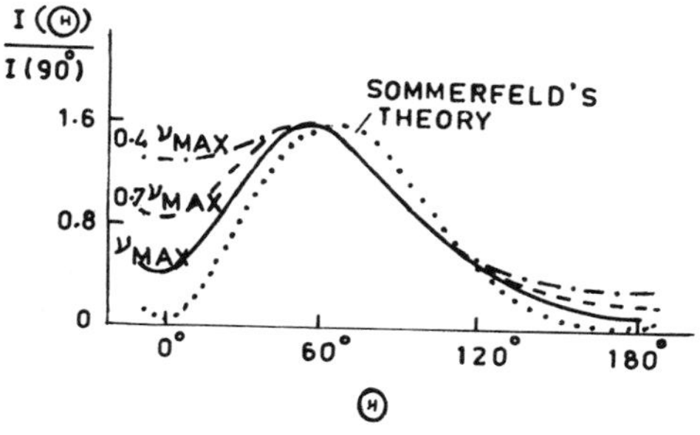

Fig. 1.22. Spatial distribution from the observed spectral distribution curves at various angles shown in Fig.1.21. The theoretical curve (dotted) is according to (1.81)

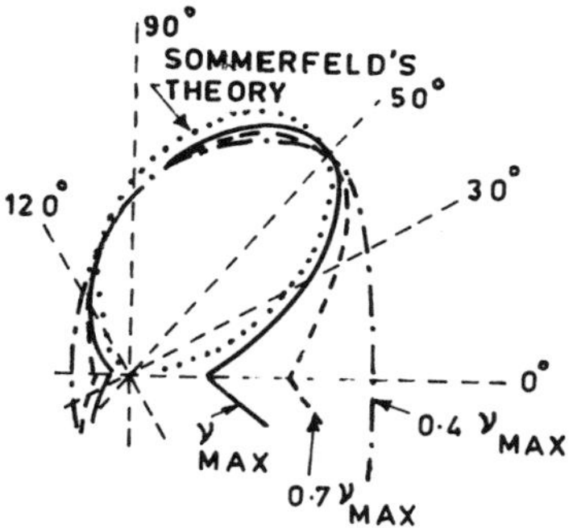

Fig. 1.23. Intensity as a function of direction in polar coordinates. The theoretical curve (dotted) is according to (1.81)

agree with the results of NICHOLAS in the range $40^{\circ} < \theta < 140^{\circ}$. Outside of this range, especially at smaller angles, the lines become slanting.

KERSCHER and KULENKAMPFF obtained distribution curves from Fig.1.21 by reading intensities at chosen values of $\nu = \nu_{max}$, 0.7 ν_{max} and 0.4 ν_{max}. The resulting curves are shown in Fig.1.22 and are in agreement with the results of HONERJÄGER [1.10]. These curves can also be drawn on a polar diagram, as shown in Fig.1.23. Experiments at higher energies (45 to 170 kV) have been made by THORDARSON [1.14] and SESEMANN [1.15]. In their experiments, the transmission target formed the window in the X-ray tube and the intensity of continuous radiation transmitted through the target was easily observed, down to $\theta = 0^{\circ}$. Such a transmission target approaches a thin target in its effect.

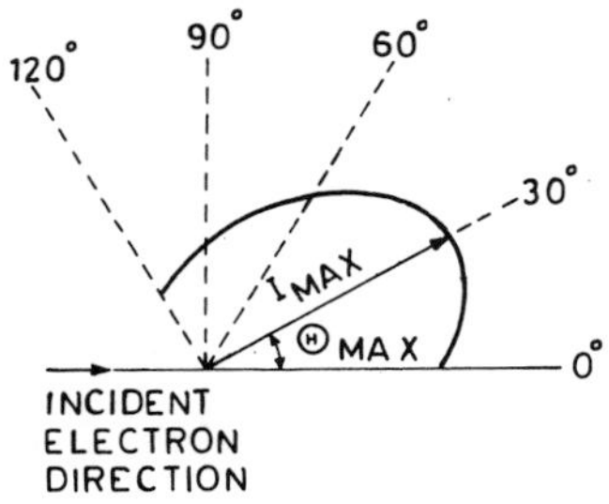

Fig. 1.24. Spatial distribution of $I_{tot}(\Theta)$ of continuous X-rays from a thick ^{4}Be target at V = 140 kV

Measurements on the thick target of ^{4}Be have been made by DETERMANN [1.16] at a voltage V = 140 kV. His results (Fig.1.24) show that even for such a light element as beryllium the distribution of intensity is nearly the same in all directions. In particular, the intensity at $\Theta = 0$ is not only not zero, as theoretically predicted by SOMMERFELD, but even exceeds the intensity at 90° and is only a little less than the maximum intensity.

Note that the mean depth of X-ray production inside a solid target is of the order of 10^4 Å [1.17].

1.17 Shortcomings of Classical Theory

There is a strong resemblance between the theoretical curves of Fig.1.17 and experimental curves of Fig.1.23. However, there are significant differences that cannot be ignored. In Table 1.1, we compare the experimental situation of spectral and spatial distributions of continuous X-rays with the predictions from the classical theory.

It is clear from Table 1.1 that the classical theory can give only qualitative agreement with experiment, in respect to low-frequency spectral distribution, spatial distribution and polarization. However, it utterly fails to explain the observed cutoff frequency ν_{max}. Although the former aspects require a moving electron that is brought to rest with a constant negative acceleration, the latter (ν_{max}) aspect requires an oscillating electron. So there is a conflict; to resolve it we must move out of the classical theory and use quantum ideas.

Table 1.1

Classical theory	Experimental data	Remarks
1) Θ_{max} is ~65° for $\frac{v}{c} \sim \frac{1}{3}$ (Fig.1.17)	Θ_{max} is ~55° for $\frac{v}{c} \sim \frac{1}{3}$ for thin targets (Fig.1.23)	SOMMERFELD assumed that the electrons are brought completely to rest in the target, whereas, in thin targets, the electrons are not greatly retarded while passing through them
2) Intensity is zero at $\Theta = 0°$ and $\Theta = 180°$ (Fig.1.17) for all frequencies	Intensity is not zero at $\Theta = 0°$, and $\Theta = 180°$ (Figs.1.23, 24). For thin targets, the intensity at $\Theta = 0°$ is minimum for the cutoff frequency ν_{max} and increases for lower frequencies. For thick targets, intensity is almost as great at $\Theta = 0°$ as at $\Theta = 90°$	SOMMERFELD assumed that the electrons are decelerated in the direction of motion. It appears that not all electrons do so and that some transverse motion takes place owing to collisions with the target atoms. Such transverse motions are least likely to occur in the first few atomic layers of the target from where the radiation near the ν_{max} is emitted. As electrons enter deeper in the target, they emit radiation of lower frequencies and also are more and more deflected
3) Polarization should be complete in the plane of the incident electron beam	Polarization is partial	All electrons are not decelerated in the direction of motion, as assumed by SOMMERFELD
4) No sharp high-frequency limit is predicted	There is sharp high-frequency limit, ν_{max}, determined by the voltage on the X-ray tube	Quantum theory is needed to explain this cutoff frequency in a natural way. The maximum energy that an electron can give up is eV. When all of it is converted into a quantum of radiation energy, we get $eV = h\nu_{max}$, where h is Planck's constant

1.18 Semiclassical Quantum Theory of Kramers

A semiclassical quantum theory of the continuous spectrum was developed by KRAMERS [1.18] and extended by WENTZEL [1.19]. KRAMERS first considered a classical theory of bremsstrahlung in nonrelativistic Coulomb collisions, then, using the Bohr's correspondence principle, he introduced quantum theory in the calculation.

At low electron energies ($v/c \sim 1/3$) Sommerfeld's equation (1.81) gives the approximate intensity of soft (low-energy) X-rays, if empirical values of $\dot{v}$

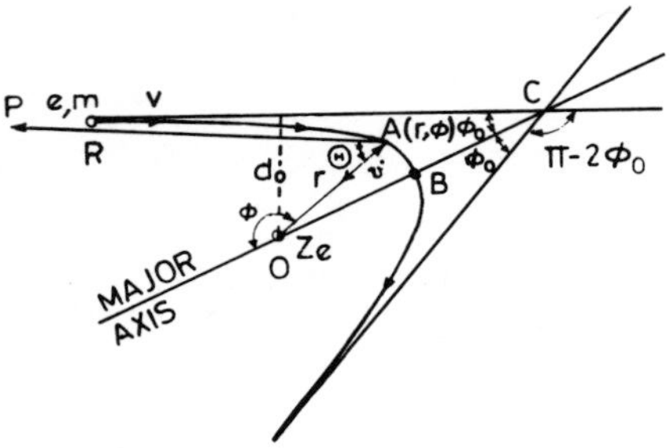

Fig. 1.25. The deflection of an electron along a parabolic path in the Coulomb field of the nucleus Ze. The impact parameter is d_0

are substituted. In this case, the collisions of incident electrons with nuclei are rare. The main source of energy loss and momentum transfer is electron-electron collision. However, single collisions between particles of like e/m do not produce dipole radiation[3] (1.69). Therefore, for low energies, where electron-electron collisions predominate, the radiation is due to the collective effects of the atomic electrons in retarding the incident electron. At higher energies, and particularly for heavy targets, the X-ray yield is radiation due to the deflection of the incident electron in the Coulomb field of a nucleus. For electron collisions with nuclei, dipole radiation is possible; in these circumstances the contribution mainly comes from single collisions.

As a model for this single-collision process, KRAMERS considered the collision of a fast, but nonrelativistic ($v/c \sim 1/3$), electron of charge e, mass m, and velocity v with a fixed nucleus of point charge Ze. Let d_0 be the impact parameter and $\pi - 2\phi_0$ the angle of scattering (Fig.1.25). The electron will describe a hyperbolic orbit with the nucleus at one of the foci such that (see Appendix A)

$$\tan\phi_0 = \frac{md_0v^2}{Ze^2} \quad . \tag{1.92}$$

[3]Consider a system of two charged particles. Uniform motion of the system as a whole (motion of center of mass) does not lead to radiation. Therefore, consider only the relative motion of the particles. Let the center of mass be the origin. Then the dipole moment of the system $\underline{D} = e_1\underline{r}_1 + e_2\underline{r}_2$ has the form

$$\underline{D} = \frac{e_1m_2 - e_2m_1}{m_1+m_2}\,\underline{r} = \mu\left(\frac{e_1}{m_1} - \frac{e_2}{m_2}\right)\underline{r} \quad ,$$

where $\underline{r} = \underline{r}_1 - \underline{r}_2$ and $\mu = m_1m_2/(m_1+m_2)$.

The deflection of the electron in the Coulomb field of the nucleus will always imply an accelerated motion and hence a possibility of radiation.

If ε is the eccentricity and r, ϕ the polar coordinates of any point A on the orbit, we have for an attractive potential

$$\frac{1}{r} = \frac{1-\varepsilon\cos\phi}{d_0\tan\phi_0} = \frac{1-\varepsilon\cos\phi}{d_0^2v^2}\cdot\frac{Ze^2}{m} \quad , \quad \varepsilon = \frac{1}{\cos\phi_0} \quad , \tag{1.93}$$

where all the angles are measured from the major axis. We also have

$$\dot{v} = \frac{dv(T)}{dT} = \frac{Ze^2}{mr^2} \quad , \tag{1.94}$$

$$mr^2\dot{\phi} = d_0mv = \text{constant (angular momentum is conserved)} \quad , \tag{1.95}$$

$$dU = \frac{2}{3}\frac{e^2}{c^3}\dot{v}^2\,dT \quad \text{[from (1.68)]} \quad . \tag{1.96}$$

The total loss of energy is found by integrating (1.96) over the time and using $r^2d\phi = d_0vdT$

$$U = \int_{-\infty}^{+\infty}\frac{2}{3}\frac{e^2}{c^3}\dot{v}^2\,dT = \frac{2Z^2e^6}{3c^3m^2}\int_{-\infty}^{+\infty}\frac{dT}{r^4} = \frac{2Z^2e^6}{3c^3m^2d_0v}\int_{\phi_0}^{2\pi-\phi_0}\frac{d\phi}{r^2}$$

$$= \frac{2Z^4e^{10}}{3c^3m^4d_0^5v^5}\int_{\phi_0}^{2\pi-\phi_0}(1-\varepsilon\cos\phi)^2d\phi$$

$$= \frac{2Z^4e^{10}}{3c^3m^4d_0^5v^5}\left[(2\pi-2\phi_0)\left(1+\frac{1}{2}\frac{1}{\cos^2\phi_0}\right)+3\tan\phi_0\right] \quad , \tag{1.97}$$

where in the last step we have used $\varepsilon = 1/\cos\phi_0$.

If d_0 is large, ϕ_0 is large and the angle of scattering (deflection from the original direction) $\pi - 2\phi_0$ is small. For d_0 larger than the radius of the electron cloud of the atom, the deflection of the incident electron is negligible and hence no radiation is emitted. We would therefore be interested in the case of small d_0 or *small* ϕ_0. For this case (1.97) reduces to

$$U \simeq \frac{2Z^4e^{10}}{3c^3m^4d_0^5v^5}[3\pi] = \frac{2\pi Z^4e^{10}}{c^3m^4d_0^5v^5} \quad , \tag{1.98}$$

and

$$\varepsilon = \frac{1}{\cos\phi_0} \simeq 1 \quad \text{(orbit is a parabola)} \quad . \tag{1.99}$$

The acceleration of the electron, $\dot{\underline{v}}$, can be resolved into a component $\dot{v}_{\parallel} = -\dot{v}\cos\phi$ parallel to the axis of the parabola and a component $\dot{v}_{\perp} = -\dot{v}\sin\phi$ perpendicular to the axis.

To simplify further discussion we reduce the problem to a particular parabola for which

$$d_0 v = 1 \quad , \quad \frac{Ze^2}{m} = 1 \quad \text{and so} \quad U = \frac{2\pi e^2}{c^3} \quad . \tag{1.100}$$

To obtain the spectral distribution, we recall, (1.69), that the energy of the electromagnetic radiation can be expressed in terms of the dipole moment of an electron harmonically oscillating about some point. To find the frequency distribution of the energy radiated by an electron moving along a parabola, we can resolve the component accelerations $\dot{v}_{\parallel}$ and $\dot{v}_{\perp}$ according to the frequencies of such oscillators, the aggregate of which (Fourier integral) gives the same effect as does the electron under study. Following KRAMERS, we define our Fourier transforms for motion along the *reduced parabola* (1.100) in terms of the *reduced frequency* γ, as

$$\begin{aligned} \dot{v}_{\parallel}(t) &= \int_0^{\infty} \Psi(\gamma)\cos\gamma t\, d\gamma \quad , \\ \dot{v}_{\perp}(t) &= \int_0^{\infty} \Phi(\gamma)\sin\gamma t\, d\gamma \quad , \end{aligned} \tag{1.101}$$

and the inverse

$$\begin{aligned} \Psi(\gamma) &= \frac{1}{\pi}\int_{-\infty}^{+\infty} \dot{v}_{\parallel}(\tau)\cos\gamma\tau\, d\tau \quad , \\ \Phi(\gamma) &= \frac{1}{\pi}\int_{-\infty}^{+\infty} \dot{v}_{\perp}(\tau)\sin\gamma\tau\, d\tau \quad . \end{aligned} \tag{1.102}$$

Note that it does not matter what variable is used in (1.102) on the right-hand side, as it is ultimately integrated out.

The total energy emitted is

$$U = \int_0^{\infty} U(\gamma)d\gamma = \int_{-\infty}^{+\infty} \frac{2}{3}\frac{e^2}{c^3}\dot{v}^2\, dT = \frac{2e^2}{3c^3}\int_{-\infty}^{+\infty}\left(\dot{v}_{\parallel}^2 + \dot{v}_{\perp}^2\right) dT = \frac{2\pi e^2}{c^3} \quad , \tag{1.103}$$

where in the last step we have used (1.100). This shows that

$$\int_{-\infty}^{+\infty}\left(\dot{v}_{\parallel}^2 + \dot{v}_{\perp}^2\right) dT = 3\pi = \pi\int_0^{\infty}\left[\Psi^2(\gamma) + \Phi^2(\gamma)\right]d\gamma \quad , \tag{1.104}$$

where we have used Parseval's formula.

We introduce the function

$$P(\gamma) \equiv \frac{1}{3}\left[\Psi^2(\gamma)+\Phi^2(\gamma)\right] = \frac{c^3}{2\pi e^2}\,U(\gamma) = \frac{U(\gamma)}{U} \tag{1.105}$$

with the normalization

$$\int_0^\infty P(\gamma)d\gamma = 1 \quad . \tag{1.106}$$

We can now write

$$U = \frac{2\pi e^2}{c^3}\int_0^\infty P(\gamma)d\gamma = \frac{2\pi e^2}{c^3} \quad . \tag{1.107}$$

According to (1.105), the function $P(\gamma)$ determines the distribution of energy with reduced frequency for the reduced parabola. $P(\gamma)d\gamma$ gives the relative amount of radiation energy for which the reduced frequency lies between γ and $\gamma + d\gamma$. KRAMERS has estimated $P(\gamma)$ and found

$$\int \frac{P(\gamma)}{\gamma}\,d\gamma = \text{constant} = \frac{4}{\pi\sqrt{3}} \quad . \tag{1.108}$$

We can translate our results from the reduced parabola to the actual parabola. From (1.93) we find that the distance OB of the perihelion B of the parabola from the focus O (Fig.1.25) for the case ϕ_0 small is

$$OB = \left[\frac{m}{Ze^2}\,\frac{d_0^2v^2}{1-\varepsilon\cos\phi}\right]_{\varepsilon=1,\phi=180^\circ} = \frac{md_0^2v^2}{2Ze^2} \quad .$$

The angular velocity of the electron at the perihelion, from (1.95), is

$$\dot{\phi}_B = \frac{d_0v}{(OB)^2} = \frac{4Z^2e^4}{m^2d_0^3v^3} \quad \text{(actual)}$$

$$= 4 \quad \text{(reduced)} \quad . \tag{1.109}$$

Therefore, if $\omega = 2\pi\nu$ is the frequency for the actual parabola, we can write

$$\frac{2\pi\nu}{\gamma} = \frac{4Z^2e^4}{m^2d_0^3v^3}\cdot\frac{1}{4} \quad ,$$

or

$$\gamma = 2\pi\nu\,\frac{m^2d_0^3v^3}{Z^2e^4} \quad . \tag{1.110}$$

We can replace the reduced distribution function $P(\gamma)$ by the actual distribution function, such that

$$P(\gamma)d\gamma = P(\nu)d\nu \quad . \tag{1.111}$$

Using (1.110), we obtain

$$P(\nu) = P(\gamma)\,\frac{d\gamma}{d\nu} = \frac{2\pi m^2 d_0^3 v^3}{Z^2 e^4}\,P(\gamma) \quad . \tag{1.112}$$

As a result of each collision of the electron with the target atom, the electron radiates energy $U(\gamma)d\gamma$ between γ and $\gamma + d\gamma$. Using (1.105), we can write

$$U(\gamma)d\gamma = UP(\gamma)d\gamma = UP(\nu)d\nu = U(\nu)d\nu = UP(\gamma)\,\frac{d\gamma}{d\nu}\,d\nu \quad . \tag{1.113}$$

For a large number N of collisions, with the same impact parameter d_0, the energy of radiation between ν and $\nu + d\nu$ is given by

$$NU(\nu)d\nu = 2\pi N\,\frac{m^2 d_0^3 v^3}{Z^2 e^4}\,U \cdot P\left(2\pi\nu\,\frac{m^2 d_0^3 v^3}{Z^2 e^4}\right) \cdot d\nu \quad , \tag{1.114}$$

where U is given by (1.98).

Following KRAMERS, we apply the *correspondence principle* to introduce the quantum ideas. As a result of the collision of an electron with a target atom, a photon of frequency ν and energy $h\nu$ is emitted. Let $q(\nu)d\nu$ be the *quantum probability* of a collision in which a photon of energy $h\nu$ with frequency between ν and $\nu + d\nu$ is emitted. For a large number N of collisions, the total energy emitted between ν and $\nu + d\nu$ is

$$NU(\nu)d\nu = Nq(\nu)h\nu d\nu \quad . \tag{1.115}$$

Using the correspondence principle, we can equate it with (1.114)

$$Nq(\nu)h\nu d\nu = NUP(\gamma)\,\frac{d\gamma}{d\nu}\,d\nu \tag{1.116}$$

$$= 2\pi N\,\frac{m^2 d_0^3 v^3}{Z^2 e^4}\,U \cdot P\left(2\pi\nu\,\frac{m^2 d_0^3 v^3}{Z^2 e^4}\right) \cdot d\nu$$

$$= N\,\frac{4\pi^2 Z^2 e^6}{c^3 m^2 d_0^2 v^2} \cdot P\left(2\pi\nu\,\frac{m^2 d_0^3 v^3}{Z^2 e^4}\right) \cdot d\nu \quad \text{for} \quad \nu < \nu_{max} \quad ,$$

$$= 0 \quad \text{for} \quad \nu > \nu_{max} \quad , \tag{1.117}$$

where ν_{max} is determined by the condition that the entire energy of the incident electron is emitted as a photon

$$eV = \frac{1}{2} mv^2 = h\nu_{max} \quad . \tag{1.118}$$

Here $h = 6.626 \times 10^{-27}$ erg-s is Planck's constant.

We can calculate the energy distribution in the continuous spectrum from a *thin* target. Let the thin target contain A atoms per cm^2 of atomic number Z. It is bombarded by a beam of electrons of velocity v in which n electrons in 1 s pass across a cross section of 1 cm^2. We can take $N = nA$. The spectral density $I(\nu)$ of the continuous spectrum between ν and $\nu + d\nu$ is then obtained by integrating (1.117) over all directions of the impact parameter d_0 and over all the values of this parameter, $0 < d_0 < \infty$, for the thin target surface. Thus

$$\begin{aligned} I(\nu)d\nu &= Nd\nu \int_0^\infty q(\nu)h\nu \cdot 2\pi d_0 dd_0 \\ &= Nd\nu \frac{8\pi^3 Z^2 e^6}{c^3 m^2 v^2} \int P\left(2\pi\nu \frac{m^2 d_0^3 v^3}{Z^2 e^4}\right) \cdot \frac{dd_0}{d_0} \quad . \end{aligned} \tag{1.119}$$

From (1.110), we have

$$\begin{aligned} d_0^3 &= \frac{Z^2 e^4}{2\pi m^2 v^3} \cdot \frac{\gamma}{\nu} \quad , \\ \frac{dd_0}{d_0} &= \frac{Z^2 e^4}{2\pi m^2 v^3} \cdot \frac{d\gamma}{\nu 3 d_0^3} = \frac{1}{3} \frac{d\gamma}{\gamma} \quad . \end{aligned} \tag{1.120}$$

Therefore, for thin targets,

$$\begin{aligned} I(\nu)d\nu &= Nd\nu \frac{8\pi^3 Z^2 e^6}{3c^3 m^2 v^2} \int_0^\infty \frac{P(\gamma)}{\gamma} d\gamma = N \frac{32\pi^2 Z^2 e^6}{3\sqrt{3} c^3 m^2 v^2} d\nu \\ &= \beta \frac{Z^2}{v^2} d\nu = \beta' \frac{Z^2}{V} d\nu \qquad \text{for} \quad \nu < \nu_{max} \quad , \\ &= 0 \qquad\qquad\qquad\qquad\quad \text{for} \quad \nu > \nu_{max} \quad , \end{aligned} \tag{1.121}$$

where we have used (1.108) and

$$\beta = N \frac{32\pi^2 e^6}{3\sqrt{3} c^3 m^2} \quad , \quad \beta' = N \frac{16\pi^2 e^5}{3\sqrt{3} c^3 m} \quad . \tag{1.122}$$

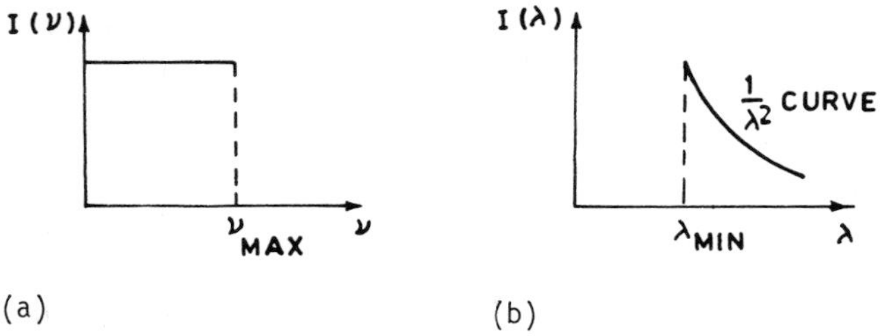

Fig. 1.26. (a) $I(\nu)$ vs ν plot, and (b) $I(\lambda)$ vs λ plot, for a thin target

It is seen that $I(\nu)$ does not depend upon the frequency ν (Fig.1.26a). This is in agreement with the experiments of NICHOLAS [1.8] (Fig.1.19). For a given Z and V, we can write (1.121) as

$$\begin{aligned} I(\nu) &= \text{constant} \quad \text{for} \quad \nu < \nu_{max} \\ &= 0 \quad \text{for} \quad \nu > \nu_{max} \quad . \end{aligned} \tag{1.123}$$

For X-rays $\nu = c/\lambda$, where λ is the wavelength. We can go from $I(\nu)$ vs ν (Fig. 1.26a) curve to $I(\lambda)$ vs λ (Fig.1.26b) curve by using the prescription $I(\nu)d\nu = I(\lambda)d\lambda$, which is necessary if the area under both curves is to be proportional to the energy in the spectrum. Since $-d\nu/d\lambda = c/\lambda^2$, we can write

$$\begin{aligned} I(\lambda) = \frac{c}{\lambda^2} I(\nu) &= \frac{b}{\lambda^2} \quad , \quad \lambda > \lambda_{min} \quad , \\ &= 0 \quad , \quad \lambda < \lambda_{min} \quad , \end{aligned} \tag{1.124}$$

where b is a constant and λ_{min} is the shortest wavelength emitted. This is plotted in Fig.1.26b. DUANE and HUNT [1.20] experimentally discovered the *short-wavelength limit*, also called *Duane-Hunt limit*, and used it to determine the value of h/e,

$$\frac{h}{e} = \frac{V}{\nu_{max}} = \frac{V\lambda_{min}}{c} \quad \text{(Duane-Hunt limit)} \quad . \tag{1.125}$$

It is found that the following relationships hold

$$\lambda_{min}\nu_{max} = c \quad , \quad \lambda_{min}[\text{Å}] = \frac{hc}{eV} = \frac{12398.1}{V[\text{volts}]} \quad .$$

Accurate measurements of h/e have been made by OHLIN [1.21,22], BEARDEN et al. [1.23], and FELT et al. [1.24]. Use is made of the fact that the graph between ν_{max} and V is a straight line.

The dependence of $I(\nu)$ on the tube voltage V, $I(\nu) \propto 1/V$, predicted by KRAMERS (1.121) has been experimentally verified by HARWORTH and KIRKPATRICK [1.12]. They used a 199 Å nickel foil and varied the voltage from 8 to 180 kV. A minimum voltage is necessary to emit the radiation of a given frequency ν (or wavelength λ),

$$V_{min} = \frac{h\nu}{e} \quad . \tag{1.126}$$

For voltages $V > V_{min}$ the $I(\nu)$ vs V curve is shown in Fig.1.27 for a given ν. It is called an *isochromat*.

Let us now calculate the spectral distribution for a *thick target* as is usually found in an X-ray tube. KRAMERS takes into account the decrease of velocity of the electrons with depth x in the target by applying the Thomson-Whiddington law [1.25] $dE/dx = -\text{const}/E$ or

$$\frac{dv^4}{dx} = -bZ \quad . \tag{1.127}$$

Bohr's theoretical value of the constant is (see Appendix B)

$$b = \frac{16\pi e^4 L}{m^2}\, l \quad , \tag{1.128}$$

where L is the number of atoms in 1 cm^3 of target and $l \simeq 6$.

To obtain the total radiation $I_{tot}(\nu)$ of frequency ν, we can integrate $I(\nu)$ over the depth x in the limits 0 to some x_ν, to which the velocity v_ν given by

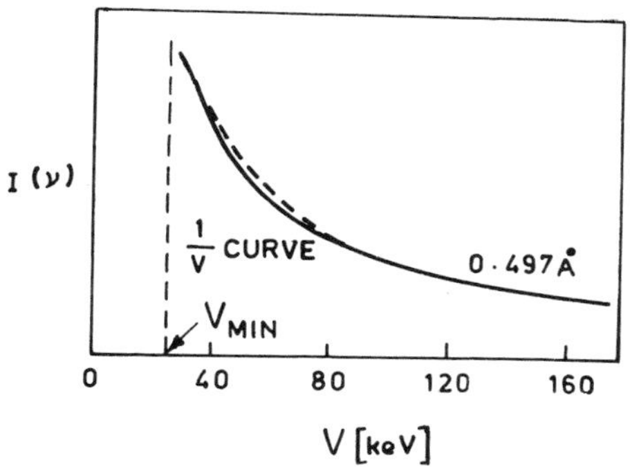

Fig. 1.27. Isochromat for a thin Ni target. Solid line is theoretical and dotted line is experimental

$$\frac{1}{2} m v_\nu^2 = h\nu \tag{1.129}$$

corresponds. For $x > x_\nu$ the velocity will decrease and the kinetic energy of the electron will no longer be sufficient to emit a photon of frequency ν. Thus, for a thick target,

$$I_{tot}(\nu) = \int_0^{x_\nu} I(\nu)dx = \int_{v_0}^{v_\nu} I(\nu) \frac{dx}{dv} dv \quad , \tag{1.130}$$

where v_0 is the initial velocity of electrons, $I(\nu)$ is given by (1.121), and $dx/dv = -4v^3/bZ$ by (1.127). Remembering that (1.130) gives the total intensity of radiation from a thick target, we replace the number of atoms A per 1 cm^2 of thin target surface by the number of atoms L in 1 cm^3 of the target. Then

$$I_{tot}(\nu) = -\frac{4}{bZ}\int_{v_0}^{v_\nu} I(\nu)v^3 dv = -\frac{4}{Z} \cdot \frac{m^2}{16\pi e^4 L1} \cdot nL\, \frac{32\pi^2 Z^2 e^6}{3\sqrt{3}c^3 m^2}\int_{v_0}^{v_\nu} v dv$$

$$= n\alpha Z\left(v_0^2 - v_\nu^2\right) \quad , \quad \alpha = \frac{4\pi e^2}{3\sqrt{3}1c^3} \quad . \tag{1.131}$$

Thus, layers that cause equal electron-energy loss give equal intensities. At the Duane-Hunt limit $(1/2)mv_0^2 = h\nu_{max}$, so that

$$I_{tot}(\nu) = n\alpha' Z(\nu_{max} - \nu) \quad , \quad \alpha' = \frac{8\pi e^2 h}{3\sqrt{3}1c^3 m} = 495 \times 10^{-50} \quad . \tag{1.132}$$

This is in good agreement with the most complete empirical work of KULENKAMPFF [1.26] who found the following formula

$$I_{tot}(\nu) = CZ\left[(\nu_{max} - \nu) + aZ\right] \simeq CZ(\nu_{max} - \nu) \quad , \tag{1.133}$$

where C and a are constants, and aZ term is much smaller in magnitude, $aZ \ll \nu_{max} - \nu$. It is of the same form as Kramers' result (1.132) with $C = n\alpha'$.

From (1.132) we can find the total intensity I of a continuous spectrum at voltage V

$$I = n\alpha' Z \int_0^{\nu_{max}} (\nu_{max} - \nu)d\nu = n\alpha' Z \frac{1}{2}\nu_{max}^2 = n\alpha' Z\left(\frac{eV}{h}\right)^2$$

$$= n\, \frac{4\pi e^4}{3\sqrt{3}1c^3 mh} ZV^2 \quad . \tag{1.134}$$

This is in agreement with I given by the area under the experimental curve for the energy distribution of the continuous spectrum (Fig.1.28a). If $I_{tot}(\nu)$

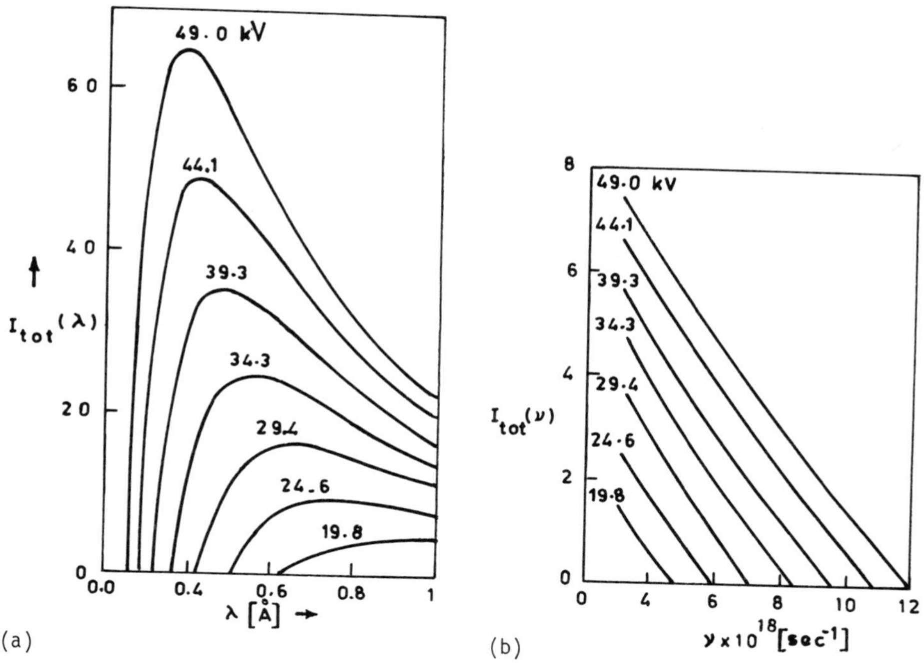

Fig. 1.28. (a) The spectral intensity $I_{tot}(\lambda)$ for a thick tungsten target versus λ, at various voltages V [1.27]. (b) $I_{tot}(\nu)$ vs ν plot

is plotted against ν (Fig.1.28b), a straight line results, down to a frequency $\nu = \nu_{max}/2$, in agreement with the empirical formula (1.133). The slope is independent of the voltage. The constant C varies by 10 percent at higher voltages (30 to 50 kV for W target).

Such curves were first obtained by ULREY [1.28]. In later work (Fig.1.28) several corrections were made to determine the actual distribution of the continuous X-ray spectrum as it is emitted in the target itself. These corrected curves then can be compared with the theory. The main corrections are: incomplete absorption in the detector; absorption in detector window, the atmosphere, the X-ray tube wall, and the target itself; reflectivity of the crystal analyzer; and removal of effects of order higher than the first. An empirical expression for $I_{tot}(\lambda)$ can be obtained from (1.133), by use of (1.124),

$$I_{tot}(\lambda) = \frac{c}{\lambda^2} I_{tot}(\nu) = \frac{CcZ}{\lambda^2}\left(c\,\frac{\lambda-\lambda_{min}}{\lambda\lambda_{min}} + aZ\right) \simeq \frac{Cc^2Z}{\lambda^3\lambda_{min}}\,(\lambda-\lambda_{min}) \quad , \qquad (1.135)$$

which represents the experimental data (Fig.1.28a) fairly well. Figure 1.28a shows that the wavelength of maximum intensity (hump), λ_m, depends on V, and that, approximately,

$$\lambda_m = \frac{3}{2}\lambda_{min} \quad . \tag{1.136}$$

To calculate λ_m, we can use $dI_{tot}(\lambda)/d\lambda = 0$ to get from (1.135)

$$\lambda_m = \frac{3}{2}\lambda_{min}\frac{1}{1+(a/c)Z\lambda_{min}} \simeq \frac{3}{2}\lambda_{min} \quad , \tag{1.137}$$

in agreement with (1.136).

The value of λ_{min} (or ν_{max}) corresponds to the maximum energy that an electron can emit in a single collision. In fact, it is more likely that the incident electrons give up less energy in individual collisions, and in a thick target undergo several collisions before coming to rest. This explains the occurrence of other X-ray photons of lower energies which give rise to the rest of the continuous spectrum (Fig.1.28). The hump at $\lambda_m = 1.5\lambda_{min}$ merely reflects the statistically most probable energy loss. The nature of the curve changes to Fig.1.26b, when the target is so thin that multiple collisions are not likely. Then, in most of the collisions, either total energy is radiated or none at all. This explains the intensity maximum at λ_{min} (Fig.1.26b). WEBSTER [1.29] suggested that because the electrons quickly lose velocity as they enter the target, the radiation of maximum frequency can be emitted only from the surface layer of a thick target. A spectrum from such a target can then be thought of as a superposition of a succession of thin-target curves, like Fig.1.26b, modified by self-absorption in the target of incident electrons and of emitted X-ray photons. Webster's analysis is shown in Fig.1.29.

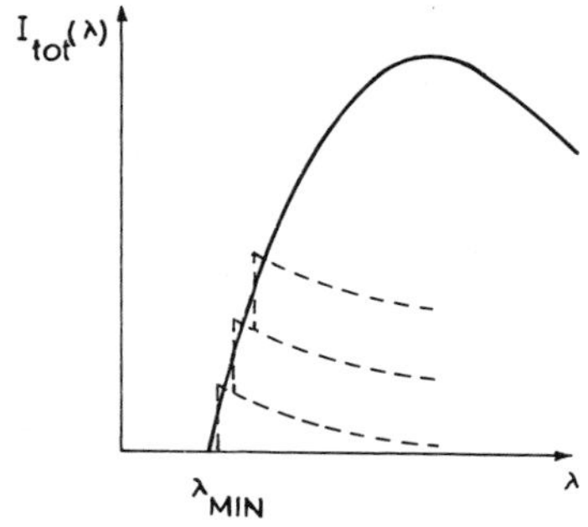

Fig. 1.29. Webster's analysis of a thick-target spectrum into a succession of thin-target spectra

Another quantitative test for Kramers' theory is provided by calculation of the *efficiency* of production of X-rays

$$\eta = \frac{\text{X-ray energy}}{\text{Cathode-ray energy}} = \frac{I}{neV} = \frac{4\pi e^3}{3\sqrt{3}\,1c^3mh} ZV = 9.2 \times 10^{-10} ZV \quad , \qquad (1.138)$$

where V is in volts. Calorimetric determination of η has been made by RUMP [1.30]. The average empirical equation for the various data is $\eta = 11 \times 10^{-10}$ ZV. Thus Kramers' theory is in good agreement with experiments.

For a tungsten (Z=74) target operated at 50 kV, the efficiency is about 0.4 percent. The remainder of the energy appears as heat, which makes the cooling of targets necessary.

The process responsible for conversion of the incident electron energy into heat energy is the large impact parameter, small scattering angle, and multiple elastic scattering of electrons by atoms. An electron that traverses a thick target undergoes a large number of such small-angle (*glancing*) deflections and generally emerges at a small angle that is the cumulative statistical superposition of a large number of deflections. In most of these glancing deflections, radiation is not emitted; they lead to elastic processes that involve momentum transfers to recoiling target atoms. Only rarely is the electron deflected through a large angle; because these events are rare, such an electron makes only one such collision that leads to the emission of an X-ray photon. This circumstance permits us to divide the angular range into two regions — one of relatively large scattering angles ($\theta = \pi - 2\phi_0$, where ϕ_0 is shown in Fig.1.25) which contains only the single scatterings that emit photons, and one region of very small scattering angles, which contains the heat-producing multiple elastic scatterings.

Quantum considerations show that the classical result, that radiation is emitted in *every* collision in which an electron is deflected, is incorrect. In quantum theory, there is a small but finite probability that a photon will be emitted each time the electron suffers a deflection. However, this probability is so small that usually no photon is emitted. In the few collisions that lead to photon emission, a relatively large amount of energy is radiated. Thus, the quantum theory replaces the multitude of small classical energy losses by a much smaller number of large energy losses; the averages are nearly equal in the two theories.

KRAMERS has not estimated the polarization of X-rays. However, a qualitative discussion based on his parabolic orbits is possible. Consider Fig.1.12 with $\theta = \pi/2$ as measured from the direction of initial velocity $\underline{v}$. At the point of observation P, the vector $\underline{E}$ is perpendicular to $\underline{R}$ and lies in a plane that

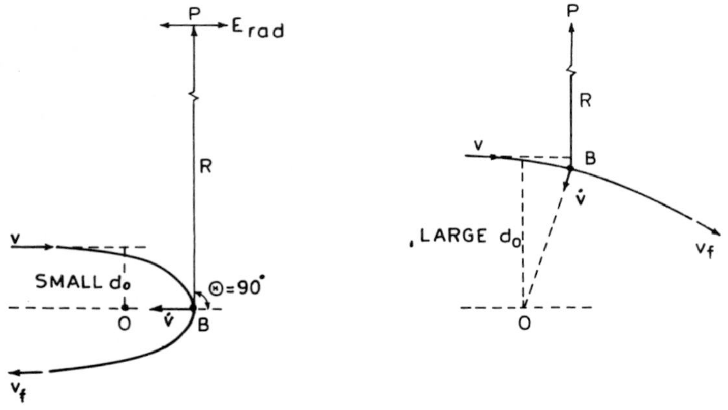

(a) (b)

Fig. 1.30. (a) Orbit for small d_0 and large angle of scattering ($\pi - 2\phi_0 \simeq \pi$), and (b) for large d_0 and small angle of scattering ($\pi - 2\phi_0 \simeq 0$)

contains $\underline{R}$ and $\dot{\underline{v}}$. If the impact parameter d_0 is small, the electron will approach the nucleus at 0 very closely (Fig.1.30a). The angle of scattering, that is, the angle between the direction of the initial velocity $\underline{v}$ and the direction of the final velocity $\underline{v}_f$, in this case is close to π. This is the case of *strong deflection*. The acceleration $\dot{\underline{v}}$ is almost parallel to the axis of $\underline{v}$ and perpendicular to $\underline{R}$. Therefore, the vector $\underline{E}$ of the radiation field attains its maximum value according to (1.66), $E_{rad} = (e/c^2R)\dot{v}\sin\Theta$. Also, the electron radiates the maximum frequency ν_{max} at the instant when it passes the parihelion B (point of shortest distance from 0) of its orbit, in this case.

On the other hand, in the case of large d_0 (Fig.1.30b) the deflection of the electron is small (weak deflection), the distance 0B is large, $\dot{\underline{v}}$ is almos perpendicular to $\underline{v}$ and parallel to $\underline{R}$ making $E_{rad} \to 0$ due to (1.66), and the electron emits radiation of lower frequency (longer wavelength). Thus, at $\nu \simeq \nu_{max}$, the vector $\underline{E}$ is of great magnitude and parallel to $\underline{v}$; at $\nu \simeq 0$ the vector $\underline{E}$ is small in magnitude and perpendicular to $\underline{v}$. In both cases, the radiation is completely polarized. At $0 < \nu < \nu_{max}$ the polarization is only partial, in agreement with observations, (Sect.1.12).

WENTZEL [1.19] applied the correspondence principle in a different way than KRAMERS. KRAMERS has simply cut off the classical radiation at ν_{max} by introducing quantum ideas. WENTZEL has retained all frequencies and proposed a correspondence principle for the transition probability between hyperbolic (open) electron orbits, in analogy with the Bohr correspondence principle for elliptical (closed) orbits in atoms. By summing up for the different open orbits, along which electrons move in the target, WENTZEL obtained results in agreement with experiments. We can conclude that the quantum theories of KRAMERS and WENTZEL have been justified by their general agreement with observed data.

1.19 Quantum Mechanical Considerations

Instead of using the correspondence principle on an ad hoc basis, full quantum mechanical calculations can be used. Early work on these lines is due to SOMMERFELD [1.31]. This work was refined by later workers. In particular, KIRKPATRICK and WIEDMANN [1.32] expressed the theory in such a way that its results can be compared directly with experiment.

Nonrelativistic theory of electron-nucleus bremsstrahlung based on Sommerfeld's approach has been described by BERESTETSKII et al. [1.33]. They also give the quasi-classical relativistic theory of bremsstrahlung. Full relativistic field-theory treatment is given in HEITLER [1.34] and other recent books on the subject.

To illustrate the type of results obtained in quantum mechanics, we present here a simple nonrelativistic perturbation calculation based on the Born approximation. We consider the dipole radiation during collision of two particles having different e/m values.

The dipole moment of two particles that have charges e_1, e_2 and masses m_1, m_2 in their center-of-mass system is (see p.36)

$$\underline{D} = \mu\alpha\underline{r} \quad , \quad \alpha = \frac{e_1}{m_1} - \frac{e_2}{m_2} \quad , \tag{1.139}$$

where μ is the reduced mass, and $\underline{r} = \underline{r}_1 - \underline{r}_2$. The matrix element is

$$\underline{D}_{\underline{p}'\underline{p}} = \int \Psi^*_{\underline{p}'}\underline{D}_{\text{operator}}\Psi_{\underline{p}}\, d\tau = -\frac{1}{\omega^2}\int \Psi^*_{\underline{p}'}\ddot{\underline{D}}_{\text{operator}}\Psi_{\underline{p}}\, d\tau \quad , \tag{1.140}$$

because in quantum mechanics $(v_x)_{\underline{p}'\underline{p}} \equiv (\dot{x})_{\underline{p}'\underline{p}} = \left[-\frac{i}{\hbar}(E_i - E_f)\right]x_{\underline{p}'\underline{p}} = -i\omega x_{\underline{p}'\underline{p}}$ and similarly $(\ddot{x})_{\underline{p}'\underline{p}} = -\omega^2 x_{\underline{p}'\underline{p}}$, where $\omega = (E_i - E_f)/\hbar = (\underline{p}^2 - \underline{p}'^2)/2\mu$, with $\underline{p} = \mu\underline{v}$, $\underline{p}' = \mu\underline{v}'$ as the momenta of relative motion[4].

---

[4] $\hbar = h/2\pi$ (normalized Planck's constant).

Because a force can be expressed as the negative gradient of the potential, we can write, in natural units ($\hbar$=c=1),

$$\ddot{\underline{D}} = \mu\alpha\ddot{\underline{r}} = -\alpha\nabla \frac{e_1 e_2}{r} \quad . \tag{1.141}$$

Using $\Psi_{\underline{p}} = e^{i\underline{p}\cdot\underline{r}}$, $\Psi_{\underline{p}'} = e^{i\underline{p}'\cdot\underline{r}}$ and the result ($\underline{q}=\underline{p}'-\underline{p}$)

$$\left(\nabla \frac{1}{r}\right)_{\underline{p}'\underline{p}} = \int_0^\infty \int_0^\pi e^{-i\underline{q}\cdot\underline{r}} \left(\frac{\partial}{\partial r}\frac{1}{r}\right)\hat{\underline{n}}_r \, r^2 \, dr \, \sin\theta \, d\theta \, 2\pi$$

$$= -2\pi \, \hat{\underline{n}}_r \int_0^\infty \int_{-1}^{+1} e^{-iqrx} \, dx \, dr = -4\pi \, \frac{\hat{\underline{n}}}{q} \int_0^\infty \frac{\sin(qr)}{r} \, dr = -4\pi \, \frac{\underline{q}}{q^2} \quad ,$$

we have for the matrix element

$$\underline{D}_{\underline{p}'\underline{p}} = -\frac{1}{\omega^2} \ddot{\underline{D}}_{\underline{p}'\underline{p}} = + \frac{\alpha e_1 e_2}{\omega^2} \left(\nabla \frac{1}{r}\right)_{\underline{p}'\underline{p}} = - \frac{4\pi\alpha e_1 e_2}{\omega^2} \frac{\underline{q}}{q^2} \quad . \tag{1.142}$$

The probability of dipole radiation is (for example, see [1.35])

$$dw = \frac{\omega^3}{2\pi} \left|\underline{e}\cdot\underline{D}_{\underline{p}'\underline{p}}\right|^2 d\Omega_{\underline{k}} = \frac{16\pi^2\alpha^2 e_1^2 e_2^2}{2\pi\omega q^4} (\underline{e}\cdot\underline{q})(\underline{e}^*\cdot\underline{q}) \, d\Omega_{\underline{k}} \quad , \tag{1.143}$$

where $\underline{e}$ is the unit polarization vector. Assuming that the wave functions are normalized to one particle per unit volume (V=1), we can write the differenti cross section as

$$d\sigma_{\underline{k}\underline{p}'} = \frac{V}{v} \frac{d^3p'}{(2\pi)^3} dw = \frac{e_1^2 e_2^2}{\pi^2} \alpha^2 \frac{v'}{v} \frac{\mu^2}{q^4} (\underline{e}\cdot\underline{q})(\underline{e}^*\cdot\underline{q}) \frac{d\omega}{\omega} d\Omega_{\underline{p}'} \, d\Omega_{\underline{k}} \quad , \tag{1.144}$$

for the emission of a photon $\underline{k}$ into the solid angle $d\Omega_{\underline{k}}$ with the scattering o the electron into the range of states $d^3p' = p'^2 dp' d\Omega_{\underline{p}'}$, where $(p^2/2\mu)+(p'^2/2\mu$ $= \omega$ gives $dp' = \mu d\omega/p'$ or $d^3p' = \mu p' \, d\omega \, d\Omega_{\underline{p}'}$.

To sum over polarizations, we note that the two unit polarization vectors together with $\underline{k}/k$ form a mutually orthogonal system, so

$$\sum_{\text{pol}} (\underline{e}\cdot\underline{q})^2 + \frac{1}{k^2} (\underline{k}\cdot\underline{q})^2 = q^2 \quad ,$$

which gives

$$\sum_{\text{pol}} (\underline{e}\cdot\underline{q})^2 = q^2 \sin^2\theta \quad , \tag{1.145}$$

where θ is the angle between the direction of the photon $\underline{k}$ and the vector $\underline{q}$, which lies in the scattering plane. Thus, integrating over the directions of photons,

$$\int_0^{\pi} \sin^2\theta d\Omega_{\underline{k}} = 2\pi \int_0^{\pi} \sin^3\theta d\theta = 2\pi \cdot \frac{4}{3} \quad ,$$

we get

$$d\sigma_{\omega\theta} = \frac{8}{3\pi} e_1^2 e_2^2 \alpha^2 \frac{v'}{v} \frac{d\omega}{\omega} \mu^2 \frac{2\pi \sin\theta \, d\theta}{p'^2 + p^2 - 2pp' \cos\theta} \quad , \tag{1.146}$$

where θ is the scattering angle. Integrating[5] over θ, we finally get

$$d\sigma_{\omega} = \frac{16}{3c^3} e_1^2 e_2^2 \alpha^2 \frac{\mu^2}{p^2} \log \frac{p+p'}{p-p'} \frac{d\omega}{\omega} \quad . \tag{1.147}$$

For radiation in the field of a fixed center of Coulomb force (nucleus) this is equivalent to the bremsstrahlung cross section in which v and v' are the initial and final velocities of the electron, $e_1 = -e$, $e_2 = Ze$, $m_1 = \mu$ and $m_2 = \infty$. With this interpretation, $T = p^2/2\mu$ is the kinetic energy of the primary electron and $T - \omega = p'^2/2\mu$ is the kinetic energy of the scattered electron. Then we can write

$$d\sigma_{\omega} = \text{const.}\ Z^2 \frac{d\omega}{\omega} \frac{1}{T} \log \frac{[T^{\frac{1}{2}} + (T-\omega)^{\frac{1}{2}}]^2}{\omega} \quad . \tag{1.148}$$

Because intensity = energy × probability, $I(\omega)d\omega = \hbar\omega d\sigma_{\omega}$, we get the familiar result that intensity per unit frequency interval is independent of frequency (Fig.1.26a). Further, at the Duane-Hunt limit $v' = 0$, or $T = \hbar\omega_{max}$, so $d\sigma_{\omega}$ vanishes. Thus we get the ω_{max} cutoff in the intensity in a natural way and not in an ad hoc manner as in Kramers' semiclassical quantum theory.

---

[5] $\int_{-1}^{+1} \frac{dx}{a-bx} = \frac{1}{-b} [\log(a-bx)]_{-1}^{+1} = \frac{1}{b} \log \frac{a+b}{a-b}$.

2. Characteristic X-Rays

In an atom, a centrally charged nucleus is surrounded by Z electrons. The electrons most tightly bound to the nucleus are called K-shell electrons, the next group the L-shell electrons, and so on. This is the picture of Bohr's atom which was given around 1913.

2.1 Line Emission

When fast cathode-ray electrons strike a target in the X-ray tube, mainly two things happen. 1) The incident electron is deflected in the collision process with the target atom, giving rise to the continuous radiation, as seen in Chap.1. 2) The incident electron, if it has sufficient energy, may knock out one of the inner bound electrons in the target atom. If such a hole (or *vacancy*) is created (say) in the inner-most (K) shell, then according to the Bohr theory, an electron of the L or M shell can fall into the K shell, and a quantum of radiation is emitted. In this case the lines $K\alpha$, $K\beta$, ... will appear in the emission spectrum (Fig.2.1). This reasoning was given by KOSSEL [2.1,74,75], following a suggestion of BARKLA. A similar result follows in the case when an electron is knocked out from the L shell of the target atom.

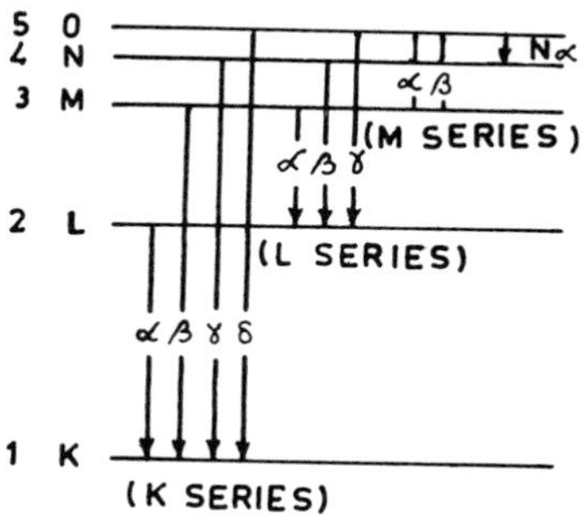

Fig. 2.1. The scheme of shells in an atom, after Barkla and Kossel. The electronic transitions correspond to the X-ray emission lines in the quantum theory of Bohr

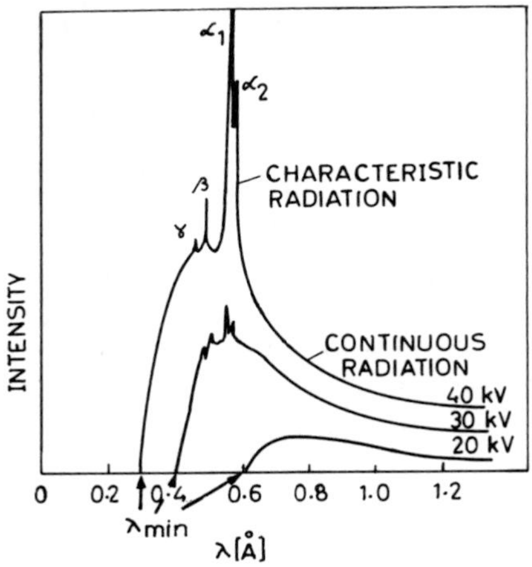

Fig. 2.2. X-ray spectrum of ^{47}Ag target at various tube voltages

Both of the processes 1) and 2) can compete with each other simultaneously In some collisions, the incident electrons are deflected, and in others the inner electron from the target atom is knocked out. Experiments confirm this, because the line emission, *characteristic of the target atom*, appears as superimposed on the continuous emission.

Consider the X-ray spectrum emitted by a silver (^{47}Ag) target X-ray tube as the voltage is increased systematically (Fig.2.2). At 20 kV only a continuous spectrum is observed. The ionization potential (Appendix C) for ^{47}Ag K shell is 25.511 kV. At this voltage, according to the Barkla-Kossel hypothesis, a K electron is knocked out, creating a *hole* in the K shell. As a result, three faint lines appear at 0.4860, 0.4962 and 0.5597 Å, respectively As the voltage is increased further to 30 kV, these lines become stronger and the one at 0.5597 Å begins to appear as a resolved *doublet* ($K\alpha_1,\alpha_2$). At 40 kV the second peak also begins to show a doublet of two lines at 0.4950 and 0.4967 Å (not shown, owing to the large scale of Fig.2.2). These lines arise due to the transitions of electrons in the inner shells of the target atom and so represent the radiation *characteristic* of the atom or material of the target.

Figure 2.3 shows the spectrum for ^{42}Mo and ^{74}W at 35 kV. This voltage is not enough to excite the K lines of ^{74}W, and L lines would be off the diagram at longer wavelengths.

WEBSTER [2.2] increased the voltage on a rhodium (^{45}Rh) target X-ray tube and observed the intensity of the Kα line at 0.613 Å. His experimental plot is shown in Fig.2.4. It is an *isochromat*. The curve shows that the intensity

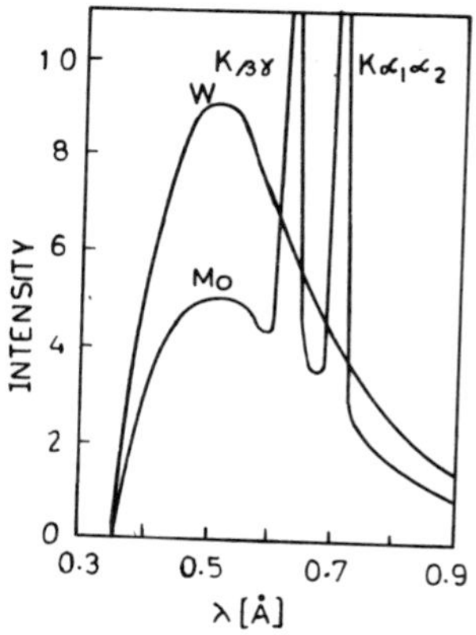

Fig. 2.3. X-ray spectrum from ^{42}Mo and ^{74}W at 35 kV

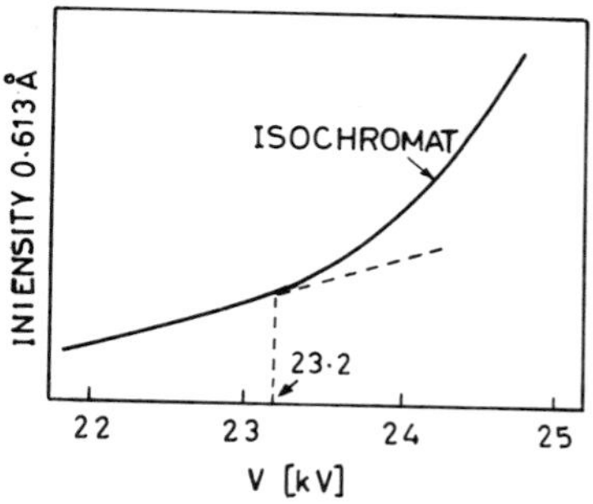

Fig. 2.4. Isochromat of rhodium Kα line

increases gradually as the voltage increases, until at 23.2 kV the slope increases suddenly. This can be understood in the following way. Below 23.2 kV the X-ray intensity recorded at 0.613 Å is entirely due to the background continuous radiation, which increases slowly with voltage (see Fig.1.28a). This radiation is not characteristic of ^{45}Rh, because a similar intensity would be observed for any target of comparable Z. At 23.2 kV, the incident electrons have just enough energy to eject the K-shell electrons from ^{45}Rh atoms. As soon as holes in the K-shell are produced, transitions from other shells occur. In particular, electrons from the L-shell undergo transitions to the K level (hole in K-shell is transferred to the L-shell), and give rise to the ^{45}Rh Kα line of energy according to the Bohr frequency relation

$$E(K\alpha) = E_K - E_L = h\nu_{(K\alpha)} = hc/\lambda_{(K\alpha)} \quad . \tag{2.1}$$

Here E_K, E_L,... are the energies required to remove a K, L,... electron, respectively. As the tube voltage is further increased, more K electrons are ejected, yielding a more intense Kα line.

From Fig.2.1 we find that the observed lines can be arranged in a term scheme by using the difference relations

$$h\nu_{(K\beta)} = h\nu_{(K\alpha)} + h\nu_{(L\alpha)} \quad ,$$

$$h\nu_{(K\gamma)} = h\nu_{(K\alpha)} + h\nu_{(L\beta)} = h\nu_{(K\beta)} + h\nu_{(M\alpha)} \quad , \text{etc.} \tag{2.2}$$

These relations correspond to Ritz's combination principle and are in agreement with experiments.

An important point to be noted is that before any of the K-characteristic X-rays are emitted, a K electron first must be ejected from the atom under study. The Kα line emission will transfer this *hole* (initially in the K-shell) to the L-shell. We can now obtain the Lα line when this hole in the L-shell is transferred to the M-shell (Fig.2.1). Thus, the entire line-emission spectrum is produced because of an initial hole in the K-shell. Of course, the frequencies (or wavelengths) of these lines will lie in quite different range

2.2 Moseley Law

The discovery of Bragg diffraction by crystals enabled X-ray spectroscopists to measure wavelengths precisely. MOSELEY [2.3] discovered that the frequency of the strongest (unresolved Kα) characteristic line emitted by an element increases systematically as its atomic number Z increases. It was known that according to the quantum theory of Bohr the energies of quantized atomic states are given by

$$E_n = -R_\infty hc\left(\frac{M}{M+m}\right)\frac{Z^2}{n^2} \quad , \quad n = 1,2,3,\dots \quad , \tag{2.3}$$

where m = electron mass, M = nucleus mass, and $R_\infty = 2\pi^2 me^4/ch^3 = 109737{\cdot}31\ \mathrm{cm}^{-1}$ is the *Rydberg constant* for an atom of infinite mass. One Rydberg unit of energy, expressed in terms of electron volts, is 13.605 eV. The frequency of a spectral line is given by

$$\nu = R_\infty c\left(\frac{M}{M+m}\right)Z^2\left(\frac{1}{n_f^2} - \frac{1}{n_i^2}\right) \quad , \tag{2.4}$$

where n_i and n_f are the principal quantum numbers of the initial and final states of the atom involved in the transition, respectively.

The result (2.4) holds good for an atom that consists of a nucleus with one external electron. In a complex atom the nuclear charge will not be exactly Ze but modified by the presence of other external electrons in the atom. Because of this screening of the nuclear charge, the effective charge that attracts the electron is less than Ze. It is usually written as $Z_{eff} = Z - \sigma$, where σ is the *screening constant*. When this correction is included, the factor $M/(M+m)$ can be ignored; it represents a very small correction due to the motion of the nucleus about the center of mass of the system.

In the Barkla-Kossel theory, the origin of a K series line is a state in which one electron is missing in the innermost shell (n=1). The *hole* can then be filled by the transition of an electron from one of the outer shells (n>1). Thus, the Kα line involves the transition of an electron from the initial L-shell (or state) $n_i = 2$ to the final K state $n_f = 1$ (Fig.2.1)

$$\nu_K = R_\infty c(Z-\sigma_K)^2\left(\frac{1}{1^2} - \frac{1}{2^2}\right) = R_\infty c \frac{3}{4} (Z-\sigma_K)^2 \quad . \tag{2.5}$$

This gives

$$\left(\frac{\tilde{\nu}}{R_\infty}\right)_K^{\frac{1}{2}} = \left(\frac{\nu}{cR_\infty}\right)^{\frac{1}{2}} = 0.866\ (Z-\sigma_K) = a_K(Z-\sigma_K) \quad , \tag{2.6}$$

where $\tilde{\nu} = \nu/c = 1/\lambda$ is the *wave number*. For the Lα line, due to a transition from M state to L state, $n_i = 3$ and $n_f = 2$,

$$\nu_L = R_\infty c \frac{5}{36} (Z-\sigma_L)^2 \quad .$$

Guided by the result (2.6) from Bohr's theory, MOSELEY plotted $(\tilde{\nu}/R_\infty)^{\frac{1}{2}}$ of the measured Kα line against Z. He obtained a straight line (Fig.2.5) that could be expressed as

$$\left(\frac{\tilde{\nu}}{R}\right)_{K\alpha}^{\frac{1}{2}} = 0.874\ (Z-1.13) \qquad (\textit{Moseley law}) \quad . \tag{2.7}$$

Precision measurements have shown that $(\tilde{\nu}/R_\infty)^{\frac{1}{2}}$ is not exactly a linear function of Z for the resolved $K\alpha_1$ and $K\alpha_2$ lines. This can be seen by plotting large scale graphs. To avoid this inconvenience, the *modified Moseley diagram* has been introduced (Fig.2.6). To obtain it, the constants a_K and σ_K in (2.6) are given values that make the *basic line* of $(\tilde{\nu}/R_\infty)^{\frac{1}{2}}$ versus Z follow as closely as possible the experimental points. The differences between the true

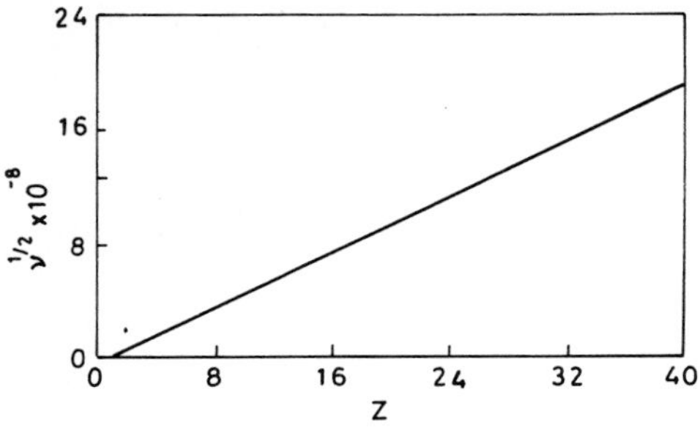

Fig. 2.5. Moseley plot for the $K\alpha$ line

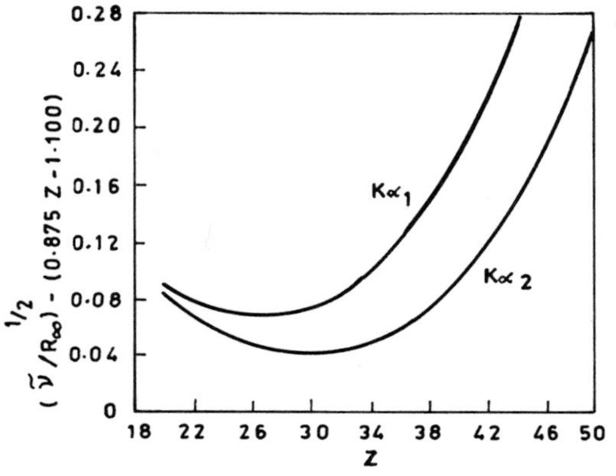

Fig. 2.6. Modified Moseley diagram of the $K\alpha_1$ and $K\alpha_2$ lines. The basic line is $(\tilde{\nu}/R_\infty)^{\frac{1}{2}} = 0.875Z - 1.100$

$(\tilde{\nu}/R_\infty)^{\frac{1}{2}}$ values and those represented by the basic straight line are then plotted against Z (Fig.2.6). These diagrams have helped to systematize the X-ray spectra and to fix the atomic numbers of newly discovered elements.

Moseley plots can be drawn for all lines in K, L, M, ... series. For the $L\alpha_1$ lines of elements with atomic number $Z > 62$ the values of constants in (2.6) are $a = 0.376$ and $\sigma = 7.90$. Knowing the $\tilde{\nu}/R_\infty$ value from (2.6) for a photon, we can calculate the wavelength by using

$$\lambda(\text{Å}) = \frac{911.27}{\tilde{\nu}/R_\infty} \quad . \tag{2.8}$$

In measurements by use of crystal spectrographs, the λ is usually given in the "eks" unit, XU, or in "kiloeks" unit, kXU. By agreement [2.4],

$$1 \text{ kXU} \equiv 1000 \text{ XU} = (1.00202 \pm 0.00003) \text{ Å} \quad . \tag{2.9}$$

Recent work of BEARDEN [2.5] gives

$$1\ \mathrm{kXU} = (1.002076 \pm 0.000005)\ \text{Å} \quad . \tag{2.10}$$

Thus, 1 kXU $\simeq$ 1 Å. We shall use both of these units of λ. Note that $\tilde{\nu}/R_\infty$ in (2.8) is a true dimensionless quantity, with both R_∞ and $\tilde{\nu}$ measured in cm^{-1}. Some authors have used $(\mathrm{XU})^{-1}$ for $\tilde{\nu}$ and cm/XU for $\tilde{\nu}/R_\infty$. The various conversion factors can be found from the following example [2.5]:

Line	[Å]	[keV]	$(\tilde{\nu}/R_\infty)$	[XU]	$(\tilde{\nu}/R_\infty)$ [cm/XU]
$^{42}\mathrm{MoK}\alpha_1$	0.709300	17.47934	1284.74	707.831	1287.40 .

The energy in [keV] is inversely proportional to the corresponding wavelength

$$\lambda E = E/\tilde{\nu} = 12.3981\ \text{Å} - \mathrm{keV} = 12372.42\ \mathrm{XU} - \mathrm{keV} \quad . \tag{2.11}$$

Thus, $ch/e = 1.23981 \times 10^{-4}$ eV cm, (1.125), gives the energy associated with the wavelength.

In the case of continuous spectra, the observed cutoff frequency found a natural explanation in the quantum theory. Here, also, quantum ideas are helpful in understanding the *discrete* frequencies of the characteristic emission lines. Therefore, we can hope, once again, to obtain some insight into the problem by first developing a classical model based on oscillators, and then applying the correspondence principle to bring in the quantum ideas.

2.3 Classical Oscillator Model

To explain the discrete lines of a spectrum, the classical theory provides the atom with a set of electron oscillators, each of which corresponds to a definite line. In this model, the electron in an atom undergoes harmonic vibrations around the nucleus to which it is bound. The electron is decelerated as it moves away from the nucleus and accelerated when it moves toward it. We, therefore, expect it to emit radiation energy. When a line is strong, it is provided with a number of identical oscillators (the so-called *oscillator strength*).

For one-dimensional harmonic oscillator (Fig.2.7) the equation of motion is

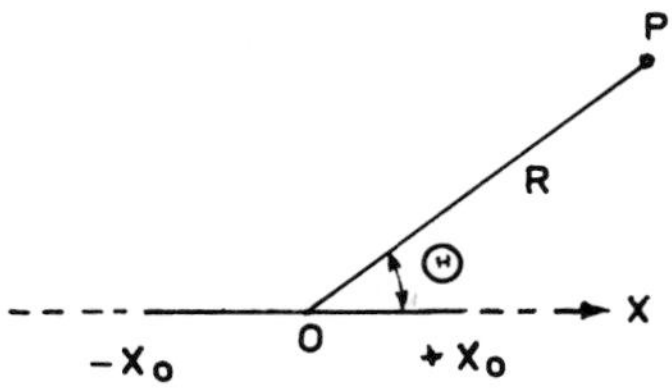

Fig. 2.7. The classical oscillator or dipole. P is the point of observation of the radiation from such a dipole

$$x = x_0 \cos(\omega t) = x_0 \cos(2\pi\nu t) \quad , \tag{2.12}$$

where x_0 is the amplitude, and ω the angular frequency. The acceleration at any point x is

$$a = d^2x/dt^2 = -4\pi^2\nu^2 x \quad . \tag{2.13}$$

From (1.67) we find the radiation intensity at the point P, that is at a distance R from O (Fig.2.7), to be

$$I(\Theta,x) = \frac{4\pi^3 e^2}{c^3} \frac{\sin^2\Theta}{R^2} \nu^4 x^2 \quad . \tag{2.14}$$

Suppose there are N atoms per 1 cm^3 with Z electrons per atom. The Z electrons can be divided into groups specified by the index j, with f_j electrons that have the same frequency ν. The number f_j is called the *oscillator strength* of the jth oscillator, and it satisfies the obvious sum rule

$$\sum_j f_j = Z \quad . \tag{2.15}$$

Since $N_\nu = Nf_j$ is the number of oscillators per 1 cm^3 with frequency ν and amplitude x_0, the rate at which the energy is lost is obtained by integrating over a spherical surface and multiplying by N_ν,

$$-\frac{dU}{dt} = P = N_\nu \int_0^\pi I(\Theta,x) \cdot 2\pi R \sin\Theta \cdot R\, d\Theta = N_\nu \frac{32\pi^4 e^2}{3c^3} \nu^4 x^2 \quad . \tag{2.16}$$

The electron makes many oscillations per second. We wish to find the average value $\langle x^2 \rangle$ of x^2 for the period $T = 1/\nu$,

$$\langle x^2 \rangle = \frac{\oint x^2 dt}{\oint dt} = \frac{\oint x_0^2 \cos^2 2\pi\nu t\, dt}{T} = \frac{1}{T} x_0^2 \int_0^T \cos^2 2\pi\nu t\, dt = \frac{1}{T} x_0^2 \frac{1}{2\nu} = \frac{1}{2} x_0^2 \quad . \tag{2.17}$$

Using this result, we can write the power averaged over the period T, or intensity I, as

$$I \equiv \left\langle -\frac{dU}{dt}\right\rangle = \langle P\rangle = N_\nu \frac{32\pi^4 e^2}{3c^3}\nu^4 \langle x^2\rangle = N_\nu \frac{16\pi^4 e^2}{3c^3}\nu^4 x_0^2 \quad . \tag{2.18}$$

In general, the oscillations of the dipole occur along some direction $r(x,y,z)$ with an amplitude $r_0(x_0,y_0,z_0)$. Then we can write

$$\langle P\rangle = N_\nu \frac{32\pi^4 e^2}{3c^3}\nu^4(\langle x^2\rangle + \langle y^2\rangle + \langle z^2\rangle) = N_\nu \frac{16\pi^4 e^2}{3c^3}\nu^4 r_0^2 \quad , \tag{2.19}$$

where

$$r_0^2 = x_0^2 + y_0^2 + z_0^2 = 2(\langle x^2\rangle + \langle y^2\rangle + \langle z^2\rangle) \quad . \tag{2.20}$$

2.4 Quantum Theory

We can invoke the correspondence principle to introduce the quantum ideas. Such an approach lacks rigor, but leads to the same result as would be given by a correct quantum mechanical derivation of the transition probability.

To convert the classical expression (2.19) into the corresponding expression in the quantum theory, we take the following three steps:

Step 1: We replace the classical frequency ν by the quantum frequency ν_{fi} of photons that are radiated during the transitions of the atom from the initial state i of energy E_i to the final state f of energy E_f

$$\nu \rightarrow \nu_{fi} = (E_i - E_f)/h \quad , \quad (i \rightarrow f) \quad . \tag{2.21}$$

Step 2: We replace the amplitude x_0 of the oscillations of the dipole by the *matrix element* x_{fi}

$$x_0 \rightarrow x_{fi} = (f|x|i) = \int \psi_f x \psi_i \, d\tau \quad , \quad d\tau = dx\,dy\,dz \quad , \tag{2.22}$$

where ψ_i and ψ_f are the normalized initial- and final-state wave functions, respectively. Similarly, $y_{fi} = (f|y|i)$ and $z_{fi} = (f|z|i)$.

Step 3: We replace [2.6] the time average of the coordinate of the oscillator $\langle x^2\rangle$ by the square of the matrix element of the same quantity for the transition $i \rightarrow f$,

$$<x^2> \rightarrow 2|x_{fi}|^2 \quad . \tag{2.23}$$

The factor 2 occurs on the right-hand side to give us the final result in agreement with the quantum mechanical result. The difference occurs because we have taken $\cos\omega t$ rather than the full exponential form to describe the oscillator.

Equation (2.23) gives the well-known connection between the classical quantities and the quantum mechanical quantities according to the *correspondence principle*. We can write, from (2.20,23),

$$r_0^2 \rightarrow 4\left(|x_{fi}|^2+|y_{fi}|^2+|z_{fi}|^2\right) \equiv 4r^2(i,f) \quad . \tag{2.24}$$

For the emission process, the initial state consists of an atom excited (one electron missing) in an inner shell without the presence of quanta. In the final state, an X-ray quantum is emitted and the hole in the inner level is filled (*spontaneous emission*).

Let w_{fi} be the transition probability per unit time of an atom going from the state i to the state f. The correspondence principle allows us to write, from (2.19,24), the total energy radiated per unit time, or the intensity of the line, in spontaneous emission as

$$I(i,f) = <P>_{fi} = N_\nu\, h\nu_{fi}\, w_{fi} = N_\nu \frac{64\pi^4 e^2}{3c^3} \nu_{fi}^4\, r^2(i,f) \quad . \tag{2.25}$$

We have interpreted the power as the product of the spontaneous rate of transition from i to f and the emitted photon energy $h\nu_{fi}$. Equation (2.25) gives

$$w_{fi} = \frac{64\pi^4 e^2}{3c^3 h} \nu_{fi}^3\, r^2(i,f) \quad . \tag{2.26}$$

This is just the quantum mechanical result for spontaneous emission (see, for example, [2.7]).

We can easily evaluate the order of magnitude of w_{fi}. If E is the energy of the atom, the Bohr radius a_0, or the dimension of the atom, is roughly $a_0 \sim e^2/E$. Putting $r_0 \sim a_0$, we can write (use $e^2/\hbar c = 1/137$)

$$w_{fi} \sim \frac{e^2}{\hbar c^3}\, \omega^3 a_0^2 \sim \frac{1}{137}\left(\frac{\omega e^2}{c\hbar\omega}\right)^2 \omega \sim \frac{\omega}{137^3} \quad , \tag{2.27}$$

which is $\sim 10^8\ s^{-1}$ for the optical region and $\sim 10^{11}\ s^{-1}$ for X-rays.

In (2.25,26), we should include two more factors before the theory can be compared with experiments. Before any X-ray emission line appears, an inner bound electron must be ejected from the atom, that is, the atom must be *ionized* in an inner shell. This ionized state is the initial excited state. In an X-ray tube, this occurs when the incident cathode electrons collide with the target atoms and ionize them. Therefore, we must introduce a function F_i that determines the *ionization probability* of an inner bound level i of the atom. The second factor is the so-called *statistical weight* g_i of the state i, which gives the number of different possible orientations of the underlying configuration that gives rise to the state under consideration.

For the emission of X-ray photons, we therefore have the useful result

$$I(i,f) \sim \nu_{fi}^4 \, F_i \, g_i \, r^2(i,f) \quad . \tag{2.28}$$

Here F_i is the ionization probability for effectively one electron per 1 cm^3.

Inner-shell ionization will also occur when a target is bombarded by protons or heavier ions. The topic of X-ray production by heavy charged particles has been reviewed by MERZBACHER and LEWIS [2.8], and by OGURTSOV [2.9].

2.5 Ionization Function

We are interested in the relative intensity of a given (say K) line that arises in a particular transition $i \to f$ in a thin target, when the energy E of the incident electrons is varied. In such a case, we can treat all the quantities in (2.28) as constants except F_i, and write

$$I(i,f) \sim F_i \quad (\textit{thin target}) \quad . \tag{2.29}$$

A satisfactory theory of ionization function is not available. DAVIS [2.10] made an *ad hoc* assumption that the relative excess of energy of the incident electrons over the binding energy E_i is a measure of the ionization probability. He took

$$F_i = (V_x - V_i)/V_x \quad , \tag{2.30}$$

where eV_x is the energy of the incident electron at a depth x below the target surface, and eV_i is the ionization energy of the level i. The number of atoms that are excited to the ith level (hole created in the ith level) in 1 cm^2 of the layer dx per second is

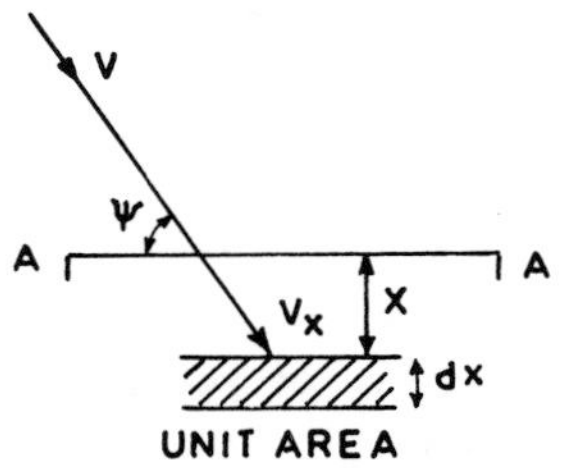

Fig. 2.8. A target layer at a depth x

$$dN_i = n\ NF_i \frac{dx}{\sin\psi} = nN \frac{V_x - V_i}{V_x} \frac{dx}{\sin\psi} \quad , \tag{2.31}$$

where n is the number of incident electrons that cross a unit-area layer between x and x + dx below the target surface, N is the number of collisions with target atoms per unit path length of the incident electrons, and ψ is the angle of incidence of the electrons from the target face (Fig.2.8). The path length in the layer dx is $dx/\sin\psi$.

For a thick target, we can use the Thomson-Whiddington law (1.127) to get a relation between n and V_x, if we express the velocity v_x at depth x in terms of V_x and replace dx by $dx/\sin\psi$

$$\frac{1}{2} mv_x^2 = eV_x \quad , \quad \frac{dv_x^4}{dx} = -bZ \quad ,$$

$$dV_x^2 = 2V_x dV_x = -a \frac{dx}{\sin\psi} \quad ; \quad a = \frac{4e^2bZ}{m^2} \quad . \tag{2.32}$$

Substituting dx from (2.32) in (2.31) and integrating from the tube voltage V to V_i (corresponding to 0 to x, after which there is not enough energy left for ionization to be possible), we get

$$N_i = -\frac{2nN}{a} \int_V^{V_i} (V_x - V_i) dV_x = \frac{2nN}{a}\left[V\left(\frac{1}{2}V - 1\right) - V_i\left(\frac{1}{2}V_i - 1\right)\right] \simeq \frac{nN}{a}\left(V^2 - V_i^2\right) \ . \tag{2.33}$$

The total intensity $I_{tot}(i,f)$ can now be written as

$$I_{tot}(i,f) = \frac{16\pi^4 e^2}{3c^3} \nu_{fi}^4 \ N_i \ g_i \ r^2(i,f) = \mathrm{const.}\left(V^2 - V_i^2\right) \quad , \tag{2.34}$$

in the simplified classical calculation of F_i presented here. This formula is found to be reliable up to $V = 3V_i$ for the intensity measurements of K line made by WOOTEN [2.11], and by HILL and TERREY [2.12]. For $V > 3V_i$ the depth of penetration of electrons into the target increases and the neglected factor of the absorption of X-rays within the target begins to predominate. Consequently, a lower intensity is observed than predicted by (2.34) (Fig.2.9).

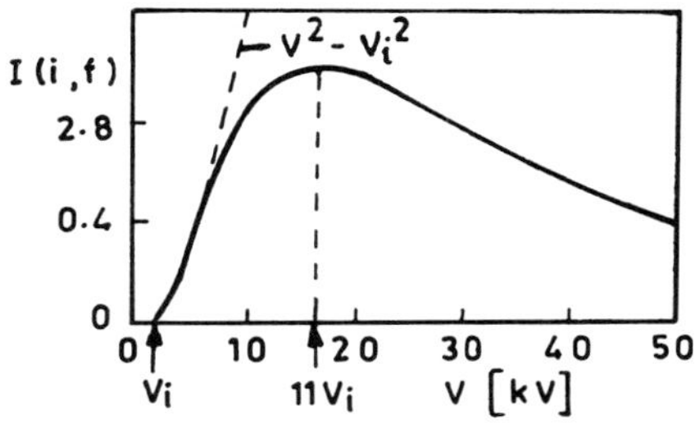

Fig. 2.9. The observed intensity I(i,t) of the $K\alpha_{1,2}$ line of aluminium versus the voltage V in kV

Several other formulas have been developed, which incorporate various corrections [2.13,14] and hold up to about $10V_i$. Even this agreement is due to the fact that some of the corrections not included tend to cancel each other. For experimental details and discussion of corrections see, for example, the papers of WEBSTER et al. [2.15], and POCKMAN et al. [2.16]. The L_{III} ionization of silver has been studied by McCUE [2.17]. For recent work, see GREEN and COSSLETT [2.18].

2.6 X-Ray Terms

The stationary states of the hydrogenlike atom are defined by the well-known wave functions

$$\psi_{nlm} = R_{nl}(r) \cdot Y_l(\theta) \cdot Z_m(\phi) = N_{nlm}\, e^{-\rho/2} \rho^l L_{n+l}^{2l+1}(\rho) \cdot P_l^m(\cos\theta) \cdot e^{im\phi} ,$$
$$(n=1,2,3,\dots;\ 0 \leqq l < n;\ -l \leqq m \leqq l;\ l,m \text{ both integers}) \quad , \qquad (2.35)$$

where N_{nlm} is the normalization factor, and $\rho = Zr/a_0$ with a_0 as the radius of first Bohr orbit. The energy of a state or level is given by

$$E_n = -R_\infty hc \frac{Z^2}{n^2} \quad , \quad n = 1,2,3, \dots \quad , \qquad (2.36)$$

which depends only on n. We have neglected the term $M/(M+m) \simeq 1$.

If we include spin, (2.35) suggests that the electrons in an atom can be described by *four* quantum numbers. The *principal quantum number* n (=1,2,3,...) specifies the shell

n	: 1	2	3	4	5	6	7
Shell	: K	L	M	N	O	P	Q

and determines the energy. The *azimuthal* quantum number l (=0,1,2,...n-1) determines the orbital angular momentum

l	: 0	1	2	3	4	5	6
Symbol	: s	p	d	f	g	h	i

Each shell is subdivided into subshells according to the values of l

Shell n:	1	2		3			4
Subshell l:	0	0	1	0	1	2	0
Designation:	1s	2s	2p	3s	3p	3d	4s ...

The *magnetic* quantum number m (=0,±1,±2,...,±l) specifies the orientation of the electron's orbital[1] in a magnetic field. The *spin* quantum number m_s (=±1/2) describes the orientation of the spin.

A specification of n and l for all the electrons bound to a nucleus is called the *electronic configuration* of the atom (or ion). The assignment of the electrons to electronic quantum states is determined by 1) the thermodynamic considerations that require the lowest energy states to be filled up first, and 2) the exclusion principle of Pauli, according to which *no two electrons in an atom can have the same four quantum numbers*. Any subshell can contain up to $2(2l+1)$ electrons, because m can range from the value l down through zero to $-l$, making $2l+1$ different values; m_s can be either 1/2 or -1/2. Thus, several states ψ_{nlmm_s} have the same energy E_n (degeneracy).

Subshell l:	0	1	2	3	4	5	6
Symbol :	s	p	d	f	g	h	i
Number :	2	6	10	14	18	22	26

The normal configuration of electrons in an atom is expressed as $1s^2\ 2s^2\ 2p^6\ 3s^2\ 3p^6\ 3d^{10}\ 4s^2$.... The total number of electrons in a shell (given n) is

$$\sum_{l=0}^{n-1} 2(2l+1) = 2n^2 \quad \text{(total degeneracy)} \quad . \tag{2.37}$$

For discussing atomic spectra, it is necessary to take into account the *spin-orbit* interactions. For this, it is convenient to define two different

[1]The wave function for a single electron can be plotted and is called an orbital. Often m is written as m_l.

Table 2.1. Allowed combinations of quantum numbers

Shell	K	L			M				
n	1	2			3				
l	0	0	1		0	1		2	
j	$\frac{1}{2}$	$\frac{1}{2}$	$\frac{1}{2}$	$\frac{3}{2}$	$\frac{1}{2}$	$\frac{1}{2}$	$\frac{3}{2}$	$\frac{3}{2}$	$\frac{5}{2}$
m_j	$\pm\frac{1}{2}$	$\pm\frac{1}{2}$	$\pm\frac{1}{2}$	$\pm\frac{1}{2}\pm\frac{3}{2}$	$\pm\frac{1}{2}$	$\pm\frac{1}{2}$	$\pm\frac{1}{2}\pm\frac{3}{2}$	$\pm\frac{1}{2}\pm\frac{3}{2}$	$\pm\frac{1}{2}\pm\frac{3}{2}\pm\frac{5}{2}$
Number of	2	2	2	4	2	2	4	4	6
electrons	2	8			18				

quantum numbers: the *total angular momentum* quantum number $j = l \pm m_s$, and the quantum number m_j that can have all integrally spaced values from $+j$ to $-j$. Each subshell with $l \geq 1$ can now be further subdivided. For example, the 2p ($n=2, l=1$) subshell can be subdivided into $2p_{1/2}$ ($n=2, l=1, j=1/2$) and $2p_{3/2}$ ($n=2, l=1, j=3/2$). Thus, the L ($n=2$) shell has in all three subshells $2s_{1/2}$ ($n=2, l=0, j=1/2$), $2p_{1/2}$ and $2p_{3/2}$. These are called the *spectroscopic terms*. When the spin-orbit interaction is included, we can think of each subshell, as defined by a given nl, being split into two nlj subshells, each composed of all electrons that have the same values of n, l, and j. Each closed subshell contains $2j+1$ electrons, one for each value of m_j. The distribution of quantum numbers among the lowest energy states is shown in Table 2.1. We have not shown m, because it is unimportant in determining the energy. The electron distribution among the various energy states of free atom is given in Appendix D. A useful complete set of four quantum numbers is $nljm_j$. We shall use this set rather than the set $nlmm_s$.

The electrons in any completely filled shell or subshell contribute *zero* resultant angular momentum, both orbital and spin, to the angular momentum of the whole atom. In a closed subshell, for every electron with z component of orbital (spin) angular momentum $+m\hbar$ ($+m_s\hbar$) there is another electron with z component $-m\hbar$ ($-m_s\hbar$). The resulting charge distribution for a closed subshell is spherically symmetric. All such inner electrons form the *core*. The charge density for a closed shell has the form

$$e \sum_{m=-l}^{+l} |R_{nl}(r)|^2 |Y_{lm}(\theta,\phi)|^2 \quad , \quad Y_{lm}(\theta,\phi) = Y_l(\theta) Z_m(\phi) \quad ;$$

this is spherically symmetric because of the property of spherical harmonics,

$$\sum_{m=-1}^{+1} |Y_{1m}(\theta,\phi)|^2 = \frac{2l+1}{4\pi} \quad .$$

Electrons that have the same n and l values are called *equivalent electrons*. The vacancy principle of PAULI [2.19] states that *from a configuration of z electrons that have the same nl values, the same spectroscopic terms will arise as in a configuration of the same nl electrons that lack z electrons to complete the closed shell.*

We have seen that the X-ray characteristic spectra originate from the lack of one electron (a *hole*) in an inner closed shell. By the Pauli vacancy principle, the set of *X-ray terms* that arise from this circumstance will correspond to the set of terms of a hydrogenlike atom. The single hole behaves just like an "anti-electron" (not positron) with spin 1/2 outside the core in a hydrogenlike atom. Therefore, we can base our discussion of X-ray characteristic spectra on the theory of the spectra of hydrogenlike atoms, which is well understood. Within this approximation, we can assign the quantum numbers $nljm_j$ to the single hole.

The binding energy of the inner or core electrons in the atom is large. If such an electron is ejected, the *excited* atom becomes mechanically unstable with respect to ionization. This is accompanied by the readjustment of the electron cloud, and the formation of a stable ion. However, as the interaction between the electrons themselves in the atom is relatively weak, the probability of such an occurrence is comparatively small. Consequently, the lifetime τ of the excited state is long. This makes the *width* $\Delta E \sim \hbar/\tau$ of the level so small that it is reasonable to regard the energies of an atom with a hole as discrete energy levels of quasistationary states of a hydrogenlike atom. These levels are called *X-ray terms* because the transitions between these levels give rise to X-ray line emission. It is important to realize that *X-ray states are hole states.*

We have seen that the total angular momentum of the set of electrons that occupy any closed shell is zero. When one electron has been removed (to where on the outside does not matter in the first approximation), the shell acquires some angular momentum j and very high energy. If the hole occurs in the 1s shell ($1s^{-1}$, superscript indicates one missing electron), we get a high positive energy (equal in magnitude to the ionization, or binding, energy of 1s electron) K-level or X-ray term. Similarly, if an electron is removed from the 2s shell ($2s^{-1}$) we obtain a level, not as high as the K-level, that is called the L_I level in the X-ray notation. In general, for the nl shell, the angular momentum j can take the value $l \pm (1/2)$. Thus, we obtain levels that

Table 2.2. X-ray levels

n	l = 0	l = 1		l = 2		l = 3	
1	$1s_{1/2}^{-1}$ K	-	-	-	-	-	-
2	$2s_{1/2}^{-1}$ L_I	$2p_{1/2}^{-1}$ L_{II}	$2p_{3/2}^{-1}$ L_{III}	-	-	-	-
3	$3s_{1/2}^{-1}$ M_I	$3p_{1/2}^{-1}$ M_{II}	$3p_{3/2}^{-1}$ M_{III}	$3d_{3/2}^{-1}$ M_{IV}	$3d_{5/2}^{-1}$ M_V	-	-
4	$4s_{1/2}^{-1}$ N_I	$4p_{1/2}^{-1}$ N_{II}	$4p_{3/2}^{-1}$ N_{III}	$4d_{3/2}^{-1}$ N_{IV}	$4d_{5/2}^{-1}$ N_V	$5f_{5/2}^{-1}$ N_{VI}	$5f_{7/2}^{-1}$ N_{VII}
5	$5s_{1/2}^{-1}$ O_I	$5p_{1/2}^{-1}$ O_{II}	$5p_{3/2}^{-1}$ O_{III}	$5d_{3/2}^{-1}$ O_{IV}	$5d_{5/2}^{-1}$ O_V	-	-
6	$6s_{1/2}^{-1}$ P_I	$6p_{1/2}^{-1}$ P_{II}	$6p_{3/2}^{-1}$ P_{III}	-	-	-	-

can be denoted by $nl_j^{-1} = 1s_{1/2}^{-1}(K)$ $2s_{1/2}^{-1}(L_I)$, $2p_{1/2}^{-1}(L_{II})$, $2p_{3/2}^{-1}(L_{III})$, ..., where the value of j appears as a suffix to the letter that gives the position of the hole. The symbol in brackets that follows the spectroscopic notation is the X-ray notation. The X-ray levels are given in Table 2.2, in the spectroscopic and the Bohr-Coster X-ray notations. The table ends with O_V and P_{III} because 5f and 6d shells do not get filled in the usual elements; see Appendix E. It has become a common practice to drop the superscript -1 from the spectroscopic notation. We shall do so from now on. Thus, a transition $K \rightarrow M_{III}$, equivalent to $1s_{1/2}^{-1} \rightarrow 3p_{3/2}^{-1}$, will be written simply as $1s_{1/2} \rightarrow 3p_{3/2}$. They all indicate the transition of the hole.

From Table 2.2 we see that for L (n=2) shell we get three X-ray levels L_I (n=2,l=0,j=1/2), L_{II} (n=2,l=1,j=1/2), L_{III} (n=2,l=1,j=3/2). For $l > 0$, there are two values of j ($=l \pm m_s$) namely, $j = l - (1/2)$ and $j = l + (1/2)$. All such levels (same nl values but different j values) will therefore be *spin doublets* like L_{II} and L_{III}, or M_{IV} and M_V, etc. Levels with same nj values but different l values are called *screening doublets*, such as L_I and L_{II}, or M_{III} and M_{IV}, etc.

2.7 Energies of Atomic X-Ray Levels and Energy-Level Diagram

We have just noted that the discussion of the X-ray emission spectra can be based on the theory of the optical spectra of hydrogenlike atoms. The nonrelativistic Schrödinger equation gives

$$H_0 = \frac{p^2}{2m} - \frac{Ze^2}{r} \quad , \quad H_0\psi_{nlm} = E_n\psi_{nlm} \quad , \tag{2.38}$$

$$E_n = -R_\infty hc \frac{Z^2}{n^2} = -\frac{m\alpha^2c^2}{2}\frac{Z^2}{n^2} \quad , \tag{2.39}$$

where $\alpha = e^2/\hbar c$ is the *fine-structure constant* and ψ_{nlm} is of the form (2.35). In this nonrelativistic theory, the wave function involves the quantum numbers, n,l,m, but the energy depends only on n. There is Coulomb (accidental) degeneracy[2] with respect to l, and also degeneracy with respect to spin direction if spin is included. In a full theory, we should include relativistic and spin effects and see if these degeneracies can be removed to some extent, which would yield a *fine structure*.

The relativistic expression for the energy of a free electron is

$$E_{rel} = (c^2p^2+m^2c^4)^{\frac{1}{2}} = mc^2\left(1+\frac{p^2}{m^2c^2}\right)^{\frac{1}{2}} = mc^2\left(1+\frac{p^2}{2m^2c^2}-\frac{p^4}{8m^4c^4}+\ldots\right) \quad . \tag{2.40}$$

It involves the rest-mass energy mc^2. To a first approximation, the Hamiltonian for the electron in a Coulomb field can be written as

$$H = mc^2\left(1+\frac{p^2}{m^2c^2}\right)^{\frac{1}{2}} - mc^2 - \frac{Ze^2}{r} = \left(\frac{p^2}{2m}-\frac{Ze^2}{r}\right) - \frac{p^4}{8m^3c^2} = H_0 + H' \quad , \tag{2.41}$$

where the perturbation term is

$$H' = -\frac{p^4}{8m^3c^2} = -\frac{1}{2mc^2}\left(\frac{p^2}{2m}\right)^2 = -\frac{1}{2mc^2}\left(H_0+\frac{Ze^2}{r}\right)^2 \quad . \tag{2.42}$$

The resulting *relativistic correction* of the energy E_n is given by the expectation value of H' over the unperturbed normalized wave functions $\psi_{nlm}=R_{nl}Y_lZ_m$, (2.35),

$$\begin{aligned} E'_{rel} &= \int \psi^*_{nlm}\, H'\, \psi_{nlm}\, d\tau \\ &= -\frac{1}{2mc^2}\left(E_n^2+2Ze^2E_n\int R^2_{nl}\frac{1}{r}r^2dr+Z^2e^4\int R^2_{nl}\frac{1}{r^2}r^2dr\right) \quad . \end{aligned} \tag{2.43}$$

By use of the generating functions for the associated Laguerre polynomials, integrals of the form

---

[2]The nonrelativistic accidental degeneracy in a Coulomb field is due to the conservation law for the operator

$$\frac{\underline{r}}{r} + \frac{1}{2me^2}\,(\underline{l}\times\underline{p}-\underline{p}\times\underline{l}) \quad ,$$

where $\underline{l} = \underline{r}\times\underline{p}$ is the angular-momentum operator.

$$\langle r^{-s} \rangle = \int_0^\infty R_{nl}^2 \frac{1}{r^s} r^2 \, dr \quad ,$$

have been evaluated [2.20]. We get

$$\langle r^{-1} \rangle = (Z/a_0)\, n^{-2} \quad ,$$
$$\langle r^{-2} \rangle = (Z/a_0)^2 \left[n^3 \left(l + \frac{1}{2} \right) \right]^{-1} \quad ,$$
$$\langle r^{-3} \rangle = (Z/a_0)^3 \left[n^3 l(l+1) \left(l + \frac{1}{2} \right) \right]^{-1} \quad , \tag{2.44}$$

where $a_0 = \hbar^2/me^2 = \hbar/mc\alpha$. By use of them, we get

$$E'_{rel} = - E_n (\alpha Z)^2 \frac{1}{n} \left(\frac{3}{4n} - \frac{1}{l + \frac{1}{2}} \right) \quad . \tag{2.45}$$

With this correction,

$$E_{nl} = E_n + E'_{rel} = E_n \left[1 - (\alpha Z)^2 \frac{1}{n} \left(\frac{3}{4n} - \frac{1}{l + \frac{1}{2}} \right) \right] \quad . \tag{2.46}$$

The energy now depends both on n and l. This *removes the degeneracy between states of the same n and different l.*

We shall now consider the effect of the electron *spin*. This effect in one-electron spectra arises from the interaction of the magnetic moment of the electron with the effective magnetic field set up by its own motion around the nucleus. It is called the *spin-orbital interaction.*

From (1.46) we see that $\underline{A} = (1/c)\underline{v}\phi$. Therefore, from (1.18a) for a particle moving along (say) the x axis,

$$\underline{H} = \nabla \times \underline{A} = \frac{1}{c} \nabla \times (\underline{v}\phi) = - \frac{1}{c} \underline{v} \times \nabla\phi \quad . \tag{2.47}$$

Because $\underline{v}$ is along the x axis, only the y and z components of gradϕ are important in the cross product. For $\underline{A}$ independent of time, these components are just the negatives of the y and z components of $\underline{E}$, (1.49a). Thus,

$$\underline{H} = \frac{1}{c} \underline{v} \times \underline{E} \quad . \tag{2.48}$$

A Dirac electron has a magnetic moment $\underline{\mu}_s = (e\hbar/2mc)\underline{s}$, where $\underline{s}$ is the spin vector. When it moves in the central electric field, it sees the effective magnetic field (2.48). The resulting interaction term is

$$H'_{\text{spin-orbit}} = -\underline{\mu}_s \cdot \underline{H} = \frac{e\hbar}{2mc^2}\,\underline{s} \cdot (\underline{E}\times\underline{v}) = \frac{e\hbar}{2mc^2}\,\underline{s} \cdot \left(-\frac{\partial V}{\partial r}\,\hat{\underline{r}}_0\times\underline{v}\right) \quad , \tag{2.49}$$

where $\hat{\underline{r}}_0$ is a radial unit vector. Using $\hat{\underline{r}}_0 = \underline{r}/r$ and $V = Ze^2/r$, we get the spin-orbit interaction energy,

$$H'_{\text{spin-orbit}} = -\frac{e\hbar}{2m^2c^2}\,\frac{1}{r}\,\frac{\partial V}{\partial r}\,\underline{s} \cdot (\underline{r}\times\underline{p}) = \frac{Ze^2\hbar^2}{2m^2c^2}\,\frac{1}{r^3}\,\underline{l} \cdot \underline{s} = \zeta(r)\underline{l} \cdot \underline{s} \quad , \tag{2.50}$$

where $\underline{l} = \underline{r} \times \underline{p}$ and $\underline{s}$ are in units of $\hbar$, and $\zeta(r) = Ze^2\hbar^2/2m^2c^2r^3$. The total unperturbed wave function is

$$\psi_{nlmm_s} = \psi_{nlm}\psi_{\text{spin}} = |nlm_l)|m_s) = |nlm_lm_s) \quad . \tag{2.51}$$

The correction due to $H'_{\text{spin-orbit}}$ is given by

$$(nlm_lm_s|H'_{\text{spin-orbit}}|nlm_lm_s) = (nl|\zeta(r)|nl)(lm|\underline{l}|lm) \cdot (m_s|\underline{s}|m_s) \quad , \tag{2.52}$$

because $\zeta(r)$ operates on the radial part $|nl) = R_{nl}(r)$, $\underline{l}$ on angular part $|lm) = Y_l(\theta)Z_m(\phi)$ and $\underline{s}$ on the spin part $|m_s) =$ Pauli spinor.

Using (2.44), we have

$$\begin{aligned}(nl|\zeta(r)|nl) &= \frac{Ze^2\hbar^2}{2m^2c^2}\int_0^\infty \frac{1}{r^3}\,R_{nl}^2\,r^2\,dr \\ &= \frac{e^2\hbar^2Z}{2m^2c^2}\left(\frac{Z}{a_0}\right)^3 \frac{1}{n^3l(l+1)\left(l+\frac{1}{2}\right)} \\ &= R_\infty\,ch\,\alpha^2Z^4/\left[n^3l(l+1)\left(l+\frac{1}{2}\right)\right] \quad .\end{aligned} \tag{2.53}$$

The other matrix elements in (2.52) can be written immediately if we use the vector model to calculate the scalar product $\underline{l} \cdot \underline{s}$ (Fig.2.10). Because of the magnetic coupling of $\underline{l}$ and $\underline{s}$, we get the total quantum number $\underline{j} = \underline{l} + \underline{s}$, which is now a constant of motion, and the states can be designated by jm_j in place of m_lm_s, where $\underline{j}^2 = j(j+1)$ and $j_z = m_j\hbar$. We have the operator equation

Fig. 2.10. Owing to the spin-orbit interaction, $\underline{l}$ and $\underline{s}$ precess around $\underline{j} = \underline{l} + \underline{s}$, which remains fixed in the absence of external torques

$$\underline{j}^2 = (\underline{l}+\underline{s})^2 = \underline{l}^2 + \underline{s}^2 + 2\underline{l}\cdot\underline{s} \quad . \tag{2.54}$$

Because l, j and s are all good quantum numbers, $\underline{l}\cdot\underline{s}$ also has diagonal elements given by

$$(lj|\underline{l}\cdot\underline{s}|lj) = \frac{1}{2}\left[j(j+1)-l(l+1)-s(s+1)\right] \quad , \tag{2.55}$$

where $s(s+1) = 3/4$. When $l > 0$, the value of j can be either $l + (1/2)$ or $l - (1/2)$. Thus, the spin-orbit interaction gives the correction to the energy

$$E'_{\text{spin-orbit}} = (nl|\zeta(r)|nl)\begin{cases} \frac{1}{2}\,l & \text{if } j = l + \frac{1}{2} \quad , \\ -\frac{1}{2}\,(l+1) & \text{if } j = l - \frac{1}{2} \quad . \end{cases} \tag{2.56}$$

We find that *the spin-orbit interaction removes the* $m_l m_s$ *degeneracy within each configuration* (same nl). Consequently, for the same n,l, the states of different j have different energies. (We are still left with a (2j+1)-fold degeneracy due to m_j which can take all integrally spaced values from -j to +j. This can be removed by applying an external magnetic field.) *The state of lower j lies lowest*, as required by the Hund rule. The doublet separation $\zeta_{nl}(r)[l+(1/2)]$ is of magnetic origin and smaller than electronic energies by a factor of order α^2.

The fully corrected energy in the first approximation [we neglect higher terms in (2.40)] is

$$\begin{aligned} E_{nlj} &= E_n + E'_{\text{rel}} + E'_{\text{spin-orbit}} \\ &= -R_\infty hc\left[\frac{Z^2}{n^2} + \frac{\alpha^2 Z^4}{n^3}\left(\frac{1}{l+\frac{1}{2}} - \frac{3}{4n}\right) - \frac{\alpha^2 Z^4}{n^3}\,\frac{(lj|\underline{l}\cdot\underline{s}|lj)}{l(l+1)\left(l+\frac{1}{2}\right)}\right] \quad . \end{aligned} \tag{2.57}$$

Using (2.56), we can write

$$E_{nlj} = -R_\infty hc\left[\frac{Z^2}{n^2} + \frac{\alpha^2 Z^4}{n^4}\left(\frac{n}{j+\frac{1}{2}} - \frac{3}{4}\right)\right] \quad . \tag{2.58}$$

which holds for both $j = l + (1/2)$ and $j = l - (1/2)$. The relativistic Dirac theory provides the correct basis for this result. We now have an interesting situation. The spin-orbit interaction gives a correction of the same order of magnitude as the relativity correction. The net result is that the two levels that have the same j value but different values of l, now have the same enery (are degenerate) (Fig.2.11) in the hydrogen atom. The same situation persists even if the exact relativistic correction is included when the Dirac

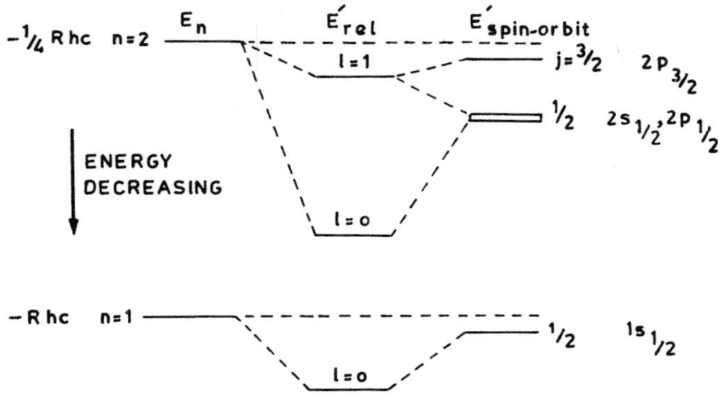

Fig. 2.11. The effect of various corrections on the energy level diagram of the hydrogen atom for *optical* levels with $n = 1,2$

equation is solved. In the hydrogen atom, this degeneracy is removed by the radiation correction (Lamb shift). In hydrogenlike terms and in X-ray terms ($Z > 4$) this degeneracy is removed by the dominating effect of *screening* that depends on the l value and produces a large separation.

The (one or) two *levels* in which each configuration is split are together said to constitute a *doublet term*. The *terms* are designated as ^2S (doublet S), ^2P, ^2D ... according to the l value. The separate levels are designated by adding the value of j as a subscript, thus $^2S_{\frac{1}{2}}$; $^2P_{\frac{1}{2}}$, $^2P_{3/2}$; $^2D_{3/2}$, $^2D_{5/2}$; etc. The quantum number n or the configuration label nl is sometimes prefixed to these symbols, and to specify an individual state the m_j value given as a superscript; thus $2\ ^2P^{3/2}_{3/2}$ or $2p\ ^2P^{-\frac{1}{2}}_{3/2}$.

In view of the Pauli vacancy principle, the X-ray levels are analogous to the levels of hydrogenlike atoms. Therefore, the energies of X-ray states (one inner electron missing) can at once be written from (2.58) provided that the following three modifications are made:

1) The electrons bound in an atom have negative energies, (2.58). For inner hole states, *the energy is taken to be positive.*

2) In the principal energy term, $E_n (\propto Z^2/n^2)$, the Z is replaced by an effective Z, $Z_{eff} = Z - \sigma_1(n,l)$, where σ_1 is called the *total screening constant*. The reason is that the actual nuclear charge seen by an inner shell electron is not Z but is screened by the other electrons that come in between. It is also affected by the other electrons in the external region. This screening will change as we go from one shell to another. Because *all* of the electrons of the atom affect this screening, σ_1 depends on n,l and Z. It separates the

terms with same n but different l, like $L_I(2s_{1/2})$ and $L_{II}(2p_{1/2})$. They form the *screening doublets* (L_I-L_{II}, M_I-M_{II}, ...).

3) In the second term of (2.58), the Z is replaced by $Z_{eff} = Z - \sigma_2(n,l)$, where σ_2 is called the *internal screening constant*. This second term contains the interaction of the magnetic moments l and s of a given electron with itself and with the magnetic moments of other electrons of the atom. However, the total magnetic moment of a closed shell is zero. Therefore, only the electrons of the excited shell (one electron missing) matter. Because the distribution of electrons in any such given shell remains the same for all elements, σ_2 does not depend on Z. In general, it depends on n,l.

With these modifications in (2.58), we can write the energy of *atomic X-ray levels*, as

$$
\begin{aligned}
E_{nlj} &= E_n(\sigma_1) + E'_{rel}(\sigma_2) + E'_{spin\text{-}orbit}(\sigma_2) \\
&= +R_\infty hc \left\{ \frac{\left[Z-\sigma_1(n,l)\right]^2}{n^2} + \frac{\alpha^2\left[Z-\sigma_2(n,l,j)\right]^4}{n^3} \left(\frac{1}{l+\frac{1}{2}} - \frac{3}{4n}\right) \right. \\
&\quad \left. - \frac{\alpha^2\left[Z-\sigma_2(n,l,j)\right]^4}{n^3} \frac{(lj|\underline{l}\cdot\underline{s}|lj)}{l(l+1)\left(l+\frac{1}{2}\right)} \right\} \\
&= +R_\infty hc \left\{ \frac{\left[Z-\sigma_1(n,l)\right]^2}{n^2} + \frac{\alpha^2\left[Z-\sigma_2(n,l,j)\right]^4}{n^4} \left(\frac{n}{j+\frac{1}{2}} - \frac{3}{4}\right) \right\} \quad , \qquad (2.59)
\end{aligned}
$$

to a first approximation. Because of the presence of $\sigma_1(n,l)$ and $\sigma_2(n,l,j)$, the terms $2s_{1/2}(L_I)$ and $2p_{1/2}(L_{II})$, that were degenerate in (2.58), will have separate energies. The (2j+1)-fold degeneracy with respect to m_j remains in (2.59) and determines the *statistical weight* of the level.

A qualitative representation of the X-ray atomic energy levels is shown in Fig.2.12. In X-rays it is customary *to choose the neutral atom in its ground state as the zero for energy measurements*. Then all other excited (hole) states have *positive* energy. Although one-electron symbols are retained, each level denotes the energy of the entire system when one electron is missing from the particular electronic shell.

The photon energies of emitted X-ray lines can be shown as *hole transitions* from a higher level to a lower level on this energy level diagram with the *selection rules*

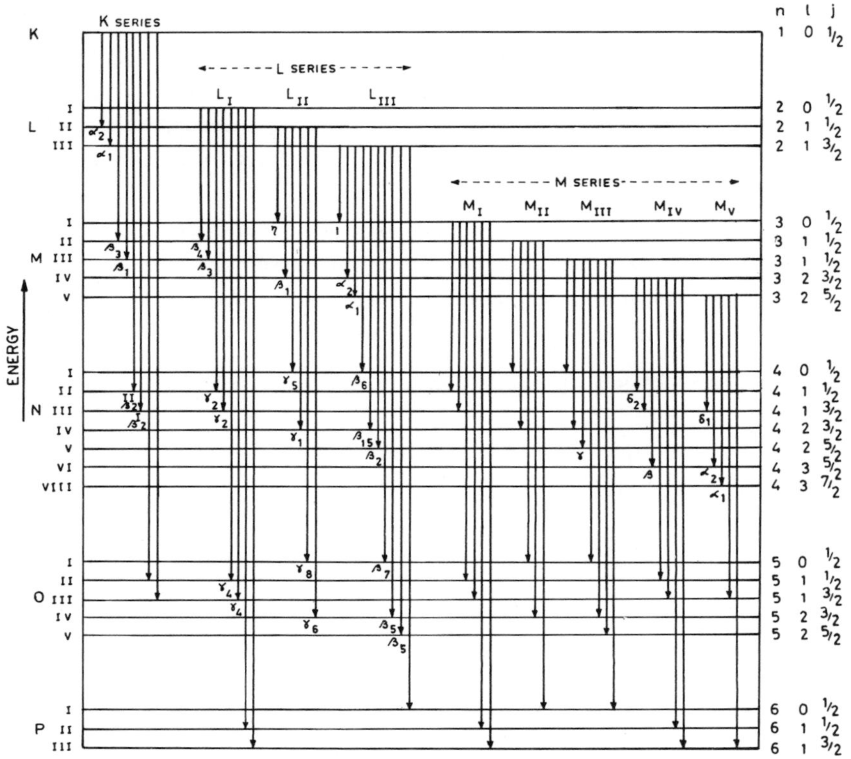

Fig. 2.12. Qualitative X-ray atomic energy level diagram. Some of the allowed electric-dipole hole transitions are shown by arrows. Selection rules are $\Delta l = \pm 1$, $\Delta j = 0$ or ± 1 (the j-transition $0 \to 0$ is forbidden)

$$\left.\begin{array}{l} \Delta n \neq 0 \quad , \\ \Delta l = \pm 1 \\ \Delta j = 0, \pm 1 \end{array}\right\} \text{electric-dipole selection rules} \quad . \tag{2.60}$$

The observed characteristic X-ray emission lines that obey these selection rules are called *diagram lines*. This nomenclature emphasizes that the condition $h\nu_{fi} = E_i - E_f$, where E_i and E_f are given on the energy level diagram, gives the frequencies of all the observed emission lines. In general, lines that arise from one-electron (or one-hole) jumps between singly ionized states, with the inactive electrons retaining their original quantum numbers as if frozen, are known as *diagram lines*; all others are *satellites* (see Chap.4). The notation $K \to L_{III}$ indicates the transition of the hole. The conventional X-ray nomenclature of the type $K\alpha_1$ is adequate for the diagram lines.

The above calculations have been done for an isolated atom. For such free atoms, the width of a level may be determined solely by the lifetime of the corresponding energy state. In the study of X-ray spectra, we usually have solid targets. In the solid state, the levels will be affected by the field of neighboring atoms. However, because only the outer-valence-electron wave functions overlap (hybridize), we can assume that the inner levels remain intact, in considering both their relative energy positions and their widths. Therefore, X-ray spectra can be used to determine the X-ray atomic-energy levels. Another way of saying the same thing is to note that X-ray lines arise from transitions between excited states of energies of the order of thousands of Rydberg units (in the range of several keV). On the other hand, the interaction energy between atoms in a solid is of the order of only a Rydberg unit or less. Hence, to a first approximation, such interactions can be ignored and the spectra interpreted as due to isolated atoms.

There is no universally accepted nomenclature for the X-ray lines. The designations given in Fig.2.12 are commonly used.

Whereas most of the features of X-ray spectra can be explained by a one-electron (or one-hole) jump model, using hydrogenlike wave functions, an adequate description of X-ray processes should, in general, take into account the manner in which the relatively inactive electrons are affected by the change of potential brought about either by the creation of the initial hole or by its decay. For this we need the many-electron wave functions of the initial and final states. The most accurate common representation of such wave functions is given by the Slater determinant of orthonormal spin orbitals

$$\Psi = (N!)^{-1} \det\left[\psi_{1s\alpha}(1),\psi_{1s\beta}(2),\psi_{2s\alpha}(3),\ldots,\psi_{\eta\lambda\beta}(N)\right] \quad .$$

Minimizing the total energy of the product yields the Hartree-Fock equations, which can be solved to give ψ's that are good approximations to the spatial wave functions. To calculate state energies, BAGUS [2.21] has used wave functions of the Slater type. Calculations of this kind for hole states are accurate (±1 eV in 3000) but not available for many elements. In such cases, KOOPMANS' [2.22] theorem is often used, in which the one-electron orbital energy obtained from Hartree-Fock equations is used as the level energy. It implies that the inactive electron orbitals are identical in the initial and final states. It has been shown that inner-shell ionization energies calculated from Koopmans' theorem are related to a weighted-average energy for singly and multiply excited states that are created by a sudden ionization process [2.23].

2.8 Electric-Dipole Selection Rules

The transition probability per unit time for the spontaneous emission is given by (2.26)

$$w_{fi} = \frac{64\pi^4}{3c^3h} \nu_{fi}^3 \; |(f|\underline{D}(r)|i)|^2 \quad , \tag{2.61}$$

where $\underline{D} = e\underline{r}$ is the *electric-dipole moment*. The transitions between the initial states $\psi_{n,l,m}$ and final states $\psi_{n',l',m'}$ will be possible if

$$\int \psi^*_{n'l'm'} |\underline{D}(r)| \psi_{nlm} \, d\tau \neq 0 \qquad (\textit{dipole integral}) \quad . \tag{2.62}$$

The components of $\underline{D}$ along x,y,z axes are

$$D_x = D(r) \sin\theta \cos\phi \quad ,$$
$$D_y = D(r) \sin\theta \sin\phi \quad ,$$
$$D_z = D(r) \cos\theta \quad . \tag{2.63}$$

As $D(r) = er$ is a function of r alone, we can write the dipole integral (2.62) as a product of three factors:

$$\int_0^\infty R^*_{n'l'}(r) |D(r)| R_{nl}(r) \, r^2 \, dr \quad ,$$
$$\int_0^\pi Y^*_{l'm'}(\theta) \begin{vmatrix} \sin\theta \\ \sin\theta \\ \cos\theta \end{vmatrix} Y_{lm}(\theta) \sin\theta \, d\theta \quad ,$$
$$\int_0^{2\pi} Z^*_{m'}(\phi) \begin{vmatrix} \cos\phi \\ \sin\phi \\ 1 \end{vmatrix} Z_m(\phi) \, d\phi \quad . \tag{2.64}$$

The integral in r is nonvanishing and so may be ignored for discussing the selection rules.

Let us consider the integral for D_z,

$$\int_0^\pi P^{m'}_{l'} |\cos\theta| P^m_l \sin\theta \, d\theta \int_0^{2\pi} e^{i(m-m')\phi} \, d\phi \quad . \tag{2.65}$$

The ϕ integral is zero for $m \neq m'$. Therefore, we must have

$$m = m' \quad . \tag{2.66}$$

For the θ integral, we use the recursion formula

$$|\cos\theta| P_l^m = \frac{l-m+1}{2l+1} P_{l+1}^m + \frac{l+m}{2l+1} P_{l-1}^m \quad , \tag{2.67}$$

and the orthonormality of associated Legendre polynomials. Because $m = m'$, we must have, for (2.62) to be nonzero,

$$m = m' \quad , \quad l \pm 1 = l' \quad , \quad (\text{for } D_z) \quad . \tag{2.68}$$

We consider D_y and D_z in the combinations

$$\begin{aligned} D(x+iy) &= D(r) \sin\theta \, e^{i\phi} \quad , \\ D(x-iy) &= D(r) \sin\theta \, e^{-i\phi} \quad . \end{aligned} \tag{2.69}$$

For $D(x+iy)$, the integrals are

$$\int_0^\pi P_{l'}^{m'} |\sin\theta| P_l^m \sin\theta \, d\theta \int_0^{2\pi} e^{i(m-m'+1)} \, d\phi \quad . \tag{2.70}$$

The ϕ integral is not zero only if $m+1 = m'$. Using the recursion formula

$$(1-\cos^2\theta)^{\frac{1}{2}} P_l^m = |\sin\theta| P_l^m = \frac{1}{2l+1} \left(P_{l+1}^{m+1} - P_{l-1}^{m+1} \right) \quad , \tag{2.71}$$

we note that (2.62) is not zero only when

$$m + 1 = m' \quad , \quad l \pm 1 = l' \quad , \quad \text{for } D(x+iy) \quad . \tag{2.72}$$

Similarly, for $D(x-iy)$, we get

$$m - 1 = m' \quad , \quad l \pm 1 = l' \quad , \quad \text{for } D(x-iy) \quad . \tag{2.73}$$

The *electric-dipole-moment selection rules* can therefore be expressed as

$$\Delta l = \pm 1 \quad ; \quad \Delta m = 0, \pm 1 \quad . \tag{2.74}$$

The observed emission lines shown in Fig.2.12 obey these selection rules.

2.9 Relative Intensities of Emission Lines in a Multiplet

The relative intensities of spectral lines can be defined either by means of their peak intensities or by the area below their intensity distribution curves.

A *multiplet* of the X-ray spectrum is *a group of lines whose initial states arise from a single configuration* nl, *and whose final states likewise arise from another single configuration* n'l'. A simple example is provided when the initial state or the final state is already a single level. Thus, $K\alpha_1\alpha_2$, $L\beta_3\beta_4$, $Ll\eta$, being simple doublets, are multiplets (Fig.2.13). Another example is the $L\beta_1\alpha_1\alpha_2$ triplet.

The intensity of the emission line is given by $I(if) \sim \nu_{fi}^4\ F_i\ g_i\ r^2(i,f)$, (2.28). The factors ν_{if}^4 and F_i can be taken to be identical for the lines of any multiplet that are close to each other. Thus, effectively the relative intensity of the lines of a multiplet depends on two factors,

$$I(i,f) \sim g_i\ r^2(i,f) \quad . \tag{2.75}$$

To calculate the relative intensities of the lines of any one multiplet, the sum rule, discovered empirically by BURGER et al. [2.24,25], is used. It states that *if the separation of the splitting of either the initial or final levels of a multiplet is imagined to be reduced to zero, then the sum of the quantities* $g_i\ r^2(i,f)$ *for the lines, whose frequencies would as a result ap-*

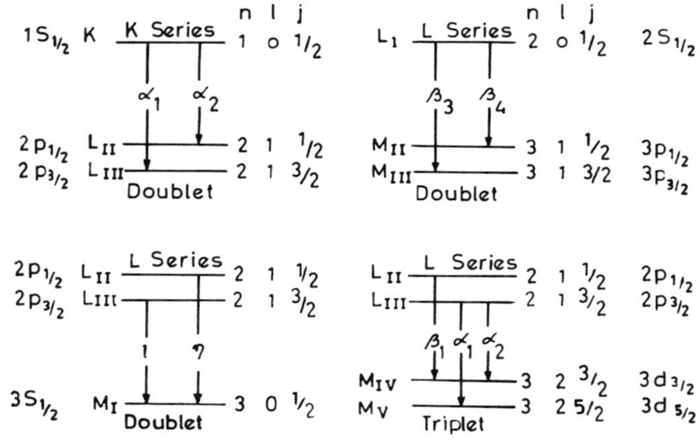

Fig. 2.13. Examples of multiplets

pear to be the same, are proportional to the statistical weights of the unreduced levels. This sum rule can be obtained from the quantum mechanical considerations (see, for example, [2.26]).

We have seen (Sect.2.7) that the statistical weight g of a term characterized by the quantum number j is 2j + 1. Let us apply the *Burger-Dorgelo-Ornstein* rule to the $K\alpha_1\alpha_2$ doublet (Fig.2.13). Here one level is already common and so the reducing process is not necessary. We have

$$\frac{I(K\alpha_1)}{I(K\alpha_2)} = \frac{g_K\, r^2(\alpha_1)}{g_K\, r^2(\alpha_2)} = \frac{g_{L_{III}}}{g_{L_{II}}} = \frac{2\times\frac{3}{2}+1}{2\times\frac{1}{2}+1} = \frac{2}{1} \quad . \tag{2.76}$$

This is in good agreement with experiments [2.27] over a wide range of atomic numbers:

Element	: ^{26}Fe	^{29}Cu	^{40}Zr	^{47}Ag
$I(K\alpha_2)/I(K\alpha_1)$	: 0.500	0.497	0.502	0.499 .

Let us now apply the above rule twice to the triplet $L\beta_1\alpha_1\alpha_2$ (Fig.2.13). Imagine the separation of M_{IV} and M_V to be zero. Then the rule gives

$$\frac{I(L\alpha_1)+I(L\alpha_2)}{I(L\beta_1)} = \frac{g_{L_{III}}\, r^2(\alpha_1)+g_{L_{III}}\, r^2(\alpha_2)}{g_{L_{II}}\, r^2(\beta_1)} = \frac{g_{L_{III}}}{g_{L_{II}}} = \frac{2}{1} \quad . \tag{2.77}$$

Now imagine the separation of L_{II} and L_{III} to be zero. Then

$$\frac{I(L\alpha_2)+I(L\beta_1)}{I(L\alpha_1)} = \frac{g_{L_{III}}\, r^2(\alpha_2)+g_{L_{II}}\, r^2(\beta_1)}{g_{L_{III}}\, r^2(\alpha_1)} = \frac{g_{M_{IV}}}{g_{M_V}} = \frac{2\times\frac{3}{2}+1}{2\times\frac{5}{2}+1} = \frac{2}{3} \quad . \tag{2.78}$$

From (2.77,78), we get

$$I(L\alpha_1) + I(L\alpha_2) = 2I(L\beta_1) \quad , \quad I(L\alpha_2) + I(L\beta_1) = \frac{2}{3}\, I(L\alpha_1) \quad ,$$
$$I(L\alpha_1) : I(L\alpha_2) : I(L\beta_1) = 9:1:5 = 100:11:56 \quad , \tag{2.79}$$

again in close agreement with experiments [2.28,29]:

Element	: ^{42}Mo	^{47}Ag	^{74}W	^{78}Pt
$L\alpha_1:\alpha_2:\beta_1$	: 100:13:62	100:12:59	100:11.5:52	100:11.4:51 .

Similar calculations yield the following theoretical values:

$$
\begin{aligned}
&\text{K series}: \alpha_1:\alpha_2 = \beta_1:\beta_2 = 2:1 \quad ,\\
&\text{L series}: \gamma_3:\gamma_2 = \beta_3:\beta_4 = 1:\eta = \beta_6:\gamma_5 = 2:1 \quad ,\\
&\qquad \alpha_1:\alpha_2:\beta_1 = \beta_2:\beta_{15}:\gamma_1 = 9:1:5 \quad ,\\
&\qquad \beta_5:\beta_6 = 2:1 \quad ,\\
&\text{M series}: \alpha_1:\alpha_2:\beta_1 = 20:1:14 \quad . \qquad (2.80)
\end{aligned}
$$

These values are in fair agreement with the experimental values, so long as we use the quantities $g_i\ r^2(i,f)$ to calculate the intensity ratios. However, if we use the full expression $\nu_{fi}^4\ F_i\ g_i\ r^2(i,f)$, (2.28), the agreement is spoiled. For example, if the ν_{fi}^4 correction is included, the value of $L\beta_1$ would be decreased by a factor 0.57 in ^{74}W and 0.39 in ^{92}U, which completely upsets the agreement.

Again, $F(L_{II})$ and $F(L_{III})$ need not be equal even for high voltages on the X-ray tube. A rough classical estimate would give $F_i = E_i^{-1}$, where E_i is the energy transferred to the atom in the impact when it is put in the excited state i. If this factor is included, the intensity of $L\beta_1$ should be increased by the ratio $E(L_{II})/E(L_{III})$, that is, 1.13 for ^{74}W and 1.22 for ^{92}U.

Thus, with both of the factors $\nu_{fi}^4\ F_i$ included, the total factor to be applied to the $L\beta_1$ line is 0.64 in ^{74}W and 0.48 in ^{92}U. This will change the theoretical $g_i\ r^2(i,f)$ ratios $L\alpha_1:\alpha_2:\beta_1 = 100:15:56$ to $\nu_{fi}^4\ F_i\ g_i\ r^2(i,f)$ ratios 100:11:23.5. The new ratios are not in agreement with the experimental ratios 100:11.5:52 for ^{74}W and 100:11:49.5 for ^{92}U. This clearly shows that the theoretical discussion of the problem is inadequate.

Realistic calculations for relative line intensities have been made only recently in the active-electron approximation, including the effects of retardation, by use of wave functions obtained by self-consistent relativistic Hartree-Fock methods with a Slater approximation to the exchange term [2.30, 31]. Relative intensities, corrected for self-absorption effects and for instrumental response, have also been measured recently [2.32,33].

For the ratio of characteristic to continuous radiation see TOTHILL [2.34], and UNSWORTH and GREENING [2.35]. The ratio varies approximately as $1/Z^3$.

2.10 Screening and Spin Doublets

The energy E_{nlj} of the neutral hydrogenlike atoms is given by (2.58). This equation is strictly valid only for the optical levels of hydrogen atoms (Z=1)

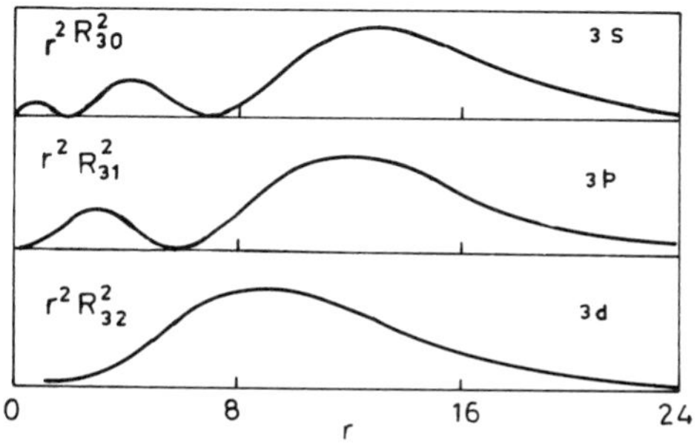

Fig. 2.14. Radial probability density for $n = 3$, $l = 0,1,2$, states of the hydrogenlike atoms. r is in units of a_0

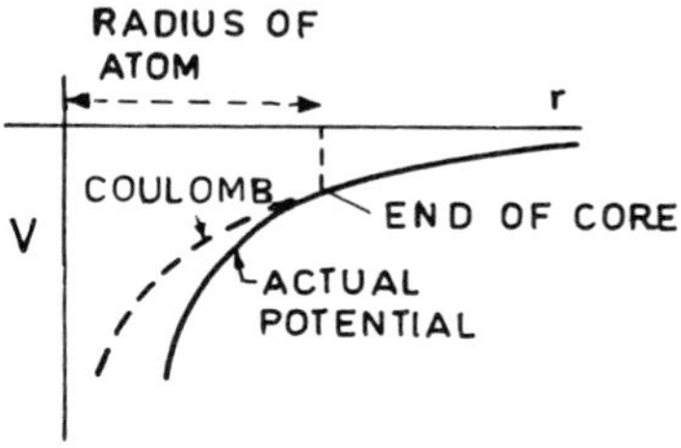

Fig. 2.15. The screening effect on the potential

For hydrogenlike atoms ($Z > 1$) with one (valence) electron outside the core of $Z - 1$ electrons, this equation will hold strictly only if the valence electron remains always outside the core and sees a shielded or screened effective nuclear charge eZ_{eff}, which is exactly $eZ_{eff} = e[Z-(Z-1)] = e$, as in the case of the hydrogen atom. However, $Z_{eff} > 1$ for hydrogenlike atoms because quantum mechanically there is a certain small but finite probability that the valence electron will be within the charge cloud of the core electrons (Fig.2.14). During this time, the valence electron is closer to the nucleus and so is not screened by all of the $Z - 1$ core electrons but only by a part of them, giving $Z_{eff} > 1$. In fact, as the distance r of the valence electron from the nucleus approaches zero, the valence electron begins to see the full charge $eZ_{eff} = eZ$. As it moves away from the nucleus, this is decreased, till it finally reduces to just e outside the core. We can, therefore, write $Z_{eff}(r) = Z - \sigma_1(r)$, with $\sigma_1(r) \to 0$ and $r \to 0$, $\sigma_1(r) \to Z - 1$ as $r \to$ "radius of the atom". We can say that $\sigma_1(r)$ is a measure of the incompleteness of screening. For the configuration nl we can write simply $Z_{eff} = Z - \sigma_1(n,l)$, because the radial probability density has a strong maximum at the appropriate value of r. Clearly $\sigma_1(n,l) < Z - 1$ and $Z_{eff} > 1$, which makes the electron more strongly attracted to the nucleus than for hydrogen (Fig.2.15). Because $\sigma_1(r)$ increases with r, we expect $\sigma_1(n,l)$ to increase with increase of n and l.

For the hydrogenlike atoms, we can get some idea of the change of energy levels by writing $V = -(e^2/r) - \delta V$, where $-\delta V$ is the correction of the Coulomb potential. The correction is taken to be negative because the effect of incomplete screening in the inner shells is always to increase the force of binding of the electron to the nucleus. The first-order correction of the nth energy level will be

$$\delta E_{nl} = -\int R^*_{nl}(r)\delta V\, R_{nl}(r)\, r^2\, dr \quad . \tag{2.81}$$

Now δE_{nl} will be large when $R_{nl}(r)$ is large, in the same place that δV is large, namely at small r. If we plot the radial probability density $R^*_{nl}R_{nl}r^2$ for the same n (=3) and different l (0,1,2) (Fig.2.14), we find that *the higher* l *value becomes, the smaller is the probability density near the origin*. Intuitively, we can see the reason for this behavior. In the radial part of the Schrödinger equation for the hydrogen problem (see, for example, [2.7]) the effective potential is $-(e^2/r) + [l(l+1)/r^2]$, where the second term, called the "centrifugal term", keeps the electron away from the nucleus when l is large. We, therefore, conclude that δE_{nl} given by (2.81) will be largest for the lowest value of l. The (l=0) s optical state is thus depressed the most, the (l=1)p optical state the next, and so on. This means that levels of the same n and different l (such as the 3s and 3p levels) are separated by an amount, with the s level more strongly bound, that increases with the increase of the deviation from the Coulomb potential. The separated levels of same n and different l (2s from 2p, 3s from 3p, etc.) are the *screening doublets*. They will now be separated even if they have the same j values, such as $2s_{\frac{1}{2}}$ and $2p_{\frac{1}{2}}$. Without $\sigma_1(n,l)$, they coincide according to (2.58), which is strictly true only for the hydrogen atom ($Z=1$, $\sigma_1=0$).

If we adhere to the idea that the X-ray terms (of positive energy) arise from the hole states in a manner analogous to the hydrogenlike states, then we can replace the Bohr term in $-R_\infty hc\, Z^2/n^2$ by $+R_\infty hc[Z-\sigma_1(n,l)]^2/n^2$, where the *total screening constant* $\sigma_1(n,l)$ separates out the X-ray levels with the same n and different l, such as L_I and $L_{II,III}$. Because the L_I state requires the removal of the more strongly bound $2s_{\frac{1}{2}}$ electron and L_{II} the removal of the comparatively less strongly bound $2p_{\frac{1}{2}}$ electron, the L_I level will have higher positive energy than the L_{II} level, and an *X-ray screening doublet* will be formed. The ordering of levels is obviously reversed when we go from optical spectra to X-ray spectra.

If we ignore the corrections that involve α^2 in (2.59), we have

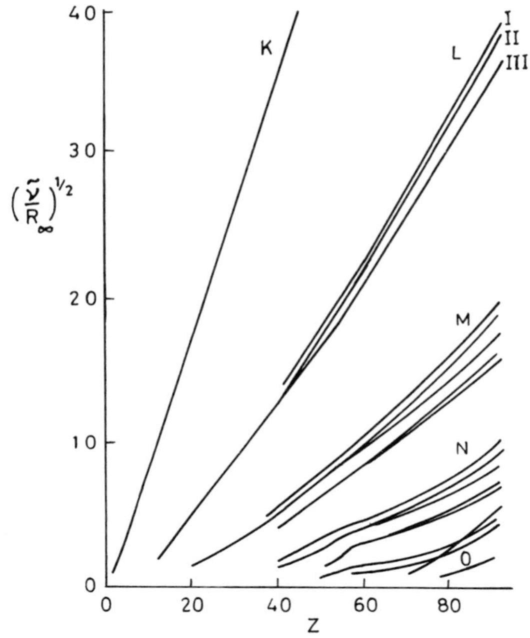

Fig. 2.16. Moseley diagram for the X-ray levels K, L, M,... where $\tilde{\nu}$ is wave number and $(\tilde{\nu}/R_\infty)^{1/2} = (E/13.6)^{1/2}$, E being in electron volts

$$E_{nlj} = +R_\infty\, hc\, \frac{Z_{eff}^2}{n^2} = +R_\infty\, hc\, \frac{[Z-\sigma_1(n,l)]^2}{n^2} \quad , \tag{2.82}$$

or, using the relation

$$\frac{\tilde{\nu}}{R_\infty} = \left|\frac{E_{nlj}}{R_\infty hc}\right| \quad , \tag{2.83}$$

we can write

$$\left(\frac{\tilde{\nu}}{R_\infty}\right)^{1/2}_{nlj} = \frac{Z-\sigma_1(n,l)}{n} \quad . \tag{2.84}$$

For a given n (=1,2,3,...), we can express (2.84) as

$$\left(\frac{\tilde{\nu}}{R_\infty}\right)^{1/2} = aZ + b \quad ; \quad a = 1/n \quad , \quad b = -\sigma_1/n \quad . \tag{2.85}$$

This is called *Moseley's law for X-ray atomic levels*. A plot of experimental data against Z is nearly a straight line over significant intervals of Z (Fig.2.16). It is called the Bohr-Coster diagram [2.36]. Sharp breaks in some of the straight lines are associated with changes of the filling of corresponding levels with electrons. For example, filling of the $N_{VI,VII}$ (4f) level in the N-shell begins after $Z = 57$, giving breaks in the lines for the M, N and O levels at $Z = 57$.

HERTZ [2.37] noted that on a Moseley plot (Fig.2.16) the L_I and L_{II} are separated and run parallel, preserving a $\Delta(\tilde{\nu}/R_\infty)^{\frac{1}{2}}$ separation independent of Z. This is consistent with (2.84), because for $n = 2$ we have for the L_I, L_{II} separation

$$\Delta\left(\frac{\tilde{\nu}}{R_\infty}\right)^{\frac{1}{2}} = \left(\frac{\tilde{\nu}}{R_\infty}\right)^{\frac{1}{2}}_{20j} - \left(\frac{\tilde{\nu}}{R_\infty}\right)^{\frac{1}{2}}_{21j} = \frac{Z-\sigma_1(L_I)}{2} - \frac{Z-\sigma_1(L_{II})}{2}$$

$$= \frac{\sigma_1(L_{II})-\sigma_1(L_I)}{2} = \frac{\Delta\sigma_1}{n} \quad , \tag{2.86}$$

which is independent of Z.

The finding of HERTZ [2.37] that *the difference between the square roots of the energies of two screening-doublet levels is a constant essentially independent of* Z, is called the *screening-doublet law* (or the *irregular-doublet law*). It is approximately true for screening doublets like L_IL_{II}, M_IM_{II}, $M_{III}M_{IV}$, etc. Because σ_1 takes into account the effect of the entire electron cloud, it depends on Z, and it is difficult to calculate it accurately[3]. Moseley plots provide a rough method for experimental determination of it. It is found that σ_1 increases with Z, as we expect. It increases with increasing Z, n and l.

We have seen that two levels with same n,l but different directions of spin, that is, with different values of j are called *spin doublets* (L_{II}-L_{III}, M_{II}-M_{III} ...). HERTZ also noted that when the Moseley diagram for the X-ray atomic levels is drawn (Fig.2.16), *the difference between the square roots of the energies between two spin-relativity doublets is proportional to the*

[3] If a nucleus of charge +Ze is surrounded by a spherical electron cloud of density $\rho(r)$, then the potential at a distance r_0 from the nucleus is

$$V(r_0) = \frac{Ze}{r_0} - \frac{1}{r_0}\int_0^{r_0} \rho(r)4\pi r^2 dr - \int_{r_0}^{\infty} \rho(r)4\pi r dr = \frac{Ze}{r_0} - \frac{q_{r<r_0}}{r_0} - \frac{q_{r>r_0}}{r_{ext}} \quad ,$$

where r_{ext} is the average radius for the charge distribution for $r > r_0$. The inside screening effect (second term) does not depend on the details of the charge distribution. The external screening effect (third term) depends on it; $\sim 1/r_{ext}$.

square of $Z-\sigma_2$, where $\sigma_2(n,l)$ is the *internal screening constant* appropriate to the spin-orbit interaction. This is called the *spin-relativity doublet law* (or the *regular-doublet law*). Because only electrons inside the shell in question influence σ_2, following BOHR and COSTER [2.36], we expect σ_2 to be independent of Z. But σ_2 increases for subshells farther from the nucleus. We can determine σ_2 from experiments. The σ_2 is smaller than σ_1 for a given (n,l). The reason is that, for the spin-orbit interaction, it is the net electric intensity experienced by the electron that is significant, not the energy. A symmetric closed shell of electrons, external to the electron in question, takes no part in reducing the effective nuclear charge, insofar as the electric intensity is concerned, but it does insofar as energy is concerned.

From (2.59) we get two separate terms.
For $j = l + (1/2)$,

$$E_{n,l,l+\frac{1}{2}} = +R_\infty\, hc\, \frac{[Z-\sigma_1(n,l)]^2}{n^2} + R_\infty\, hc\, \alpha^2\, \frac{[Z-\sigma_2(n,l)]^4}{n^4} \left(\frac{n}{l+1} - \frac{3}{4}\right) \quad . \tag{2.87}$$

For $j = l - (1/2)$,

$$E_{n,l,l-\frac{1}{2}} = +R_\infty\, hc \frac{[Z-\sigma_1(n,l)]^2}{n^2} + R_\infty\, hc\, \alpha^2\, \frac{[Z-\sigma_2(n,l)]^4}{n^4} \left(\frac{n}{l} - \frac{3}{4}\right) \quad . \tag{2.88}$$

These equations give the *spin-doublet formula*,

$$\Delta E = R_\infty\, hc\, \alpha^2\, \frac{[Z-\sigma_2(n,l)]^4}{n^3\; l(l+1)} \quad , \quad (|\Delta j| = 1 \text{ but same } nl) \quad , \tag{2.89}$$

for the separation of spin-doublets like $L_{II}L_{III}$, $M_{II}M_{III}$, $M_{IV}M_V$, $N_{II}N_{III}$, $N_{IV}N_V$ and $N_{VI}N_{VII}$. Unlike (2.86), here the separation depends strongly on Z, as observed by HERTZ.

Two *lines* are called a *spin-doublet* when: 1) they have a common initial level and the final levels are a spin-doublet, or 2) they have a common final level and the initial levels are a spin-doublet. Such lines of K and L series are given below:

K-series spin-doublets	Lines	Levels
	$\alpha_1\alpha_2$	$L_{II}L_{III}$
	$\beta_1\beta_3$	$M_{II}M_{III}$

L-series spin-doublets	Lines	Levels
	$l\eta$	$L_{II}L_{III}$
	$\alpha_2\beta_1$	$L_{II}L_{III}$
	$\beta_6\gamma_5$	$L_{II}L_{III}$
	$\beta_{15}\gamma_1$	$L_{II}L_{III}$
	$\gamma_8\beta_7$	$L_{II}L_{III}$
	$\beta_5\gamma_6$	$L_{II}L_{III}$
	$\beta_4\beta_3$	$M_{II}M_{III}$
	$\alpha_1\alpha_2$	$M_{IV}M_V$
	$\gamma_2\gamma_3$	$N_{II}N_{III}$
	$\beta_{15}\beta_2$	$N_{IV}N_V$

The spin-doublets are recognized by their having the same frequency difference all through the spectrum of one and the same element (given Z). Examples are the lines $K\alpha_1$ and $K\alpha_2$, the $L\beta_1$ and $L\alpha_2$, the $L\eta$ and Ll, and the $L\gamma_5$ and $L\beta_6$, all characterized by having the L_{II} and L_{III} levels either as initial or final levels. There are no electric-dipole lines that form screening doublets by having L_I and L_{II} levels as initial or as final levels. However, one dipole line and one multipole line can form such a doublet (see Sect.2.13).

It is useful to calculate the wavelength difference $\Delta\lambda$ between the lines of a spin-doublet. Because $\nu = 1/\lambda$ and $h\Delta\nu = \Delta E$, we have

$$|\Delta\lambda| = \frac{\Delta\tilde{\nu}}{\tilde{\nu}^2} = \frac{R_\infty}{\tilde{\nu}^2}\Delta\left(\frac{\tilde{\nu}}{R_\infty}\right) = \frac{R_\infty\alpha^2}{n^3\ l(l+1)}\ \frac{(Z-\sigma_2)^4}{\tilde{\nu}^2} \quad . \tag{2.90}$$

From the Moseley law for lines, (2.6), we get

$$(\tilde{\nu}/R_\infty)^2 = A^4(Z-\sigma)^4 \quad \text{(for lines)} \quad . \tag{2.91}$$

In general, $\sigma \neq \sigma_2$. However, for large Z we can take $\sigma \simeq \sigma_2$. Using this, we get for spin-doublets

$$|\Delta\lambda| \simeq \frac{\alpha^2}{A^4\ R_\infty\ n^3\ l(l+1)} \quad . \tag{2.92}$$

This approximate formula tells us that $\Delta\lambda$ for the lines of a spin-doublet will not depend on Z, in contrast to $\Delta\tilde{\nu}$ given by (2.89). It is supported by

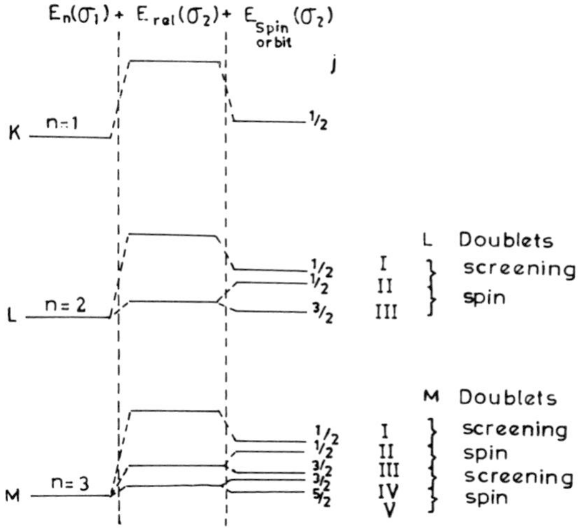

Fig. 2.17. Diagram showing the splitting of simple X-ray atomic levels as relativity and spin-orbit effects, together with screening effects, are introduced, (2.59)

the data. Let us consider the spin doublet $K\alpha_1$, $K\alpha_2$ that arises from the transitions $K \rightarrow L_{III}$, $K \rightarrow L_{II}$, respectively. The measured values of $\Delta\lambda$ in XU remain between 4 and 5, whereas $\Delta(\tilde{\nu}/R_\infty)$ varies from about 100 to 25035 for Z values that range from 26(Fe) to 92(U).

The rule (2.89) for a spin doublet, that $\Delta\tilde{\nu}$ is same for a given Z, and the rule (2.92), that $\Delta\lambda$ for a spin doublet is nearly constant for any Z, have together been very useful in systematizing X-ray spectra. There are some interesting exceptions to the constant $\Delta\lambda$ rule, which are connected with incomplete electronic shells (see, for example, [2.38]).

In Fig.2.17 we show how the simple X-ray levels of the Bohr atom are split up through the relativity and spin effects, when screening is included (2.59). When the screening effects are not included the levels with the same n,j values but different l values coincide (Fig.2.11).

Experimental values of σ_1 and σ_2 can be obtained in the following way. The spin-doublet formula (2.89) can be used to determine approximately the internal screening constant σ_2,

$$\Delta\left(\frac{\tilde{\nu}}{R_\infty}\right) = \frac{\Delta E}{R_\infty hc} = \frac{E_i - E_f}{R_\infty\, hc} = \alpha^2\,\frac{[Z-\sigma_2(n,l)]^4}{n^3\; l(l+1)} \quad . \tag{2.93}$$

For $n=2$, $l=1$, this will correspond to the separation of spin-doublet levels $L_{II}L_{III}$. If we determine $\Delta(\tilde{\nu}/R_\infty)$ from the distance between the observed $K\alpha_1$ and $K\alpha_2$ lines in Rydbergs, a value for σ_2 is obtained,

$$\Delta\left(\frac{\tilde{\nu}}{R_\infty}\right)_{\text{expt}} = \alpha^2 \, \frac{[Z-\sigma_2(n=2,l=1)]^4}{2^4} \quad (\text{for } L_{II}L_{III}) \quad . \tag{2.94}$$

For a better approximation, we should improve (2.59) by including more terms in the relativistic correction (2.40). This gives the Bohr-Sommerfeld expression for orbital energies, that involves the two screening constants σ_1 and σ_2, as

$$\begin{aligned} E_{nlj} = R_\infty hc\Bigg[&\frac{(Z-\sigma_1)^2}{n^2} + \frac{\alpha^2(Z-\sigma_2)^4}{n^4}\left(X-\frac{3}{4}\right) \\ &+ \frac{\alpha^4(Z-\sigma_2)^6}{n^6}\left(\frac{1}{4}X^3+\frac{3}{4}X^2-\frac{3}{2}X+\frac{5}{8}\right) \\ &+ \frac{\alpha^6(Z-\sigma_2)^8}{n^8}\left(\frac{1}{8}X^5+\frac{3}{8}X^4+\frac{1}{8}X^3-\frac{15}{8}X^2+\frac{15}{8}X-\frac{35}{64}\right)+\dots\Bigg] \quad , \\ X = n/\left(j+\frac{1}{2}\right) \quad . \end{aligned} \tag{2.95}$$

This equation, like (2.59), holds for $j=l+(1/2)$ and $j=l-(1/2)$. Using it, we get in place of (2.94), for the $L_{II}L_{III}$ doublet,

$$\Delta\left(\frac{\tilde{\nu}}{R_\infty}\right)_{\text{expt}} = \frac{\alpha^2(Z-\sigma_2)^4}{2^4}\left[1+\frac{5}{2}\frac{\alpha^2}{2^2}(Z-\sigma_2)^2+\frac{53}{8}\frac{\alpha^4}{2^4}(Z-\sigma_2)^4+\dots\right] \quad . \tag{2.96}$$

SOMMERFELD [2.39] has shown that this can be solved for $Z-\sigma_2$ in an approximate manner to give the result

$$\begin{aligned} [Z-\sigma_2(n=2,l=1)]^2 = &\left[\frac{4}{\alpha}\sqrt{\Delta\left(\frac{\tilde{\nu}}{R_\infty}\right)_{\text{expt}}} - 5\Delta\left(\frac{\tilde{\nu}}{R_\infty}\right)_{\text{expt}}\right] \\ &\times \left[1+\frac{19}{32}\alpha^2\,\Delta\left(\frac{\tilde{\nu}}{R_\infty}\right)_{\text{expt}}\right] \quad . \end{aligned} \tag{2.97}$$

Values of σ_2 for the $L_{II}L_{III}$ doublet, obtained by putting $K\alpha_1$, $K\alpha_2$ data for $\Delta(\tilde{\nu}/R_\infty)_{\text{expt}}$ in (2.97), are given in Table 2.3. Thus, both $\Delta\lambda$ and σ_2 remain nearly constant as predicted. Values of σ_2 for other spin-doublet levels can be found in a similar way. The average values of $\sigma_2(n,l)$ are given in Table 2.4. The value of $\sigma_2(n,l)$ increases with the increase in n and l, because the amount of screening from the nucleus also increases.

Table 2.3. The internal screening constant σ_2 for $L_{II}L_{III}$ doublet

Element	$\Delta\lambda(K\alpha_2-K\alpha_1)$ [XU]	$\Delta(\tilde{\nu}/R_\infty)$ (experiment)	$Z-\sigma_2$	σ_2
42 Mo	4.27	7.73	38.53	3.47
47 Ag	4.39	12.74	43.49	3.51
50 Sn	4.45	16.74	46.50	3.50
66 Dy	4.72	58.41	62.51	3.49
74 W	4.83	98.84	70.52	3.48
82 Pb	4.88	158.49	78.56	3.44

Table 2.4. Values of σ_2 for the X-ray spin-doublet levels

Level	n	l	σ_2
$L_{II,III}$	2	1	3.5
$M_{II,III}$	3	1	8.5
$M_{IV}\ M_V$	3	2	13.0
$N_{II,III}$	4	1	17.0
$N_{IV,V}$	4	2	24
$N_{VI,VII}$	4	3	34

The above method cannot be used to find $\sigma_2(n,l=0)$ for *single* levels like K, L_I, M_I, ... (l=0 and only a single value of j=1/2). We therefore first determine σ_1.

The formula (2.84) is much too rough for determining the X-ray total screening constant $\sigma_1(n,l)$. A better approximation is obtained by using (2.83,59), for the level characterized by n,j, σ_1, σ_2,

$$\frac{\tilde{\nu}}{R_\infty} = \frac{(Z-\sigma_1)^2}{n^2}\left[1+\frac{\alpha^2}{n^2}\frac{(Z-\sigma_2)^4}{(Z-\sigma_1)^2}\left(\frac{n}{j+\frac{1}{2}}-\frac{3}{4}\right)\right] \quad . \tag{2.98}$$

Using $(1+x)^{\frac{1}{2}} \simeq 1+\frac{1}{2}x$, we get

$$\left(\frac{\tilde{\nu}}{R_\infty}\right)^{\frac{1}{2}} = \frac{Z-\sigma_1}{n} + \frac{\alpha^2}{2n^3}\frac{(Z-\sigma_2)^4}{Z-\sigma_1}\left(\frac{n}{j+\frac{1}{2}}-\frac{3}{4}\right) \quad . \tag{2.99}$$

Similarly, for the second level of a screening doublet (like $2s_{\frac{1}{2}}$, $2p_{\frac{1}{2}}$ or L_I, L_{II}) characterized by n, j, σ_1', σ_2',

$$\left(\frac{\tilde{\nu}'}{R_\infty}\right)^{\frac{1}{2}} = \frac{Z-\sigma_1'}{n} + \frac{\alpha^2}{2n^3}\frac{(Z-\sigma_2')^4}{Z-\sigma_1'}\left(\frac{n}{j+\frac{1}{2}}-\frac{3}{4}\right) \quad . \tag{2.100}$$

For a screening doublet, n and j are the same but l differs by 1. From (2.99, 100) we have

$$\Delta\left(\frac{\tilde{\nu}}{R_\infty}\right)^{\frac{1}{2}} = \left(\frac{\tilde{\nu}}{R_\infty}\right)^{\frac{1}{2}} - \left(\frac{\tilde{\nu}'}{R_\infty}\right)^{\frac{1}{2}}$$

$$= \frac{\sigma_1'-\sigma_1}{n} + \frac{\alpha^3}{2n^3}\left(\frac{n}{j+\frac{1}{2}}-\frac{3}{4}\right)\left[\frac{(Z-\sigma_2)^4}{Z-\sigma_1} - \frac{(Z-\sigma_2')^4}{Z-\sigma_1'}\right] \quad . \tag{2.101}$$

Here $\sigma_1 = \sigma_1(n=2,l=0)$ and $\sigma_1' = \sigma_1(n=2,l=1)$, for example, for the screening doublet L_I, L_{II}. We again get (2.86), $\Delta(\tilde{\nu}/R_\infty)^{\frac{1}{2}} = \Delta\sigma_1/n$, when the second term in (2.101) is ignored.

We already know σ_2 for the L_{II} level (Table 2.4). Using this σ_2 and the experimental values of $(\tilde{\nu}/R_\infty)^{\frac{1}{2}}$, we can calculate the total screening constant σ_1 for the L_{II} levels of various elements from (2.100). These values of σ_1 (L_{II}) are plotted against Z, see Fig.2.18. Although σ_2 is independent of Z, σ_1 increases slowly with Z which changes in slope associated with the filling of shells. However, $\Delta\sigma_1$ should remain independent of Z (2.86). This fact can be exploited to find σ_2 for the level L_I, by using a method of trial and error. Experimental values of $(\tilde{\nu}/R_\infty)^{\frac{1}{2}}$ are selected for the L_I level, and various numerical values of σ_2 (not known for L_I) are substituted by hand in (2.99). This gives a set of curves for the σ_1 versus Z plot. It contains one curve for each value of σ_2 used. From this set, the curve that runs parallel to the $\sigma_1(L_{II})$ curve for the L_{II} level drawn earlier (Fig.2.18) is selected. This gives us $\sigma_1(L_I)$ for the L_I level (Fig.2.18) and also the corresponding σ_2 for the L_I level. The values of σ_2 found by trial in this way are given in Table 2.5. The value for the K level has been added for the sake of completeness.

The foregoing interpretations cannot be true in a precise sense because, quantum mechanically, no sharp distinction between internal and external shells is meaningful. Therefore, there is not much to be gained by attempting to choose σ_1 and σ_2 in such a way as to represent the data best. The Bohr-Sommerfeld result (2.59) is in accord with the main facts; that is all that can be expected. For recent calculations see [2.76,77].

In the exact Dirac relativistic theory of the hydrogen atom, the relativistic and spin-orbit effects are not treated separately as they are implied in the theory. Therefore, we can use a single screening constant σ_{nlj}, which

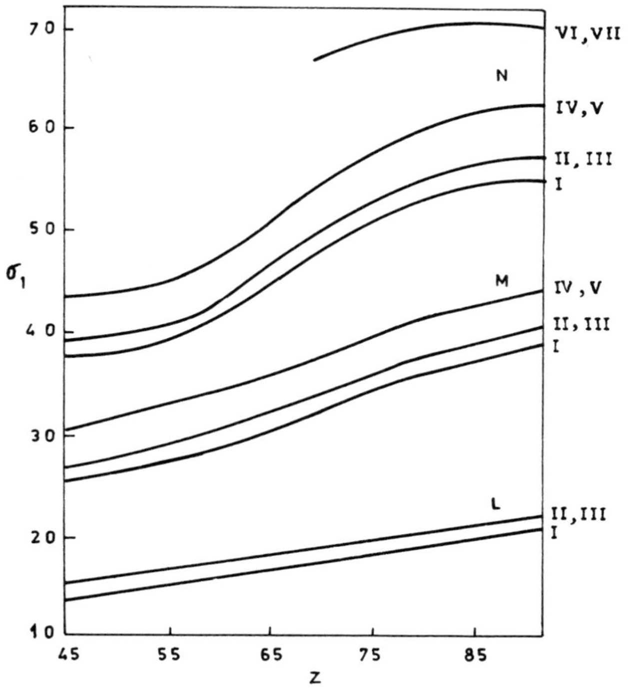

Fig. 2.18. The screening constant $\sigma_1(n,l;Z)$ as a function of Z

Table 2.5. $\sigma_2(n,l)$ for individual levels

Level	σ_2
K	0
L_I	2.0
M_I	6.8
N_I	14

allows us to replace Z by $Z-\sigma_{nlj}$. The Dirac theory gives the term-value equation, which includes both the relativistic effect and the spin-orbit effect, as

$$E_{nlj} = mc^2\left(1+(\alpha Z_{nj})^2\left\{n-j-\frac{1}{2}+\left[\left(j+\frac{1}{2}\right)^2-(\alpha Z_{nj})^2\right]^{\frac{1}{2}}\right\}^{-2}\right)^{-\frac{1}{2}} \quad ,$$

where Z_{nj} is the effective atomic number, $Z_{nj} = Z-\sigma_{nlj}$. A simple program would be to calculate a set of values for σ_{nlj} by using the experimental values of

E_{nlj} for the various shells from the X-ray data. Such a calculation has been performed by MANDE and DAMLE [2.40]. Z_{nj} can be expressed as Z minus a sum over occupied orbitals of the product of electron numbers and orbital screening constants, so that the matrix equation is

$$Z_{nj} = Z - S_{nj} N_{nj} + S_{nj} \quad .$$

Here N_{nj} is the orbital occupation matrix and S_{nj} is the screening matrix. Recently VEIGELE et al. [2.41] wrote a screening parameter program based on the Dirac theory.

2.11 Γ-Sum and Permanence Rules Applied to the X-Ray Doublet

So far, we have considered the energy of X-ray atomic levels (hole states) as analogous to the optical levels of hydrogenlike atoms, on the basis of Pauli's vacancy principle and Koopmans' theorem. It would be proper to check the correctness of this procedure by taking into account all of the electrons in the closed shell with a given configuration.

The mutual orientations of all orbital angular momenta $\underline{l}_i$ and spins $\underline{s}_i$ of separate electrons in a configuration can couple together in various ways that depend on the atom and the shell considered, to form J, the *total angular momentum of the configuration*. We shall consider two extreme types of couplings.

2.11.1 Russell-Saunders (LS) Coupling

In this coupling, the main interactions are between the $\underline{l}_i$'s to form a resultant $\underline{L} = \sum_i \underline{l}_i$, and the $\underline{s}_i$'s to form a resultant $\underline{S} = \sum_i \underline{s}_i$. The resultant vectors $\underline{L}$ and $\underline{S}$ then combine to form their resultant $\underline{J} = \underline{L} + \underline{S}$, that has the magnitude $\sqrt{J(J+1)}\hbar$. The possible values that can occur in a given LS term are all those integrally spaced values that satisfy $|L-S| \leqq J \leqq L+S$. The occurrence of atomic states characterized by quantum numbers L and S is called LS or *Russell-Saunders coupling*. It is a useful coupling for nearly closed-shell configurations.

2.11.2 jj Coupling

In this case $\underline{l}_i$ couples with the corresponding $\underline{s}_i$ to form a resultant $\underline{j}_i = \underline{l}_i + \underline{s}_i$. Then these $\underline{j}_i$'s couple together to form the resultant $\underline{J} = \sum_i \underline{j}_i$. It is called the jj-coupling and is a good approximation when the spin-orbit interaction is strong. All degrees of couplings between perfect LS and jj can occur in atoms.

Let γ_i represent the increase of the energy of a term because of spin-orbit interaction $\underline{l}_i \cdot \underline{s}_i$ of the single i^{th} electron. In view of (2.50,59) it can be written as

$$\gamma_i = \zeta \underline{l}_i \cdot \underline{s}_i = \zeta l_i s_i \cos(\underline{l}_i, \underline{s}_i) \quad , \tag{2.102}$$

where

$$\zeta = R_\infty \, hc \, \frac{\alpha^2 (Z-\sigma_2)^4}{n^3 \, l(l+1)\left(l+\frac{1}{2}\right)} \quad . \tag{2.103}$$

Let us apply a magnetic field $\underline{H}$. As the strength of this magnetic field is increased, the couplings between the $\underline{L}$ and $\underline{S}$ vectors or between the $\underline{j}_i$ vectors, depending on the type of coupling applicable, are disrupted. Further increase of $\underline{H}$ will disrupt even the couplings between the separate $\underline{l}_i$ and $\underline{s}_i$ vectors. In the end, a stage will come when there are no mutual couplings left and a *direct* coupling of the separate vectors $\underline{l}_i$ and $\underline{s}_i$ with the external field $\underline{H}$ comes into play. Then, all of the separate vectors $\underline{l}_i$ and $\underline{s}_i$ begin to precess relative to the direction of $\underline{H}$. The angle $(\underline{l}_i, \underline{s}_i)$ (Fig.2.10) will no longer be constant; in (2.102) it will be necessary to average $\cos(\underline{l}_i, \underline{s}_i)$ over all of its possible values.

Figure 2.19 describes the precession of $\underline{l}_i$ and $\underline{s}_i$ relative to $\underline{H}$. The vector $\underline{s}_i$ is extended to meet the plane, in which the end C of the vector $\underline{l}_i$

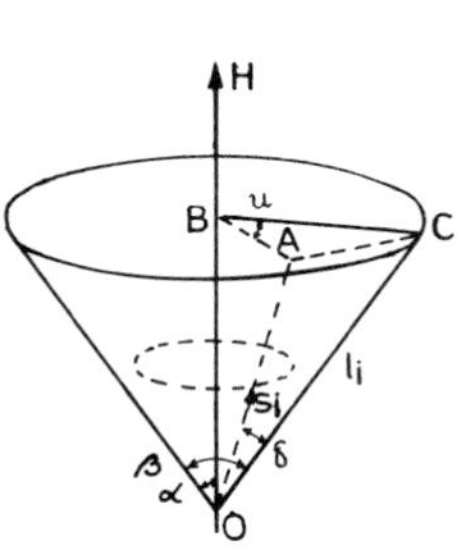

Fig. 2.19. Precession of $\underline{l}_i$ and $\underline{s}_i$ around $\underline{H}$

rotates, at the point A. The vector $\underline{H}$ intersects this plane at the point B. Consider the following angles,

$$\alpha = (\underline{s}_i,\underline{H}) \quad , \quad \beta = (\underline{l}_i,\underline{H}) \quad , \quad \delta = (\underline{l}_i,\underline{s}_i) \quad , \quad u = \text{angle ABC} \quad .$$

The angle u can vary from 0 to 2π while $\underline{l}_i$ and $\underline{s}_i$ precess around $\underline{H}$ at various velocities. The triangles OBA and OBC are right angled, so that

$$AB = l_i \cos\beta \tan\alpha \quad , \quad BC = l_i \cos\beta \tan\beta \quad , \quad OA = l_i \cos\beta/\cos\alpha \quad .$$

From the triangles ABC and AOC, we can find two values of AC by using the law of cosines. Equating them, we get

$$\cos\delta = \cos\alpha \cos\beta(1+\tan\alpha \tan\beta \cos u) \quad .$$

Because the average value of cosu is equal to zero, $\langle\cos u\rangle = 0$, we have $\langle\cos\delta\rangle = \cos\alpha \cos\beta$.

The quantization of the vectors $\underline{l}_i$ and $\underline{s}_i$ is characterized by the quantum numbers m_{l_i} and m_{s_i}. From Fig.2.19

$$m_{l_i} = |\underline{l}_i|\cos\beta \quad , \quad m_{s_i} = |\underline{s}_i|\cos\alpha \quad ,$$

whence

$$\langle\cos\delta\rangle = \frac{m_{l_i}m_{s_i}}{|\underline{l}_i||\underline{s}_i|} = \frac{m_{l_i}m_{s_i}}{l_i s_i} \quad ,$$

and (2.102) becomes

$$\gamma_i = \zeta m_{l_i}m_{s_i} \quad . \tag{2.104}$$

Let us define Γ as the sum of all of the corrections γ_i of the energy of a given configuration (same nl value) in some definite orientation with respect to $\underline{H}$. Thus, in a very strong magnetic field

$$\Gamma = \sum_i \gamma_i = \zeta \sum m_{l_i}m_{s_i} \quad . \tag{2.105}$$

In a magnetic field, the character of the orientation of the electrons is determined by the quantity

Table 2.6. Possible orientations of a $2p^5$ configuration in a very strong magnetic field. $M = \sum_{i=1}^{5} (m_{l_i} + m_{s_i})$

Orientation number	m_{l_1}	m_{l_2}	m_{l_3}	m_{l_4}	m_{l_5}	m_{s_1}	m_{s_2}	m_{s_3}	m_{s_4}	m_{s_5}	M
1	1	1	0	0	-1	$\frac{1}{2}$	$-\frac{1}{2}$	$\frac{1}{2}$	$-\frac{1}{2}$	$\frac{1}{2}$	$\frac{3}{2}$
2	1	1	0	0	-1	$\frac{1}{2}$	$-\frac{1}{2}$	$\frac{1}{2}$	$-\frac{1}{2}$	$-\frac{1}{2}$	$\frac{1}{2}$
3	-1	-1	0	0	1	$\frac{1}{2}$	$-\frac{1}{2}$	$\frac{1}{2}$	$-\frac{1}{2}$	$\frac{1}{2}$	$-\frac{1}{2}$
4	-1	-1	0	0	1	$\frac{1}{2}$	$-\frac{1}{2}$	$\frac{1}{2}$	$-\frac{1}{2}$	$-\frac{1}{2}$	$-\frac{3}{2}$
5	-1	-1	1	1	0	$\frac{1}{2}$	$-\frac{1}{2}$	$\frac{1}{2}$	$-\frac{1}{2}$	$\frac{1}{2}$	$\frac{1}{2}$
6	-1	-1	1	1	0	$\frac{1}{2}$	$-\frac{1}{2}$	$\frac{1}{2}$	$-\frac{1}{2}$	$-\frac{1}{2}$	$-\frac{1}{2}$

$$M = \sum_i \left(m_{l_i} + m_{s_i}\right) \quad . \tag{2.106}$$

In a configuration of several electrons, a particular value of M may arise from several different orientations with respect to $\underline{H}$. Such orientations belong to one and the same type; we shall denote the sum of their Γ values by $(\Sigma\Gamma)_M$.

Let us calculate the possible values of M and $(\Sigma\Gamma)_M$ that arise from a configuration of five p electrons ($2p^5$ configuration), that is, $L_{II,III}$ levels. These levels arise when one electron is missing from the $n = 2$, $l = 1$ subshell (capable of holding six electrons when closed). The Pauli exclusion principle greatly restricts the possible number of distinct orientations. The allowed arrangements are given in Table 2.6, together with their M values, which characterize each orientation.

For the various orientations given in Table 2.6, we can calculate γ_i and Γ values (Table 2.7).

From Table 2.6, we find that for M we get four numerically distinct values, $M = -3/2$, $-1/2$, $1/2$, $3/2$. Table 2.8 gives the Γ sums for the corresponding M values.

Values of Γ, corresponding to various orientations characterized by M, determine the *fine structure* of levels formed by a given nl configuration. Thus, each level is split into *six* sublevels determined by the different possible orientations (M values) of the five p electrons. As the strength of the magnetic field is decreased to zero, $H \to 0$, these sublevels gradually merge be-

Table 2.7. Values of γ and Γ of a $2p^5$ configuration in a strong magnetic field. $\gamma_i = \zeta m_{l_i} m_{s_i}$, $\Gamma = \sum_{i=1}^{5} \gamma_i$

Orientation number	γ_1	γ_2	γ_3	γ_4	γ_5	Γ
1	$\frac{1}{2}\zeta$	$-\frac{1}{2}\zeta$	0	0	$-\frac{1}{2}\zeta$	$-\frac{1}{2}\zeta$
2	$\frac{1}{2}\zeta$	$-\frac{1}{2}\zeta$	0	0	$\frac{1}{2}\zeta$	$\frac{1}{2}\zeta$
3	$-\frac{1}{2}\zeta$	$\frac{1}{2}\zeta$	0	0	$\frac{1}{2}\zeta$	$\frac{1}{2}\zeta$
4	$-\frac{1}{2}\zeta$	$\frac{1}{2}\zeta$	0	0	$-\frac{1}{2}\zeta$	$-\frac{1}{2}\zeta$
5	$-\frac{1}{2}\zeta$	$\frac{1}{2}\zeta$	$\frac{1}{2}\zeta$	$-\frac{1}{2}\zeta$	0	0
6	$-\frac{1}{2}\zeta$	$\frac{1}{2}\zeta$	$\frac{1}{2}\zeta$	$-\frac{1}{2}\zeta$	0	0

Table 2.8. Values of $(\Sigma\Gamma)_M$ for a $2p^5$ configuration in a strong magnetic field

M =	$-3/2$	$-1/2$	$1/2$	$3/2$
Γ	$-\frac{1}{2}\zeta$	0	$\frac{1}{2}\zeta$	$-\frac{1}{2}\zeta$
		$\frac{1}{2}\zeta$	0	
$(\Sigma\Gamma)_M$	$-\frac{1}{2}\zeta$	$\frac{1}{2}\zeta$	$\frac{1}{2}\zeta$	$-\frac{1}{2}\zeta$

cause the mutual interactions become significant; we are left with a certain number of limiting levels of interest. This situation corresponds to the X-ray terms envisaged on the basis of a hydrogenlike atom.

To obtain the desired $H \to 0$ limit, we shall use the following semi-empirical rules given by LANDÉ [2.42] and GOUDSMIT [2.43]:

1) Γ-*permanence rule* states that, for a given configuration, the $(\Sigma\Gamma)_M$ values are independent of the strength of the magnetic field.

2) Γ-*sum rule* states that the sum of Γ values for terms of the same J arising from a given configuration is independent of the coupling (LS or jj).

With these rules, which have been well tested in the optical spectra, we begin to reduce the strength of $\underline{H}$. In the weak-field orientations of the $2p^5$ configuration, Table 2.6 shows that the M values can be grouped into two sets, -3/2, -1/2, 1/2, 3/2 and -1/2, 1/2. Because M is a projection of the vector J on the direction of the field, these sets correspond to two values of J, J = 3/

Table 2.9. $(\Sigma\Gamma)_M$ values for the terms $J = 3/2$, $1/2$, in a weak magnetic field

M =	- 3/2	- 1/2	1/2	3/2
J = 3/2	$\Gamma_{3/2}$	$\Gamma_{3/2}$	$\Gamma_{3/2}$	$\Gamma_{3/2}$
J = 1/2		$\Gamma_{1/2}$	$\Gamma_{1/2}$	
$(\Sigma\Gamma)_M$	$\Gamma_{3/2}$	$\Gamma_{3/2}+\Gamma_{1/2}$	$\Gamma_{3/2}+\Gamma_{1/2}$	$\Gamma_{3/2}$

and $J = 1/2$. When $H \to 0$, we will be left with only two levels characterized by these two values of J. We wish to find the Γ values, $\Gamma_{3/2}$ and $\Gamma_{1/2}$, for these J values. In Table 2.9 we calculate the Γ sums for a weak-magnetic-field case, irrespective of the type of the coupling (LS or jj) involved. For the simple $2p^5$ configuration, only one term of the same J value is present. So we have no Γ sums for a given J.

The Γ-permanence rule allows us to equate the $(\Sigma\Gamma)_M$ values of Tables 2.8,9. We get

$$\Gamma_{3/2} = -\frac{1}{2}\zeta \quad , \quad \Gamma_{3/2} + \Gamma_{1/2} = \frac{1}{2}\zeta \quad . \tag{2.107}$$

Because for $2p^5$ configuration there are no Γ sums for constant J, we obtain uniquely the values

$$\Gamma_{3/2} = -\frac{1}{2}\zeta \quad , \quad \Gamma_{1/2} = \zeta \quad . \tag{2.108}$$

These equations remain unchanged when we take $H = 0$. The levels characterized by $J = 3/2$ and $J = 1/2$ are a spin doublet. Their separation is given by

$$\Delta E_{spin} = \Gamma_{1/2} - \Gamma_{3/2} = \frac{3}{2}\zeta \quad . \tag{2.109}$$

Because $\Gamma_{3/2} < 0$ and $\Gamma_{1/2} > 0$, the $J = 3/2$ level is lower on the energy scale than the $J = 1/2$ level.

With the hydrogenlike model, the spin-doublet formula (2.89) gave

$$\Delta E_{spin} = R_\infty hc\, \alpha^2 \frac{(Z-\sigma_2)^4}{n^3\, l(l+1)} = \zeta\left(1 + \frac{1}{2}\right) = \frac{3}{2}\zeta \quad , \quad \text{for } \begin{array}{l} n = 2, \\ l = 1, \end{array} \tag{2.110}$$

which agrees with (2.109).

If the correction to the principal energy E_n is determined from (2.58) for the single electron configuration $2p^1$, then for the two values of j of an optical electron, $j = 3/2$ and $j = 1/2$, we would get negative corrections. The

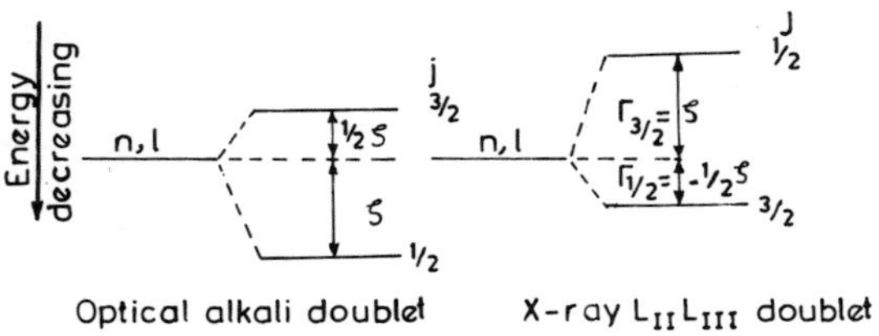

Fig. 2.20. Inversion of the X-ray spin doublet

absolute value of the correction for j = 3/2 is smaller. Therefore, the j = 3/2 level is higher in the spin-doublet of optical levels (Fig.2.20). In other words, compared to the *regular* spin doublet of optical levels, the X-ray spin-doublet is *inverted* (Fig.2.20).

The Γ-sum rule allows us to say that the spin-doublet formula will apply to the X-ray levels for all values of Z, because a change of Z may imply a change of the type of coupling. For low Z, usually, LS coupling is applicable. For high Z, the spin-orbit interaction becomes significant, which makes jj coupling appropriate.

The above discussion shows that our use of Pauli's vacancy principle was justified. Both $2p^5$ (or $2p^{-1}$) and $2p^1$ configurations give spin doublets of the same separation.

2.12 Quantum Theory of Spontaneous Emission of X-Ray Lines and Multipoles

In an X-ray tube, the target atom is excited by bombardment with cathode electrons. As a result, an atom emits an X-ray photon without electromagnetic radiation being present initially to induce the radiative process. Such a process is called *spontaneous emission*. Einstein calculated the transition probability for spontaneous emission on semiclassical consideration, using the results of induced processes. For a full quantum mechanical treatment, we have to consider the quantization of the pure radiation field.

The interaction between the atom and the radiation field, treated as one dynamical system, can bring about radiative transitions even if *in the initial state no light quanta exist*. When the entire system (atom + radiation field) is treated as a single entity, the Hamiltonian is independent of time, and a transition between atomic levels is to be thought of as a redistribution of the energy between the two parts of the coupled system.

The interaction term can be found by considering the Hamiltonian

$$H = \left[\frac{1}{2m}\left(\underline{p} - \frac{e}{c}\underline{A}\right)^2 + e\phi\right] + H_{\text{radiation}} \quad , \tag{2.111}$$

where $\underline{A}$ and ϕ are the vector and scalar potentials of the radiation field. The interaction term is

$$H' = -\frac{e}{mc}\,\underline{p}\cdot\underline{A} = \frac{ie\hbar}{mc}\,\nabla\cdot\underline{A} \quad . \tag{2.112}$$

We can quantize the radiation field by enclosing it in a certain volume V with $\underline{A}$ periodic on the walls. The radiation field can then be regarded as an infinite set of harmonic oscillators, one for each value of the propagation vector $\underline{k}_1$ and a particular direction of polarization $\underline{e}_1$. We can therefore express $\underline{A}$ as (see Appendix F),

$$\underline{A} = \left(\frac{4\pi c^2}{V}\right)^{\frac{1}{2}} \sum_1 \underline{e}_1\left(q_1\, e^{i\underline{k}_1\cdot\underline{r}} + q_1^\dagger\, e^{-i\underline{k}_1\cdot\underline{r}}\right) \quad , \tag{2.113}$$

where the q's have the nonvanishing matrix elements

$$q_{n-1,n} = \left(\frac{n\hbar}{2\omega}\right)^{\frac{1}{2}} e^{-i\omega t} \quad , \quad q^\dagger_{n+1,n} = \left[\frac{(n+1)\hbar}{2\omega}\right]^{\frac{1}{2}} e^{i\omega t} \quad . \tag{2.114}$$

Here $q^\dagger_{n+1,n}$ means that initially we have n photons and finally $n+1$ photons. In our case $n=0$ and $q^\dagger_{1,0}$ is the *only* nonvanishing matrix element for the emission of one photon. We can write the matrix element for the spontaneous emission as

$$H'_{f,1;i,0} = -\frac{e}{mc}\left(\frac{2\pi c^2\hbar}{V\omega}\right)^{\frac{1}{2}} \int \psi_f^*\, e^{-i\underline{k}\cdot\underline{r}}(\underline{e}\cdot\underline{p})\psi_i\, d\tau \quad . \tag{2.115}$$

We can expand $\exp(-i\underline{k}\cdot\underline{r})$ in (2.115) as

$$e^{-i\underline{k}\cdot\underline{r}} = 1 - i\underline{k}\cdot\underline{r} + \frac{1}{2!}(i\underline{k}\cdot\underline{r})^2 - \ldots \quad . \tag{2.116}$$

The term $(\underline{k}\cdot\underline{r})^n$ is said to give the electric-2^{n+1}-pole and the magnetic-2^n-pole transitions. Thus, if we retain only the first term, $\exp(-i\underline{k}\cdot\underline{r}) \to 1$, we get the *electric-dipole transition*.

2.12.1 Spontaneous Electric-Dipole Transition

We first show that the replacement $\exp(-i\underline{k}\cdot\underline{r}) \to 1$ is equivalent to replacing the atom by an electric dipole. Note that, because ψ_i and ψ_f are eigenstates of the unperturbed Hamiltonian of the atom H_0, we have

$$\frac{1}{m}(\underline{p})_{fi} = \frac{d}{dt}(\underline{r})_{fi} = \frac{1}{i\hbar}\left[\underline{r},H_0\right]_{fi} = \frac{1}{i\hbar}(E_i-E_f)(\underline{r})_{fi} = i\omega_{fi}(r)_{fi} \quad . \tag{2.117}$$

In this approximation, (2.115) becomes

$$H'_{f1;i0} = -i\left(\frac{2\pi\hbar}{V\omega}\right)^{\frac{1}{2}}\omega_{fi}\,\underline{e}\cdot(e\underline{r})_{fi} \quad . \tag{2.118}$$

This is the *electric-dipole matrix element for the spontaneous-emission process*. The transition probability per unit time is given by

$$w_{fi} = \frac{2\pi}{\hbar}|H'_{f1;i0}|^2\,\rho(E) = \frac{4\pi^2}{V\omega}\,\omega_{fi}^2|\underline{e}\cdot(e\underline{r})_{fi}|^2\,\rho(E) \quad , \tag{2.119}$$

where $\rho(E)$ is the density of final continuum states of the emitted photon. Phase-space considerations give (for photons $E = \hbar\omega = pc$),

$$\rho(E) = \frac{d}{dE}\left(\frac{p^2dpd\Omega}{h^3/V}\right) = \frac{d}{d\hbar\omega}\left(\frac{V\omega^2d\omega}{8\pi^3c^3}\,d\Omega\right) = \frac{V\omega^2}{8\pi^3\hbar c^3}\,d\Omega \quad . \tag{2.120}$$

Putting $\omega_{fi} = \omega$, we get

$$w_{fi} = \frac{\omega_{fi}^3}{2\pi\hbar c^3}\int|\underline{e}\cdot(e\underline{r})_{fi}|^2\,d\Omega \quad . \tag{2.121}$$

Let $\underline{e}$, $\underline{k}$ and $\underline{r}_{fi}$ be in the plane of paper, for convenience, with $\theta = (\underline{e},\underline{r}_{fi})$ and $\Theta = (\underline{k},\underline{r}_{fi})$ (Fig.2.21). Because $(\underline{e},\underline{k}) = \pi/2$, we can write $(\underline{e}\cdot\underline{r}_{fi}) = r_{fi}\cos\theta = r_{fi}\sin\Theta$, and

$$w_{fi} = \int_0^{\pi}\int_0^{2\pi}\frac{\omega_{fi}^3}{2\pi\hbar c^3}|(er)_{fi}|^2\sin^3\Theta\,d\Theta\,d\Phi = \frac{4}{3}\,\frac{e^2\omega_{fi}^3}{\hbar c^3}\,|r_{fi}|^2 \quad , \tag{2.122}$$

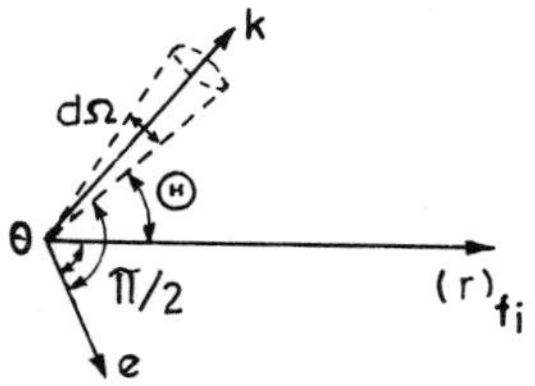

Fig. 2.21. Angles for the integration between $\underline{k}$ and $(\underline{r})_{fi}$

in exact agreement with the semiclassical result for the spontaneous-emission rate, (2.26), derived earlier. The intensity will be proportional to the energy emitted per second, $\hbar\omega_{fi}$.

2.12.2 Spontaneous Higher-Multipole Transitions

Let us consider the next term, $-i\underline{k}\cdot\underline{r}$, in the expansion (2.116). Its contribution to the matrix element is

$$h_{fi} = C\,\frac{ie}{mc}\int \psi_f^*(\underline{k}\cdot\underline{r})(\underline{e}\cdot\underline{p})\psi_i\,d\tau = h_{fi}^M + h_{fi}^Q \quad , \tag{2.123}$$

where $C=(2\pi c^2\hbar/V\omega)^{\frac{1}{2}}$ and

$$h_{fi}^M = C\,\frac{ie}{2mc}\int \psi_f^*\Big[(\underline{e}\cdot\underline{p})(\underline{k}\cdot\underline{r})-(\underline{e}\cdot\underline{r})(\underline{k}\cdot\underline{p})\Big]\psi_i\,d\tau \quad , \tag{2.124}$$

$$h_{fi}^Q = C\,\frac{ie}{2mc}\int \psi_f^*\Big[(\underline{e}\cdot\underline{p})(\underline{k}\cdot\underline{r})+(\underline{e}\cdot\underline{r})(\underline{k}\cdot\underline{p})\Big]\psi_i\,d\tau \quad . \tag{2.125}$$

Using the identity $(\underline{a}\times\underline{b})\cdot(\underline{c}\times\underline{d}) = (\underline{a}\cdot\underline{c})(\underline{b}\cdot\underline{d}) - (\underline{a}\cdot\underline{d})(\underline{b}\cdot\underline{c})$, we obtain

$$h_{fi}^M = C\,\frac{ie}{2mc}\int \psi_f^*(\underline{e}\times\underline{k})\cdot(\underline{p}\times\underline{r})\psi_i\,d\tau = C\,\frac{ie}{2mc}\int \psi_f^*(\underline{k}\times\underline{e})\cdot\underline{L}\,\psi_i\,d\tau \quad , \tag{2.126}$$

where $\underline{L}=\underline{r}\times\underline{p}$ is the orbital angular momentum. Consider an electron that moves with speed v in a circle or radius r. It creates a current $ev/2\pi r$ in a loop of area πr^2 and has orbital angular momentum $l=mvr$. Therefore, the equivalent dipole magnetic moment is

$$\mu = \frac{ev}{2\pi r}\cdot\frac{\pi r^2}{c} = \frac{e}{2mc}\,l \quad \text{or} \quad \underline{\mu} = \frac{e}{2mc}\,\underline{L} \quad . \tag{2.127}$$

These considerations suggest

$$h_{fi}^M = iC(\underline{k}\times\underline{e})\cdot\int \psi_f\,\underline{\mu}\,\psi_i\,d\tau \quad , \tag{2.128}$$

for a *magnetic-dipole transition.*

The matrix element h_{fi}^Q can be simplified by using $\underline{p}=m\dot{\underline{r}}$ so that

$$\underline{p}\underline{r} + \underline{r}\underline{p} = \frac{im}{\hbar}\Big([H,\underline{r}]\underline{r}+\underline{r}[H,\underline{r}]\Big) = \frac{im}{\hbar}\,[H,\underline{r}\underline{r}] \quad , \tag{2.129}$$

and

$$h^Q_{fi} = C \frac{e}{2\hbar c} (E_f - E_i) \int \psi^*_f (\underline{e}\cdot\underline{r})(\underline{k}\cdot\underline{r})\psi_i \, d\tau$$

$$= C \frac{\omega}{2c} |\underline{k}| (er^2)_{fi} \cos(\underline{e},\underline{r}) \cos(\underline{k},\underline{r}) \quad . \qquad (2.130)$$

As h^Q_{fi} involves r^2, it gives rise to the *electric-quadrupole transition.*

The electric-dipole transitions give most intense lines because

$$\frac{\text{magnetic-dipole moment}}{\text{electric-dipole moment}} \sim \frac{\left(\frac{e}{mc}\right) mvr}{er} = \frac{v}{c} \ll 1 \quad , \qquad (2.131)$$

$$\frac{\text{electric-quadrupole moment}}{\text{electric-dipole moment}} \sim \frac{er^2\omega/c}{er} \sim \frac{r}{\lambda} \quad . \qquad (2.132)$$

For valence electrons, r is of the order of 1 Å. Therefore, in optical spectra ($\lambda \sim 5000$ Å) we have $r/\lambda \ll 1$; consequently the intensities of electric-quadrupole lines are low. Even for soft X-rays ($\lambda \sim 100$ Å), $(r/\lambda)^2 \sim 10^{-4}$, which indicates low transition probability. For hard X-rays ($\lambda \sim 1$ Å), inner-core electrons are involved, with $r \sim 0.1$ Å. At these distances our approximation is not very good, but we still find $(r/\lambda)^2 \lesssim 10^{-2}$. Therefore, higher-order multipole transitions are *weak*, but they exist.

2.13 Parity Selection Rules and Forbidden Lines

The wave function $\psi(x,y,z)$ is said to have *even* (+) or *odd* (-) *parity*, which depends on the sign in the equation

$$\psi(x,y,z) = \begin{cases} +\psi(-x,-y,-z) & \text{even parity} \quad , \\ -\psi(-x,-y,-z) & \text{odd parity} \quad , \end{cases} \qquad (2.133)$$

under reflection through the origin of coordinates ($x \to -x$, $y \to -y$, $z \to -z$). In our matrix elements, we have integrals of the form

$$I = \int_{-a}^{+a} F(x) dx \quad , \quad F(x) = \psi^*(x) O_{op}(x) \psi(x) \quad . \qquad (2.134)$$

Clearly, we have

$$I = \begin{cases} \text{not zero} & \text{if} \quad F(x) = +F(-x) \quad , \\ \text{zero} & \text{if} \quad F(x) = -F(-x) \quad , \end{cases} \qquad (2.135)$$

although F(x) has not been explicitly specified.

For our matrix elements, we have for the overall parity

$$F(x) = \psi^*(x)0_{op}(x)\psi(x)$$
$$= (\pm 1)\psi^*(-x) \cdot (\pm 1)0_{op}(-x) \cdot (\pm 1)\psi(-x) = \pm F(-x) \quad , \qquad (2.136)$$

where 0_{op} is any of the operators like er for electric dipole, $\underline{\mu}$ for magnetic dipole, and (er^2) for electric quadrupole. In all, we have the following eight distinct possibilities:

ψ_f^*	0_{op}	ψ_i	F
+	+	+	+
+	+	-	-
-	+	+	-
-	+	-	+
+	-	+	-
+	-	-	+
-	-	+	+
-	-	-	-

For polar coordinates, the reflection through the origin ($\underline{r} \to -\underline{r}$) implies

$$r \to r \quad , \quad \theta \to \pi - \theta \quad , \quad \phi \to \phi + \pi \quad , \qquad (2.137)$$

as limits of θ are from 0 to π. For $\psi(r,\theta,\phi) = R_{nl}(r)Y(\theta)Z(\phi)$ reflection means

$$R_{nl}(r) \to R(r) \quad ,$$
$$Y_{l|m|}(\theta) \to Y_{l|m|}(\pi-\theta) = P_l^{|m|}(-\cos\theta) \quad ,$$
$$Z_{|m|}(\phi) \to Z_{|m|}(\phi+\pi) = e^{i|m|(\phi+\pi)} \quad ,$$
$$\psi(r,\theta,\phi) \to \psi(r,\pi-\theta,\phi+\pi) \quad . \qquad (2.138)$$

In our case, $Z_{|m|}(\phi+\pi) = (-1)^{|m|} e^{i|m|\phi} = (-1)^{|m|} Z_{|m|}(\phi)$. Therefore, $Z_{|m|}(\phi)$ has the parity of $|m|$. For $Y_{l|m|}(\theta)$,

$$Y_{l|m|}(\pi-\theta) = P_l^{|m|}(-\cos\theta) = (-1)^{l-|m|}P_l^{|m|}(\cos\theta) = (-1)^{l-|m|}Y_{l|m|}(\theta) \quad .$$

The explanation is that $P_l^m(x)$ is equal to an even part $(1-x^2)^{|m|/2}$ times a polynomial in x that has the parity of $l-|m|$ for $x \to -x$. Thus, for the whole function $\psi(r,\vartheta,\phi) \to (-1)^{|m|}(-1)^{l-|m|}\psi(r,\theta,\phi) = (-1)^l\psi(r,\theta,\phi)$; that is, it is multiplied by $(-1)^l$. In other words, *all states with even* l *are even, and all those with odd* l *are odd.*

We now get the following table of parities and matrix elements I for the electric-dipole transition:

ψ_f	$\underline{r}$	ψ_i	F	$I_{(er)}$	Remark
+	-	+	-	zero	forbidden
+	-	-	+	nonzero	allowed
-	-	+	+	nonzero	allowed
-	-	-	-	zero	forbidden

Because, for even $l = 0,2,4,\ldots$, ψ has even parity and for odd $l = 1,3,5,\ldots$, ψ has odd parity, we are led to the familiar selection rule

$$\Delta\pi,\ \text{yes} \quad ; \quad \Delta l(\text{yes}) = \pm 1 \quad (\textit{electric dipole}) \quad ,$$
$$(\Delta j = 0 \text{ or } \pm 1, \text{ the j-transition } 0 \to 0 \text{ forbidden}) \quad , \qquad (2.139)$$

where $\Delta\pi$ means change in parity. This agrees with the result (2.74) explicitly derived earlier. It follows that *all spectral lines due to electric-dipole radiation arise from transitions between states of opposite parity*. This is called the *Laporte rule*.

Orbital angular momentum $\underline{L} = \underline{r} \times \underline{p}$ is a *pseudovector*, because it does not change its sign under parity transformation, as do other *polar* or ordinary vectors (r or p); $\underline{r} \times \underline{p} \to (-\underline{r}) \times (-\underline{p}) = \underline{r} \times \underline{p}$. Therefore, unlike $\underline{r}$, the dipole magnetic moment is a pseudovector; it gives the following table:

ψ_f	$\underline{\mu}$	ψ	F	$I_{(\mu)}$
+	+	+	+	nonzero
+	+	-	-	zero
-	+	+	-	zero
-	+	-	+	nonzero

In this case, we get nonvanishing matrix components only between states of the *same parity*, which is just the opposite of the Laporte rule for electric-dipole radiation. The selection rules are

$$\Delta\pi,\ \text{no} \quad ; \quad \Delta l = 0 \quad (\textit{magnetic dipole}) \quad ,$$
$$(\Delta j = 0 \text{ or } \pm 1, \text{ the j-transition } 0 \to 0 \text{ is forbidden}) \quad . \qquad (2.140)$$

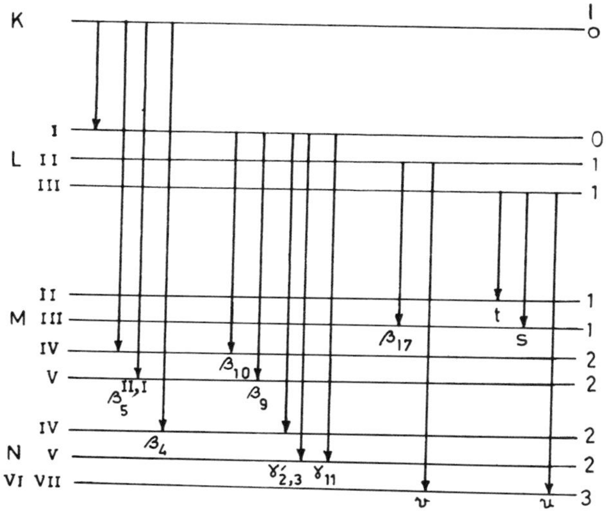

Fig. 2.22. Energy-level diagram showing the observed forbidden lines. Energy levels are not to scale

Because $\underline{r}^2$ is a scalar, $\underline{r}^2 \to \underline{r}^2$ under parity transformation, for *electric-quadrupole radiation*, we get the same table as for the magnetic-dipole case. The selection rules are,

$$\Delta\pi, \text{ no} \quad ; \quad \Delta l = 0 \text{ or } \pm 2 \quad (\textit{electric quadrupole}) \quad ,$$
$$(\Delta j = 0, \pm 1 \text{ or } \pm 2, \text{ the j-transitions } 0 \to 0, \tfrac{1}{2} \to \tfrac{1}{2} \text{ and } 0 \rightleftarrows 1 \text{ are forbidden}) \quad . \tag{2.141}$$

Diagram lines that are not allowed by the electric-dipole selection rules (2.139) are called *forbidden lines*. Weak forbidden lines are observed, which obey selection rules like (2.140) or (2.141), for higher poles. In Fig.2.22, we show transitions for some of the important (forbidden) quadrupole lines that have been observed in spite of their small intensities. The forbidden lines are *diagram lines* in the sense that their transitions can be shown on the energy-level diagram similar to that used for the electric-dipole allowed lines (Fig.2.12).

The forbidden lines $K\beta_5$ ($K \to M_{IV,V}$ unresolved) and $K\beta_4$ ($K \to N_{IV,V}$ unresolved) have been observed by IDEI [2.44], DUANE [2.45], ROSS [2.46,47], and CARLSSON [2.48]. ROSS has estimated that the intensity of the forbidden line $K\beta_5$ in ^{42}Mo and ^{42}Rh is of the order of 1/1000 of that of the allowed $K\alpha_1$ ($K \to L_{III}$) line. However, in ^{46}Pd it is nearer 1/250. Resolved lines $K\beta_5^{II}$ ($K \to M_{IV}$;

Table 2.10. Observed forbidden lines

Transition	Symbol	Parity (i-f)	Transition	Symbol	Parity (i-f)
$K \to L_I$	-	even-even	$L_{II} \to M_{III}$	$L\beta_{17}$	odd-odd
$K \to M_{IV,V}$	$K\beta_5^{II,I}$	even-even	$L_{II} \to N_{VI,VII}$	Lv	odd-odd
$K \to N_{IV,V}$	$K\beta_4^{II,I}$	even-even	$L_{III} \to M_{II}$	Lt	odd-odd
$L_I \to M_{IV}$	$L\beta_{10}$	even-even	$L_{III} \to M_{III}$	Ls	odd-odd
$L_I \to M_V$	$L\beta_9$	even-even	$L_{III} \to N_{VI,VII}$	Lu	odd-odd
$L_I \to N_{IV,V}$	$L\gamma'_{2,3}$	even-even			
$L_I \to N_V$	$L\gamma_{11}$	even-even			

625.78 XU); $K\beta_5^I$ ($K \to M_V$; 625.62 XU); $K\beta_4^{II}$ ($K \to N_{IV}$; 618.95 XU) and $K\beta_4^I$ ($K \to N_V$; 618.73 XU) have been reported for ^{42}Mo by HULUBEI [2.49]. For other such measurements see the wavelength tables of CAUCHOIS and HULUBEI [2.50], SANDSTRÖM [2.38], and BEARDEN [2.5]. The $K \to L_I$ line has been observed by BECKMAN [2.51]. It is a magnetic-dipole line with an intensity of 0.02 compared to the $K\alpha_1$ line.

In the L series, the most intense observed forbidden lines are $L\beta_{10}$ ($L_I \to M_{IV}$) and $L\beta_9$ ($L_I \to M_V$). ALLISON and ARMSTRONG [2.52] found that the intensity ratio of these lines to the allowed line $L\alpha_1$ ($L_{III} \to M_V$) is about 1/300 at 30.7 kV in ^{74}W. In Table 2.10, we give the list of some of the important observed forbidden lines in the K series and the L series.

It may happen that a forbidden line coincides with an allowed line. An unexpected increase of the ratio $K\beta_2/K\beta_1$ led MEYER [2.53] to show that for elements with $Z < 32$, the quadrupole line $K\beta_5$ takes the place of the β_2 line.

Relative intensities of lines arising from magnetic-dipole and electric-quadrupole transitions have been carefully measured recently by SMITHER et al. [2.32], and RAO et al. [2.33]. The angular correlation of X-ray emission has been studied by CATZ [2.54] from the point of view of admixed magnetic-quadrupole with electric-dipole radiation.

2.14 Absorption Discontinuities

We have seen (Chap.1) that in an X-ray tube continuous X-rays are also produced. When these X-rays are passed through an absorbing material, changes

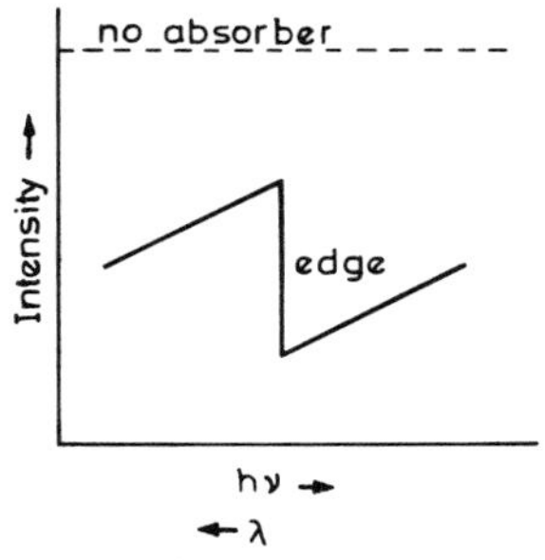

Fig. 2.23. Effect of the absorber on the distribution of transmitted intensity in a continuous spectrum

occur in the distribution of intensity with wavelength (or frequency). These changes are gradual, owing to the scattering of X-rays within the material but become abrupt when the incident photon has just enough energy hν to remove one of the inner electrons in the absorbing atom to *outside* the atom or at least outside the system of complete shells. At this energy, heavy absorption occurs and a sudden change of the intensity transmitted through the material is observed. This is the so-called *absorption discontinuity* or the *absorption edge* (Fig.2.23). On the high-frequency side of this edge, the continuous spectrum is considerably weakened. A measurement of the wavelength of the edge can be used to calculate the critical energy of the corresponding level, that is, the minimum energy of a photon absorbed by a given level. The absorption edge is thus a characteristic feature of the absorbing element.

We cannot have an absorption line in normal atoms corresponding to a transition of the electron from the K-level to, say, L_{III} level (or of the hole from the L_{III} level to K level) because in normal atoms, the L_{III}, or the other inner levels, are full (no vacancy to which the electron can go) according to the Pauli exclusion principle. The shift of emphasis from electron to hole is useful in this sense in discussing X-ray absorption spectra. We can say that, in normal atoms, the holes are on the outside, so the X-ray absorption spectrum is produced by the transition of one of these outer holes to an inside shell. The absorption spectrum is, as a result, directly affected by the external structure of the atoms and the way in which it is influenced by the physical and chemical state in the absorbing material. It is therefore necessary to state carefully what we mean by the outside of the atom in discussing the absorption spectra.

In early days, it was proposed by BARKLA, on the basis of his absorption studies, that in an X-ray absorption process the atom is completely ionized, that is, an inner electron is removed from a given inner shell to "infinity". This explains that there is one edge for each spectral series or subseries

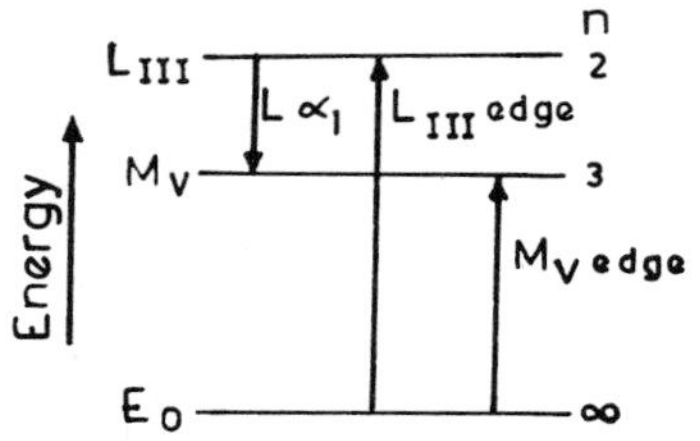

Fig. 2.24. Origin of L_{III} edge, M_V edge and $L\alpha_1$ line, with the outside of the atom defined as $n=\infty$

(one K; three L: L_I, L_{II}, L_{III}; five M edges; etc.). However, it is an oversimplification of the problem as can be seen from the following argument. If a frozen orbital model is adapted (Koopmans' theorem approximation), the energies of the states can be associated with the active electron or hole. According to this view, Fig.2.24 shows the hole transitions for L_{III} edge, M_V edge, and the $L\alpha_1$ emission line. It gives relationships of the type

$$h\nu_{L\alpha_1} = L_{III} - M_V \quad ,$$
$$h\nu_{L_{III}\ \mathrm{edge}} = L_{III} - E_0 \quad ,$$
$$h\nu_{M_V\ \mathrm{edge}} = M_V - E_0 \quad . \tag{2.142}$$

Here L_{III} and M_V designate the energies of corresponding levels and E_0 of the *zero level*, that is, the energy of an electron removed to rest at an infinite distance from the atom. From (2.142), we obtain

$$h\nu_{L\alpha_1} = h\nu_{L_{III}\ \mathrm{edge}} - h\nu_{M_V\ \mathrm{edge}} \quad . \tag{2.143}$$

This is *not* in agreement with experiment.

Later on, SIEGBAHN proposed the following explanation. In the absorption process, the atom is not ionized, but instead is excited by the transition of an inner shell electron to an unfilled external level. These transitions also *obey the electric-dipole selection* rule, $\Delta l = \pm 1$, that has been established for the emission lines. The absorption edge is determined by a transition to the first such allowed unfilled level.

In ^{42}Mo, the level $N_{IV,V}$ is partially filled, and the $N_{VI,VII}$ level is completely unoccupied. All of the shells within $N_{IV,V}$ are fully occupied. The transitions allowed by the selection rules for the L_{III} edge, the M_V edge, and the $L\alpha_1$ line in this theory are now drawn as shown in Fig.2.25 (compare with Fig.2.24) and described by the relations

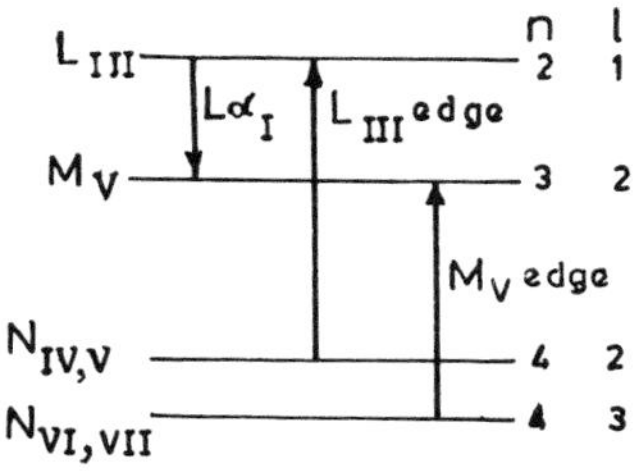

Fig. 2.25. Origin of L_{III} edge, M_V edge, and $L\alpha_1$ line in the Siegbahn model

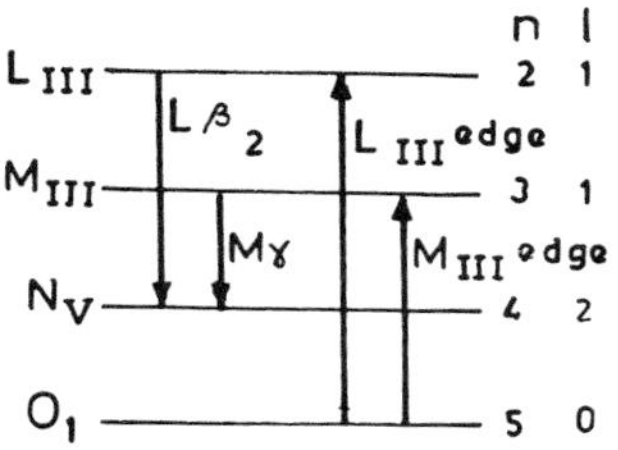

Fig. 2.26. Origin of edges L_{III}, M_{III}, and lines $L\beta_2$, $M\gamma$, in the Siegbahn model

$$h\nu_{L\alpha_1} = L_{III} - M_V \quad ,$$

$$h\nu_{L_{III}\ \mathrm{edge}} = L_{III} - N_{IV,V} \quad ,$$

$$h\nu_{M_V\ \mathrm{edge}} = M_V - N_{VI,VII} \quad . \tag{2.144}$$

It is clear from (2.144) that the equality (2.143) is not valid. In fact, we get

$$h\nu_{L\alpha_1} > h\nu_{L_{III}\ \mathrm{edge}} - h\nu_{M_V\ \mathrm{edge}} \quad , \tag{2.145}$$

in agreement with experiments.

As another example, consider the palladium (^{46}Pd) atom. All levels up to $N_{IV,V}$ are completely occupied. The remaining levels are empty. In Fig.2.26 are shown the levels L_{III}, M_{III}, N_V and O_I of ^{46}Pd and transitions for the L_{III} edge, M_{III} edge, and the emission lines $L\beta_2$ ($L_{III} \to N_V$) and $M\gamma$ ($M_{III} \to N_V$). The first allowed empty level is O_I for the excitation of the atom from both the levels L_{III} and M_{III}. Thus, we get

$$h\nu_{L\beta_2} = L_{III} - N_V \quad ,$$

$$h\nu_{M\gamma} = M_{III} - N_V \quad ,$$

$$h\nu_{L_{III}\ \text{edge}} = L_{III} - O_I \quad ,$$

$$h\nu_{M_{III}\ \text{edge}} = M_{III} - O_I \quad , \tag{2.146}$$

whence

$$h\nu_{L\beta_2} - h\nu_{M\gamma} = h\nu_{L_{III}\ \text{edge}} - h\nu_{M_{III}\ \text{edge}} \quad . \tag{2.147}$$

This equality is in agreement with experiments.

IDEI [2.44] pointed out that it is possible to calculate the energy-level system of an atom from the $h\nu$ (or $\tilde{\nu}/R$) values of one of its absorption limits (say, L_{III} edge) and its emission lines. In early work [2.55,56], the L_{III} level was chosen as the fundamental level, because it could be measured with high precision. Ascribing to the L_{III} edge an energy identical with that of the L_{III} level, the relative energies of all of the other levels can be deduced by following the transitions for the emission lines back and forth. Thus, from (2.147) the M_{III} level can be determined relative to the L_{III} leve by using the $h\nu$ values of the observed lines $L\beta_2$, $M\gamma$. In fact, more than four such combinations are possible, which increases the reliability of the result It is easy to identify these lines with specific transitions by using the Siegbahn-Grotrian diagram shown in Fig.2.27. It is obtained from Figs.2.12,17 by separating the levels according to their l quantum numbers.

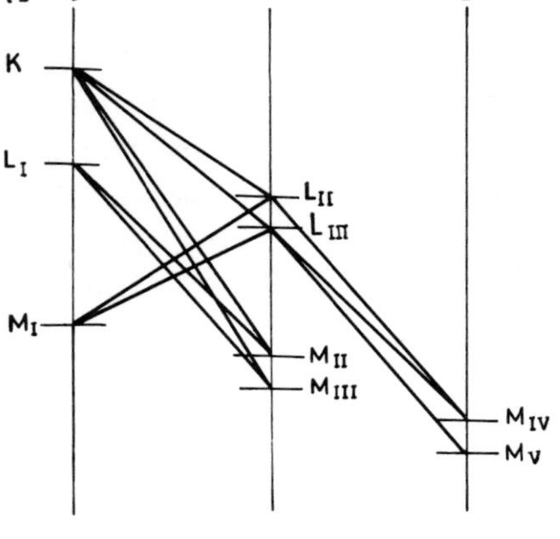

Fig. 2.27. The Siegbahn-Grotrian energy level diagram for an atom

If Siegbahn's idea is correct, the X-ray absorption spectrum of a free atom ought to show a series of absorption lines, corresponding to higher empty optical levels, following the edge. This structure has been found for the inert gases. The absorbing material is in solid form in the usual experiments. In the solid state, the optical (outer) levels will be distorted and broadened to form a *band* due to the hybridization of neighboring-atom valence orbitals. Some structure still remains, according to KOSSEL [2,57a,b], although the conditions are complicated. It is called the *Kossel structure* close to the main absorption edge on the high-energy side, and has been observed experimentally.

An emission line will *not* appear if there are normally no electrons in the energy level, that is the initial level for the transition of the electron. Accordingly, elements with low atomic number Z are poor in lines and spectral series, compared to those with high Z.

Let us now see what emission lines can possibly appear for each atom as Z changes and what is the closest external (optical) level to which the inner-shell electrons can be excited in the absorption process. In Appendix D the distribution of electrons in the various levels of the free atoms is given. We should compare this with the energy level and transition diagram (Fig.2.12). It gives meaningful information only for atoms of monoatomic gases or of vapors, where the atoms are free. Usually, the targets and absorbing materials used in experimental studies are solids. Therefore, a comparison of the diagram of Fig.2.12 and of the table of Appendix D with the experimental results shows how the distribution of electrons in a solid material varies.

We make this comparison with the K series as an example (Fig.2.28) to see how the X-ray spectra vary with increase of Z in free atoms and in solids. The first electron appears at $L_{II,III}$ level with boron (^5B). Therefore, the

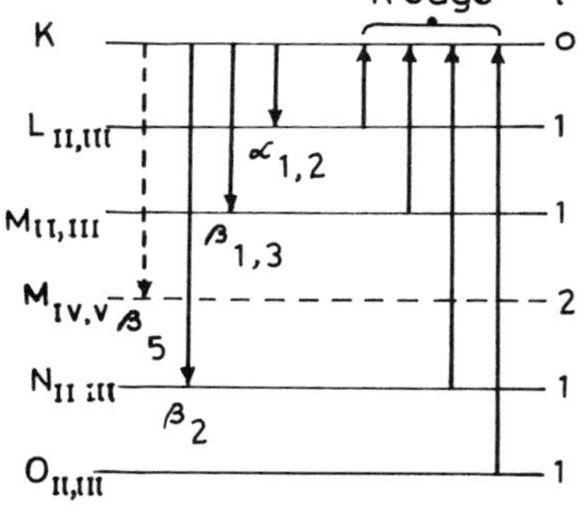

Fig. 2.28. Possible lines of K series and the K absorption edges of different elements. $K\beta_5$ is a quadrupole line. Other transitions obey the dipole rule $\Delta l = \pm 1$

Table 2.11. Appearance of the first line of the K series with an increase in Z

Level	K line	First appearance		Element for which the level is first filled in free atom
		Free atoms	Solids	
$L_{II,III}$	$\alpha_{1,2}$	5 B	3 Li	10 Ne
$M_{II,III}$	$\beta_{1,3}$	13 Al	11 Na	18 Ar
$M_{IV,V}$	β_5	21 Sc	17 Cl	29 Cu
$N_{II,III}$	β_2	31 Ga	30 Zn	36 Kr
$O_{II,III}$	-	49 In	-	54 Xe

$K\alpha_{1,2}$ line can first appear in a free boron atom. However, in solids it is observed even for ^{3}Li and ^{4}Be. In fact, all lines of the K series appear sooner than the first electron appears at the corresponding level of the free atom, see Table 2.11. This indicates that, as a result of chemical bonding in a solid, the external electrons of the atom are excited to the next optical level, from where these electrons can complete the transition to the hole left in the K level. Such lines are called semioptical for the elements ^{17}Cl, ^{18}Ar, ^{19}K and ^{20}Ca.

In Table 2.11 we also show the elements in which the levels $L_{II,III}$, $M_{II,III}$, ... are first filled up. So long as the level $N_{II,III}$ is partially filled, the line $K\beta_2$ occurs during the transition $K \rightarrow N_{II,III}$, whereas the K edge is due to the same transition but in the opposite direction, $N_{II,III} \rightarrow K$. With increasing Z, when the level $N_{II,III}$ is filled, the absorption process removes the K electron to the next level $O_{II,III}$ allowed by the selection rule, whereas the $K\beta_2$ line is still due to the transition $K \rightarrow N_{II,III}$. As a result, the K edge energy now exceeds the photon energy of the $K\beta_2$ line. SANDSTRÖM [2.58] studied the change of the difference of these two energies by plotting the difference between their $\tilde{\nu}/R_\infty$ values against Z (Fig.2.29). As expected, this difference is nearly zero up to $Z = 35$; then it begins to increase sharply.

For $Z = 36$ (^{36}Kr), the level $N_{II,III}$ is filled up and the absorption and emission transitions cease to have common levels. Because ^{36}Kr is a noble gas, the closed shell $N_{II,III}$ is not disrupted by chemical bonding. A similar plot for the $K\beta_5$ ($K \rightarrow M_{IV,V}$) line (Fig.2.30) shows that the level $M_{IV,V}$ is first filled up *in a solid* at the element ^{32}Ge, although in a free atom it is filled up at ^{29}Cu (Table 2.11). This implies that the chemical bonds promote the external (valence) electrons to the next optical levels. This creates a vacancy

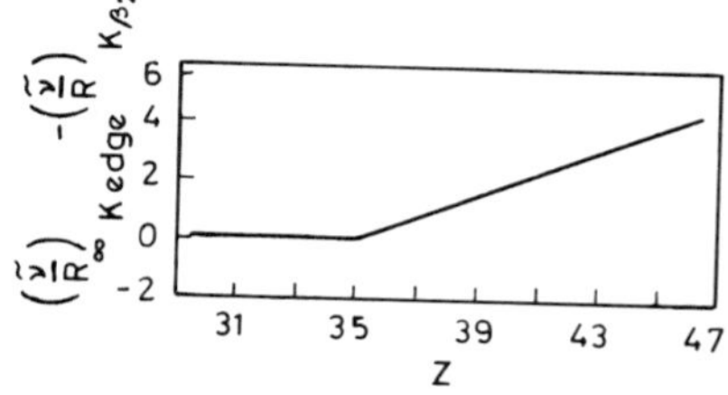

Fig. 2.29. $\Delta\tilde{\nu}/R_\infty$ for the K edge and $K\beta_2$ line showing the effect of the electron configuration

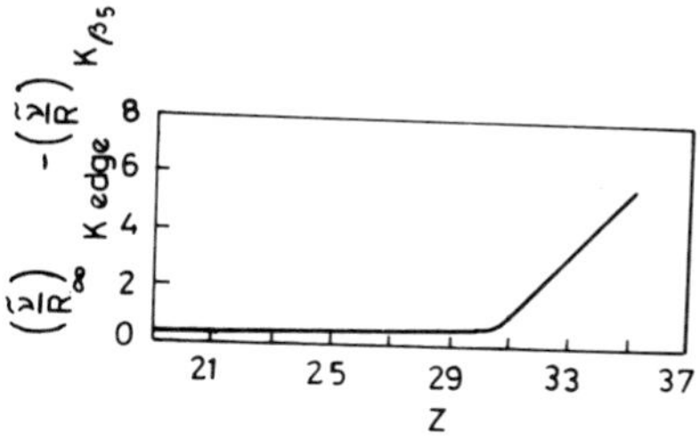

Fig. 2.30. $\Delta\tilde{\nu}/R_\infty$ for the K edge and $K\beta_5$ line versus Z

in the $M_{IV,V}$ level (full in free atoms ^{29}Gu, ^{30}Zn and ^{31}Ga) and enables it to participate in the K absorption processes up to ^{31}Ga. Analogous L-series plots have been given by INGELSTAM [2.59].

Thus, in solids, the external (valence) electrons are redistributed and filling of the external levels no longer corresponds to Table 2.11. The intensity of an observed line depends on the number of electrons in the outer level that can participate in the transition and also on the probability of the transition. Both of these factors change with the increase of the atomic number Z, when chemical bonding is present. Therefore, the variation of the relative intensities of the lines with Z is a topic of interest. This variation has been studied experimentally by several workers [2.53,60-63].

In Fig.2.31 we show a plot of the relative intensity of the $K\beta_2$ line with respect to the line $K\beta_1$ of unchanging intensity, versus Z. We also show by a dotted curve the number of electrons N(4p) present in the level 4p ($N_{II,III}$) as a function of Z. Clearly, the line $K\beta_2$ appears first at ^{30}Zn but attains a constant relative intensity only near ^{50}Sn. The initial rise up to ^{36}Kr can be attributed mainly to the gradual filling up tp the $N_{II,III}$ level. We have seen that, even in solids, this level is filled up at ^{36}Kr. Therefore, beyond ^{36}Kr the increase of intensity is because of a change of the probability of the transition $K\to N_{II,III}$.

In Fig.2.32, we show the change of the relative intensity of the quadrupole line $K\beta_5$ with Z. For $Z>30$ the level $M_{IV,V}$ is an internal level; therefore the $K\beta_5$ has the usual low intensity of a quadrupole line. With decrease of Z the level $M_{IV,V}$ becomes external, that is, a level of valence electrons.

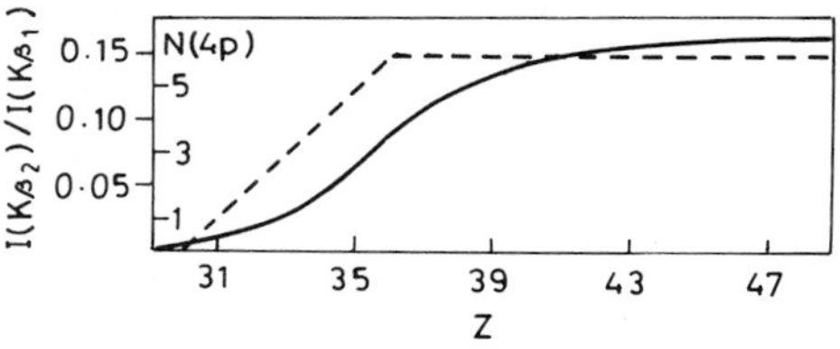

Fig. 2.31. Solid curve: the relative intensity of the $K\beta_2$ line. Dotted curve: the number of electrons N(4p) at the level 4p ($N_{II,III}$) of free atoms

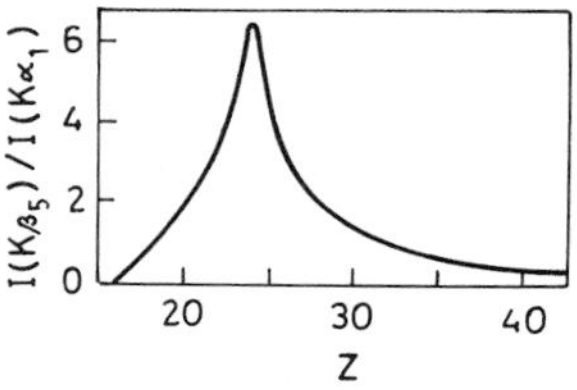

Fig. 2.32. Relative intensity of the line $K\beta_5$ versus Z

In metals, it ceases to be a sharp atomic level and becomes a broad band that overlaps the neighboring levels. The electrons of the overlapping levels are indistinguishable; they acquire common properties, and the quadrupole line becomes a dipole line. This, of course, brings about a sharp increase of intensity. With a further decrease of Z, the level $M_{IV,V}$ becomes unoccupied, resulting in the decrease of intensity of the line $K\beta_5$ to zero at ^{16}S.

It is now evident that an inner-shell hole can be created either during collision of bombarding cathode electrons with the target atoms in an X-ray tube, or during absorption of X-ray photons of sufficient energy by the atoms of the absorbing material. In the former case, we get both continuous and characteristic spontaneous emission of X-rays; in the later case, the induced-absorption process is followed by induced emission of characteristic lines alone (fluorescence radiation). Early experimental work of BARKLA [2.64] was based on the absorption of X-rays, and the accompanying K and L fluorescence X-rays.

2.15 Comparison of Optical and X-Ray Spectra

Following are some of the essential features in which the optical spectra differ from the X-ray spectra:

1) Optical spectra states are truly stationary states of a neutral atom.	1) X-ray spectra states are quasi-stationary states of an inner-shell ionized atom.
2) The valence electron in the outermost part of the atom is excited to a higher unoccupied stationary state; the electron emits radiation when it goes back to the ground state, remaining inside the atom throughout.	2) First, one inner hole is created to excite the atom. Then, an electron from an occupied state of higher quantum number fills this hole, to emit an X-ray line. We thus have a transfer of hole (or electron) between inner levels which are occupied in the normal atom.
3) The optical terms depend on both the initial and final states of the outer electron. Therefore, optical spectra change their characteristics as we go from one element to another and form compounds because of the changes of energies of the outer electron groups. The frequencies of lines, and also their excitation potentials, increase periodically with Z.	3) The inner electron groups are similarly arranged for all except the very light elements. Therefore, the only spectral differences to be expected are those that result from the different stability of these inner-shell electrons for different values of the effective nuclear charge, as we go from one element to another and form compounds. The frequencies of the lines, and also their excitation potentials, increase steadily, not periodically, with increasing Z (Moseley law).
4) Generally, incomplete shells (less than half full) participate.	4) Complete shells participate, from which one electron is missing.
5) The optical terms may or may not be hydrogenlike.	5) In the frozen orbital approximation, the X-ray terms are given by the Dirac electron theory of hydrogenlike states applicable to a hole in the inner shell.
6) Optical atomic absorption spectra usually show line absorption due to the transition of the valence electron from the ground state to a higher vacant state. Both the initial and final states are stationary states and so have sharply defined energy values.	6) X-ray spectra usually show absorption edges due to the removal of an inner electron to an empty state above the Fermi level, because all of the other states below the Fermi level are full, by the Pauli exclusion principle.

2.16 Nomenclature of X-Ray Lines

The literature contains several systems of labeling energy levels and emission lines. The Bohr and Coster notation, Table 2.12, is widely accepted for energy levels. We have also used this notation.

Table 2.12. Various notations for the energy levels

Sommerfeld	K	L_{11}	L_{21}	L_{22}	M_{11}	M_{21}	M_{22}	M_{32}	M_{33}
Early papers	K	L_3	L_2	L_1	M_5	M_4	M_3	M_2	M_1
Mott and others	K	L_1	L_2	L_3	M_1	M_2	M_3	M_4	M_5
Some journals	K	L1	L2	L3	M1	M2	M3	M4	M5
Bohr and Coster	K	L_I	L_{II}	L_{III}	M_I	M_{II}	M_{III}	M_{IV}	M_V
Optical symbols	$1^2S_{1/2}$	$2^2S_{1/2}$	$2^2P_{1/2}$	$2^2P_{3/2}$	$3^2S_{1/2}$	$3^2P_{1/2}$	$3^2P_{3/2}$	$3^2D_{3/2}$	$3^2D_{5/2}$

For X-ray lines, the designations given by SIEGBAHN [2.55] (Fig.2.12) are commonly used. Because there is no agreed system of names, except for the most prominent lines, it becomes necessary to state the transition in terms of the initial and final levels involved. The general basis of Siegbahn's nomenclature is as follows. First the series is given the name K, L, M, ... that corresponds to $n = 1,2,3, \ldots$ for the initial X-ray state (hole state). Then, the strongest unresolved line in the series is named α, the next strongest β, and so on. Under high resolution the α line is in general a doublet: the more-intense component is called α_1 and the weaker α_2. In the L series the β_1 line is next strongest. There are many weaker lines in the same wavelength region; these are named β_2, β_3, ..., β_{15}. A similar system is followed for the γ lines. The relative positions of the less-prominent lines vary from element to element. A nomenclature that seems logical for one element may appear arbitrary for an element of much different Z. In many cases, the lines took names in the order of their discovery, and often the names given by the discoverers have persisted.

In 1952 AGARWAL [2.65] suggested that a systematic approach would be to name each line K, L, M, ... corresponding to $n = 1,2,3, \ldots$ for the initial state, and α, β, γ, ... corresponding to $\Delta n = 1,2,3, \ldots$ for the difference of n from the initial to the final state. Superscripts (subscripts) 1,2,3 might be used for the initial (final) energy levels involved in the transition [2.66]. Thus, the $K\alpha_1$ ($K \to L_{III}$) line might be written as $K\alpha_3^1$, $K\alpha_2$ ($K \to L_{II}$) as $K\alpha_2^1$, $K\beta_1$ ($K \to M_{III}$) as $K\beta_3^1$, $L\alpha_1$ ($L_{III} \to M_V$) as $L\alpha_5^3$, $L\beta_1$ ($L_{II} \to M_{IV}$) as $L\alpha_4^2$, $L\eta$ ($L_{II} \to M_I$) as $L\alpha_1^2$, $L\gamma_1$ ($L_{II} \to N_{IV}$) as $L\beta_4^2$, $L\gamma_8$ ($L_{II} \to O_I$) as $L\gamma_1^2$, and so on. Obviously, an α line would be softer (of lower frequency) than a β line, and $L\alpha_2^1$ would be softer than $L\alpha_3^1$, etc. For electric-dipole (allowed) lines, the difference between the supercript and the subscript can be only 1 or 2, with following combinations forbidden:

Superscript	2	3	4	5	...
Subscript	3	2	5	4	

These are the superscripts and subscripts that, along with combinations differing by 0, 3, 4, ..., etc., occur for the forbidden lines. Thus, $L\beta_{10}$ ($L_I \to M_{IV}$) is $L\alpha_4^1$, $L\beta_{17}$ ($L_{II} \to M_{III}$) is $L\alpha_3^2$, Ls ($L_{III} \to M_{III}$) is $L\alpha_3^3$, etc.

This systematic nomenclature, if accepted, would introduce some order in the tables and diagrams of the emission lines in X-ray spectra. We give here the identifications of characteristic lines according to BEARDEN [2.67] and in this nomenclature (in parentheses):

K series

KL_{II} α_2 (α_2^1)

KL_{III} α_1 (α_3^1)

KM_{II} β_3 (β_2^1)

KM_{III} β_1 (β_3^1)

$KN_{II,III}$ β_2 ($\gamma_{2,3}^1$)

$KN_{IV,V}$ β_4 ($\gamma_{4,5}^1$)

L series

L_IM_{II} β_4 (α_2^1)

L_IM_{III} β_3 (α_3^1)

L_IM_{IV} β_{10} (α_4^1)

L_IM_V β_9 (α_5^1)

L_IN_{II} γ_2 (β_2^1)

L_IN_{III} γ_3 (β_3^1)

$L_IO_{II,III}$ γ_4 ($\gamma_{2,3}^1$)

$L_{II}M_I$ η (α_1^2)

$L_{II}M_{III}$ β_{17} (α_3^2)

$L_{II}M_{IV}$ β_1 (α_4^2)

$L_{II}N_I$ γ_5 (β_1^2)

$L_{II}N_{IV}$ γ_1(β_4^2)

$L_{II}N_{VI}$ ν (β_6^2)

$L_{II}O_I$ γ_8 (γ_1^2)

$L_{II}O_{IV}$ γ_6 (γ_4^2)

$L_{III}M_I$ l (α_1^3)

$L_{III}M_{II}$ t (α_2^3)

$L_{III}M_{III}$ s (α_3^3)

$L_{III}M_{IV}$ α_2 (α_4^3)

$L_{III}M_V$ α_1 (α_5^3)

$L_{III}N_I$ β_6 (β_1^3)

$L_{III}N_{IV}$ β_{15} (β_4^3)

$L_{III}N_V$ β_2 (β_5^3)

$L_{III}N_{VI,VII}$ u ($\beta_{6,7}^3$)

$L_{III}O_I$ β_7 (γ_1^3)

$L_{III}O_{IV,V}$ β_5 ($\gamma_{4,5}^3$)

M series

$M_{III}N_V$ γ (α_5^3)

$M_{IV}N_{II}$ ζ_2 (α_2^4)

$M_{IV}N_{VI}$ β (α_6^4)

M_VN_{III} ζ_1 (α_3^5)

M_VN_{VI} α_2 (α_6^5)

M_VN_{VII} α_1 (α_7^5)

3. Interaction of X-Rays with Matter

In optics, the term *dispersion* means the separation of electromagnetic waves into their component wavelengths. The first treatment in the X-ray region was given by COMPTON. He adapted the theory developed earlier for the optical region by LORENTZ [3.1]. He assumed that the electrons in atoms can act like oscillators.

3.1 Free, Damped, Oscillator

Let a classical oscillator consist of an electron that oscillates along the x axis, about a central point, under a restoring force F(x). From (2.13),

$$F(x) = m\ddot{x} = -m\omega_0^2 x \quad , \tag{3.1}$$

where m is the electron mass and $\omega_0 = 2\pi\nu_0$ is the initial angular frequency of the oscillations. The general solution is

$$x = x_0 \, e^{i\omega_0 t} \quad . \tag{3.2}$$

In (2.12), we used only the real part of this solution.

An oscillating electron is accelerated. Therefore, it emits radiation at a rate given by Larmor's formula (1.68),

$$-\frac{dU}{dt} = \frac{2e^2}{3c^3}\ddot{x}^2 \quad . \tag{3.3}$$

Because of this loss of energy by radiation, the amplitude of oscillations will gradually decrease, as if the electron were being retarded by a frictional force ϕ. We can find the fictitious force ϕ by equating the work done by it in unit time to the negative of the energy radiated in that time,

$$\phi\dot{x} = -\frac{2e^2}{3c^3}\ddot{x}^2 \quad . \tag{3.4}$$

Thus ϕ is equivalent to the radiation damping.

In this equation, $\dot{x}$ and $\ddot{x}$ are basically uncorrelated; consequently, there is no solution for ϕ that would hold for all times. We seek a solution that represents an average over a long time interval t_1 to t_2,

$$\int_{t_1}^{t_2} \phi\dot{x}\,dt = -\frac{2e^2}{3c^3}\int_{t_1}^{t_2} \ddot{x}^2 dt = -\frac{2e^2}{3c^3}\left[(\dot{x}\ddot{x})_{t_1}^{t_2} - \int_{t_1}^{t_2} \dot{x}\dddot{x}\,dt\right] \quad . \tag{3.4}$$

The motion being periodic, we can choose $\ddot{x} = 0$ at $t = t_1$ and $t = t_2$, so that

$$\int_{t_1}^{t_2} \left(\phi - \frac{2e^2}{3c^3}\dddot{x}\right)\dot{x}\,dt = 0 \quad . \tag{3.5}$$

This suggests that we can identify the damping force as

$$\phi = \frac{2e^2}{3c^3}\dddot{x} \quad . \tag{3.6}$$

It is called *radiation damping* or *Lorentz frictional force*. It describes the reaction of the radiation on the charge.

The motion of the electron under the combined effect of the restoring force $F(x)$ and the Lorentz frictional force $\phi(\dddot{x})$ is given by Newton's law,

$$m\ddot{x} = F(x) + \phi(\dddot{x}) \quad . \tag{3.7}$$

From (3.1,6), we have for the free, damped, oscillations of an electron,

$$\ddot{x} - \alpha\dddot{x} + \omega_0^2 x = 0 \quad , \tag{3.8}$$

where

$$\alpha = 2e^2/(3mc^3) = 6.2 \times 10^{-24}\ \mathrm{s} \quad . \tag{3.9}$$

This is called the *Abraham-Lorentz equation* of motion. It includes, in some approximate and time-average way, the reactive effects of the emission of radiation.

3.2 Form and Width of Lines

As α is a small quantity in (3.8), we can make the approximation,

$$\dddot{x} = -\omega_0^2 \dot{x} \quad . \tag{3.10}$$

Then (3.8) becomes

$$\ddot{x} + \gamma\dot{x} + \omega_0^2 x = 0 \quad , \tag{3.11}$$

where

$$\gamma = \alpha\omega_0^2 = \frac{2e^2\omega_0^2}{3mc^3} = \frac{8\pi^2 e^2}{3mc\lambda_0^2} \quad . \tag{3.12}$$

For $\lambda_0 = 1$ Å, $\omega_0 = 1.8 \times 10^{19}$ s^{-1} and $\gamma/\omega_0 = \alpha\omega_0 = 1.14 \times 10^{-4}$, that is, $\gamma/\omega_0 \ll 1$ for the X-ray region. For *small* γ, the solution of (3.11) is

$$x = \left(x_0\, e^{-\gamma t/2}\right) e^{i\omega_0 t} = x_0\, e^{\left(i\omega_0 - \frac{1}{2}\gamma\right)t} \quad , \tag{3.13}$$

as can be verified by direct substitution. It has the desired form because the amplitude $x_0 \exp(-\gamma t/2)$ is gradually (exponentially) damped as time passes.

For an accelerated charge, the field is given by (1.66),

$$E = \frac{e}{c^2 R}\, \ddot{x} \sin\theta \quad , \tag{3.14}$$

at the point P (Fig.2.7) that is situated at a distance R and polar angle θ. From (3.13), we have for small γ,

$$\ddot{x} = \left(i\omega_0 - \frac{1}{2}\gamma\right)^2 x_0\, e^{\left(i\omega_0 - \frac{1}{2}\gamma\right)t} \simeq -\omega_0^2 x_0\, e^{\left(i\omega_0 - \frac{1}{2}\gamma\right)t} \quad . \tag{3.15}$$

Because the electromagnetic field at the point P, at the time of observation t, is determined by the state of the oscillator R/c seconds earlier, that is, at the instant $t - (R/c)$, we should write from (3.14,15),

$$E = \frac{e}{c^2 R} \sin\theta \left[-\omega_0^2 x_0\, e^{\left(i\omega_0 - \frac{1}{2}\gamma\right)\left(t - \frac{R}{c}\right)}\right] = \left(E_0\, e^{-\frac{1}{2}\gamma t}\right) e^{i\omega_0\left(t - \frac{R}{c}\right)} \quad , \tag{3.16}$$

where

$$E_0 = -\frac{e}{c^2R}\sin\theta\,\omega_0^2 x_0\, e^{\gamma R/2c} \quad , \tag{3.17}$$

is a constant for a fixed point P. Thus the energy, like the amplitude, decreases exponentially at a rate proportional to ω_0^2.

The radiation emitted by the oscillator is not monochromatic, because of the radiation damping. The *linewidth* can be obtained by making a Fourier analysis of the field,

$$E(t) = \int_0^\infty E_\omega\, e^{i\omega t}\, d\omega = E_0\, e^{-\frac{1}{2}\gamma t}\, e^{i\omega_0\left(t-\frac{R}{c}\right)} \quad . \tag{3.18}$$

If we assume $E(t) = 0$ for $-\infty < t < 0$, we have

$$\begin{aligned}
E_\omega &= \frac{1}{\pi}\int_{-\infty}^{+\infty} E(t)\, e^{-i\omega t}\, dt = \frac{1}{\pi}\int_0^\infty E(t)\, e^{-i\omega t}\, dt \\
&= \frac{E_0}{\pi}\int_0^\infty e^{-\frac{1}{2}\gamma t}\, e^{i\omega_0\left(t-\frac{R}{c}\right)}\, e^{-i\omega t}\, dt \\
&= \frac{E_0}{\pi}\, e^{-i\omega_0 R/c}\int_0^\infty e^{-\left[i(\omega-\omega_0)+\frac{1}{2}\gamma\right]t}\, dt \\
&= \frac{E_0}{\pi}\, e^{-i\omega_0 R/c}\,\frac{1}{i(\omega-\omega_0)+\frac{1}{2}\gamma} \\
&= \frac{E_0}{\pi}\,\frac{-i(\omega-\omega_0)+\frac{1}{2}\gamma}{(\omega-\omega_0)^2+\frac{1}{4}\gamma^2}\, e^{-i\omega_0 R/c} \quad .
\end{aligned} \tag{3.19}$$

The modulus of E_ω is given by

$$|E_\omega| = \frac{E_0}{\pi}\left[\frac{1}{(\omega-\omega_0)^2+\frac{1}{4}\gamma^2}\right]^{\frac{1}{2}} \quad . \tag{3.20}$$

Substitution of E_ω in (3.18) gives

$$E(t) = \frac{E_0}{\pi}\, e^{-i\omega_0 R/c}\int_0^\infty \frac{e^{i\omega t}}{i(\omega-\omega_0)+\frac{1}{2}\gamma}\, d\omega \quad . \tag{3.21}$$

This shows that the field of a classical oscillator can be represented in the form of a continuous set of simple harmonic oscillators of the type (3.19).

The spectral intensity (a real quantity) is given by

$$I_\omega = \frac{c}{4\pi}|E_\omega|^2 = \frac{c}{4\pi^3}\,\frac{E_0^2}{(\omega-\omega_0)^2+\frac{1}{4}\gamma^2} \quad . \tag{3.22}$$

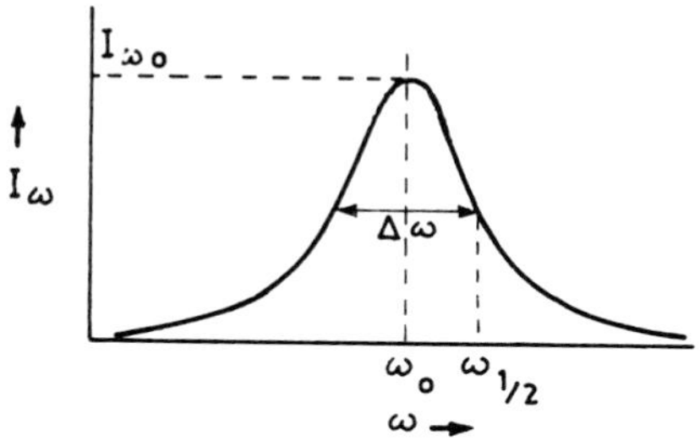

Fig. 3.1. The dispersed shape of a spectral line at ω_0

This expression has a maximum at $\omega = \omega_0$,

$$(I_\omega)_{max} = I_{\omega_0} = \frac{cE_0^2}{\pi^3\gamma^2} \quad , \tag{3.23}$$

and we can write

$$I_\omega = \frac{I_{\omega_0}}{\left(\frac{\omega-\omega_0}{\gamma/2}\right)^2+1} \quad . \tag{3.24}$$

This distribution of intensity as a function of frequency is called the *dispersion formula*; it determines the shape of the spectral line (Fig.3.1). It is in agreement with experiments in which symmetrical lines are observed.

The value of frequency at half intensity, $\omega_{1/2}$, is determined by

$$I_{\omega_{1/2}} = \frac{I_{\omega_0}}{\left(\frac{\omega_{1/2}-\omega_0}{\gamma/2}\right)^2+1} = \frac{1}{2} I_{\omega_0} \quad . \tag{3.25}$$

This gives

$$|\omega_{1/2}-\omega_0| = \frac{1}{2}\gamma \quad . \tag{3.26}$$

Therefore, the full frequency width at half maximum is

$$\Delta\omega = \gamma = \frac{8\pi^2 e^2}{3mc\lambda_0^2} \quad , \tag{3.27}$$

where we have used (3.12). The corresponding width expressed in wavelength is,

$$\Delta\lambda = \frac{\lambda_0^2}{2\pi c}|\Delta\omega| = \frac{4\pi e^2}{3mc^2} = \frac{4\pi}{3} r_e = 0.118 \text{ XU} \quad , \tag{3.28}$$

where we have used $\omega = 2\pi c/\lambda$, $|\Delta\omega| = (2\pi c/\lambda^2)\Delta\lambda$. The factor $r_e = e^2/mc^2 = 2.818 \times 10^{-13}$ cm is the classical electron radius (calculated from $e^2/r = mc^2$). Obviously, $\Delta\lambda$ is a universal constant, independent of the frequency of the oscillator and of the wavelength λ_0 corresponding to the intensity maximum of the line. This conclusion is not supported by experiments, because radiation damping is not the only source of the width of a spectral line. The full width of the $K\alpha_1$ line at half intensity varies from 1.50 XU for ^{20}Ca (λ_0 = 3351 XU) to 0.16 XU for ^{74}W (λ_0 = 208.6 XU).

If I_0 is the intensity at $t = 0$, then from (3.16),

$$I(t) = \frac{c}{4\pi} |E(t)|^2 = I_0 e^{-\gamma t} \quad . \tag{3.29}$$

We define the effective *lifetime* τ of the oscillator as the time at which the intensity is decreased by the factor 1/e, where e is the base of natural logarithms. Thus, we have

$$I(\tau) = I_0 e^{-\gamma\tau} = I_0 \exp(-1) \quad . \tag{3.30}$$

This gives,

$$\gamma\tau = 1 \quad , \quad \gamma = \Delta\omega = \frac{1}{\tau} \quad , \quad \Delta\nu = \frac{\Delta\omega}{2\pi} = \frac{1}{2\pi\tau} \quad . \tag{3.31}$$

If τ is short, the oscillations are rapidly damped, the motion is no longer simple harmonic, and the width is large. This is in qualitative agreement with observations.

The relation $\Delta\omega\tau = 1$ is equivalent to the relation

$$\Delta E \Delta t \sim \hbar \quad , \tag{3.32}$$

where $\Delta E = \hbar\Delta\omega$. Equation (3.32) is the quantum mechanical relation between the lifetime and the energy width of a state.

3.3 Forced, Damped, Oscillator

We assume that the velocity of the electron is nonrelativistic. Therefore, the effect of the magnetic vector in the external incoming radiation can be neglected.

The equation of motion of a bound electron that belongs to the q shell and is in an external (polarized) plane wave field, is

$$\ddot{x} - \alpha\dddot{x} + \omega_q^2 x = -\frac{e}{m} E_0 e^{i\omega\left(t - \frac{y}{c}\right)} . \tag{3.33}$$

Here ω_q is a natural frequency of the electron and q refers to K, L, M, etc. We have used the Abraham-Lorentz equation (3.8) because we intend to deal with steady-state oscillations. We have considered an electromagnetic wave $E(y,t)$ that propagates along the y axis and is polarized in the xy plane. The negative sign means that the force exerted on the electron of charge $-e$ is $-eE$. The electric vector of this wave is parallel to the x axis. At $y = 0$, $t = 0$, the field strength is E_0; at $y = 0$, $t = t$, it is $E_0 \exp(i\omega t)$; at $y = y$, $t = t$, E is equal to a value that the electric field had y/c second earlier, at the point $y = 0$. Thus,

$$E(y,t) = E_0 e^{i\omega\left(t - \frac{y}{c}\right)} .$$

We seek a solution of (3.33) in the form of oscillations with the frequency and phase of the inducing field,

$$x = x_0 e^{i\omega\left(t - \frac{y}{c}\right)} . \tag{3.34}$$

Substitution of this in (3.33) gives for a forced, damped, oscillator,

$$\begin{aligned} x(t) &= -\frac{e}{m} E_0 \frac{1}{\left(\omega_q^2 - \omega^2\right) + i\alpha\omega^3} e^{i\omega\left(t - \frac{y}{c}\right)} \\ &= -\frac{eE}{m\left(\omega_q^2 - \omega^2\right) + i\,\frac{2e^2}{3c^3}\,\omega^3} , \end{aligned} \tag{3.35}$$

where $\alpha = 2e^2/3mc^3$, by (3.9).

3.4 Complex Dielectric Constant

Because x represents the displacement of the electron under the action of the electric field, it is connected with the *polarization* of the medium (Fig.3.2). If n_q is the number of q-type electrons per unit volume, the volumetric polarization of the medium is given by

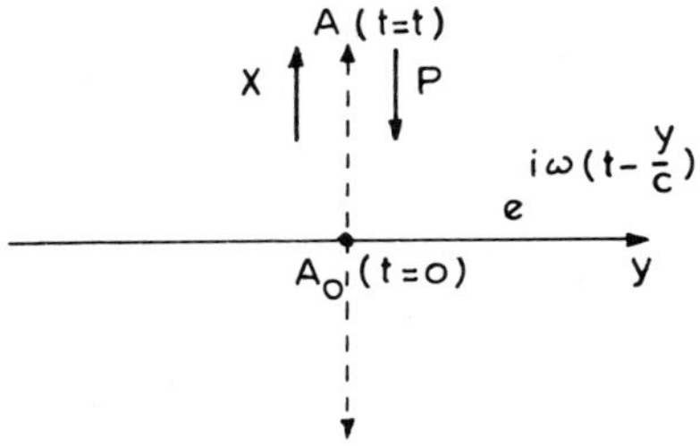

Fig. 3.2. The electromagnetic wave moves along the y axis with the electric vector parallel to the x axis. It displaces the electron initially (t=0) at A_0 to A by an amount x

$$P_q = -n_q\, ex \quad . \tag{3.36}$$

The dielectric permeability of a medium ε_q is related to P by

$$\varepsilon_q = 1 + 4\pi \frac{P_q}{E} \quad . \tag{3.37}$$

From (3.35-37), we obtain

$$\varepsilon_q = 1 + \frac{4\pi n_q e^2}{m\left(\omega_q^2 - \omega^2\right) + i\,\frac{2e^2}{3c^3}\omega^3} \quad . \tag{3.38}$$

Separating the real and imaginary parts, we get for the complex dielectric constant

$$\varepsilon_q = 1 - 2\delta_q - i2\beta_q \quad , \tag{3.39}$$

where

$$\delta_q = -\frac{2\pi n_q e^2 m\left(\omega_q^2-\omega^2\right)}{m^2\left(\omega_q^2-\omega^2\right)^2 + \frac{4e^4}{9c^6}\omega^6} \quad , \tag{3.40}$$

$$\beta_q = \frac{4\pi n_q e^4 \omega^3}{3c^3\left[m^2\left(\omega_q^2-\omega^2\right)^2 + \frac{4e^4}{9c^6}\omega^6\right]} \quad . \tag{3.41}$$

In Fig.3.3 we have plotted δ_q, which occurs in the real part of ε_q, and β_q, which occurs in the imaginary part of ε_q, as a function of the frequency ω of the incident radiation. We find that δ_q is zero at $\omega = \omega_q$, and β_q has a maximum at this frequency (resonance absorption). Also, δ_q is strongly frequency dependent near this frequency and undergoes a change in sign. This is the phenomenon of *anomalous dispersion*.

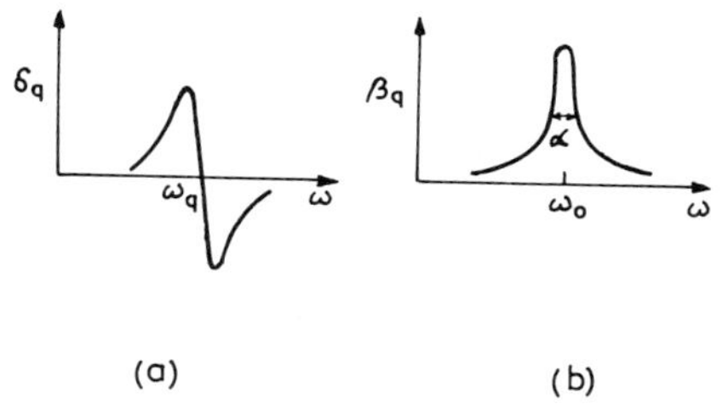

Fig. 3.3. The dependence of (a) δ_q on ω, and of (b) β_q on ω

In addition to the primary electromagnetic wave, we have the electromagnetic wave radiated by q-type electrons that are under forced oscillations. The resultant wave is propagated through the medium with a phase velocity v given by the well-known relation

$$v = c/\varepsilon^{\frac{1}{2}} \quad . \tag{3.42}$$

Because of ε, it is a complex quantity.

It is useful to define the so-called complex-wave slowness, S_q, by

$$\begin{aligned} S_q &= v_q^{-1} = \frac{1}{c}\,\varepsilon^{\frac{1}{2}} = \frac{1}{c}\,(1-2\delta_q-i2\beta_q)^{\frac{1}{2}} \\ &\simeq \frac{1}{c}\,(1-\delta_q-i\beta_q) \equiv S_r - iS_i \quad , \end{aligned} \tag{3.43}$$

where we have used the fact that δ_q and β_q are small quantities ($\sim 10^{-6}$). We can write, for the wave in the medium,

$$E_q(y,t) = E_{0,q}\, e^{i\omega\left(t-\frac{y}{v_q}\right)} \quad . \tag{3.44}$$

From (3.43,44),

$$E_q(y,t) = \left(E_{0,q}\, e^{-\omega S_i y}\right) e^{i\omega(t-S_r y)} \quad . \tag{3.45}$$

Thus, the amplitude is exponentially damped as y in the medium increases. If $|E_q(y)|$ is the amplitude at a depth y in the medium,

$$|E_q(y)| = E_{0,q}\, e^{-\omega S_i y} \quad , \tag{3.46}$$

we can say that the complex nature of the dielectric constant, and hence of the phase velocity, leads to attenuation (or absorption) of the wave. The intensity is

$$I_q(y) = I_{0,q}\, e^{-(2\omega S_i)y} = I_{0,q}\, e^{-(\tau_q)y} \quad , \tag{3.47}$$

where

$$(\tau_q) = 2\omega S_i = 2\omega v_{i,q}^{-1} = \frac{2\omega}{c}\,\beta_q = \frac{4\pi}{\lambda}\,\beta_q \quad , \tag{3.48}$$

is the *partial linear absorption coefficient* for the q-type electrons.

We have shown that the real part of the complex dielectric constant is related to the phase velocity of waves in the medium, and the imaginary part, β_q, with the absorption. This is a general result. The amount of energy abstracted from the incident wave is converted to energy of the bound-electron system.

3.5 Refractive Index

The *refractive index* is defined as the ratio of the phase velocities in vacuum and in the medium, c/v_q. Because v_q is complex in our theory, we get a complex refractive index, μ_c. The *partial refractive index* for the q-type electrons is

$$\mu_{c,q} = \mu_q - i\mu_{i,q} = \varepsilon_q^{\frac{1}{2}} = cv_q^{-1} = 1 - \delta_q - i\beta_q \quad , \tag{3.49}$$

where we have used (3.43), and $\mu_q = 1 - \delta_q$ is the real part of $\mu_{c,q}$. The *unit decrement* δ_q gives the departure of μ_q from 1.

For X-ray frequencies, we have from (3.40),

$$\delta_q = \frac{2\pi n_q e^2 m\left(\omega^2 - \omega_q^2\right)}{m^2\left(\omega^2 - \omega_q^2\right)^2 + \frac{4e^4}{9c^6}\omega^6} \simeq \frac{2\pi n_q e^2}{m}\,\frac{1}{\omega^2 - \omega_q^2} \quad , \tag{3.50}$$

where we have neglected the small term $(4e^4/9c^6)\omega^6$ that appears in the denominator because of the radiation damping. This is easily justified. For quartz (SiO_2) the largest value of ω_q is obtained for the K electrons of silicon (^{14}Si),

$$\omega_{(Si)K} = 2\pi c/\lambda_{(Si)K} = 2.8 \times 10^{18}\ s^{-1} \quad ,$$

where $\lambda_{(Si)K} = 6.73$ Å. For incident X-rays of wavelength 1 Å,

$$\omega = 2\pi c/\lambda = 18.9 \times 10^{18}\ s^{-1} > \omega_{(Si)K} \quad ,$$
$$\left(\omega^2 - \omega^2_{(Si)K}\right)^2 = 1.2 \times 10^{77} \quad ,$$
$$(4e^4/9c^6)\omega^6 m^{-2} = \alpha^2\omega^6 = 1.8 \times 10^{69} \ll \left(\omega^2 - \omega^2_{(Si)K}\right)^2 \quad .$$

Thus, the approximation is good for all the electrons of the quartz atom.

If the medium contains electrons of various natural frequencies ω_q, we can write for the refractive index

$$\mu_c = \mu - i\mu_i = \sum_q \mu_{e,q} = 1 - \sum_q (\delta_q + i\beta_q) = 1 - \delta - i\beta \quad , \tag{3.51}$$

where

$$\delta = \sum_q \delta_q \quad , \quad \beta = \sum_q \beta_q \quad .$$

Now δ gives the unit decrement of the refractive index μ.

For the X-ray frequencies ($\omega \gg \omega_q$), we can write

$$\delta = \sum_q \delta_q = \frac{2\pi e^2}{m} \sum_q \frac{n_q}{\omega^2 - \omega_q^2} \simeq \frac{2\pi e^2}{m\omega^2} \sum_q n_q = \frac{e^2 n\lambda^2}{2\pi mc^2} \quad , \tag{3.52}$$

where n is the total number of electrons per unit volume of medium. For X-rays being dispersed by crystals like quartz, calcite, etc., in a spectrograph, the unit decrement δ is a small positive quantity, and

$$\mu = 1 - \delta \lesssim 1 \quad \text{(X-rays)} \quad . \tag{3.53}$$

(For optical frequencies, in a glass prism, $\omega < \omega_q$, so that $\mu > 1$). It follows that *for X-rays the medium of a quartz crystal is less dense than a vacuum.*

If $N_A = 6.0225 \times 10^{23}$ is Avogadro's number, A the atomic weight, Z the atomic number and ρ the density, then (3.52) gives

$$n = ZN_A\rho/A \quad ,$$

$$\frac{\delta}{\lambda^2} = \frac{e^2 N_A}{2\pi mc^2}\frac{Z\rho}{A} = 2.7 \times 10^{10}\frac{Z\rho}{A} = \text{constant} \quad , \quad (\lambda \ll \lambda_k) \quad . \tag{3.54}$$

For ^{13}Al (A=27, ρ=2.7), $\delta = 1.74 \times 10^{-6}$ for the MoKα (0.708 Å) radiation. The experimental value is 1.68×10^{-6}. Therefore, the Lorentz theory is satisfactory for $\lambda \ll \lambda_K$.

LARSSON [3.2] has compared his experiments on calcite ($CaCO_3$) with (3.50), when the damping factor $(4e^4/9c^6)\omega^6$ is neglected,

$$\delta = \frac{2\pi e^2}{m} \sum_q \frac{n_q}{\omega^2-\omega_q^2} = \frac{e^2}{2\pi mc^2} \lambda^2 \sum_q \frac{n_q\lambda_q^2}{\lambda_q^2-\lambda^2} \quad . \tag{3.55}$$

For a compound,

$$n_q = (N_A\rho/M)\, z_q \quad ,$$

where z_q is the number of electrons of the q-type and M is the molecular weight. If λ_K is the K-edge wavelength of Ca, we can separate out the term for the z_K electrons in the K level of Ca, and write

$$\frac{\delta}{\lambda^2} = \frac{e^2N_A}{2\pi mc^2} \frac{\rho}{M} \left(\frac{z_K\lambda_K^2}{\lambda_K^2-\lambda^2} + \sum_{q\neq K} \frac{z_q\lambda_q^2}{\lambda_q^2-\lambda^2} \right) \quad .$$

Near the absorption edge, $\lambda \approx \lambda_K$, it is the first term in the parentheses that will cause departure from the value given by (3.54) for the case $\lambda << \lambda_K$. Therefore, for the remaining electrons $\lambda < \lambda_q$, and

$$\sum_{q\neq K} \frac{z_q\lambda_q^2}{\lambda_q^2-\lambda^2} = \frac{1}{1-(\lambda/\lambda_q)^2} \sum_{q\neq K} z_q \simeq \sum_{q\neq K} z_q = Z_M - z_K \quad ,$$

$$\frac{\delta}{\lambda^2} = \frac{e^2N_A}{2\pi mc^2} \frac{\rho}{M} \left[\frac{z_K\lambda_K^2}{\lambda_K^2-\lambda^2} + (Z_M-z_K) \right] = 2.68 \times 10^{-6} \frac{\rho}{M} \left[\frac{2\lambda_K^2}{\lambda_K^2-\lambda^2} + (Z_M-2) \right] \quad . \tag{3.56}$$

For $CaCO_3$, $\rho = 2.71$ gcm^{-3}, $M = 100.07$, $Z_M = 50$, $z_K = 2$, $\lambda_K = 3.06$ Å, so that,

$$\frac{\delta}{\lambda^2} = 0.0732 \times 10^{-6} \left(\frac{18.78}{9.390-\lambda^2} + 48 \right) \quad ,$$

where λ is in Å.

In Fig.3.4, we plot this result. For $\lambda << \lambda_K$, the quantity in the parentheses tends to 50, and for $\lambda >> \lambda_K$, to 48. Therefore, $\delta/\lambda^2 \to 3.65 \times 10^{-6}$ for $\lambda \to$ small and $\delta/\lambda^2 \to 3.50 \times 10^{-6}$ for $\lambda \to$ large. As $\lambda \to \lambda_K$ from the short (long) wavelength side, $\delta/\lambda^2 \to +\infty(-\infty)$. Experimental points, obtained with the crystal-wedge method for known emission lines between 1.5 Å and 3.7 Å, show that the Lorentz theory gives correct results for $\lambda << \lambda_K$. Near the absorption edge, $\lambda \approx \lambda_K$, the agreement breaks down. The reason is that we have neglected the

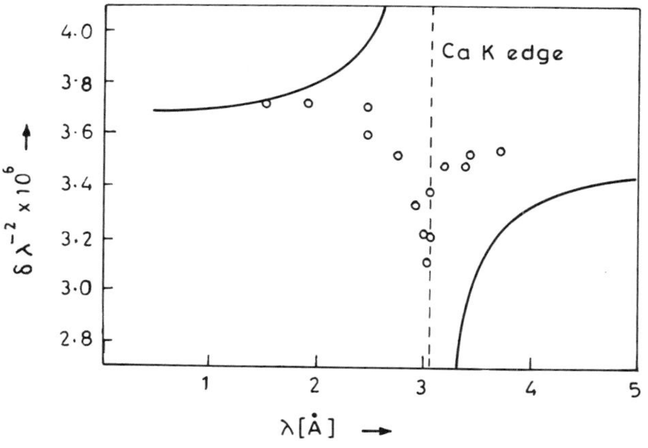

Fig. 3.4. Plot of δ/λ^2 versus λ(Å) for calcite. The curves are from the Lorentz theory and the experimental points are from [3.2]

damping factor, which becomes important in this region. Also, the Lorentz theory does not include the effect of absorption near the edge (see Sect.3.10). For $\lambda >> \lambda_K$ the effects of other absorption edges appear and spoil the agreement with this simple theory. Near the absorption edge δ/λ^2 is not constant and anomalous dispersion occurs (δ/λ^2 varies with λ).

3.6 Correction of the Bragg Equation

BRAGG [3.3] gave a simple explanation of the observed angles of the beams diffracted from a crystal. It is based on the idea that in the diffraction of X-rays by the atoms of a crystal lattice there occurs, *as it were*, specular (mirrorlike) reflection of the X-rays from the internal atomic planes. Each plane reflects only a very small fraction of the radiation, as with a very lightly silvered mirror, such that the angle of incidence is equal to the angle of reflection.

Consider a set of parallel lattice planes spaced equal distances d apart (Fig.3.5). The radiation of wavelength λ is incident at a glancing angle θ. The path difference for rays reflected from the adjacent planes is 2d sinθ. It is the extra distance travelled *inside* the crystal. The radiation reflected from successive planes will arrive at PP'P" in phase, to give constructive interference (strong reflection), whenever the path difference is an integral

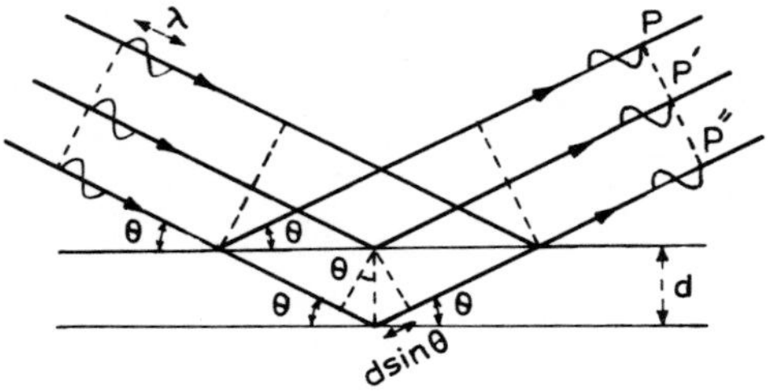

Fig. 3.5. Derivation of the Bragg equation

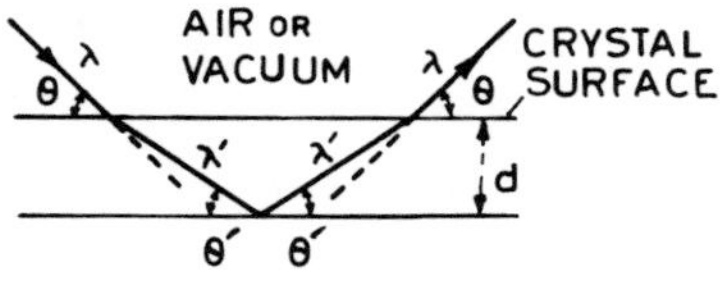

Fig. 3.6. Construction for the modified Bragg law. Rays bend away from the normal because $\mu < 1$

multiple of wavelength, $n\lambda$ (n=0,1,2,3,...). Thus the condition for constructive reflection of the incident radiation is simply

$$2d \sin\theta = n\lambda \quad (n = \text{order of reflection}) \quad . \tag{3.57}$$

This is the *Bragg law*, and θ is the *Bragg angle*. Bragg reflection can occur only for $\lambda \leqq 2d$. For the visible light $\lambda \gg 2d$.

We should correct (3.57) for the effect of refraction of X-rays when they enter the crystal ($\mu \lesssim 1$) (Fig.3.6). This will alter the extra distance travelled inside the crystal, and therefore the required path difference. Inside the crystal the wavelength λ becomes λ' and the corresponding Bragg angle is θ'. The refractive index can be defined, as in optics, by the well-known relation

$$\mu = 1 - \delta = \lambda/\lambda' = \cos\theta/\cos\theta' \quad . \tag{3.58}$$

Within the crystal the Bragg law is

$$n\lambda' = 2d \sin\theta' \quad . \tag{3.59}$$

From (3.58,59),

$$\mu = \frac{\lambda}{\lambda'} = \frac{\lambda}{2(d/n)\sin\theta'} = \frac{n\lambda}{2d\left(1 - \frac{\cos^2\theta}{\mu^2}\right)^{\frac{1}{2}}} \quad . \tag{3.60}$$

Rearrangement gives

$$n\lambda = 2d \sin\theta \left(1 + \frac{\mu^2 - 1}{\sin^2\theta}\right)^{\frac{1}{2}} . \tag{3.61}$$

Because μ is nearly one, we can write $\mu^2 - 1 = (\mu+1)(\mu-1) \simeq 2(-\delta)$ and

$$n\lambda \simeq 2d \sin\theta \left(1 - \frac{2\delta}{\sin^2\theta}\right)^{\frac{1}{2}} = 2d \sin\theta \left(1 - \frac{\delta}{\sin^2\theta} + \ldots\right) ,$$

or

$$n\lambda = 2d \sin\theta \left(1 - \frac{\delta}{\sin^2\theta}\right) , \tag{3.62}$$

to the first power in the small quantity δ. This is the *modified Bragg law.*

We can write (3.62) in terms of a fictitious angle θ_B as

$$n\lambda = 2d \sin\theta_B , \tag{3.63}$$

where

$$\sin\theta_B = \sin\theta \left(1 - \frac{\delta}{\sin^2\theta}\right) = \sin\theta - \frac{\delta}{\sin\theta} . \tag{3.64}$$

Then

$$\sin\theta - \sin\theta_B = \Delta(\sin\theta) = \cos\theta(\Delta\theta) ,$$

$$\Delta\theta = \theta - \theta_B = \frac{\sin\theta - \sin\theta_B}{\cos\theta} = \frac{\delta}{\sin\theta \cos\theta} = \frac{2\delta}{\sin 2\theta} > 0 . \tag{3.65}$$

For small θ,

$$\Delta\theta \approx \delta/\theta . \tag{3.66}$$

Thus, a significant deviation from the Bragg law will occur only for small θ, because $\delta \sim 10^{-6}$.

It may be noted that no modification of the Bragg equation is needed if the symmetrical reflection occurs from the planes that are perpendicular to the surface of a thin crystal (like mica or quartz sheet). To see this, consider two parallel rays A and B that fall at angles θ to such planes (Fig.3.7). In this case, the path difference between A and B arises entirely from the

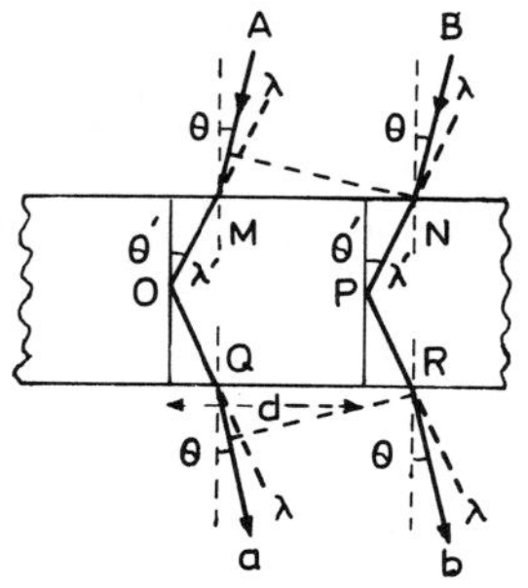

Fig. 3.7. Symmetrical reflection from planes perpendicular to the crystal surface

distances travelled *outside* the crystal, and is equal to MN $\sin\theta = 2d \sin\theta 2$ as shown in Fig.3.7. Therefore, the condition for constructive interference is $n\lambda = 2d \sin\theta$. This is just the usual Bragg equation.

3.7 Measurement of Refractive Index

3.7.1 The Method of Critical Angle of Reflection

The real refractive index is given by

$$\mu = 1 - \delta = \cos\theta/\cos\theta' \quad , \tag{3.67}$$

where the angles θ and θ' are shown in Fig.3.8a. For X-rays, $0 < \delta \ll 1$; therefore,

$$\Delta\theta = \theta - \theta' > 0 \quad , \tag{3.68}$$

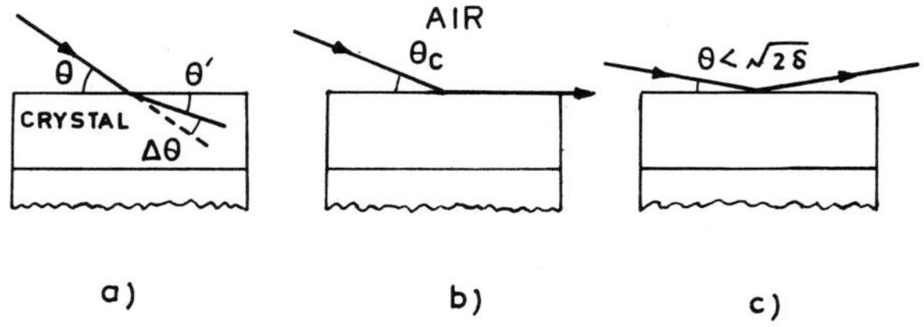

Fig. 3.8. (a) $\theta > (2\delta)^{\frac{1}{2}}$; (b) $\theta_c = (2\delta)^{\frac{1}{2}}$; (c) $\theta < (2\delta)^{\frac{1}{2}}$

Table 3.1. Some values of the critical angle θ_c

Substance	λ [Å]	$\delta \times 10^6$	θ_c
Glass	0.7078	1.64	6'10"
Calcite	1.537	8.80	14'25"
Quartz	10.0	356	91'40"

$$\cos\theta' = (1-\delta)^{-1} \cos\theta \simeq (1+\delta) \cos\theta \quad , \tag{3.69}$$

$$\cos\theta - \cos\theta' \equiv \Delta(\cos\theta) \equiv - \sin\theta \, \Delta\theta = -\delta \cos\theta \quad , \tag{3.70}$$

$$\Delta\theta = \delta \cot\theta \quad . \tag{3.71}$$

Obviously, the angle of deviation $\Delta\theta = 0$ for $\theta = \pi/2$. As θ decreases, $\Delta\theta$ increases and is maximum at $\theta' = 0$ (Fig.3.8b). The angle θ for this value of θ' is the minimum (*critical*) angle, θ_c, at which the X-ray beam can enter a medium through a surface of separation. From (3.67),

$$\mu = 1 - \delta = \cos\theta_c = 1 - \frac{1}{2!} \theta_c^2 + \frac{1}{4!} \theta_c^4 - \dots \quad , \tag{3.72}$$

and for small values of θ_c,

$$\theta_c = (2\delta)^{\frac{1}{2}} \quad . \tag{3.73}$$

For a given substance, δ/λ^2 is nearly constant. Therefore, we have $\theta_c \propto \lambda$. From the known values of δ we can compute θ_c, see Table 3.1, or we can reverse the procedure.

For glancing angles $\theta < (2\delta)^{\frac{1}{2}}$, none of the rays are transmitted; that is, *all* of the rays incident at $\theta < \theta_c$ are reflected at the surface (*total reflection*) (Fig.3.8c).

COMPTON [3.4], DOAN [3.5], and SMITH [3.6] have used (3.73) to measure δ. It is difficult to measure accurately the small angle θ_{min}.

3.7.2 The Method of Symmetrical Reflection

For a given emission line of wavelength λ, measured in two different *orders of reflection* n_1 and n_2, we have from the modified Bragg law (3.62),

$$\frac{\lambda}{2d} = \frac{1}{n_1} \left(\sin\theta_1 - \frac{\delta}{\sin\theta_1}\right) = \frac{1}{n_2} \left(\sin\theta_2 - \frac{\delta}{\sin\theta_2}\right) \quad ,$$

or

$$\delta = \frac{n_1^{-1} \sin\theta_1 - n_2^{-1} \sin\theta_2}{(n_1 \sin\theta_1)^{-1} - (n_2 \sin\theta_2)^{-1}} \quad . \tag{3.74}$$

STENTSTRÖM [3.7] measured δ from this equation by measuring θ at various n. It is not a very accurate method, because the deviations from the Bragg angle are very small in symmetrical reflections.

3.7.3 The Method of Unsymmetrical Reflection

The deviation is enhanced in the unsymmetrical reflection. For most glasses $\delta \approx 2\times 10^{-6}$, so that $\theta_{min} = (\Delta\theta)_{max} = 2\times 10^{-3} \approx 7'$. At $\theta = 0.1 = 5^\circ.73'$ we get $\Delta\theta = \delta \cot\theta = 2\times 10^{-5} \approx 4''$. Thus, a decrease from $5^\circ.73'$ to $7'$ enhances the angle of deviation by a factor of 100. At $\theta = 45^\circ$, we have $\Delta\theta = \delta = 2\times 10^{-6} \approx 0.4''$.

Let a portion of the crystal be ground to form a surface that makes an angle α with the reflecting atomic planes (Fig.3.9). The ray S is reflected in a symmetrical way and the ray U in the unsymmetrical way. The glancing angle for U is much smaller, and so the deviation caused by refraction is much larger than for S.

The crystal is mounted in a spectrometer and the spectral lines are recorded for the two cases S and U. From (3.67) and Fig.3.9, we have

$$\mu = 1 - \delta = \frac{\cos\theta}{\cos\theta'} = \frac{\cos\theta_1}{\cos(\theta'-\alpha)} = \frac{\cos\theta_2}{\cos(\theta'+\alpha)} \quad , \tag{3.75}$$

$$\theta = \theta' + \Delta \quad , \quad \theta_1 = \theta' - \alpha + \Delta_1 \quad , \quad \theta_2 = \theta' + \alpha + \Delta_2 \quad . \tag{3.76}$$

Because Δ and Δ_2 are small, we can write

$$1 - \delta = \frac{\cos\theta' \cos\Delta - \sin\theta' \sin\Delta}{\cos\theta'} \simeq 1 - \tan\theta' \sin\Delta \quad ,$$

or

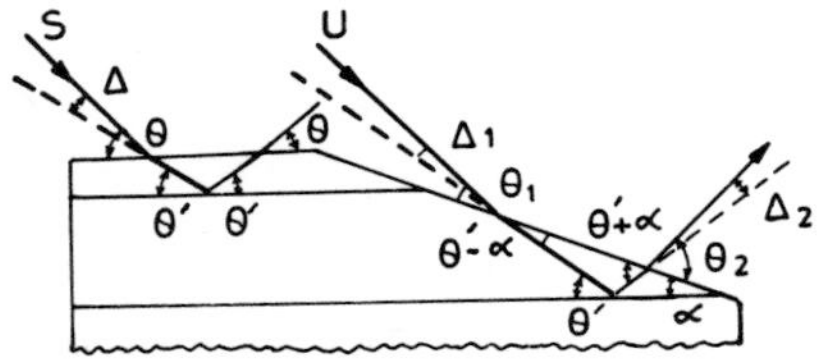

Fig. 3.9. Symmetrical, S, and unsymmetrical, U, reflections from a crystal wedge

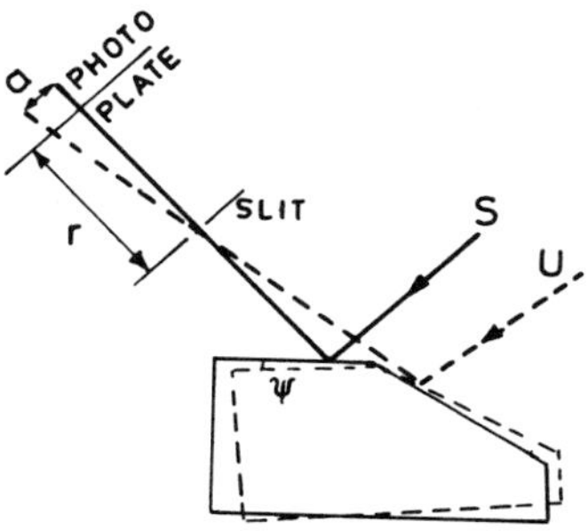

Fig. 3.10. Determination of δ by the method of rotation of the crystal wedge

$$\sin\Delta = \delta\cot\theta' \approx \delta\cot\theta \quad . \tag{3.77}$$

Similarly,

$$\sin\Delta_2 \simeq \delta\cot(\theta'+\alpha) \approx \delta\cot(\theta+\alpha) \quad . \tag{3.78}$$

Δ_1 is also a small quantity but greater than Δ and Δ_2. Therefore, in a slightly better approximation,

$$1-\delta = \frac{\cos(\theta'-\alpha+\Delta_1)}{\cos(\theta'-\alpha)} = \cos\Delta_1 - \tan(\theta'-\alpha)\sin\Delta_1$$
$$= \left(1-\frac{1}{2!}\Delta_1^2+\dots\right) - \tan(\theta'-\alpha)\sin\Delta_1 \quad ,$$

or

$$\delta \simeq \frac{1}{2}\Delta_1^2 + \tan(\theta-\alpha)\sin\Delta_1 \quad . \tag{3.79}$$

From (3.76), we get

$$\Delta_1 = \theta_1 + \theta_2 - 2\theta + 2\Delta - \Delta_2 \quad . \tag{3.80}$$

We can measure θ, θ_1 and θ_2. From (3.80), we obtain $\Delta_1 \approx \theta_1 + \theta_2 - 2\theta$. This is put in (3.79) to give an approximate value for δ. We can now determine Δ and Δ_2 from (3.77,78), and then a better value for Δ_1 from (3.80). Thus, δ can be found from (3.79) by successive approximations.

Another way of making the measurements is shown in Fig.3.10. The crystal wedge is placed on the cross table of a spectrometer and a particular line is recorded through the symmetrical reflection, S. The crystal is now rotated by a small angle ψ and the same line is recorded through an unsymmetrical reflection, U. The angular distance between the two lines is a/r (Fig.3.10). Because of the refraction effect, this angle will be less than ψ,

$$\psi - \frac{a}{r} = \varepsilon > 0 \quad . \tag{3.81}$$

From Fig.3.9,

$$\varepsilon = \Delta_1 - \Delta = \Delta_1 - (\theta - \theta') \quad . \tag{3.82}$$

From (3.75,82),

$$1 - \delta = \frac{\cos(\theta' - \alpha + \Delta_1)}{\cos(\theta' - \alpha)} = \frac{\cos(\theta + \varepsilon - \alpha)}{\cos(\theta' - \alpha)} = \frac{\cos\theta}{\cos\theta'} \quad . \tag{3.83}$$

These relations give

$$\tan\theta' = \frac{\cos(\theta + \varepsilon - \alpha)}{\sin\alpha \, \cos\theta} - \cot\alpha \quad , \tag{3.84}$$

$$\delta = 1 - \frac{\cos(\theta + \varepsilon - \alpha)}{\cos(\theta' - \alpha)} \quad . \tag{3.85}$$

We can measure θ, α, and ε. Thus, δ can be calculated from (3.84,85).

The methods based on unsymmetrical reflections have been used to find δ by HATLEY and DAVIS [3.8,9], and DAVIS and NARDOFF [3.10,11]. In these experiments, α was chosen close to θ' so that θ_1 is small and the deviation Δ_1 is large.

3.7.4 The Method of Refraction in a Prism

A narrow X-ray beam falls on the prism at a point A close to its edge (Fig. 3.11). This reduces the absorption of X-rays in the crystal. The reflected ray is AB, and the refracted ray is AC. With the prism removed, the direct ray is recorded at D. The three rays recorded at B, C and D provide enough data for the determination of δ.

For large deviation $\Delta\theta$, the angle θ is chosen to be small and close to the critical angle. Then the refracted ray emerges from the prism base at almost a right angle and its deviation at this point can be neglected. With $\theta = \theta' + \Delta\theta$, we have

$$\delta = 1 - \frac{\cos\theta}{\cos\theta'} = \frac{\cos(\theta - \Delta\theta) - \cos\theta}{\cos\theta'} = \frac{2\sin(\theta - \frac{1}{2}\Delta\theta)\,\sin\frac{1}{2}\Delta\theta}{\cos\theta'} \simeq (\theta - \frac{1}{2}\Delta\theta)\Delta\theta \quad . \tag{3.86}$$

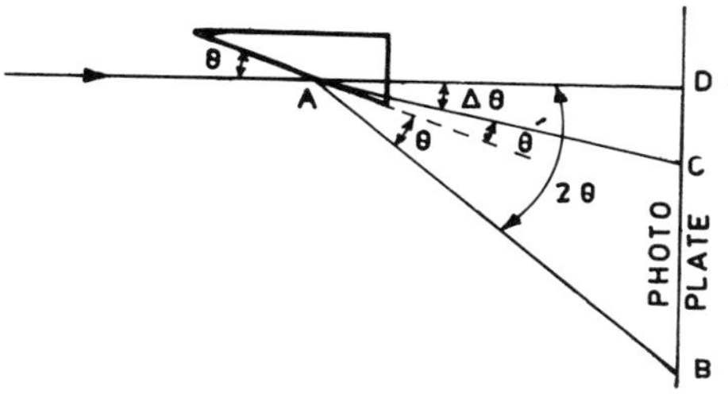

Fig. 3.11. Refraction in a prism

From the experiment, we can measure $2\theta = BD/AD$ and $\Delta\theta = DC/AD$. Therefore, δ can be found from (3.86).

This method has been used by LARSSON et al. [3.12]. DAVIS and SLACK [3.13-15] placed the prism between two calcite crystals in the parallel position of a double-crystal spectrometer.

3.8 Absorption of X-Rays and Dispersion Theory

We can calculate the amount of energy abstracted from the incident X-rays and converted to energy of the bound electrons in an atom, which form a system of oscillators.

3.8.1 Absorption by an Undamped Oscillator

In the first approximation, we neglect the damping factor. Under the influence of the electric field of the incident X-rays, the electron in an atom is accelerated from rest to

$$\ddot{x} = -\frac{e}{m}E \quad , \quad E = E_0 \sin\frac{2\pi}{T}t \quad , \tag{3.87}$$

where T is the time period. If we average over one-half the period,

$$<\ddot{x}> = -\frac{e}{m}<E> \quad . \tag{3.88}$$

We assume the acceleration to be uniform. Then the path s travelled by the electron in time T/2 is

$$s = \frac{1}{2}<\ddot{x}>\left(\frac{1}{2}T\right)^2 = -\frac{1}{8}\frac{e}{m}\frac{<E>}{\nu^2} = -\frac{1}{8}\frac{e<E>}{mc^2}\lambda^2 \quad . \tag{3.89}$$

In the second half of the period, the electron is brought back to rest. The total path travelled in the whole period is

$$x_T = 2s = -\frac{1}{4}\frac{e\lambda^2}{mc^2}<E> \quad . \tag{3.90}$$

The restoring force is, by (3.1),

$$F(x) = -m\omega_q^2 x \quad , \quad \omega_q = 2\pi\nu_q = 2\pi c/\lambda_q \quad , \tag{3.91}$$

where ω_q is the natural frequency of the free oscillations of the absorbing electron of the q-type. The energy absorbed is given by the work done over the path x_T,

$$W_T = -\int_0^{x_T} F(x)\, dx = \frac{1}{2} m\omega_q^2 x_T^2 = \frac{e^2\lambda^4}{32mc^4}\omega_q^2 <E>^2 \quad . \tag{3.92}$$

Therefore, the rate of energy absorption is

$$\dot{W} = \frac{1}{T} W_T = \frac{c}{\lambda} W_T = \frac{\pi^2 e^2 \lambda^3}{8mc\lambda_q^2} <E>^2 \quad . \tag{3.93}$$

The intensity I represents the energy of radiation that crosses unit area, held normal to the beam, in 1 s, and is

$$I = \frac{c}{4\pi} <E^2> \quad . \tag{3.94}$$

We define the absorption coefficient τ_e as the ratio of the energy $\dot{W}$ absorbed (abstracted) per unit time to the intensity I,

$$\tau_e = \frac{\dot{W}}{I} = \frac{\pi^3 e^2 \lambda^3}{2mc^2\lambda_q^2}\frac{<E>^2}{<E^2>} \quad . \tag{3.95}$$

Because $E(t) = E_0 \sin\frac{2\pi}{T}t$, we have

$$<E> = \frac{1}{(T/4)}\int_0^{T/4} E(t)\, dt = \frac{2}{\pi}E_0 \quad ,$$

$$<E^2> = \frac{1}{(T/4)}\int_0^{T/4} [E(t)]^2\, dt = \frac{1}{2}E_0^2 \quad ,$$

$$<E>^2/<E^2> = 8/\pi^2 \simeq 0.81 \quad .$$

Therefore for one electron of the q-type, we obtain

$$\tau_e = \frac{4\pi e^2}{mc^2\lambda_q^2}\lambda^3 \quad . \tag{3.96}$$

If there are z_q electrons at the q level, the partial atomic absorption coefficient τ_q is given by

$$\tau_q = z_q\tau_e = 4\pi r_e\left(z_q/\lambda_q^2\right)\lambda^3 \quad , \tag{3.97}$$

where $r_e = e^2/mc^2$. The *atomic absorption coefficient* is equal to

$$\tau_a = \sum_q \tau_q \equiv 4\pi r_e\lambda^3 \sum_{q=1}^{Z}\left(z_q/\lambda_q^2\right) \quad . \tag{3.98}$$

For the K level, $z_K = 2$, $E_K = R_\infty hc Z^2$, $\lambda_K = \lambda_{Kedge} = c/\nu_K = hc/E_K = 1/(R_\infty Z^2)$. Therefore, from (3.97)

$$\tau_K = 8\pi r_e R_\infty^2 Z^4\lambda^3 = \begin{cases} 8.6\times 10^{-2}\, Z^4\lambda^3 & (\lambda \text{ in cm}) \\ 8.6\times 10^{-26}\, Z^4\lambda^3 & (\lambda \text{ in Å}) \end{cases} \quad . \tag{3.99}$$

There is some loss of radiation from the incident beam because of the scattering. We can define the *atomic attenuation coefficient* μ_a as

$$\mu_a = \tau_a + \sigma_a \quad , \tag{3.100}$$

where τ_a is the *true* atomic absorption coefficient and σ_a the *atomic scattering coefficient*, $\sigma_a \ll \tau_a$.

In Fig.3.12, we show the observed variation of μ_a for ^{78}Pt as a function of λ of incident X-rays. The discontinuities are called *absorption edges*, and

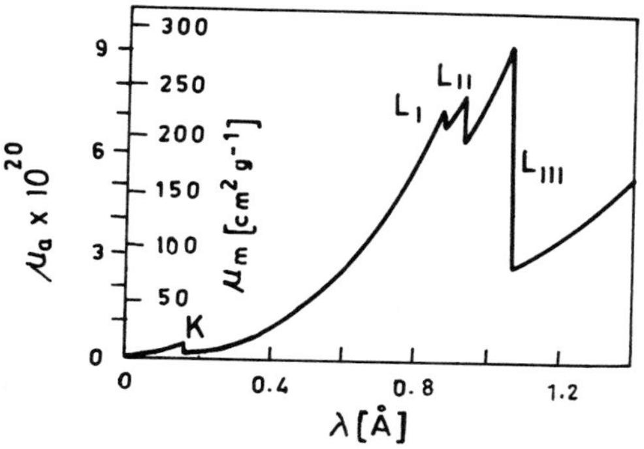

Fig. 3.12. Variation of μ_a with λ for platinum

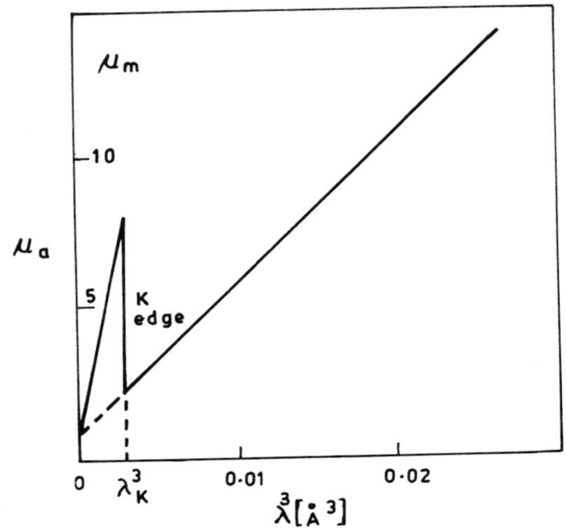

Fig. 3.13. Observed λ^3 dependence of μ_a for lead

occur at wavelengths that are characteristic of the absorbing element. Their presence led BARKLA to suggest, in early days, that they are related to the energies required to knock K, L, M, etc., electrons out of the atom, and labeled the edges accordingly. For this reason, the true atomic absorption coefficient τ_a is called the *photoelectric atomic absorption coefficient*. The gradual rise of the absorption curve is proportional to λ^3. A plot of μ_a as a function of λ^3 gives straight-line sections (Fig.3.13).

When similar measurements are made for other elements, similar curves are obtained, with edges at different characteristic wavelengths. A plot of μ_a against Z^4 gives a linear relationship. This suggests that we can write the observed dependence as

$$\mu_a = \tau_a + \sigma_a = \begin{cases} c_a\lambda^3 Z^4 + \sigma_a & \text{for} \quad \lambda < \lambda_K \\ c_a'\lambda^3 Z^4 + \sigma_a & \text{for} \quad \lambda > \lambda_K \end{cases} , \tag{3.101}$$

where $c_a = 2.25\times 10^{-2}$ cm and $c_a' = 0.32\times 10^{-2}$ cm for λ in cm. We can write for $\lambda < \lambda_K$,

$$\tau_a = \tau_K + \tau_L + \tau_M + \dots = c_a\lambda^3 Z^4 \quad , \tag{3.102}$$

where

$$\tau_K = (c_a - c_a')Z^4\lambda^3 = 1.92\times 10^{-2}\ Z^4\lambda^3 \quad , \tag{3.103}$$

$$\tau_L + \tau_M + \dots = c_a'\lambda^3 Z^4 \quad . \tag{3.104}$$

Thus, the theoretical numerical factor in (3.99) is large by a factor of about four. The reason is that we have neglected the radiation damping effect.

3.8.2 Absorption by a Damped Oscillator

When the radiation damping is included, we have to follow the Lorentz treatment (Sect.3.3). From (3.41,48),

$$(\tau_q) = \frac{2\omega}{c}\beta_q = \frac{4\pi n_q e^2}{mc}\frac{\gamma_\omega \omega^2}{\left(\omega_q^2-\omega^2\right)^2+\gamma_\omega^2\omega^2} \quad , \tag{3.105}$$

where

$$\gamma_\omega = \frac{2e^2}{3mc^3}\omega^2 \quad . \tag{3.106}$$

For ω close to ω_q, we can write $\omega_q^2 - \omega^2 \simeq 2\omega_q(\omega_q-\omega)$ and replace ω by ω_q elsewhere in (3.105), to get

$$(\tau_q) \simeq \frac{6\pi n_q c^2}{\omega_q^2}\frac{1}{1+\left[(\omega_q-\omega)^2/\frac{1}{2}\gamma\right]^2} \quad . \tag{3.107}$$

In view of (3.12), we have replaced γ_ω by γ in this region of ω.

Comparison of (3.107) with (3.25) shows that the improved classical theory leads to a line absorption, and not to a simple λ^3 dependence as given by (3.97) or by the experimental result (3.103). In X-rays, we observe an absorption edge at ω_q and not an absorption line.

This discrepancy between the Lorentz theory and experiments requires a radical modification of the theory. It can be done only in a highly artificial and formal manner, as shown below.

The natural frequency ω_q(q=K,L,M,...) is related to the absorption edges (K,L,M,...). The ejection of an electron from a K, L, M,... shell is accompanied by a transition to a higher energy state. As other inner shells are already full, the most likely transitions are to the *continuum* of states that have positive energies (Fig.3.14). Each such transition corresponds to a possible value of ω_q $(\omega_q,\omega_q',\omega_q'',\ldots,\omega_j\ldots,\infty)$; therefore, we can associate it with a *virtual oscillator*. Obviously, the frequency range of these virtual oscillators should be from ω_q to ∞. This amounts to distributing the z_q electrons of the q-type (originally of frequency ω_q) over *all* frequencies ω_j in the range $\omega_q < \omega_j < \infty$.

Let us assign to each frequency ω_j a certain fraction $f_q(\omega_j)$ of the classical oscillators. It represents the *oscillator strength* (Sect.2.3). To fre-

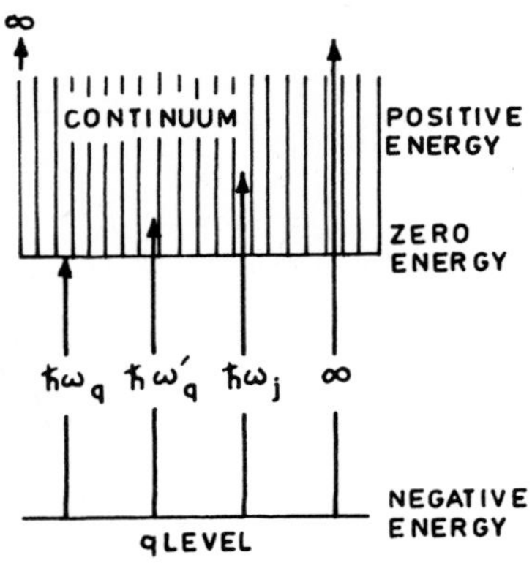

Fig. 3.14. Frequencies of virtual oscillators

quencies ω_j of the continuous spectrum corresponds, more appropriately, a spectral density of oscillator strength, $df_q/d\omega_j$. It satisfies the condition

$$z_q = \int_{\omega_q}^{\infty} f_q(\omega_j)\, d\omega_j \quad , \quad f_q(\omega_j) = \frac{df_q}{d\omega_j} \quad . \tag{3.108}$$

Here $(df_q/d\omega_j)d\omega_j$ is the number of virtual oscillators with frequencies between ω_j and $\omega_j + d\omega_j$.

Let n_q be the number of q-type electrons in 1 cm^3. Because z_q is the number of q-type electrons in one atom, the number of atoms in 1 cm^3 is n_q/z_q. The *partial atomic absorption coefficient* τ_q is defined in terms of the partial linear absorption coefficient (τ_q), (3.105), as

$$\tau_q = \frac{(\tau_q)}{n_q/z_q} = \frac{6\pi c^2}{\omega^2} \frac{z_q}{1+\left[(\omega_q-\omega)^2(\omega_q+\omega)^2/\gamma_\omega^2\omega^2\right]} \quad . \tag{3.109}$$

We use (3.108) for z_q, replace ω_q by ω_j for the virtual oscillators, and integrate from ω_q to ∞ for the spectral continuum,

$$\tau_q = \frac{6\pi c^2}{\omega^2} \int_{\omega_q}^{\infty} \frac{f_q(\omega_j)d\omega_j}{1+\left[(\omega_j-\omega)^2(\omega_j+\omega)^2/\gamma_\omega^2\omega^2\right]} \quad . \tag{3.110}$$

The expression under the integral has a sharp maximum at $\omega_j \approx \omega$. This means that only virtual oscillators of frequencies close to the frequency ω of the incident X-rays will absorb effectively. Therefore, we can replace ω_j by ω everywhere except in the difference $(\omega_j-\omega)$ term. We can also neglect the variation of $f_q(\omega_j)$ over the narrow region in which effective contributions to the integral occur, and merely use $f_q(\omega)$. Thus,

$$\tau_q(\omega) = \frac{6\pi c^2}{\omega^2} f_q(\omega) \int_{\omega_q}^{\infty} \frac{d\omega_j}{1+\left[(\omega_j-\omega)/\frac{1}{2}\gamma_\omega\right]^2}$$

$$= \frac{6\pi c^2}{\omega^2} f_q(\omega)\left[\frac{1}{2}\gamma_\omega\left(\tan^{-1}\frac{\omega_j-\omega}{\frac{1}{2}\gamma_\omega}\right)_{\omega_q}^{\infty}\right]$$

$$= \frac{3\pi c^2\gamma_\omega}{\omega^2} f_q(\omega)\left(\tan^{-1}\infty - \tan^{-1}\frac{\omega_q-\omega}{\frac{1}{2}\gamma_\omega}\right) \quad . \tag{3.111}$$

Because $\omega > \omega_q$ and $\gamma_\omega/2$ is very small, the term $(\omega_q-\omega)/(\gamma_\omega/2)$ will be a very large negative number for all ω not very close to ω_q. In this region, we can put

$$\frac{\omega_q-\omega}{\frac{1}{2}\gamma_\omega} \approx -\infty \quad , \quad \tan^{-1}\frac{\omega_q-\omega}{\frac{1}{2}\gamma_\omega} \approx -\frac{\pi}{2} \quad .$$

Thus, for ω not close to ω_q, (3.111) becomes

$$\tau_q(\omega) = \frac{3\pi^2c^2\gamma_\omega}{\omega^2} f_q(\omega) = \frac{2\pi^2e^2}{mc} f_q(\omega) = 2\pi^2 \, cr_e \, f_q(\omega) \quad , \tag{3.112}$$

where we have used (3.106) for γ_ω. This shows that *the measurement of the photoelectric (true) absorption coefficients provides a direct knowledge of oscillator strengths.*

If we use (3.108) and remember that ω and ω_j are the same, for our purpose here, we get from (3.112),

$$\frac{1}{2\pi^2 cr_e}\int_{\omega_q}^{\infty} \tau_q(\omega)d\omega = \int_{\omega_q}^{\infty} f_q(\omega)d\omega = z_q \quad , \tag{3.113}$$

or

$$\frac{1}{\pi cr_e}\int_{\nu_q}^{\infty} \tau_q \, d\nu = z_q \quad . \tag{3.114}$$

To check the *sum rule* (3.114), we can evaluate the left-hand side either by the graphical integration of the actual observed absorption curve, or we can use the empirical law,

$$\tau_q = c_q Z^4 \lambda^3 \quad , \quad 0 < \lambda < \lambda_q \quad ,$$

Table 3.2. Calculated values of z_K, with $c_K = 1.92 \times 10^{-2}$ cm from experiment

Element	λ_K [Å]	$Z^4\lambda_K^2$	z_K (calculated)
13 Al	7.9481	1.80×10^6	1.96
29 Cu	1.3806	1.35	1.46
42 Mo	0.6198	1.20	1.30
47 Ag	0.4859	1.15	1.25
50 Sr	0.4247	1.13	1.22
79 Au	0.1536	0.92	1.00

where c_q is given by experiments. Both methods give nearly the same result. Therefore, we can write

$$z_q = \frac{c_q Z^4}{\pi c r_e} \int_{\lambda_q}^{0} \lambda^3 \left(-\frac{c}{\lambda^2}\right) d\lambda = \frac{c_q Z^4}{2\pi r_e} \lambda_q^2 \quad , \tag{3.115}$$

where λ is in cm. From the Moseley law, we expect $\lambda_q^2 Z^4$ to be nearly constant. Table 3.2 gives some values of z_K according to (3.115), with $c_K/(2\pi r_e) = 1.92 \times 10^{-2}/(2\pi \times 2.82 \times 10^{-13}) \simeq 10^{10}$; $z_K \simeq Z^4\left[\lambda_K^2 \text{ (in Å)}\right] \times 10^{-6}$. We find that for the K level, the left-hand side of the sum rule (3.114) is consistently much less than the right-hand side value, $z_K = 2$, which is known to be correct from the Pauli principle. Thus, the classical approach needs to be revised [3.16], especially regarding the number of oscillators per cm^3.

From $\tau_q = c_q Z^4 \lambda^3$ and (3.115), we have

$$\tau_q = 2\pi r_e \frac{\lambda^3}{\lambda_q^2} z_q = 4\pi^2 c r_e \frac{\omega_q^2}{\omega^3} z_q \quad . \tag{3.116}$$

Comparison with (3.97) shows that tne inclusion of the effect of radiation damping has decreased the value of τ_q by a factor of two. Even now it is higher than the observed value. The reason is that, in the photoelectric absorption process, the atom gets excited, and the fluorescent radiation that results returns part of the absorbed energy to the electromagnetic field. We have not corrected for this, so far.

Equation (3.116) gives the general law of absorption, obtained in a semiempirical manner, in that we have used the experimental λ^3 law. The atomic absorption coefficient is

$$\tau_a = 2\pi r_e \lambda^3 \sum_q \left(z_q/\lambda_q^2\right) = 1.77 \times 10^{-12} \lambda^3 \sum_q \left(z_q/\lambda_q^2\right) \quad , \tag{3.117}$$

where λ is in cm. The experimental results [3.17] give

$$\tau_a = 1.71 \times 10^{-12} \lambda^3 \sum_q \left(z_q/\lambda_q^2\right) \quad .$$

From (3.112,116), we get for the oscillator-strength density,

$$f_q(\omega) = \frac{1}{2\pi^2 c r_e} \tau_q = 2 \frac{\omega_q^2}{\omega^3} z_q \quad , \quad (\omega \simeq \omega_j) \tag{3.118}$$

$$f_q(\lambda) = - f_q(\omega) \frac{d\omega}{d\lambda} = \frac{2\pi c}{\lambda^2} f_q(\omega) = 2 \frac{\lambda}{\lambda_q^2} z_q \quad , \tag{3.119}$$

where $f_q(\lambda)$ is such that for the continuous absorption

$$\int_0^{\lambda_q} f_q(\lambda) d\lambda = z_q \quad . \tag{3.120}$$

3.9 Kramers-Kallmann-Mark Theory of Refractive Index

From (3.38,106), we obtain

$$\mu_{c,q} = 1 + \frac{2\pi n_q e^2}{m} \frac{1}{\left(\omega_q^2 - \omega^2\right) + i\gamma_\omega \omega} = 1 - \delta_q - i\beta_q \quad . \tag{3.121}$$

The virtual oscillator distribution (3.118), that leads to the correct absorption law, can now be profitably introduced in the form

$$f_q(\omega_j)/z_q = 2\omega_q^2/\omega_j^3 \quad , \quad \omega_q < \omega_j < \infty \quad . \tag{3.122}$$

We have used the subscript j in order to distinguish between the incident radiation frequency ω and the virtual-oscillator frequency ω_j, in the following calculation.

Each electron of the q-type is to be replaced by a distribution of virtual oscillators, (3.122), with frequencies in the range ω_q to ∞ (Fig.3.14). This means that the distribution per electron, $f_q(\omega_j)/z_q$, will now replace unity in the numerator of the second term in (3.121),

$$\mu_{c,q} = 1 + \frac{4\pi e^2 n_q \omega_q^2}{m} \int_{\omega_q}^{\infty} \frac{d\omega_j}{\omega_j^3 \left(\omega_j^2 - \omega^2 + i\gamma_\omega \omega\right)} \quad . \tag{3.123}$$

Integration[1] gives

$$\mu_{c,q} = 1 + \frac{4\pi e^2}{m} n_q\omega_q^2 \left\{ \frac{1}{2\omega(i\gamma_\omega-\omega)} \left[\frac{1}{\omega_q^2} + \frac{1}{\omega(i\gamma_\omega-\omega)} \ln \frac{\omega_q^2}{\omega_q^2-\omega^2+i\gamma_\omega\omega} \right] \right\} \quad . \tag{3.124}$$

Let us put

$$s = \omega/\omega_q \quad , \quad d = \gamma_\omega/\omega_q \quad , \tag{3.125}$$

where d characterizes the damping. Then

$$\delta_q + i\beta_q = \frac{2\pi e^2}{m} \frac{n_q}{\omega_q^2} \frac{\ln(1-s^2+isd)+(s^2-isd)}{(s^2-isd)^2} \quad . \tag{3.126}$$

To separate the real and imaginary parts, we put, for the case $s<1$,

$$\ln(1-s^2+isd) = a + ib \quad ,$$

$$1 - s^2 + isd = e^{a+ib} = e^a \cos b + ie^a \sin b \quad ,$$

$$1 - s^2 = e^a \cos b \quad , \quad sd = e^a \sin b \quad ,$$

$$a = \frac{1}{2} \ln\left[(1-s^2)^2+s^2d^2\right] \quad , \quad b = \tan^{-1} \frac{sd}{1-s^2} = \tan^{-1}y \quad ,$$

$$(s^2-isd)^2 = s^2(s^2-d^2) - i2s^3d \quad .$$

Thus, for $s<1$, or $\omega<\omega_q$, we find,

$$\delta_q + i\beta_q = \frac{2\pi e^2}{m} \frac{n_q}{\omega_q^2} \frac{\left\{\frac{1}{2} \ln\left[(1-s^2)^2+s^2d^2\right]+s^2\right\}-i\left(sd-\tan^{-1}y\right)}{s^2(s^2-d^2)-i2s^3d} \quad ,$$

whence

$$\delta_q = \frac{2\pi e^2}{m} \frac{n_q}{\omega_q^2} \frac{\frac{1}{2}(s^2-d^2) \ln\left[(1-s^2)^2+s^2d^2\right]-2sd \tan^{-1}y+s^2(s^2+d^2)}{s^2(s^2+d^2)^2} \quad ,$$

$$\beta_q = \frac{2\pi e^2}{m} \frac{n_q}{\omega_q^2} \frac{sd \ln\left[(1-s^2)^2+s^2d^2\right]+(s^2-d^2)\tan^{-1}y+sd(s^2+d^2)}{s^2(s^2+d^2)^2} \quad .$$

---

[1] Use $\int \frac{dx}{x^3(a+bx^2)} = \int\left[\frac{1}{ax^3} - \frac{b}{ax(a+bx^2)}\right]dx = -\frac{1}{2ax^2} - \frac{b}{a}\left(\frac{1}{2a} \ln \frac{x^2}{a+bx^2}\right)$.

Because of the complex nature of the logarithm, we get a different result for the case $s > 1$. We now put

$$\ln\left[-(s^2-1-isd)\right] = a + ib \quad ,$$

whence, with $e^{-i\pi} = -1$,

$$s^2 - 1 - isd = e^a e^{i(b-\pi)}$$

$$a = \frac{1}{2}\ln\left[(s^2-1)^2+s^2d^2\right] \quad , \quad b = \pi - \tan^{-1}(-y) \quad , \quad y = \frac{sd}{1-s^2} \quad .$$

Proceeding as before, for $s > 1$, or $\omega > \omega_q$, we find

$$\delta_q = \frac{2\pi e^2}{m}\frac{n_q}{\omega_q^2}\frac{\frac{1}{2}(s^2-d^2)\ln\left[(s^2-1)^2+s^2d^2\right]+2sd\tan^{-1}(-y)+s^2(s^2+d^2)-2\pi sd}{s^2(s^2+d^2)^2}$$

$$\beta_q = \frac{2\pi e^2}{m}\frac{n_q}{\omega_q^2}\frac{sd\ln\left[(s^2-1)^2+s^2d^2\right]-(s^2-d^2)\tan^{-1}(-y)+sd(s^2+d^2)+\pi(s^2-d^2)}{s^2(s^2+d^2)^2} \quad .$$

Because γ_ω is a small quantity, we can neglect d^2. Then

$$\delta_q \simeq \frac{2\pi e^2}{m}\frac{n_q}{\omega_q^2}\frac{1}{s^2}\cdot\begin{cases} 1 + s^{-2}\ln(1-s^2) \quad , & \text{for } s < 1 \quad , \\ 1 + s^{-2}\ln(s^2-1) - s^{-3}\,2\pi d \quad , & \text{for } s > 1 \quad . \end{cases} \tag{3.127}$$

In the limit $s \gg 1$, we get[2] from (3.127)

$$\delta_q \simeq \frac{2\pi e^2}{m}\frac{n_q}{\omega_q^2}\frac{1}{s^2} = \frac{2\pi e^2}{m}\frac{n_q}{\omega^2} \quad , \quad (\omega \gg \omega_q) \quad , \tag{3.128}$$

$$\delta = \sum_q \delta_q \simeq \frac{2\pi e^2}{m\omega^2}\sum_q n_q = \frac{e^2\lambda^2}{2\pi mc^2}n \quad , \quad (\lambda \ll \lambda_q) \quad . \tag{3.129}$$

This agrees with the simple Lorentz-theory result (3.52), as it should, because we are away from the absorption edge and so corrections that arise from the continuum absorption are not important.

---

[2]Use L'Hôpital's rule: Let f and g be differentiable in an open interval and continuous on [a,b]. Assume that $g'(x) \neq 0$ for x in (a,b). Assume also that $\lim_{x\to a+} f(x) = 0 = \lim_{x\to a+} g(x)$, or that $\lim_{x\to a+} f(x) = \infty = \lim_{x\to a+} g(x)$. Then if $\lim[f'(x)/g'(x)]$ exists, so does $\lim_{x\to a+}[f(x)/g(x)]$, and the two limits are equal. The theorem also holds if $\lim_{x\to a+}[f'(x)/g'(x)] = \infty$ or $-\infty$. *Warning:* When applying this rule to f/g do not differentiate as a quotient. Just differentiate f and g individually, form the quotient of derivatives, and then take the limit of this quotient.

Let us neglect damping, d = 0. Then, from (3.127),

$$\delta_q \simeq \frac{2\pi e^2}{m}\frac{n_q}{\omega^2}\begin{cases}1+\left(\omega_q^2/\omega^2\right)\ln\left[1-\left(\omega^2/\omega_q^2\right)\right] & , \quad \text{for } \omega < \omega_q \quad ,\\ 1+\left(\omega_q^2/\omega^2\right)\ln\left[\left(\omega^2/\omega_q^2\right)-1\right] & , \quad \text{for } \omega > \omega_q \quad ,\end{cases} \tag{3.130}$$

whence

$$\frac{\delta}{\lambda^2} = \frac{\sum_q \delta_q}{\lambda^2} = \frac{e^2}{2\pi mc^2}\left(\sum_q n_q + \lambda^2 \sum_q \frac{n_q}{\lambda_q^2}\ln\left|1-\frac{\lambda_q^2}{\lambda^2}\right|\right) \quad . \tag{3.131}$$

For a compound: $n_q = (N_A\rho/M)z_q$, $\sum_q z_q = Z_M$, $\sum_{q\neq K} z_q = Z_M - z_K$. When λ is close to λ_K, we have $\lambda \ll \lambda_{q\neq K}$ for all of the remaining levels and we can set $\left(\lambda^2/\lambda_q^2\right)\ln\left|1-\left(\lambda_q^2/\lambda^2\right)\right| = 0$ for them. Then

$$\begin{aligned}\frac{\delta}{\lambda^2} &= \frac{e^2N_A\rho}{2\pi mMc^2}\left(Z_M + z_K\frac{\lambda^2}{\lambda_K^2}\ln\left|1-\frac{\lambda_K^2}{\lambda^2}\right|\right)\\ &= 2.68\times 10^{-6}\,\frac{\rho}{M}\left(Z_M + 2\frac{\lambda^2}{\lambda_K^2}\ln\left|1-\frac{\lambda_K^2}{\lambda^2}\right|\right) \quad ,\end{aligned} \tag{3.132}$$

where we have used $z_K = 2$ and λ is in Å.

For calcite ($CaCO_3$), we finally get [3.18],

$$\frac{\delta}{\lambda^2} = 0.0732\times 10^{-6}\left(50 + \frac{2\lambda^2}{9.390}\ln\left|1-\frac{9.390}{\lambda^2}\right|\right) \quad . \tag{3.133}$$

This is different from the Lorentz-theory result (3.56). For $\lambda \gg \lambda_K$ and $\lambda \ll \lambda_K$ it has, however, the same limits as the Lorentz theory.

In Fig.3.15, we compare this theory [3.19,20], (3.133), with the data of LARSSON [3.2] on calcite. It is in closer agreement with the data than the simple Lorentz theory, especially for $\lambda \leqq \lambda_K$. For $\lambda \simeq \lambda_K$, we cannot expect good agreement because we have neglected the damping term; it becomes important in this region. At $\lambda \sim \lambda_K/2$, the present theory predicts a broad maximum, unlike the Lorentz theory. LARSSON [3.2] has found evidence for its existence in quartz. We find that the theory of Kramers, Kallmann and Mark succeeds in giving qualitatively the observed shape of the X-ray dispersion curve. For further improvement we have to consider the quantum theory of dispersion.

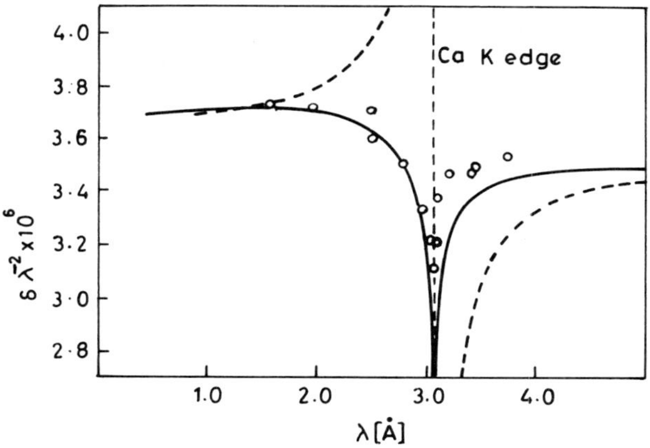

Fig. 3.15. Plot of δ/λ^2 versus λ (Å) for calcite. The solid curve is from the theory of Kramers, Kallmann and Mark (3.133). For comparison, the Lorentz-theory curve (dashed), see Fig.3.4, is also given. The experimental points are from [3.2]

3.10 Oscillator Strength and Quantum Theory of Dispersion

Suppose that $\rho(\omega_q)$ is the energy density of the incident radiation in the neighborhood $\omega = \omega_q$. The definition of the atomic absorption coefficient implies that the energy absorbed per atom in unit time is

$$\left(\frac{dU}{dt}\right)_{\text{classical}} = c\tau_q(\omega)\rho(\omega_q) = \frac{2\pi^2 e^2}{m} f_q(\omega)\rho(\omega_q) \quad , \tag{3.134}$$

where we have used (3.112) for $\tau_q(\omega)$. It represents the *induced* absorption of energy per second by the classical oscillators. This formula was first obtained by LADENBURG [3.16]. If we use $\nu_q = \omega_q/2\pi$, we get

$$\left(\frac{dU}{dt}\right)_{\text{classical}} = \frac{\pi e^2}{m} f_q(\nu)\rho(\nu_q) \quad , \quad f_q(\nu) = \frac{df_q}{d\nu} \quad . \tag{3.135}$$

In the quantum theory, the energy absorbed per atom per second, in the electronic transition $i \to f$, can be expressed as

$$\left(\frac{dU}{dt}\right)_{\text{quantum}} = h\nu_{fi}\, b_{i\to f}\, \rho_{fi} \quad , \tag{3.136}$$

where ν_{fi} is the absorbed frequency, $b_{i\to f}$ is the transition probability for the *induced* absorption, and ρ_{fi} is the density of the incident radiation of frequency ν_{fi}. Once in the excited state f, a *spontaneous* emission of radiation of frequency ν_{fi} occurs because of the electronic transition $f \to i$.

If N_i atoms per cm^3 are in the initial quantized state i and N_f in the final state f, then, according to Einstein, at equilibrium,

$$N_i h\nu_i b_{i\to f}\rho_{fi} = N_f h\nu_{fi}(a_{f\to i} + b_{f\to i}\rho_{fi}) \quad . \tag{3.137}$$

Here $a_{f\to i}$ is the probability for a spontaneous-emission transition $f \to i$, and $b_{f\to i}$ is the probability of an induced-emission transition in the presence of the radiation. N_i and N_f are related by the Boltzmann factor

$$N_i = N_f\, e^{(E_f - E_i)/kT} = N_f\, e^{h\nu_{fi}/kT} \quad ,$$

where k = Boltzmann constant and T = temperature ($^\circ$K). From (3.137),

$$\rho_{fi} = \frac{a_{f\to i}}{b_{i\to f}\, e^{h\nu_{fi}/kT} - b_{f\to i}} \quad .$$

Comparison with Planck's law for the blackbody-radiation density,

$$\rho_\nu = \frac{8\pi h\nu^3\, d\nu}{c^3(e^{h\nu/kT} - 1)} \quad ,$$

reveals that

$$b_{i\to f} = b_{f\to i} \quad , \quad a_{f\to i} = \frac{8\pi h\nu_{fi}^3}{c^3}\, b_{f\to i} \quad . \tag{3.138}$$

Therefore

$$\left(\frac{dU}{dt}\right)_{\text{quantum}} = \frac{c^3}{8\pi\nu_{fi}^2}\, a_{f\to i}\rho_{fi} \quad . \tag{3.139}$$

By the correspondence principle, we can equate (3.135,139) to get

$$f_q(\nu) = \frac{mc^3}{8\pi^2 e^2 \nu_{fi}^2}\, a_{f\to i} = f_{i\to f} \quad . \tag{3.140}$$

Here we have identified ν_q with ν_{fi}, $\rho(\nu_q)$ with ρ_{fi}, and $f_{i\to f}$ is analogous to the classical-oscillator strength $f_q(\nu)$.

From (2.25), for a single oscillator,

$$\frac{dU}{dt} = \frac{64\pi^4 e^2}{3c^3}\, \nu_{fi}^4\, r^2(i,f) = h\nu_{fi} a_{f\to i} \quad . \tag{3.141}$$

Introducing $f_{i\to f}$ from (3.140), we obtain

$$f_{i\to f} = \frac{8\pi^2 m}{3h}\,\nu_{fi}\left(|x_{i\to f}|^2+|y_{i\to f}|^2+|z_{i\to f}|^2\right) \quad . \tag{3.142}$$

If the transitions involve the continuum, we can write

$$\frac{df_{i\to f}}{dE} = \frac{8\pi^2 m}{3h}\,\nu_{fi}\,\frac{d}{dE}\left(|x_{i\to f}|^2+|y_{i\to f}|^2+|z_{i\to f}|^2\right) \quad , \tag{3.143}$$

where $df_{i\to f}/dE$ is the *oscillator-strength density* of the atom, in the energy interval dE at ν_{fi}, for the absorption transition from state i to state f.

Quantum theory defines the ν_{fi} and $f_{i\to f}$ in terms of the eigenvalues and eigenfunctions of the Schrödinger equation. Calculation of $f_{i\to f}$ is a well-defined mathematical task. However, it can be accomplished only approximately for many-electron atoms. Analytical solutions are possible when the one-electron approximation is made.

The ground-state wave function of the hydrogen atom is spherically symmetric. Therefore, the dipole-matrix elements are equal[3], $|x_{i\to f}|^2 = |y_{i\to f}|^2 = |z_{i\to f}|^2$, and we can write (3.142) as

$$f_{i\to f} = \frac{2m}{\hbar^2}\,(E_f-E_i)\,|(f|x|i)|^2 \quad . \tag{3.144}$$

Equation (3.141) determines intensity and so will get contributions from all of the final states. KUHN [3.21] and THOMAS [3.22a,b] showed (see Appendix G) that, for a one-electron system

$$\frac{2m}{\hbar^2}\sum_f (E_f-E_i)\,|(f|x|i)|^2 \equiv \sum_f f_{i\to f} = 1 \quad . \tag{3.145}$$

For an atom that contains Z electrons, we can generalize this *sum rule* to

$$\sum_f f_{i\to f} = Z \quad . \tag{3.146}$$

When both discrete and continuous absorption are present,

$$\sum_f^{\text{discrete}} f_{i\to f} + \int_{\substack{\text{ionization}\\ \text{threshold}}}^{\infty} (df_{i\to f}/dE)\,dE = Z \quad . \tag{3.147}$$

Broad spectral data are now used to investigate these sum rules [3.23-26].

[3] $|(f|x|i)|^2 = (i|x|f)(f|x|i) = (i|x^2|i) = \int|\psi_i|^2x^2dx$, because $|f\rangle\langle f| = \int|\psi_f|^2\,dx = 1$.

Table 3.3. Oscillator strengths in the Lyman series

n	Term		$f_{1\to n}$
2	Lyman α		0.4162
3	Lyman β		0.0791
4	.		0.0290
5	.		0.0139
6	.		0.0078
7	.		0.0048
8	.		0.0032
9	.		0.0022
$\sum_{10}^{\infty}$	10th Lyman line to series limit		0.0079
			0.5641
$\int_{\text{threshold}}^{\infty} df$	continuous spectrum		0.437
$\sum + \int$	line + continuous	Sum	1.0011

SUGIURA [3.27] has calculated the oscillator strength for the normal state of hydrogen as the initial state, and checked (3.147) (Table 3.3). For the (optical) Lyman series we need $|\int \psi_{n10}\, x\, \psi_{100}|^2$. This requires evaluation of the radial integral $\int^{\infty} r^2\, dr\, R_{n10}(r)\, r\, R_{100}(r)$, where the radial wave functions are known for the hydrogen atom. The result is (for example, [3.23])

$$|(n10)|x|100|^2 = \frac{1}{3}\,\frac{2^8 n^7 (n-1)^{2n-5}}{(n+1)^{2n+5}}\, a_0^2 \quad , \quad \text{(Lyman series)} \quad . \tag{3.148}$$

An estimate for $n = 2$ gives

$$f_{1\to 2} = \frac{2mc^2}{(\hbar c)^2}\,(E_2 - E_1)\,\frac{1}{3}\,\frac{2^{15}}{3^9}\,a_0^2$$

$$= \frac{2 \times 0.51\ \text{MeV}}{(197.3\times 10^{-5})^2 [\text{MeV}]^2 [\text{Å}]^2}\,(10.18\,[\text{eV}])\,\frac{1}{3}\,\frac{2^{15}}{3^9}\,(0.53)^2 [\text{Å}]^2 = 0.4162 \quad .$$

The strengths $f_{1\to n}$ of discrete transitions can be shown as a histogram, that is, as a set of rectangular blocks whose respective areas are equal to the $f_{1\to n}$ and whose bases correspond to the energies E_n. Figure 3.16 shows the positions and oscillator strengths of the discrete spectrum of hydrogen, together with the adjoining positive-energy continuum, in a manner applicable to the Lyman series of other atoms. Plot each energy level E_n (E_2=3.4 eV, E_3=1.51 eV,...) against the principal quantum number n of the corresponding

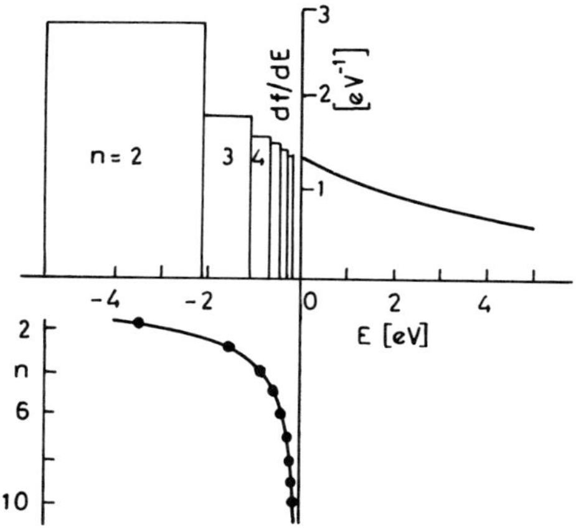

Fig. 3.16. Oscillator-strength distribution in the bound states and part of the continuum of the hydrogen atom [3.26]

line in the Lyman series and draw a smooth curve through these points. The base of the histogram block that represents the oscillator strength $f_{1\to n}$ should equal the slope of the curve, dE_n/dn, at n. The top of the histogram should then equal $f_{1\to n}(dn/dE_n)$, to give the area as $f_{1\to n}$. These tops form a staircase that constitutes an extrapolation of the positive-energy continuous spectrum. Because $(df/dE)_K$ is proportional to $\tau_K(E)$, such a plot represents the absorption spectrum.

Calculation of the continuous-spectrum contribution (Table 3.3) is rather involved, because it requires the construction of the correct positive-energy function in the Coulomb field. SUGIURA [3.27] showed that, for a single electron bound in a Coulomb field in its ground state, the oscillator strength density $df/d\omega$, for frequency ω in the K oscillator continuum, is given by (see Appendix H),

$$d\left(|x_{i\to f}|^2+|y_{i\to f}|^2+|z_{i\to f}|^2\right) = 2^7 a_0^2 s^{-5} g(s)dE \quad , \tag{3.149}$$

$$\left(\frac{df}{d\omega}\right)_K d\omega = \frac{2^7}{3}\,\frac{g(s)}{s^4}\,\omega_K^{-1} \quad , \tag{3.150}$$

where we have used (3.143), $s=\omega/\omega_K$, ω_K being the K absorption frequency, and

$$g(s) = \exp\left[-\frac{4}{(s-1)^{\frac{1}{2}}}\tan^{-1}(s-1)^{\frac{1}{2}}\right]\left(1-e^{-2\pi/(s-1)^{\frac{1}{2}}}\right)^{-1} \quad . \tag{3.151}$$

By graphical integration over the K continuum, from $s=1$ to $s=\infty$, he found a value 0.437 for the oscillator strength (Table 3.3).

The case of atoms with more than one electron has been discussed by KRONIG and KRAMERS [3.28]. We have seen that a hydrogenlike (one-hole) model is reasonable for the X-ray spectra. For the X-ray K absorption, the K electrons cannot go to L, M,... levels, because they are full. Therefore, for the X-ray case, the oscillator strength for a K electron is more nearly that for the continuous absorption in the hydrogen Lyman series, $f(\mathrm{cont})_K = 0.437$. This is considerably less than unity. For $N_i = 2$, the theoretical total oscillator strength for the K level in the one-electron approximation is then nearly 0.88.

The experimental value of $f(\mathrm{cont})_K$ can be estimated with the help of the empirical formula for the atomic absorption coefficient,

$$\begin{aligned} \tau_K(\omega) &= C_K\,\omega^{-n} = (\omega_K/\omega)^n \tau_K(\omega_K) \quad , \quad \text{for } \omega > \omega_K \quad , \\ \tau_K(\omega) &= 0 \quad , \quad \text{for } \omega < \omega_K \quad , \end{aligned} \tag{3.152}$$

$$\int \frac{df_K}{d\omega}\,d\omega = f(\mathrm{cont})_K = \frac{mcC_K}{2\pi^2e^2}\int_{\omega_K}^{\infty}\frac{d\omega}{\omega^n} = \frac{mc}{2\pi^2e^2}\,\frac{\omega_K}{n-1}\,\tau_K(\omega_K) \quad , \tag{3.153}$$

where we have used (3.112). From the knowledge of n and $\tau_K(\omega_K)$ from the experiments, we can estimate $f(\mathrm{cont})_K$ (Table 3.4). For ^{42}Mo, this value is 1.25. In fact, it is consistently higher than the theoretical value 0.88 obtained above. This shows that the approximation of an atom with one electron (hole) is not satisfactory. The effect of screening must be included. WOLF [3.29] has made calculations for the L-shell electrons.

HÖNL [3.30a,b] has modified Sugiura's quantum-mechanical calculation to include the effect of screening. According to HÖNL, we can write (3.150,151), with sufficient accuracy within the limits $\omega_K < \omega_j < 4\omega_K$, as

$$g(s) \simeq 3^{-1}\exp(-4)(4s-1) \quad , \tag{3.154}$$

Table 3.4. Oscillator strengths according to Hönl's theory

Element	λ_K [Å]	$1-\Delta_K$	$f(\mathrm{cont})_K$	
			Theoretical	Experimental
13 Al	7.94	0.710	1.53	1.40
29 Cu	1.38	0.795	1.31	1.35
42 Mo	0.618	0.818	1.24	1.25
50 Sn	0.424	0.844	1.18	1.15
74 W	0.178	0.857	1.15	-
79 Au	0.153	0.890	1.07	1.02

$$\left(\frac{df}{d\omega_j}\right)_K \simeq \frac{2^7}{9} \exp(-4) \frac{\omega_K^2}{\omega_j^4} (4\omega_j-\omega_K) \quad , \tag{3.155}$$

whence

$$f(\text{cont})_K = \int_{\omega_K}^{\infty} \left(\frac{df}{d\omega_j}\right)_K d\omega_j = \frac{2^7}{9} \exp(-4)\omega_K^2\left(\frac{2}{\omega_K^2} - \frac{1}{3\omega_K^2}\right) = 0.434 \quad .$$

The screening constants are taken to be equal, $\sigma_1 = \sigma_2 = 0.3$. For $n = 1$, $j = 1/2$, (2.59) gives E_K,

$$W_K = E_K - \varepsilon = R_\infty hc\left[(Z-\sigma)^2+(\alpha^2/4)(Z-\sigma)^4\right] - \varepsilon \quad . \tag{3.156}$$

The small correction ε takes care of the effect of the external electrons on the K level. We can write,

$$\hbar\Omega_K = W_K = E_K(1-\Delta_K) = \hbar\omega_K(1-\Delta_K) \quad ,$$

where

$$\begin{aligned}\Delta_K &= \varepsilon/E_K = 1 - (W_K/E_K) = 1 - (\hbar\Omega_K/E_K) \\ &= 1 - \frac{1/\lambda_K}{R_\infty\left[(Z-\sigma)^2+(\alpha^2/4)(Z-\sigma)^4\right]} \\ &= 1 - \frac{911/\lambda_K}{(Z-0.3)^2+1.33\times10^{-5}(Z-0.3)^4} \quad ,\end{aligned} \tag{3.157}$$

with λ_K in Å. Calculated values are given in Table 3.4.

Because $\omega_K = \Omega_K/(1-\Delta_K)$, we can write for any atom

$$f_K(\omega_j) = \left(\frac{df}{d\omega_j}\right)_K = \frac{2^7}{9} \exp(-4) z_K \frac{\Omega_K^2}{(1-\Delta_K)^2\omega_j^4}\left(4\omega_j - \frac{\Omega_K}{1-\Delta_K}\right) \quad , \tag{3.158}$$

$$f(\text{cont})_K = \int_{\Omega_K}^{\infty} \left(\frac{df}{d\omega_j}\right)_K d\omega_j = \frac{2^8}{9} \exp(-4)\left[\frac{2}{(1-\Delta_K)^2} - \frac{1}{3(1-\Delta_K)^3}\right] \quad , \tag{3.159}$$

where we have put $z_K = 2$. Values calculated from this formula by HÖNL are given in Table 3.4. They are in good agreement with the estimated experimental values.

WHEELER and BEARDEN [3.31] have evaluated $f(\mathrm{cont})_K$ in a simple way. Let p be an occupied level of an atom, other than K. Then, by the sum rule, for the K electrons,

$$f(\mathrm{cont})_K = 2\left(1 - \sum_p^{\mathrm{discrete}} f_{K\to p}\right) \quad , \tag{3.160}$$

where $f_{K\to p}$ is the oscillator strength of the virtual oscillator associated with the transition $K\to p$. The sum we require now extends over the relatively few occupied quantized states, instead of the continuum of the Hönl method. We can estimate $f_{K\to p}$ from (3.144), by the use of Hartree wave functions. The calculated values are very close to those given by Hönl's method. In recent years, more realistic wave functions, and atomic wave functions tabulated by HERMAN and SKILLMAN [3.32], have been used (for details and references see [3.26]). LASSETTRE and FRANCIS [3.33] have defined a *generalized oscillator strength*. It depends on both the momentum transferred to the oscillator and the energy involved.

HÖNL developed a quantum dispersion theory by the use of (3.158) for $f_q(\omega_j)$. From (3.122,123),

$$\mu_{c,q} = 1 + \frac{2\pi e^2 n_q}{m z_q}\int_{\omega_q}^{\infty} f(\omega_j)\,\frac{1}{\omega_j^2-\omega^2+i\gamma_\omega\omega}\,d\omega_j = 1 - \delta_q - i\beta_q \quad , \tag{3.161}$$

$$\delta_q = -\frac{2\pi e^2 n}{m}\int_{\omega_q}^{\infty} f(\omega_j)\,\frac{\left(\omega_j^2-\omega^2\right)d\omega_j}{\left(\omega_j^2-\omega^2\right)^2+\gamma_\omega^2\omega^2} \simeq -\frac{2\pi e^2 n}{m}\int_{\omega_q}^{\infty} f(\omega_j)\,\frac{1}{\omega_j^2-\omega^2}\,d\omega_j \ , \tag{3.162}$$

where, in (3.162), $n = n_q/z_q$ is the number of atoms per unit volume of the medium that contains the q electrons in question, and γ_ω^2 is a small term.

From (3.158,162), with $z_K = 2$,

$$\delta_K = -\frac{2\pi e^2 n}{m}\,\frac{2^8\exp(-4)\Omega_K^2}{9(1-\Delta_K)^2}\left[4\int_{\Omega_K}^{\infty}\frac{d\omega_j}{\omega_j^3\left(\omega_j^2-\omega^2\right)} - \frac{\Omega_K}{1-\Delta_K}\int_{\Omega_K}^{\infty}\frac{d\omega_j}{\omega_j^4\left(\omega_j^2-\omega^2\right)}\right]$$

$$= \frac{e^2 n}{2\pi m(c/\lambda)^2}\,\frac{2^8\exp(-4)}{9}\left[\frac{2}{(1-\Delta_K)^2}\left(\frac{1}{\Lambda_K^2}\ln|\Lambda_K^2-1|+1\right)\right.$$

$$\left. -\frac{1}{(1-\Delta_K)^3}\left(\frac{1}{2\Lambda_K^3}\ln\left|\frac{\Lambda_K-1}{\Lambda_K+1}\right| + \frac{1}{\Lambda_K^2}+\frac{1}{3}\right)\right] \quad , \tag{3.163}$$

where $\Lambda_K = \omega/\omega_K = \lambda_K/\lambda$, and we have used $x^{-m}(a+bx^n)^{-p-1} = (1/a)x^{-m}(a+bx^n)^{-p} - (b/a)\,x^{-m+n}(a+bx^n)^{-p-1}$ to simplify the second integration (also see p.150).

As $\Lambda_K \to \infty$, or $\lambda \to 0$, the expression has the limit

$$\left(\frac{\delta_K}{\lambda^2}\right)_{\lambda\to 0} = \frac{e^2 n}{2\pi mc^2}\,\frac{2^8\,\exp(-4)}{9}\left[\frac{2}{(1-\Delta_K)^2} - \frac{1}{3}\,\frac{1}{(1-\Delta_K)^3}\right] \quad . \tag{3.164}$$

Comparison with (3.159) gives

$$\left(\frac{\delta_K}{\lambda^2}\right)_{\lambda\to 0} = \frac{e^2 n}{2\pi mc^2}\, f(\text{cont})_K \quad . \tag{3.165}$$

The earlier calculation, (3.130), gives

$$\left(\frac{\delta_K}{\lambda^2}\right)_{\lambda\to 0} = \frac{e^2 n}{2\pi mc^2}\, z_K \quad , \quad z_K = 2 \quad . \tag{3.166}$$

When $\Lambda_K \to 0$, or $\lambda \to \infty$, the application of the L'Hôpital rule to (3.163) gives

$$\left(\delta_K/\lambda^2\right)_{\lambda\to\infty} = 0 \quad . \tag{3.167}$$

Thus, for $\lambda \gg \lambda_K$, the dispersion of K electrons is cut out.

For calcite ($CaCO_3$), from (3.157,159),

$$1 - \Delta_K = 0.760 \quad , \quad f(\text{cont})_K = 1.41 \quad . \tag{3.168}$$

When $\lambda \simeq \lambda_K$, we have $\lambda \ll \lambda_q$ for $q \neq K$ (rest of the levels). Thus, for all the remaining levels $\Lambda_K \gg 1$, and for $\lambda \simeq \lambda_K$ we can write

$$\frac{\delta}{\lambda^2} = \frac{\sum_q \delta_q}{\lambda^2} = \frac{\delta_K}{\lambda^2} + \frac{1}{\lambda^2}\sum_{q\neq K}\delta_q = \frac{\delta_K}{\lambda^2} + \frac{e^2 N_A \rho}{2\pi m M c^2}\sum_{q\neq K} f_q \quad . \tag{3.169}$$

For $\lambda \to 0$, we get

$$\left(\frac{\delta}{\lambda^2}\right)_{\lambda\to 0} = \frac{e^2 N_A \rho}{2\pi m M c^2}\sum_q f_q = \frac{e^2 N_A \rho}{2\pi m M c^2}\, Z_M \quad , \tag{3.170}$$

where the sum is over all the atoms of the molecule, and $\sum_q f_q = Z_M$, even though $f_K < z_K$. The reason for this last statement is that when a K electron is ejected in an absorption process, an electron from a higher level can fill the vacancy and absorb at the K level. Thus, a transfer of part of the oscillator strength of other levels to the K level occurs. The total oscillator strength will then be equal to Z for the whole atom, and Z_M for the molecule.

For $\lambda >> \lambda_K$, the dispersion of the K electrons is cut out. Therefore,

$$\left(\frac{\delta}{\lambda^2}\right)_{\lambda\to\infty} = \frac{e^2 N_A \rho}{2\pi m M c^2}(Z_M - f_K) \quad . \tag{3.171}$$

For $\lambda \simeq \lambda_K$, we can use (3.169) with $\sum_{q\neq K} f_q = Z_M - f_K$, and (3.163) for δ_K/λ^2 together with (3.164,170), to get

$$F \equiv \frac{2\pi m M c^2}{e^2 N_A \rho}\frac{\delta}{\lambda^2} - Z_M = \frac{2^8 \exp(-4)}{9}\left[\frac{2}{(1-\Delta_K)^2}\left(\frac{1}{\Lambda_K^2}\ln|\Lambda_K^2 - 1|\right)\right.$$
$$\left. - \frac{1}{(1-\Delta_K)^3}\left(\frac{1}{2\Lambda_K^3}\ln\left|\frac{\Lambda_K - 1}{\Lambda_K + 1}\right| + \frac{1}{\Lambda_K^2}\right)\right]$$
$$= 1.82\frac{1}{\Lambda_K^2}\ln|\Lambda_K^2 - 1| - 0.60\frac{1}{\Lambda_K^3}\ln\left|\frac{\Lambda_K - 1}{\Lambda_K + 1}\right| - 1.20\frac{1}{\Lambda_K^2} \quad . \tag{3.172}$$

Fig.3.17 shows that this result of HÖNL agrees well with data on both sides of the K edge.

PARRATT and HEMPSTEAD [3.34] have calculated the dispersion of copper for the wide range $0.1 < \lambda < 100$ Å (Fig.3.18). It shows that near the absorption edges, the dispersion is essentially determined by the contribution of oscillator strengths of only one electron shell corresponding to the edge in question. Thus, Fig.3.18 reveals that normal dispersion exists only in the region $\lambda << \lambda_K$. Anomalous dispersion prevails in the rest of the spectrum.

3.11 Quantum Theory of Line Shape and Photoabsorption Curve Shape

Let w_{fi} be the transition probability per unit time, of the transition $i \to f$. The total probability per unit that the atom will leave the state i in some way and make a transition to some other state f is

$$P_i = \sum_f w_{fi} \quad . \tag{3.173}$$

When P_i is large, the time for which the atom exists in the excited state i is small. We define

$$\tau_i = 1/P_i \tag{3.174}$$

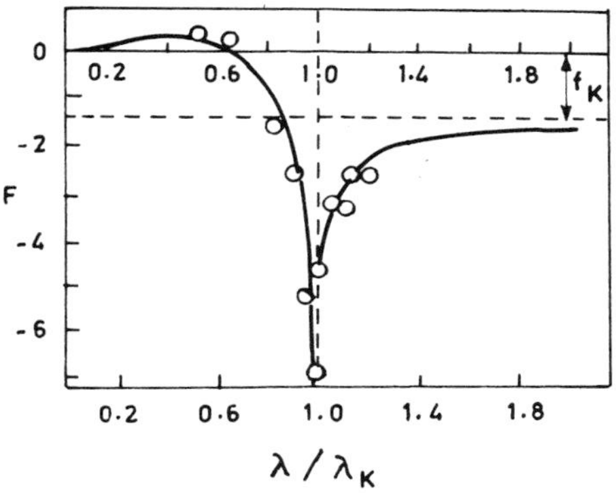

Fig. 3.17. Comparison of Hönl's dispersion theory with the data on calcite

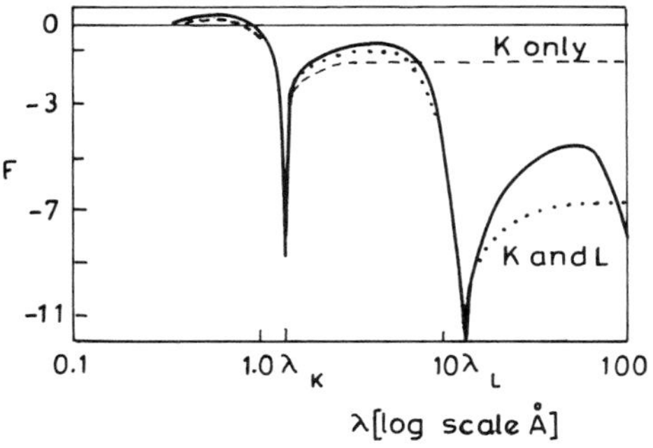

Fig. 3.18. Theoretical plot of the dispersion of copper against λ [Å] on a log scale. The result of taking only the K electrons into account is shown by the dashed curve; the result of considering both the K and L electrons is shown by the dotted curve

as the *effective lifetime* of the state i of the atom. From (3.32) we see that with the limited time τ_i there is associated some indeterminacy of the energy of the level i,

$$\Delta E_i \propto 1/\tau_i = P_i \quad . \tag{3.175}$$

This indeterminacy has a statistical meaning: The range of energy values ΔE_i at E_i will result when one atom is excited numerous times to the level i, or when many atoms are simultaneously excited to the level i.

Figure 2.1 (or Fig.2.12) shows that the closer the i shell is to the nucleus of the atom, the greater are the number of possible transitions $i \rightarrow f$, and so the larger is the value of P_i. We expect $\Delta E_K > \Delta E_L > \Delta E_M$, etc. The outer optical level f can, therefore, be taken to be sharp. For this reason, it is sometimes called the *resonance level*, and the linewidth and shape are comple-

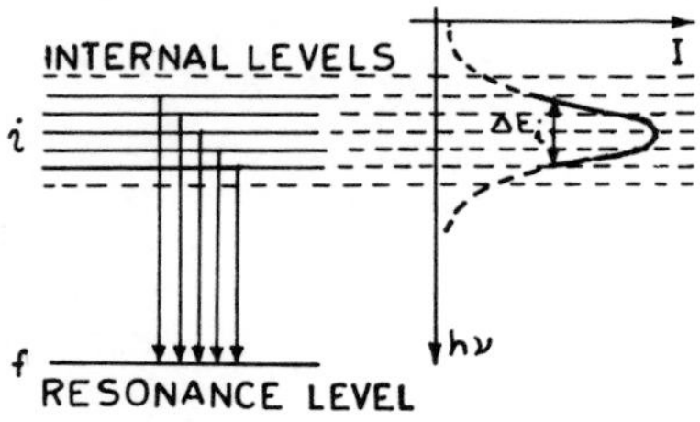

Fig. 3.19. Schematic diagram for the shape of a level

tely determined by the width and shape of the internal level i for the transition $i \rightarrow f$ (Fig.3.19).

We can find the integral intensity of the line, I_i, from (3.24). Because it is symmetrical in ω_0,

$$I_i = 2\int_{\omega_0}^{\infty} I_\omega \, d\omega = 2I_{\omega_0}\int_{\omega_0}^{\infty} \frac{d\omega}{1+\left(\frac{\omega-\omega_0}{\gamma/2}\right)^2}$$

$$= 2I_{\omega_0}\frac{1}{2}\gamma\int_0^{\infty}\frac{dx}{1+x^2} = \frac{1}{2}\pi\gamma I_{\omega_0} \quad . \tag{3.176}$$

Replacing I_{ω_0} by $(2/\pi\gamma)I_i$ in (3.24), we get the dispersion formula,

$$I_\omega \, d\omega = \frac{2}{\pi\gamma}\frac{I_i}{1+\left(\frac{\omega-\omega_0}{\gamma/2}\right)^2}\, d\omega \quad . \tag{3.177}$$

It gives the Lorentzian line shape.

By the correspondence principle, the probability $w(\omega)d\omega$ in the quantum theory, that the emitted photon of energy $\hbar\omega$ has frequency in the range ω a $\omega + d\omega$, is given by the ratio $I_\omega \, d\omega/I_i$. Thus,

$$w(E)dE = w(\omega)d\omega = \frac{1}{\hbar}w(\omega)dE = \frac{\gamma_i}{2\pi\hbar\left[\left(\frac{1}{2}\gamma_i\right)^2 + \frac{1}{\hbar^2}(E-E_i)^2\right]}\, dE \quad , \tag{3.178}$$

where γ_i, according to (3.31,174), is related to P_i,

$$\gamma_i = 1/\tau_i = P_i = \sum_f p_{fi} \quad . \tag{3.179}$$

By (3.27), γ_i characterizes the width of the level i,

$$\gamma_i = \Delta\omega = (\Delta E)/\hbar \quad , \tag{3.180}$$

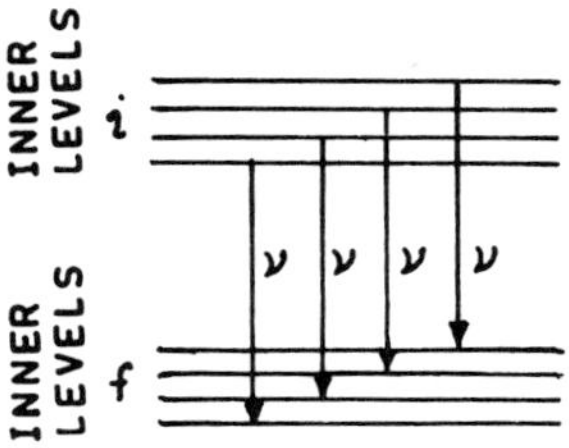

Fig. 3.20. Schematic diagram for the shape of a spectral line

where ΔE is the total width of the level at one-half of the maximum value of $w(E)$.

In the X-ray emission process $i \to f$, both the i and f levels (hole states) are inner levels. Therefore, the width of the line will depend on the sum of the widths of both the initial and final states.

If the energy of the level i is between E and $E + dE$ and of the level f between E' and $E' + dE'$ after the transition, then the energy $\hbar\omega$ of the emitted photon will lie within the limits of $\hbar\omega = E - E' - dE'$ to $\hbar(\omega + d\omega) = E - E' + dE$. This means

$$\hbar d\omega = dE + dE' \quad . \tag{3.181}$$

The line whose frequency lies between ω and $\omega + d\omega$ will arise when various transitions between the i and f levels occur (Fig.3.20). The probability for such a complex process can be expressed as

$$w(E)w(E')dEdE' = \frac{\gamma_i\gamma_f}{(2\pi\hbar)^2}\,\frac{dEdE'}{\left[\left(\frac{1}{2}\gamma_i\right)^2 + \frac{1}{\hbar^2}(E-E_i)^2\right]\left[\left(\frac{1}{2}\gamma_f\right)^2 + \frac{1}{\hbar^2}(E'-E_f)^2\right]} \quad .$$

We can put $E' = E - \hbar\omega$ and integrate over all possible energies of the level i, that is, over E. As before, this gives the ratio $I_\omega\, d\omega/I_i$,

$$\frac{I_\omega d\omega}{I_i} = \int_{-\infty}^{+\infty} w(E)w(E-\hbar\omega)dE\ dE' = \frac{b_i b_f}{\pi^2}\,da_0 \int_{-\infty}^{+\infty} \frac{da}{\left(a^2+b_i^2\right)\left[(a-a_0)^2+b_f^2\right]}$$

$$= \frac{b_i b_f}{\pi^2}\,da_0 \int_{-\infty}^{+\infty} \frac{da}{(a+ib_i)(a-ib_i)(a-a_0+ib_f)(a-a_0-ib_f)} \quad ,$$

where we have put

$$a = (E-E_i)/\hbar \quad , \quad b_i = \gamma_i/2 \quad ,$$

$$a_0 = [\hbar\omega-(E_i-E_f)]/\hbar \quad , \quad b_f = \gamma_f/2 \quad ,$$
$$\hbar\omega_{fi} = E_i - E_f \quad , \quad da_0 = d(\omega-\omega_{fi}) = d\omega = (dE')/\hbar \quad .$$

We can integrate along the real axis with poles at $a = ib_i$ and $a = a_0 + ib_f$. The result is

$$\frac{I_\omega d\omega}{I_i} = \frac{b_i b_f}{\pi^2} da_0 2\pi i\left\{\left[\frac{1}{(a+ib_i)(a-a_0+ib_f)(a-a_0-ib_f)}\right]_{a=ib_i}\right.$$
$$\left. + \left[\frac{1}{(a+ib_i)(a-ib_i)(a-a_0+ib_f)}\right]_{a=a_0+ib_f}\right\}$$
$$= \frac{b_i+b_f}{\pi} \frac{da_0}{(b_i+b_f)^2+a_0^2} = \frac{\gamma_i+\gamma_f}{2\pi} \frac{d\omega}{\left(\frac{1}{2}\gamma_i + \frac{1}{2}\gamma_f\right)^2+(\omega-\omega_{fi})^2} \quad .$$

Writing the total linewidth as

$$\gamma_{if} = \gamma_i + \gamma_f = \hbar^{-1}[(\Delta E)_i+(\Delta E)_f] \quad , \tag{3.182}$$

we get

$$I_\omega d\omega = \frac{(2\pi)^{-1}\gamma_{if}I_i}{\left(\frac{1}{2}\gamma_{if}\right)^2+(\omega-\omega_{fi})^2} d\omega \quad . \tag{3.183}$$

This has the same form as the classical dispersion formula (3.177). These results were first obtained by WEISSKOPF and WIGNER [3.35].

We can now discuss the quantum theory of the shape of the *absorption discontinuity* when the electron is removed from an inner level i to the continuous-energy region, f, of an atom (photoelectric absorption). The dispersed shape of the inner level is given by (3.178)

$$w(E_i)dE_i = \frac{\Gamma_i/2\pi}{\left(\frac{1}{2}\Gamma_i\right)^2+\left(E_i-E_{i_0}\right)^2} dE_i \quad , \tag{3.184}$$

where E_{i_0} is the energy of the distribution maximum, and Γ_i is defined by

$$\gamma_i = (\Delta E)_i/\hbar = \Gamma_i/\hbar \quad . \tag{3.185}$$

Let E_{f_0} denote the beginning of the continuous region of positive energies. This region can be regarded as a closely spaced set of resonance levels f. The

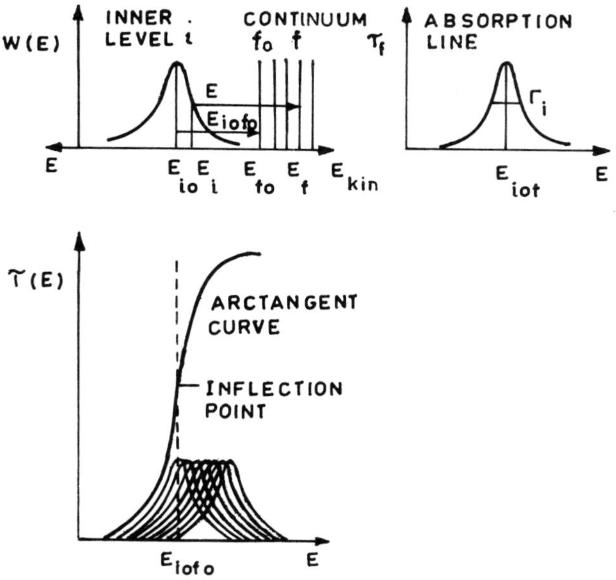

Fig. 3.21. (Top left,a) Absorption of a photon of energy E by a resonance level f. (Top right,b) τ_f for the single resonance level f. (Bottom,c) Total absorption coefficient $\tau(E)$ for the entire continuum above E_{f_0}

threshold energy of the absorbed (incident) photon is $E = E_i + E_{f_0}$. If the photon has more energy than this, the electron will make a transition from the inner level E_i to one of the resonance levels E_f of the continuous set that extends beyond E_{f_0} (Fig.3.21a). Then

$$E = \hbar\omega = E_i + E_f \quad . \tag{3.186}$$

The partial absorption coefficient $\tau_f(E)$ of the f resonance level, when the initial state has the distribution (3.184), is given by (Fig.3.21b)

$$\tau_f(E) = C\,w(E_i) = C\,w(E-E_f) = C\,\frac{\Gamma_i/2\pi}{\left(\frac{1}{2}\Gamma_i\right)^2+\left(E-E_{i_0f}\right)^2} \quad , \tag{3.187}$$

where C is a constant and

$$E_{i_0f} = E_{i_0} + E_f \quad . \tag{3.188}$$

Each resonance level f will give an absorption line of the shape (3.187), Fig.3.21b, with the same width Γ_i and constant C. The maxima of these lines will correspond to the energies E_{i_0f}, that have the range $E_{i_0f_0} \leqq E_{i_0f} \leqq \infty$, where $E_{i_0f_0} = E_{i_0} + E_{f_0}$. The set of these lines is shown in Fig.3.21c.

The total absorption coefficient $\tau(E)$ is obtained by integrating (3.187) over all possible values of E_{i_0f},

$$\tau(E) = \frac{C\Gamma_i}{2\pi} \int\limits_{E_{i_0f_0}}^{\infty} \frac{dE_{i_0f}}{\left(\frac{1}{2}\Gamma_i\right)^2 + \left(E - E_{i_0f}\right)^2} = -\frac{C}{\pi}\left(\tan^{-1}\frac{E-E_{i_0f}}{\frac{1}{2}\Gamma_i}\right)_{E_{i_0f}=E_{i_0f_0}}^{\infty}$$

$$= C\left(\frac{1}{\pi}\tan^{-1}\frac{E-E_{i_0f_0}}{\frac{1}{2}\Gamma_i} + \frac{1}{2}\right) \quad . \qquad (3.189)$$

Here C is determined by the condition $\tau(\infty) \equiv \tau_\infty = C$ at $E = \infty$. This is drawn as the *arctangent curve* in Fig.3.21c, and gives the shape of the absorption discontinuity.

When the photon energy E is equal to $E_{i_0f_0}$ (Fig.3.21a), we get $\tau(E_{i_0f_0}) = \frac{1}{2}\tau_\infty$. This point of the arctangent curve is the *inflection point* (Fig.3.21c) and is identified with the energy position of the absorption edge (or the Fermi level). For the points $E_{1/4}$ and $E_{3/4}$ we have $\tau(E)$ equal to $\tau_\infty/4$ and $3\tau_\infty/4$, respectively. They are given by

$$E_{1/4} = E_{i_0f_0} - \frac{1}{2}\Gamma_i \quad , \quad E_{3/4} = E_{i_0f_0} + \frac{1}{2}\Gamma_i \quad , \qquad (3.190)$$

so that

$$E_{3/4} - E_{1/4} = \Gamma_i \quad . \qquad (3.191)$$

This determines the width at half maximum of the initial state. Such an analysis was first given by RICHTMYER et al. [3.36]. It was developed by them for the transitions of an inner-level electron to a continuous energy band of a metal. However, their approximation works better for atoms than for solids. In fact, for metals if the correct Fermi distribution is assumed, $\tau(E)$ is given by

$$\tau(\nu) = A'\left(\tan^{-1}\frac{B}{(2\nu_e)^{\frac{1}{2}}-A} + \tan^{-1}\frac{B}{(2\nu_e)^{\frac{1}{2}}+A}\right) + \frac{1}{2}B'\ln\frac{\nu_F+D+A(2\nu_i)^{\frac{1}{2}}}{\nu_e+D-A(2\nu_e)^{\frac{1}{2}}} \quad . \qquad (3.192)$$

where $h\nu_e$ is the difference between the Fermi energy and the mean energy of electrons in the metal,

$$A = \left(D+\nu_e+\nu-\nu_{i_0f_0}\right)^{\frac{1}{2}} \quad , \quad B = \left(D-\nu_e+\nu-\nu_0\right)^{\frac{1}{2}} \quad ,$$

$$D = \left[\left(\nu_e+\nu-\nu_{i_0f_0}\right)^2+(\Gamma_i/2)^2\right]^{\frac{1}{2}} \quad , \quad A' = C'A \quad , \quad B' = C'B \quad ,$$

and C' = constant. Level widths obtained from this exact result do not differ appreciably from those given by (3.189).

3.12 Absorption Coefficients

When an X-ray beam passes through an absorber, it is attenuated. The degree of attenuation depends on scattering and various absorption processes. Without going into the details of these processes, *Lambert's law* states that *equal paths in the same absorbing medium attenuate equal fractions of the radiation*. Suppose, for the path length dx, that the intensity I is reduced by an amount dI. Then, $dI/I \propto dx$, or

$$dI/I = -\mu_l\, dx \quad , \tag{3.193}$$

where μ_l is a constant and is known as the *linear attenuation coefficient*. The negative sign indicates that I decreases as x increases. Integration gives

$$\ln I = -\mu_l\, x + a \quad , \tag{3.194}$$

where a is the constant of integration. If $I = I_0$ at $x = 0$, $a = \ln I_0$ and

$$I = I_0\, e^{-\mu_l x} \quad . \tag{3.195}$$

The dimension of μ_l is cm^{-1}. In the expression

$$\mu_l = \frac{1}{x} \ln \frac{I_0}{I} \quad , \tag{3.196}$$

all of the quantities on the right-hand side can be measured. Such measurements show that μ_l depends on the state (gas, liquid or solid) of the material. Therefore, it is useful to define the *mass attenuation coefficient* μ_m, that does not depend on the particular phase of the material.

Consider an X-ray beam of intensity I with a unit cross section. If ρ is the density of the material, a mass $dm = \text{density} \times \text{volume} = \rho dx$ will be crossed along the path length dx. The relative change of intensity dI/I along the path dx will be proportional to the mass dm in this layer,

$$dI/I = -\mu_m dm = -\mu_m \rho dx \quad , \tag{3.197}$$

$$I = I_0 e^{-\mu_m m} = I_0 e^{-\mu_m \rho x} \quad , \tag{3.198}$$

where $m = \rho x$ is the mass-per-unit-area, or *plane density*, in g cm^{-2}. In the form (3.198), it is called the *Bouguer-Lambert-Beer exponential attenuation law* [3.37-39].

Let us now find the attenuation per atom of the material. The number of atoms in dm is given by

$$dn = \frac{dm}{\text{mass of each atom}} = \frac{dm}{A/N_A} = \frac{\rho dx}{A/N_A} \quad ,$$

where A is the atomic weight and N_A Avogadro's number. Thus

$$dI/I = -\mu_a dn = -(\rho N_A/A)\mu_a dx \quad , \tag{3.199}$$

$$I = I_0 e^{-\mu_a n} \quad , \tag{3.200}$$

where μ_a is the *atomic attenuation coefficient*. Because dn is the number of atoms per cm^2 of a layer of material, the dimension of μ_a is cm^2 (or area). Clearly,

$$\mu_a = \frac{A}{\rho N_A} \mu_l = \frac{A}{N_A} \mu_m \quad . \tag{3.201}$$

The dimension of μ_a suggests that it can be interpreted as the atomic cros section σ_a of the interaction (Fig.3.22). If one photon per second is inciden on unit area, and there is one atom per unit area of material, the probabilit of absorbing or scattering this photon per second will be equal to μ_a.

The concept of μ_a is useful in that it enables us to calculate the molecular attenuation coefficient of any molecule by adding the contributions of each of the constituent atoms. Thus, for example

$$\mu_{(CaCO_3)} = \mu_{Ca} + \mu_C + 3\mu_O \quad . \tag{3.202}$$

The number of electrons per unit area is $dn_e = Zdn$. Therefore, from (3.199) we obtain

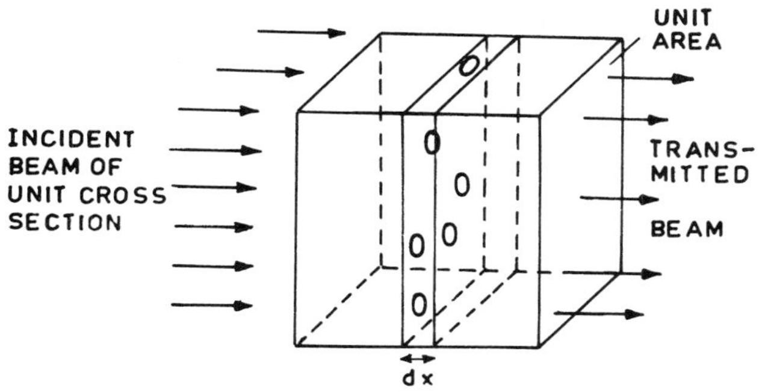

Fig. 3.22. Schematic diagram for calculation of μ_a and its relation to σ_a. The layer of thickness dx has mass-per-unit-area dm, and atom-per-unit-area dn. Each small circle shows that a cross section μ_a is associated with an atom

Table 3.5. Various attenuation coefficients

Coefficient	Dimension
Atomic μ_a	$(\text{length})^2$
Electronic $\mu_e = \mu_a/Z$	$(\text{length})^2$
Mass $\mu_m = \mu_a N_A/A$	$(\text{length})^2 (\text{mass})^{-1}$
Linear $\mu_l = \rho\mu_a N_A/A$	$(\text{length})^{-1}$

$$dI/I = -(\mu_a/Z)dn_e \quad , \quad I = I_0 e^{-\mu_e n_e} \quad , \tag{3.203}$$

where μ_e is called the *electronic attenuation coefficient*. It has the dimension of cm^2 and is equal to μ_a/Z (Table 3.5).

When X-rays pass through matter, the intensity is decreased (attenuated) because the photons are either absorbed or scattered by the atoms of the material. Photoelectric or true absorption occurs when an inner electron of the atom is completely removed from its shell. The scattering takes place mainly because of the collision of X-ray photons with the loosely bound outer electrons of the atom. If we assume these two processes to be independent, we can express μ_a as the sum of atomic coefficients of photoabsorption τ_a and of scattering σ_a,

$$\mu_a = \tau_a + \sigma_a \quad . \tag{3.204}$$

Similarly,

$$\mu_l = \tau + \sigma \quad , \quad \mu_m = \tau_m + \sigma_m \quad , \quad \mu_e = \tau_e + \sigma_e \quad , \tag{3.205}$$

where

$$\tau_e = \frac{\tau_a}{Z} = \frac{A}{N_A Z}\,\tau_m = \frac{A}{N_A Z\rho}\,\tau \quad . \tag{3.206}$$

The electronic absorption coefficient of photoabsorption, τ_e, represents an average value for *all* of the electrons (including the innermost K electrons) of the atom. Equation (3.206) is therefore valid when $\lambda < \lambda_K$.

Although σ_m remains constant, τ_m (like τ_a) increases rapidly with λ and Z. Usually, $\tau_m \gg \sigma_m$; for example, absorption exceeds scattering by a factor of 350 even for ^{24}Cr at $\lambda \sim 1$ Å. For long wavelengths and heavier elements, we can write

$$\mu_a \simeq \tau_a \quad , \quad \mu_m \simeq \tau_m \quad . \tag{3.207}$$

For the separate q levels of the atom, for $\lambda < \lambda_K$

$$\tau_a = \sum_q \tau_q = \tau_K + \left(\tau_{L_I} + \tau_{L_{II}} + \tau_{L_{III}}\right) + \dots \quad , \tag{3.208}$$

$$\tau_m = \sum_q \tau_{mq} = \tau_{mK} + \left(\tau_{mL_I} + \tau_{mL_{II}} + \tau_{mL_{III}}\right) + \dots \quad . \tag{3.209}$$

For other wavelength regions,

$$\tau_a = \tau_{L_I} + \tau_{L_{II}} + \tau_{L_{III}} + \tau_{M_I} + \dots \quad , \quad \lambda_K < \lambda < \lambda_L \quad , \tag{3.210}$$

$$\tau_a = \tau_{M_I} + \tau_{M_{II}} + \dots \quad , \quad \lambda_{L_{III}} < \lambda < \lambda_{M_I} \quad . \tag{3.211}$$

The variation of τ_a with λ and Z is given by (Fig.3.13)

$$\tau_a \simeq CZ^4\lambda^3 \quad , \tag{3.212}$$

where the constant C decreases suddenly at each absorption edge. It is obvious from (3.212) that short-wavelength X-rays are highly penetrating; they are therefore called *hard*, whereas long-wavelength X-rays are easily absorbed and are said to be *soft*.

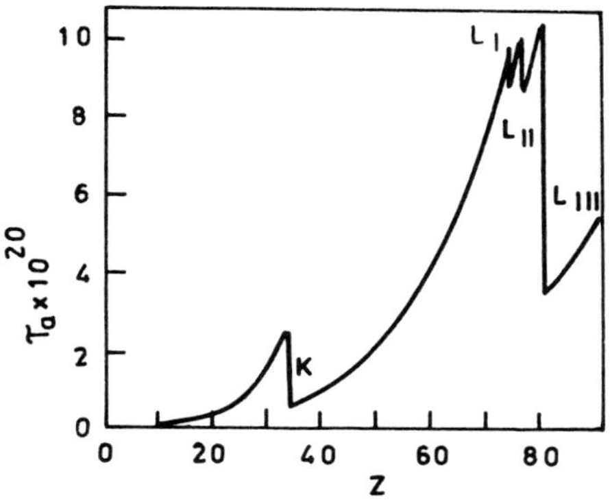

Fig. 3.23. Plot of τ_a as a function of Z for $\lambda = 1$ Å

Another useful formula for the linear absorption coefficient away from any characteristic edges is the *Victoreen formula*, $\mu_l = C\lambda^3 - D\lambda^4$ [3.40].

At constant λ, τ_a increases with Z. If X-rays of a given λ (or given energy $h\nu = hc/\lambda$) can excite the K levels of the light elements up to a certain value of Z, at which λ becomes greater than λ_K for this element, C decreases suddenly at this Z value. Consequently, a discontinuity occurs on a τ_a versus Z plot (Fig.3.23). An increase of τ_a with Z again occurs until a new discontinuity is found at another Z value, where the L_I level ceases to absorb, and so on.

3.12.1 Quantum Theory of Photoabsorption

We know that the oscillator strength f_q and photoabsorption coefficient τ_q are directly related (3.112). Therefore, from the quantum theoretical result (3.150) of SUGIURA, we can at once write

$$\tau_K = \frac{2^8 \pi e^2}{3mc} \frac{\nu_K^3}{\nu^4} \frac{e^{-4n_K \tan^{-1}(1/n_K)}}{1 - e^{-2\pi n_K}} \quad , \tag{3.213}$$

where

$$n_K = \left(\frac{\nu_K}{\nu - \nu_K}\right)^{\frac{1}{2}} \quad ; \quad \nu_K = R_\infty cZ^2 \quad . \tag{3.214}$$

This result (see Appendix H) was first derived by STOBBE [3.41]. It has the limiting cases:

1) $\nu \simeq \nu_K \quad (\eta_K >> 1)$:

$$1 - e^{-2\pi\eta_K} \simeq 1 \quad , \quad \tan^{-1}\eta_K^{-1} \simeq \eta_K^{-1} \quad ,$$

$$\tau_K \simeq \frac{2^8 \pi e^2 R_\infty^3 \exp(-4)}{3mc^2} Z^6 \lambda^4 = 1820 \; Z^6 \lambda^4 \quad , \tag{3.215}$$

where λ is in cm.

2) $\nu >> \nu_K \quad (\eta_K << 1)$:

$$1 - e^{-2\pi\eta_K} \simeq 2\pi\eta_K \simeq 2\pi(\nu_K/\nu)^{\frac{1}{2}} \quad , \quad e^{-4\eta_K \tan^{-1}\eta_K^{-1}} \simeq 1 \quad ,$$

$$\tau_K \simeq \frac{2^7 e^2 R_\infty^{5/2}}{3mc^2} Z^5 \lambda^{7/2} = 478 \; Z^5 \lambda^{7/2} \quad . \tag{3.216}$$

For the L_I and $L_{II,III}$ states, STOBBE obtained

$$\tau_{L_I} = \frac{2^{11}\pi e^2}{2mc} \frac{\nu_L^3}{\nu^4} \left(1 + 3\frac{\nu_L}{\nu}\right) \frac{e^{-8\eta_L \tan^{-1}(1/\eta_L)}}{1 - e^{-4\pi\eta_L}} \quad , \tag{3.217}$$

$$\tau_{L_{II}} + \tau_{L_{III}} = \frac{2^{12}\pi e^2}{3mc} \frac{\nu_L^4}{\nu^5} \left(3 + 8\frac{\nu_L}{\nu}\right) \frac{e^{-8\eta_L \tan^{-1}(1/\eta_L)}}{1 - e^{-4\pi\eta_L}} \quad , \tag{3.218}$$

where

$$\eta_L = \left(\frac{\nu_L}{\nu - \nu_L}\right)^{\frac{1}{2}} \quad ; \quad \nu_L = R_\infty c Z^2 \frac{1}{2^2} = \frac{c}{\lambda_L} \quad . \tag{3.219}$$

The corresponding limiting expressions are,

1) $\nu \simeq \nu_L \quad (\eta_L >> 1)$:

$$\tau_{L_I} \simeq \frac{2^5 \pi e^2 R_\infty^3 \exp(-8)}{3mc^2} Z^6 \lambda^4 \left(1 + 3\frac{\lambda}{\lambda_L}\right) = 4.17 \; Z^6 \lambda^4 (1 + 8.25 \times 10^4 \; Z^2 \lambda) \quad , \tag{3.220}$$

$$\tau_{L_{II}} + \tau_{L_{III}} \simeq \frac{2^4 e^2 R_\infty^4 \exp(-8)}{3mc^2} Z^8 \lambda^5 \left(3 + 8\frac{\lambda}{\lambda_L}\right)$$

$$= 2.29 \times 10^5 \; Z^8 \lambda^5 \; (3 + 2.20 \times 10^5 \; Z^2 \lambda) \quad . \tag{3.221}$$

2) $\nu \gg \nu_L \quad (\eta_L \ll 1)$:

$$\tau_{L_I} \simeq \frac{2^4 e^2 R_\infty^{5/2}}{3mc^2} Z^5 \lambda^{7/2} = 5.98\, Z^5 \lambda^{7/2} \quad , \tag{3.222}$$

$$\tau_{L_{II}} + \tau_{L_{III}} \simeq \frac{2^3 e^2 R^{7/2}}{mc^2} Z^{7/2} \lambda^{9/2} = 9.84 \times 10^5\, Z^7 \lambda^{9/2} \quad , \tag{3.223}$$

where λ is in cm.

From these relations, we have

$$\tau_{L_I}/\left(\tau_{L_{II}}+\tau_{L_{III}}\right) = \begin{cases} 2/11 & \text{for } \nu \simeq \nu_L \\ (1/6)(\nu/\nu_L) & \text{for } \nu \gg \nu_L \end{cases} \quad , \tag{3.224}$$

$$\tau_K \simeq 8\tau_{L_I} \simeq \frac{4}{3}\frac{\lambda_L}{\lambda}\left(\tau_{L_{II}}+\tau_{L_{III}}\right) \gg \tau_{L_{II}} + \tau_{L_{III}} \quad . \tag{3.225}$$

Thus, mainly $L_{II,III}$ levels absorb at $\nu \simeq \nu_L$, L_I at $\nu \gg \nu_L$, and K levels at $\nu \gg \nu_K$. Measurements of the relative number of photoelectrons ejected from an atom by X-rays fully support these results.

To find the numerical values of the absorption coefficients from Stobbe's equations, we should be able to calculate the frequencies ν_K and ν_L for the K and L absorption edges. For this, we can use the semiempirical method of SLATER [3.42] (see Appendix I) to estimate the screening constants σ_q. Then we can use

$$\nu_q = \frac{(Z-\sigma_q)R_\infty c}{n_q^2} \quad . \tag{3.226}$$

STOBBE simplified the calculation by using $Z_{eff} = (Z-0.3)$ for the K level, and $(Z-2\times0.85-7\times0.35) = (Z-4.15)$ for the L level, as suggested by SLATER. For more details, see PINSKER [3.62].

3.12.2 Various Attenuation Processes

Equation (3.204) can be written in greater detail, by taking into account the various attenuation processes, as

$$\sigma_{tot} = \mu_a = \tau_a + \sigma_{coh} + \sigma_{incoh} + \kappa \quad , \tag{3.227}$$

where σ_{tot} is the total cross section per atom in barns/atom (1 barn = 10^{-22} cm^2), τ_a the atomic photoelectric cross section, σ_{coh} the coherent (Rayleigh) scattering cross section, σ_{incoh} the incoherent (Compton) scattering cross section, and κ the electron-positron pair-production cross section. The mass attenuation coefficient is related to σ_{tot} by

$$\mu_l/\rho = \mu_m = \sigma_{tot}N_A/A \quad . \tag{3.228}$$

As discussed by BERGER [3.43], ALLISON [3.44] and HUBBELL [3.45], it is useful to define the *mass energy absorption coefficient* μ_{en}/ρ. It is related to, and yet distinct from, the mass attenuation coefficient μ_l/ρ. The latter is a measure of the average number of interactions between incident photons and matter that occur in a given mass-per-unit-area thickness of the material traversed. On the other hand, μ_{en}/ρ is a measure of the average fractional amount of incident photon energy transferred to kinetic energy of charged particles as a result of these interactions. Analogous to (3.228), we can write

$$\mu_{en}/\rho = \sigma_{en}N_A/A \quad , \tag{3.229}$$

where the energy-absorption cross section can be expressed as

$$\sigma_{en} = \tau_a f_\tau + \sigma_{incoh} f_{incoh} + \kappa f_\kappa \quad . \tag{3.230}$$

Here the fractions f_τ, f_{incoh} and f_κ weight the corresponding cross sections.

3.13 Absorption-Jump Ratios

Figure 3.12 shows that we obtain two values of τ_m on either side of the q absorption edge at λ_q. Let us denote the mass absorption coefficient just on the short-wavelength side of λ_q by $\tau_m(\lambda_q)$ and that just on the long-wavelength side of λ_q by $\tau_m'(\lambda_q)$, $\tau_m(\lambda_q) > \tau_m'(\lambda_q)$. The ratio

$$r_q = \tau_m(\lambda_q)/\tau_m'(\lambda_q) = \tau_a(\lambda_q)/\tau_a'(\lambda_q) > 1 \quad , \tag{3.231}$$

is called the *absorption-jump ratio* of the q level.

Several empirical relations have been suggested for r_K:

1) JÖNSSON [3.46]:

$$r_K = \lambda_{L_I}/\lambda_K \quad . \tag{3.232}$$

2) RINDFLEISCH [3.47]:

$$r_K = aZ^b \quad , \quad \log_{10} a = 1.805283 \quad , \quad b = -0.6207 \quad . \tag{3.233}$$

3) LAUBERT [3.48]:

$$r_K = a\lambda_K^b \quad , \quad \log_{10} a = 0.857652 \quad , \quad b = 0.0843 \quad . \tag{3.234}$$

4) TELLEZ-PLASENCIA [3.49]:

$$r_K = (a+bZ)^{-1} \quad , \quad a = 0.051167 \quad , \quad b = 0.0024882 \quad . \tag{3.235}$$

The absorption curves can be fitted with the formulas [3.50],

$$\begin{aligned} \tau_a &= CZ^p\lambda^n \quad , \qquad \text{for } \lambda < \lambda_K \quad , \\ \tau_a' &= C'Z^{p'}\lambda^n \quad , \qquad \text{for } \lambda_K < \lambda < \lambda_L \quad , \end{aligned} \tag{3.236}$$

where, $C = 2.64 \times 10^{-26}$, $p = 3.94$, $C' = 8.52 \times 10^{-28}$, $p' = 4.30$, and $n = 3$. AGARWAL [3.51] has shown that (3.231,236) lead to the empirical formula (3.233).

From (3.208,210), we get $\tau_a = \tau_K + \tau_a'$. Therefore,

$$(r_K-1)/r_K = \tau_K/\tau_a \quad . \tag{3.237}$$

In the region $\lambda \lesssim \lambda_K$, the photoelectric absorption is mainly because of the K-shell electrons. Therefore, $(r_K-1)/r_K$ gives the fraction of the total number of ejected photoelectrons that come from the K shell.

In a similar way, we can find the jump ratios for the absorption edges L_I, L_{II}, L_{III}, M_I, etc.

A universal absorption curve ($\log\tau_m$ versus $\log\lambda$) was drawn by JÖNSSON [3.46]. He used τ_m for $\lambda < \lambda_K$, and $\tau_m r_K = \tau_m(\lambda_{L_I}/\lambda_K)$ for $\lambda_K < \lambda < \lambda_{L_I}$. It can be extended on the same principle for longer wavelengths. A similar curve has been drawn by BÖKLEN and GEILING [3.52].

3.14 Total Reflection

Let us assume that the material does not absorb X-rays, $\beta = 0$. Then $\mu = 1 - \delta$ is real and the usual laws of reflection can be applied. We have seen, (3.73),

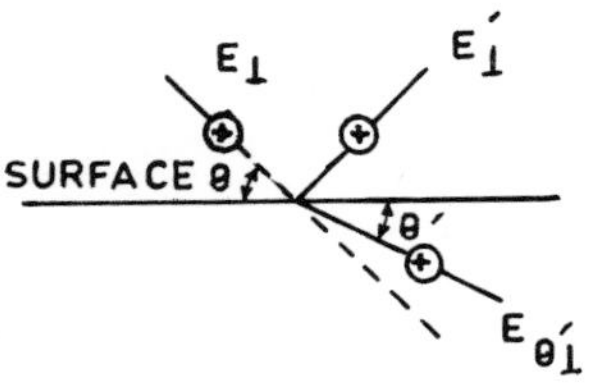

Fig. 3.24. Reflection and refraction of polarized X-rays

that X-rays are totally reflected from a plane surface at glancing angles less than a certain *critical* angle θ_c, given by

$$\theta_c = (2\delta)^{\frac{1}{2}} \quad . \tag{3.238}$$

For $\delta = 3\times10^{-6}$, we get $\theta_c = 2.4\times10^{-3}$ radians $\simeq 8'$.

COMPTON [3.4] found that, for the W L line, $\lambda = 1.279$ Å, $\theta_c = 10'$ for a sheet of plane glass and $\theta_c = 22.5'$ for a sheet of silver. These results agree with (3.238) when theoretical values of δ are used.

PRINS [3.18] pointed out that the sharpness of the limit of total reflection depends on the absorption in the material. The effect is well known in optics. Let us assume the Fresnel laws of reflection to be applicable for X-rays as well. If the electric vector $E_{\perp}$ in the incident radiation is perpendicular to the plane of incidence, and $E'_{\perp}$ is the electric vector in the reflected radiation (Fig.3.24), Fresnel's formula gives

$$E'_{\perp} = \frac{\sin\theta - \mu\sin\theta'}{\sin\theta + \mu\sin\theta'} E_{\perp} \quad . \tag{3.239}$$

Here θ and θ' are the glancing angles of incidence and refraction, respectively.

Use of $\cos\theta = \mu\cos\theta'$ and $\mu^2 = 1 - 2\delta - i2\beta$ gives

$$\frac{E'_{\perp}}{E_{\perp}} = \frac{\sin\theta - (\mu^2 - \cos^2\theta)^{\frac{1}{2}}}{\sin\theta + (\mu^2 - \cos^2\theta)^{\frac{1}{2}}} \quad , \tag{3.240}$$

$$\frac{I'_{\perp}}{I_{\perp}} \simeq \left|\frac{\theta - (\theta^2-2\delta-i2\beta)^{\frac{1}{2}}}{\theta + (\theta^2-2\delta-i2\beta)^{\frac{1}{2}}}\right|^2 = \frac{(\theta-a)^2+b^2}{(\theta+a)^2+b^2} \quad , \tag{3.241}$$

where $I'_{\perp}/I_{\perp} = R$ is the ratio of the incident and reflected intensities and

$$(\theta^2-2\delta-i2\beta)^{\frac{1}{2}} = a + ib \quad ,$$

$$a^2 - b^2 = \theta^2 - 2\delta \quad , \quad ab = -\beta \quad ,$$

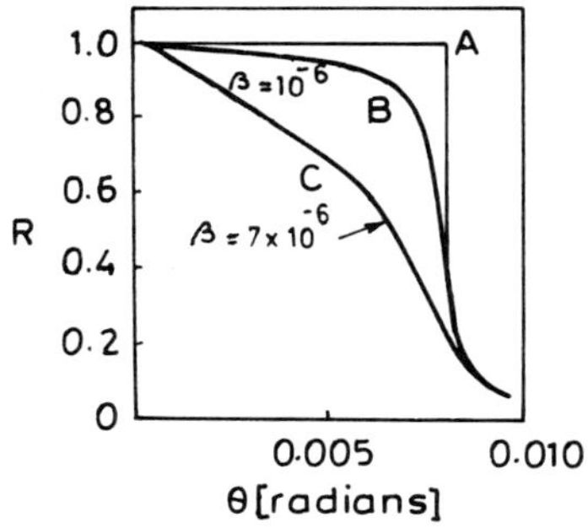

Fig. 3.25. Reflection coefficient R of iron for X-rays as a function of θ, with $\delta = 30 \times 10^{-6}$. A, negligible absorption. B, weak absorption ($\lambda \gtrsim \lambda_K$). C, strong absorption ($\lambda \lesssim \lambda_K$)

$$a^2 = \frac{1}{2}\left\{\left[(\theta^2-2\delta)^2+4\beta^2\right]^{\frac{1}{2}}+(\theta^2-2\delta)\right\} \quad ,$$

$$b^2 = \frac{1}{2}\left\{\left[(\theta^2-2\delta)^2+4\beta^2\right]^{\frac{1}{2}}-(\theta^2-2\delta)\right\} \quad .$$

If the electric vector of the incident X-rays lies in the plane of reflection, we get a result of slightly different form than (3.241). However, it gives the same numerical values, within the limits of experimental error [3.53-56].

The reflection curves given by (3.241) for zero, small and large absorption are shown in Fig.3.25. Only the curve for $\beta = 0$ (zero absorption) shows a sharp limit of total reflection. Experimental work of PRINS on an iron mirror clearly showed the difference of sharpness of total reflection for wavelengths on the two sides of the Fe K edge. This gave definite evidence for the existence of anomalous dispersion.

KIESSIG [3.57] measured reflection curves, for nickel mirrors deposited by evaporation on glass, with the help of an ionization spectrometer. For thin films, he detected maxima and minima as θ was altered. They were produced because of the *interference* between the X-rays reflected from the outer surface and those reflected from the nickel-glass surface inside the film. The value of δ is greater for nickel than for glass, and so the latter is a normal reflection from an optically denser medium. Because θ is small, the path difference between the rays reflected from the upper and lower surfaces is small enough to give interference even in the X-ray region. From the spacing of maxima and minima, the thickness of the nickel film can be determined. For the measurement of refractive index by X-ray interferometry, see PINSKER [3.62].

4. Secondary Spectra and Satellites

In an atom, ionization can be caused not only by electron impact, but also by energetic photons. In the photoionization process, the incident photon is absorbed and a single electron is ejected. It is a one-step process. Conservation of linear momentum requires that the excess energy available after ionization be partitioned in inverse proportion to the masses of the products (ejected electron and recoiled ion). The electron is about two thousand times lighter than the lightest ion (proton). Therefore, all the excess energy can be taken as the kinetic energy of the electron, with little error (less than 1 part in 10^4). We can experimentally study the process if we have a monoenergetic photon source and an electron energy analyzer.

Any other process occurring simultaneously with photoionization would be a two-step process. We shall briefly discuss all these processes in this chapter.

4.1 Photoelectric Effect

An incident quantum $h\nu_0$ of X-rays can remove a K electron, for example, out of an atom and thus cause the emission of K characteristic radiation, provided that $h\nu_0 \geqq E_K$, where E_K is the binding energy of the K electron. The incident photon is absorbed in the process. The ejected electron is called a *photoelectron* and the emitted characteristic radiation is called *fluorescent* (or *secondary*) radiation. Such a photoabsorption process is responsible for the *true* absorption of X-rays. This phenomenon is the X-ray counterpart of the photoelectric effect in the ultraviolet region; there, photoelectrons are ejected from the outer shells of a metal atom. EINSTEIN showed that a proper energy-conservation equation can be written from the viewpoint of the quantum theory,

$$h\nu_0 = \frac{1}{2} mv^2 + E_q \quad , \tag{4.1}$$

where $mv^2/2$ is the kinetic energy of the ejected electron and E_q is binding energy of the electron in the q shell.

4.2 Quantum Theory of the Photoelectric Effect

4.2.1 Born Approximation

For a K-shell electron

$$E_q = E_K = \frac{mZ^2e^4}{2\hbar^2} = \frac{1}{2} Z^2\left(\frac{e^2}{\hbar c}\right)^2 mc^2 = \frac{1}{2} Z^2 \frac{mc^2}{137^2} \quad . \tag{4.2}$$

For most of the elements $Z \ll 137$, and $E_q \ll h\nu_0 \ll mc^2$. Therefore, from (4.1), $h\nu_0 - E_q = mv^2/2$, we have $mv^2/2 \gg (1/2)Z^2mc^2/137^2$, or

$$\frac{Z}{137}\frac{c}{v} \ll 1 \quad . \tag{4.3}$$

This is the criterion for the validity of the Born approximation.

From (2.115), we can write the matrix element for the absorption of a photon as

$$H' = -\frac{e}{m}\left(\frac{\hbar}{V\nu_0}\right)^{\frac{1}{2}} \int \psi_f^*(\underline{p}_{op}\cdot\underline{e}_0)\, e^{i\underline{k}_0\cdot\underline{r}}\, \psi_i d\tau \quad , \tag{4.4}$$

where $\underline{p}_{op}$ is the momentum operator of the electron, and $\underline{k}_0$, $\underline{e}_0$ are the propagation and unit polarization vectors of the incident photon. For the K-shell electron

$$\psi_i = \left(\frac{a^3}{\pi}\right)^{\frac{1}{2}} e^{-ar} \quad , \quad a = Z/a_0 \quad , \quad a_0 = \hbar^2/me^2 \quad . \tag{4.5}$$

Because of the large volume V of the cube taken $V \gg a$, we have simply normalized ψ_i over the whole space.

In the Born approximation, we can take for ψ_f a plane wave normalized to the cube,

$$\psi_f = V^{-\frac{1}{2}} e^{i\underline{p}\cdot\underline{r}/\hbar} \quad , \tag{4.6}$$

where $\underline{p}$ is the momentum of the ejected (free) electron. Because p_{op} is Hermitian, and ψ_f is an eigenstate of momentum, we can write

$$\left\langle f|\underline{p}_{op}\cdot\underline{e}_0\, e^{i\underline{k}_0\cdot\underline{r}}|i\right\rangle = \underline{p}\cdot\underline{e}_0\left\langle f|e^{i\underline{k}_0\cdot\underline{r}}|i\right\rangle \quad ,$$

$$H' = -\frac{e}{mV}\left(\frac{a^3\hbar}{\pi\nu_0}\right)^{\frac{1}{2}} \underline{p}\cdot\underline{e}_0 \int_V e^{i(\underline{k}_0-\underline{p}/\hbar)\cdot\underline{r}}\, e^{-ar}\, d^3\underline{r} \quad , \tag{4.7}$$

where $\underline{p}$ is a number.

The factor e^{-ar} allows the integration over V to be extended over all space. Consider the integral,

$$\begin{aligned}\int d^3r\, e^{i\underline{\lambda}\cdot\underline{r}}\,\frac{e^{-ar}}{r} &= \int_0^{2\pi} d\phi \int_0^{\pi} \sin\theta d\theta \int_0^{\infty} r^2\, dr\, e^{i\lambda r\cos\theta}\,\frac{e^{-ar}}{r} \\ &= 2\pi\int_0^{\infty} rdre^{-ar}\int_{-1}^{+1} d(\cos\theta)\, e^{i\lambda r\cos\theta} \\ &= \frac{2\pi}{i\lambda}\int_0^{\infty} dr\, e^{-ar}(e^{i\lambda r}-e^{-i\lambda r}) \\ &= \frac{2\pi}{i\lambda}\left(\frac{1}{a-i\lambda}-\frac{1}{a+i\lambda}\right) = \frac{4\pi}{a^2+\lambda^2} \quad .\end{aligned} \tag{4.8}$$

Differentiation of this integral with respect to a gives

$$\int d^3\underline{r}\, e^{i\underline{\lambda}\cdot\underline{r}}\, e^{-ar} = \frac{8\pi a}{(a^2+\lambda^2)^2} \quad . \tag{4.9}$$

From (4.7,9),

$$H' = -\frac{e}{mV}\left(\frac{a^3\hbar}{\pi\nu_0}\right)^{\frac{1}{2}} (\underline{p}\cdot\underline{e}_0)\,\frac{8\pi a}{(a^2+\lambda^2)^2} \quad , \tag{4.10}$$

where $\underline{\lambda} = \underline{k}_0 - (\underline{p}/\hbar)$.

The differential cross section is given by

$$d\sigma = \frac{w}{v/V} = \frac{V}{v}\frac{2\pi}{\hbar}\,\rho_f|H'|^2\, d\Omega \quad , \tag{4.11}$$

where ρ_f is the density of the final states and the incident flux is v/V. For the photon $v = c$. For the photoelectron in the final state, $\varepsilon^2 = c^2p^2 + m^2c^4$, $\varepsilon d\varepsilon = c^2pdp$. From phase-space considerations, the number of states with momenta between p and p + dp is $\rho_p\, dp\, d\Omega = p^2\, dp\, d\Omega/(h^3/V)$. Therefore, the number of states for the photoelectron that has energy between ε and $\varepsilon + d\varepsilon$ is $\rho_\varepsilon\, d\varepsilon\, d\Omega = [(p\varepsilon)/(h^3c^2)]d\varepsilon\, d\Omega$, and we can write with total final energy $\varepsilon \simeq mc^2$

$$d\sigma = \frac{V}{c}\frac{2\pi}{\hbar}\frac{mpV}{h^3}\frac{64\pi e^2\hbar a^5(\underline{p}\cdot\underline{e}_0)^2}{m^2V^2\nu_0(a^2+\lambda^2)^4}\,d\Omega$$

$$= 32\ Z^5a_0^2\left(\frac{pc}{h\nu_0}\right)\left(\frac{\underline{p}\cdot\underline{e}_0}{mc}\right)^2\frac{1}{\left(Z^2+a_0^2\lambda^2\right)^4}\,d\Omega \quad . \tag{4.12}$$

For $h\nu_0 >> E_B$, we can write (4.1) as $h\nu_0 \simeq p^2/2m$. Therefore,

$$\frac{pc}{h\nu_0}\left(\frac{\underline{p}\cdot\underline{e}_0}{mc}\right)^2 \simeq \frac{2p}{mc}(\hat{\underline{p}}\cdot\underline{e}_0)^2 \quad ,$$

$$\lambda^2 = \hbar^{-2}(\hbar\underline{k}_0-\underline{p})^2 = \hbar^{-2}\left[(h\nu_0/c)^2-2(h\nu_0/c)p\ \hat{\underline{k}}_0\cdot\hat{\underline{p}}+p^2\right]$$

$$\simeq \hbar^{-2}\left[p^2-(p^3/mc)\hat{\underline{k}}_0\cdot\hat{\underline{p}}\right] \simeq (p/\hbar)^2\left[1-(v/c)\hat{\underline{k}}_0\cdot\hat{\underline{p}}\right] \quad , \tag{4.13}$$

where we have used $k_0 = 2\pi\nu_0/c$, $\hat{\underline{p}} = \underline{p}/p$, $\hat{\underline{k}}_0 = \underline{k}_0/k_0$, and $h\nu_0/c \simeq p^2/2mc << p$ for the nonrelativistic electron ($p << mc$). Thus, with kinetic energy $E = p^2/2m$,

$$\frac{d\sigma}{d\Omega} = \frac{64Z^5a_0^2\alpha^8(2E/mc^2)^{\frac{1}{2}}(\hat{\underline{p}}\cdot\underline{e}_0)^2}{\left[(\alpha Z)^2+\frac{2E}{mc^2}\left(1-\frac{v}{c}\hat{\underline{k}}_0\cdot\hat{\underline{p}}\right)\right]^4} \quad . \tag{4.14}$$

Let $\hat{\underline{k}}_0$ be along the z axis, and the two photon-polarization directions $\underline{e}_0^{(1)}$, $\underline{e}_0^{(2)}$ be along the x and y directions, respectively. Let $\underline{p}$ make an angle θ with the z axis, so that

$$\hat{\underline{p}} = (\sin\theta\ \cos\phi,\ \sin\theta\ \sin\phi,\ \cos\theta) \quad ,$$

$$\left(\hat{\underline{p}}\cdot\underline{e}_0^{(1)}\right)^2 = \sin^2\theta\ \cos^2\phi \quad , \quad \left(\hat{\underline{p}}\cdot\underline{e}_0^{(2)}\right)^2 = \sin^2\theta\ \sin^2\phi \quad . \tag{4.15}$$

For unpolarized photons, we have to average over the two polarization directions,

$$<(\hat{\underline{p}}\cdot\underline{e}_0)^2>_{av} = \frac{1}{2}(\sin^2\theta\ \cos^2\phi+\sin^2\theta\ \sin^2\phi) = \frac{1}{2}\sin^2\theta \quad . \tag{4.16}$$

With $\hat{\underline{k}}_0\cdot\hat{\underline{p}} = \cos\theta$, we finally get

$$\frac{d\sigma}{d\Omega} \simeq f(\alpha,a_0/Z)F(\theta) \quad ,$$

$$f(\alpha,a_0/Z) = 2^{3/2}Z^5\alpha^8a_0^2\left(\frac{E}{mc^2}\right)^{-7/2} \quad , \tag{4.17}$$

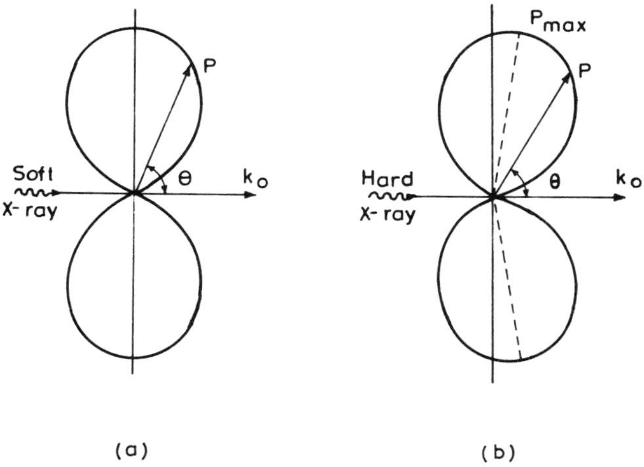

Fig. 4.1. Angular distribution of the photoelectrons for (a) soft X-rays, and (b) hard X-rays

$$F(\theta) = \frac{\sin^2\theta}{\left(1-\frac{v}{c}\cos\theta\right)^4} \simeq \sin^2\theta\left(1+4\frac{v}{c}\cos\theta\right) \quad , \tag{4.18}$$

where we have taken for light elements $h\nu_0 \gg E_B$, or equivalently $(\alpha Z)^2(mc^2/2E) \ll 1$.

Because $h\nu_0 \simeq E$, and the integrated cross section is related to the mass absorption coefficient $\tau_m = N_A\sigma/A$, we find that $\tau_m \propto \nu_0^{-7/2}Z^5$, or $\tau_m = \text{const.}\lambda_0^{3.5}Z^5$, in agreement with (3.216).

The angular distribution is given by $F(\theta)$. It is a $\sin^2\theta$ distribution for $v/c \ll 1$. No photoelectrons come out in the forward direction $\theta = 0$, and maximum number come out at $\theta = \pi/2$ (Fig.4.1a). A simple reason for this is seen if we note that the ejected electron accepts the unit of angular momentum transferred in the absorption process (dipole selection rule). In the ionization of an s electron to a p photoelectron, the axis of the p-wave function is defined by the electric vector of the radiation. If this is taken as the z axis, the angular part of the electron wave function is the same as that of an atomic p_z electron, Y_{10}, and the probability is proportional to $Y_{10}^2 = (3\pi/4)\cos^2(90-\theta) = (3\pi/4)\sin^2\theta$. For large values of v/c (hard X-ray region) the retardation term $(v/c)\cos\theta$ in the denominator gives a strong tilt in the forward direction because of the fourth power. For the relativistic case, $v \sim c$, the calculation has been done by SAUTER [4.1].

4.2.2 Dipole Approximation

Another method of calculating the photoelectric effect is to ignore the retardation term but use the correct wave function ψ_E for the continuum. When the energy $h\nu_0$ is just enough to remove, say, a K electron, the photoproduced electron continues to experience the Coulomb interaction between the nucleus and the electron. It is then said to be in the K continuum of positive energy states. In this situation, the final-state wave function ψ_E is the solution of the Schrödinger equation for an attractive Coulomb potential with $E > 0$ (see Appendix H). The corresponding result for $\tau_K = N_A\sigma/A$ is given by (3.123) The angular distribution is again given by $\sin^2\theta$.

For the L-shell electrons, the results have been obtained by SCHUR [4.2]. For a recent calculation for atoms, see the work of COOPER and ZARE [4.3].

4.2.3 Shake-up Structure

Strictly speaking, photoionization is a transition between two states characterized by N-electron wave functions. The one-electron approximation has, therefore, only a limited validity. In this approximation the primary states are the ones roughly described by KOOPMANS' [4.4] theorem, in the sense that the electron density in the ionic state resembles the original system with a hole in the concerned orbital. The most-probable continuum wave function is the one that results when the photoelectron accepts the unit of angular momentum transferred in the absorption process [4.5]. In such a simple description the relaxation of the passive electrons' orbitals toward the hole in the final state, to minimize the total energy of the system, has been ignored. This relaxation, even without explicitly involving configuration interaction (spin-orbit and spin-spin), constitutes a many-body effect, and has important consequences for both the energy and intensity of the primary states.

The perturbation of the Coulomb potential felt by the outer electrons, when a core-level electron is ejected, causes their collapse towards the positive hole, creating the relaxation energy E_q^R and allowing the possibility of excitation of the valence electrons. The *shake-up* and *shake-off* processes are manifestations of this relaxation phenomenon.

The relaxation of the valence electrons reduces the energy of the final state by E_q^R. Thus $E_q' = E_q - E_q^R$, where E_q is the orbital energy when an electron is removed from the q orbital, all other energy levels remaining undisturbed (Koopmans' theorem frozen-orbital value). Experimentally, high energy peaks the high-binding-energy side of the core level X-ray photoelectron spectroscopy (XPS) peaks have been observed in compounds of the transition elements, and in rare gases. They may be regarded as the excitation of a valence electron to an unoccupied level (shake-up) simultaneously with the core-electron ejection. The separation of the shake-up satellites in XPS from the main core-level (primary) peak corresponds to the valence-level excitation energies concerned. The shake-up structure observed in XPS is of much current theoretical [4.6,7] and experimental [4.8,9] interest.

The shake-off process can also occur when the second electron is ejected rather than excited (a double-ionization process). The partition of the kinetic energy between the two ejected electrons shows up as steps followed by a continuum, instead of the peaks found for the shake-up. Because there is always a finite value of E_q^R, it follows that there must always be shake-up and shake-off structures, even if their intensities are low. The selection rules for the shake-up transitions are $\Delta L = 0, \pm 2$. The many-electron and final-state effects in photoemission in solids have received considerable attention in recent years [4.114].

4.3 Magnetic Spectra of Photoelectrons

In the ordinary photoelectric effect, the free electrons in the metal are liberated. According to EINSTEIN, their maximum kinetic energy for incident light of frequency ν_0 is

$$\frac{1}{2} mv_m^2 = h\nu_0 - \phi_0 \quad , \tag{4.19}$$

where ϕ_0 is the work required to remove an electron from just under the surface of the metal.

The photoelectrons ejected by X-rays come from the inner shells of the atoms. When the binding energy of the q shell is written as $E_q = h\nu_q$, the energy-conservation equation is

$$E_{kin} = \frac{1}{2} mv_{m,q}^2 = h\nu_0 - h\nu_q \quad , \tag{4.20}$$

where we have neglected the small quantity ϕ_0. If $\nu_0 < \nu_K$, no photoelectrons can be ejected from the K shell.

As ν_0 increases above ν_K, the maximum energy of the photoelectrons increases linearly. The number n_q of photoelectrons ejected from the q shells of atoms of a given absorber decreases as ν_0 increases ($n_q \sim \tau_q/h\nu_0 \sim \lambda_0^4 Z^4$).

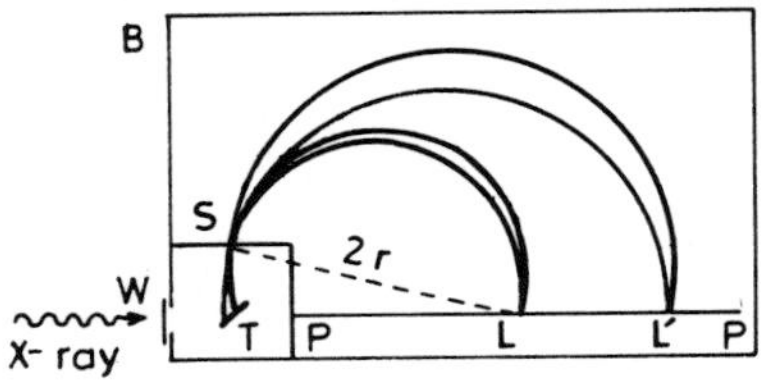

Fig. 4.2. Robinson's magnetic spectrograph

The number of photoelectrons is proportional to the intensity of the inciden X-rays.

These predictions were first verified by ROBINSON and his collaborators [4.10] with the help of a *magnetic spectrograph* (Fig.4.2). A beam of monochromatic X-rays enters, through a thin window W, an evacuated brass box B, and falls on a target T of the material under study. The whole apparatus is placed in a known magnetic field H, perpendicular to the plane of the diagra The *magnetic spectrum of photoelectrons* that are able to pass through the slit S is recorded on the photographic plate PP.

Electrons that enter S in a direction normal to $SL = 2r$, and electrons tha enter S in directions that make a small angle α with the normal, come togeth at L within a distance $\Delta = 2r(1-\cos\alpha) \simeq r\alpha^2$. Because α^2 is very small, the ele trons of *equal velocity* are said to be focused.

The radius of the circular path in the magnetic field is determined by th equation

$$Hev/c = mv^2/r \quad . \tag{4.21}$$

As the velocity of the electron is large, we express mass m as function of its rest mass m_0 and $\beta = v/c$,

$$m = m_0/(1-\beta^2)^{\frac{1}{2}} \quad . \tag{4.22}$$

Therefore, β can be calculated from the relation

$$r = \frac{\beta}{eH} \frac{m_0c^2}{(1-\beta^2)^{\frac{1}{2}}} \quad , \tag{4.23}$$

where r and H are known from the experiment. The corresponding kinetic energ of the electron is the difference between its total energy mc^2 and the rest energy m_0c^2,

$$E_{kin} = m_0c^2\left[(1-\beta^2)^{-\frac{1}{2}}-1\right] \quad . \tag{4.24}$$

We can now use (4.20) to find ν_q, for a given ν_0. Robinson's values agree with the X-ray spectroscopic values of binding energies.

The technique of XPS has been greatly refined these days (see [4.11] for a recent review). Compilations of inner atomic energy levels have been made by combining XPS data on binding energies $h\nu_q$ with X-ray emission line data (Sect.2.14). Most of the data in the tables of BEARDEN and BURR [4.12] and SIEGBAHN et al. [4.13] refer to solids. LOTZ [4.14] has tried to convert these data to free atoms.

Soft X-ray ($h\nu_0 \sim 100$ eV) or ultraviolet photoelectron spectroscopy (SXPS) is useful in surface studies and valence level studies. At this energy, the mean free path of the emitted electron is of the order of 10 Å and so can come out without further collision with atoms in the solid if it is liberated at the surface. It is then referred to as ESCA (electron spectroscopy for chemical analysis), as it has chemical applications.

For metallic samples, it is convenient to use the Fermi level in the sample as a reference level for measuring the binding energy of a core level. In general, the energy conservation equation is

$$h\nu_0 = E_q^{\text{vac}} + E_{\text{kin}}^{\text{vac}} \quad , \tag{4.25}$$

where E_q^{vac} is the binding energy of the q electron referred to the vacuum level of the sample,

$$E_q^{\text{vac}} = E^{\text{vac}} - E_q \quad . \tag{4.26}$$

Here E_q is the total energy of the initial state and $E^{\text{vac}} = E_f(q)$ is the total energy of the final state with a hole in the q subshell. If $E_{\text{kin}}^{\text{Fermi}}$ is defined as the kinetic energy that is measured in the magnetic spectrograph, then the photoelectron will behave as if it has adjusted to the spectrograph work function ϕ_{sp}, and the energy equation becomes

$$h\nu_0 = E_q^{\text{Fermi}} + E_{\text{kin}}^{\text{Fermi}} + \phi_{sp} \quad , \tag{4.27}$$

as shown in Fig.4.3. Thus, ϕ_0 does not enter in (4.27). Clearly, $E_q^{\text{vac}} - E_q^{\text{Fermi}} = \phi_0$, and $E_{\text{kin}}^{\text{vac}} - E_{\text{kin}}^{\text{Fermi}} = \phi_{sp} - \phi_0$. It is enough to measure ϕ_{sp} once. NORDLING [4.15] has used (4.27) to determine the binding energies of K and L electrons in elements like copper, zinc, and gallium. If the sample is not a metal, ϕ_0 is replaced by $\phi_0 + \phi_c$, where ϕ_c accounts for the charge build up on the sample.

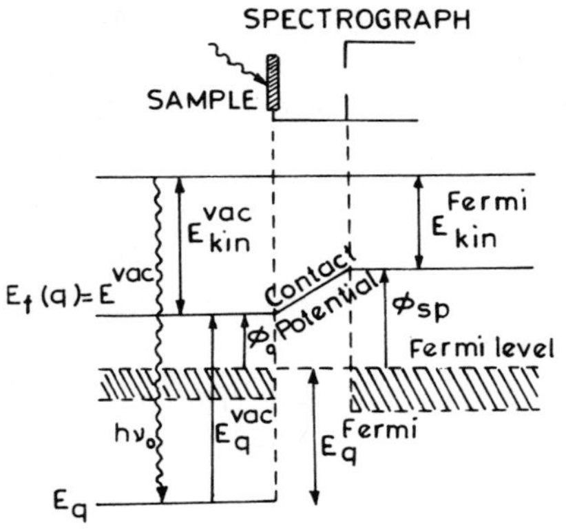

Fig. 4.3. Energy-level diagram of a metallic sample in electrical contact with the spectrograph

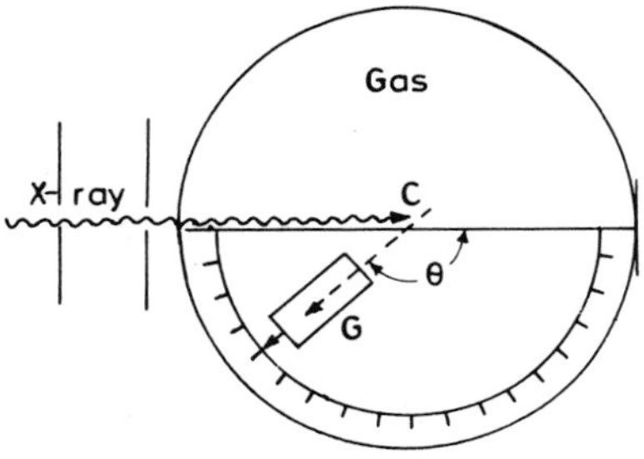

Fig. 4.4. Bothe's arrangement for the study of angular distribution of photo electrons ejected from gas atoms

The angular distribution of photoelectrons has been studied by BOTHE [4.16] in gases and by WATSON [4.17] in solids.

In Bothe's work, the X-rays passed through a gas-filled drum; some of the photoelectrons produced near the center at C were detected by a Geiger counter G placed at an angle θ with the beam (Fig.4.4).

WATSON used the method of magnetic analysis from solid samples placed at the center of twin-quadrant shaped magnetic spectrographs (Fig.4.5). The two spectrographs (I and II) can be rotated about an axis to change θ. The measurements confirm the theoretical angular distribution (4.18).

4.4 Auger Effect and Its Consequences in X-ray Spectra

MEITNER [4.18] and ROBINSON [4.19] found that atoms ionized in inner shells emit monoenergetic electrons, with energies that do no depend upon the manner

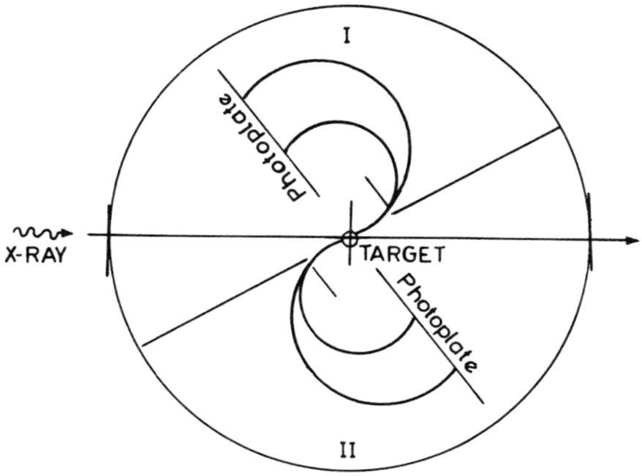

Fig. 4.5. Watson's magnetic spectrograph

in which the atoms are ionized. AUGER [4.20] used a Wilson chamber to study the photoeffect and observed a group of *slow* monoenergetic electrons, whose energies could not be explained by the energy-conservation equation (4.20). These electrons are called *Auger electrons*. Auger found that:

1) The photoelectron and the accompanying Auger electron appear at the same point (or atom).

2) The direction of ejection of the Auger electron is random and independent of the direction taken by the photoelectrons.

3) If the energy $h\nu_0$ of the incident X-rays is increased, the range of the photoelectrons increases, but that of the Auger electrons remains constant; that is, the energy of the Auger electrons does not depend on $h\nu_0$.

4) The range of the Auger electrons increases with the atomic number Z of the sample atom.

5) Not all photoelectrons are accompanied by Auger electrons.

In Auger's experiments on argon, the ejection of the photoelectron left the argon atom with a hole in the K shell (Fig.4.6a). The atom can return to the ground state by various processes. We consider here two dominant processes. The first of these is the usual *radiative process* in which an electron from one of the upper levels (say, L level) fills the vacancy and a characteristic emission line (Kα) is emitted, $h\nu_{K\alpha} = E_K - E_L$. The second process involves a *radiationless transition* in which, instead of the emission of the Ar Kα line in the electronic transition from the L shell to the K shell, the energy $E_K - E_L$ is taken up to eject a second L (or M) electron, so that the atom gets

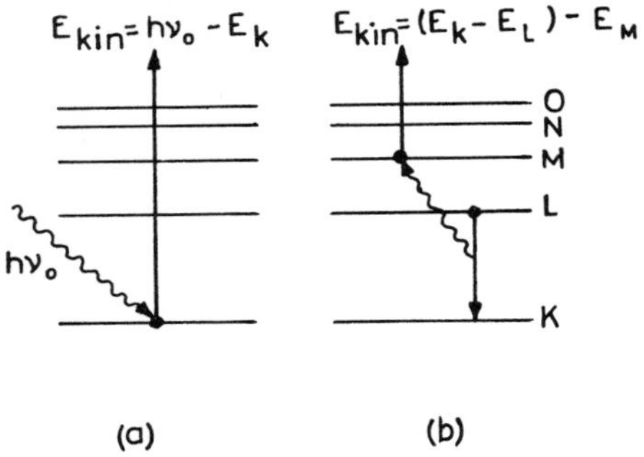

Fig. 4.6. (a) Photoelectron process and (b) Auger process (K→LM)

doubly ionized, K→LM (Fig.4.6b). Thus, in an Auger transition, an electron is emitted rather than a photon when an excited atom goes to a lower energy state.

The notation K→LM (also written as KLM) means that the initial vacancy is in the K level, and the final vacancies are in the L and M levels; the radiationless transition is K→L, and the Auger electron is ejected from the M level.

The Auger process has three important consequences in X-ray spectra:

1) The Auger process competes with the radiation process in the de-excitation of an atom. Therefore, according to the theory of Weisskopf and Wigner (Sect.3.11) it *influences the width of X-ray emission lines.*

2) The Auger process transfers a vacancy in one shell to another, and so *influences the intensity of X-ray emission lines.*

3) The Auger process leads to a state of double ionization in an atom, so it *provides one cause for the appearance of the satellite lines in X-ray spectra* that arise from transitions in atoms doubly ionized in inner shells.

The Auger effect has been recently reviewed by BURHOP and ASAAD [4.21], and by BAMBYNEK et al. [4.22].

4.4.1 Auger Effect and Widths of X-ray Emission Lines and Absorption Edges

The probability that an atom in the state i has an energy between E and E+dE is given by (3.178)

$$w_i(E)dE = \frac{\Gamma_i/2\pi}{(E_i-E)^2+(\Gamma_i/2)^2}\, dE \quad , \tag{4.2}$$

where E_i is the most-probable energy of the state, and Γ_i is the width of the state. If in a transition $i \to f$ a photon of approximate frequency $\nu_{fi} = (E_i - E_f)/h$ is emitted, then by (3.183),

$$I_{if}(\nu) = \frac{(2\pi)^{-2}(\Gamma_i+\Gamma_f)\nu}{(\nu_{fi}-\nu)^2+\left[(2h)^{-1}(\Gamma_i+\Gamma_f)\right]^2} \quad . \tag{4.29}$$

Here $I_{if}(\nu)$ is the energy radiated per transition in the frequency range ν and $\nu + d\nu$, and $\Gamma_i + \Gamma_f$ is the sum of the widths of the initial and the final states.

Let p_i^R and p_i^A be the probabilities of an atom for leaving the initial excited state i by radiative and Auger transitions, respectively. The mean lifetime τ of the state i is given by,

$$\tau_i = 1/\left(p_i^R+p_i^A\right) \quad . \tag{4.30}$$

The uncertainty principle gives for the width Γ_i of the initial state

$$\Gamma_i = \hbar/\tau_i = \hbar\left(p_i^R+p_i^A\right) = \Gamma_i^R + \Gamma_i^A \quad , \tag{4.31}$$

where $\Gamma_i^R = \hbar p_i^R$ is the *radiation width* and $\Gamma_i^A = \hbar p_i^A$ is the *Auger width*. Thus, the total width of the initial-hole state is made up of two partial widths. The absorption can be similarly treated.

4.4.2 Auger Effect and the Intensities of X-Ray Emission Lines

Let n_i be the number of atoms per unit time ionized in an inner shell i. At equilibrium, this will equal the rate at which atoms leave the state i by all possible transitions, radiative or Auger, to states of equal or lower energy. Let $p_{i\to f}^R$ be the transition rate for a radiative transition $i \to f$, and $\Gamma_{i\to f}^R = \hbar p_{i\to f}^R$ the corresponding width.

The total number $n_{i\to f}$ of radiation quanta that correspond to the transition $i \to f$ emitted per unit time is

$$n_{i\to f} = n_i \Gamma_{i\to f}^R/\left(\Gamma_i^R+\Gamma_i^A\right) \quad . \tag{4.32}$$

Thus the Auger transition probability affects the absolute intensity of the X-ray line. For the transition $a \to b$,

$$n_{a\to b} = n_a \Gamma_{a\to b}^R/\left(\Gamma_a^R+\Gamma_a^A\right) \quad . \tag{4.33}$$

The relative intensity of the two lines,

$$\frac{I_{i\to f}}{I_{a\to b}} = \frac{\nu_{fi} n_i \Gamma^R_{i\to f}\left(\Gamma^R_a + \Gamma^A_a\right)}{\nu_{ba} n_a \Gamma^R_{a\to b}\left(\Gamma^R_i + \Gamma^A_i\right)} \quad , \tag{4.34}$$

is also influenced by the Auger process.

If the initial state is the same, the relative intensity

$$\frac{I_{i\to f}}{I_{i\to b}} = \frac{\nu_{fi}\Gamma^R_{i\to f}}{\nu_{bi}\Gamma^R_{i\to b}} \quad , \tag{4.35}$$

is independent of the Auger transitions from the initial state. The effect on the satellite lines will be discussed in Sect.4.7.

4.5 Basic Theory of Auger Effect

There are two ways of looking at an Auger process:

1) It is a process of *radiationless reorganization* of an atom ionized in an inner shell because of a *direct interaction of two electrons*.

2) A quantum of X-rays is first produced, then absorbed by an outer electron of the atom in which it originates, and finally the outer electron is ejected from the atom (*internal conversion*).

In the nonrelativistic limit, both points of view lead to the same result [4.23].

4.5.1 The Nonrelativistic Theory Based on Direct Interaction of Two Electrons

Let $\psi_i(\underline{r}_1)$ and $\chi_i(\underline{r}_2)$ be the initial single-electron wave functions of the two electrons concerned, in an atom with a hole in an inner level. Let $\psi_f(\underline{r}_1)$ be the final wave function of one of the electrons in this level and $\chi_E(\underline{r}_2)$ the final wave function of the other electron in a continuum state of positive energy. The Coulomb interaction $e^2/|\underline{r}_1-\underline{r}_2|$ is the perturbation that causes this transition. The number of radiationless transitions that occur in time dt is given by the first-order perturbation theory as (see, for example, [4.24])

$$b_n dt = \frac{2\pi}{\hbar}\left|\iint \psi_f^*(\underline{r}_1)\chi_E^*(\underline{r}_2)\,\frac{e^2}{|\underline{r}_1-\underline{r}_2|}\,\psi_i(\underline{r}_1)\chi_i(\underline{r}_2)d^3\underline{r}_1 d^3\underline{r}_2\right|^2 dt \quad . \tag{4.36}$$

We shall now see qualitatively under what conditions $b_n dt$ is large. Consider the well-known Auger process $L_I \to L_{III}M_{IV,V}$, where the radiationless transition $\psi_i \to \psi_f$ corresponds to $L_I \to L_{III}$. The wave functions are

$$\begin{aligned}
L_I &: \quad \psi_i(\underline{r}_1) = u_{2s}(\underline{r}_1)/r_1 \quad , \\
L_{III} &: \quad \psi_f(\underline{r}_1) = u_{2p}(\underline{r}_1)P_1(\cos\theta_1)/r_1 \quad , \\
M_{IV,V} &: \quad \chi_i(\underline{r}_2) = u_{3d}(\underline{r}_2)P_2(\cos\theta_2)/r_2 \quad , \\
E > 0 &: \quad \chi_E(\underline{r}_2) = \sum_l (2l+1)f_l(kr)P_l(\cos\theta)/kr_2 \quad .
\end{aligned} \tag{4.37}$$

Here $u_{nl}(r)/r$ is the radial part of the wave function, and for the continuum, $f_l(kr) \sim \sin(kr - \frac{1}{2}l\pi + \delta_l)$, δ_l being the phase shift caused by the perturbing potential [4.24]. The angle θ is between $\underline{r}_2$ and the propagation vector $\underline{k}$ of the Auger electron that has the energy $E = \hbar^2k^2/2m$. We consider only the simple case of the magnetic quantum number $m = 0$. The expansion of $1/r$ is

$$1/|\underline{r}_1 - \underline{r}_2| = \sum_{n=0}^{\infty} a_n(r_1, r_2)P_n(\cos\Theta)$$

$$a_n = \begin{cases} r_1^n/r_2^{n+1} & \text{for } r_1 < r_2 \\ r_2^n/r_1^{n+1} & \text{for } r_2 > r_1 \end{cases} \quad ,$$

$$\Theta = \text{angle between } \underline{r}_1 \text{ and } \underline{r}_2 \quad . \tag{4.38}$$

From (4.36-38), we get the transition rate into unit solid angle in the direction of $\underline{k}$. If $\underline{k}$ makes an angle α with the polar axis, and β is the azimuthal angle, we can integrate over all angles of ejection to find the total rate for the Auger transition. The expansion

$$P_l(\cos\theta) = P_l(\cos\theta_2)P_l(\cos\alpha) + 2\sum_{m=1}^{l} \frac{(l-m)!}{(l+m)!} P_l^m(\cos\theta_2)P_l^m(\cos\alpha)\cos m(\phi_2 - \beta) \quad ,$$

and a similar expansion for $P_n(\cos\Theta)$, show that the angular integrations to be performed are

$$\iint P_n(\cos\Theta)P_1(\cos\theta_1)\sin\theta_1\, d\theta_1\, d\phi_1 \quad , \tag{4.39}$$

which vanishes unless $n = 1$, and

$$\iint P_1(\cos\theta_2)P_2(\cos\theta_2)P_l(\cos\theta)\sin\theta_2\, d\theta_2\, d\phi_2 \quad , \tag{4.40}$$

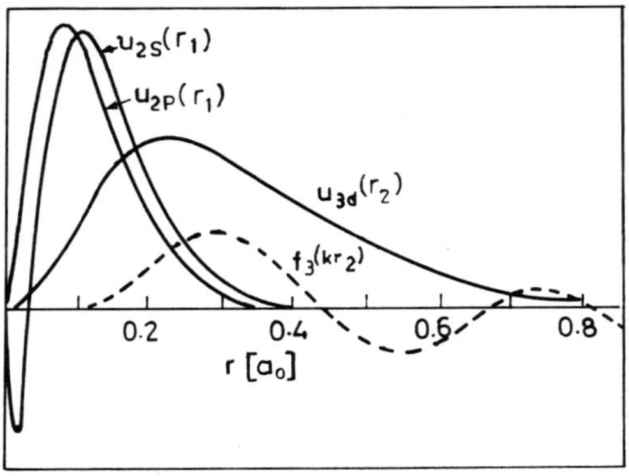

Fig. 4.7. Radial functions for the Auger-transition probability for the process $L_I \to L_{III}M_{IV,V}$

which vanishes unless $l = 1$ or 3. Therefore, the radial integrals to be evaluated are of the type

$$\iint a_n(r_1,r_2)u_{2s}(r_1)u_{2p}(r_1)u_{3d}(r_2)f_l(kr_2)\, dr_1\, dr_2 \quad , \tag{4.41}$$

with $l = 1$ or 3 and $n = 1$.

In Fig.4.7 we show the forms of the Hartree-type functions u_{nl} for $Z \sim 50$. The functions $u_{2s}(r_1)$ and $u_{2p}(r_1)$ overlap considerably to make the integral over r_1 large. The figure also shows a typical form of $f_{l=3}(kr_2)$. The asymptotic form is given by [4.25]

$$f_l = \frac{(2\pi)^{\frac{1}{2}}}{[E+2V-l(l+1)/r^2]^{1/4}} \cos\left\{ \int_R^r [E+2V(\rho)-l(l+1)/\rho^2]^{\frac{1}{2}} d\rho \right\} \quad ,$$

where V is the effective Coulomb potential and R is the lower limit chosen so as to obtain coincidence of zero points with the numerically integrated function. If f_l has a zero in the region of r_2 where $u_{3d}(r_2)$ is large, the Auger transition rate will be small. For small k (low velocity of the ejected Auger electron) only the $l = 1$ partial wave will contribute significantly and the Auger transition probability will increase with k. As k increases, the first zero of $f_l(kr_2)$ occurs for smaller values of r_2 and eventually falls in the region where $u_{3d}(r_2)$ is large. Therefore, the major contribution to (4.41) comes from $l = 3$. For larger values of k, a great deal of cancellation occurs in the r_2 integration for all l, and the Auger transition probability decreases again. The kinetic energy $\hbar^2k^2/2m$ of the ejected Auger electron will vary considerably with Z. Therefore, we expect the Auger transition probability to depend strongly on Z. Inclusion of the exchange integral does not alter these conclusions.

COSTER and KRONIG [4.26] were first to employ such arguments to clear up many puzzling anomalies in L series. Detailed calculations have been carried out for Au by MASSEY and BURHOP [4.27]. They considered transitions of the type $X_i \rightarrow X_f Y$, where X_i and X_f pertain to the same shell (say, L) but different subshells (say, L_I and L_{III}). Such transitions are said to give rise to radiationless processes of the *Coster-Kronig type*.

4.5.2 Possible Auger Transitions of the Coster-Kronig Type ($X_i \rightarrow X_f Y$)

Let E_{X_i}, E_{X_f} be the ionization energies of the two inner sublevels and E_y the energy of the outer level from which the Auger electron is ejected in the Coster-Kronig-type process $X_i \rightarrow X_f Y$. The kinetic energy of the Auger electron is

$$E_{kin} = E_{X_i} - E_{X_f} - E_Y \quad . \tag{4.42}$$

The process will occur only if $E_{kin} > 0$, or

$$E_{X_i} - E_{X_f} > E_Y \quad . \tag{4.43}$$

In Fig.4.8, we show the variation of $E_{X_i} - E_{X_f}$ and E_Y with Z for a number of levels that participate in the process. Clearly, the transition $L_I \rightarrow L_{III}M_{IV,V}$ is energetically possible for $Z < 50$ and again for $Z > 73$. Similarly, the transition $M_{III} \rightarrow M_{IV}N_{IV,V}$ can occur for $Z < 84$. In Table 4.1 we give a list of the possible Coster-Kronig transitions, according to COOPER [4.28]. In the curves of Fig.4.8 and the figures of Table 4.1, the energies of the inner levels correspond to the atomic number Z, whereas those for the outer level correspond to the atomic number $Z+1$. Because of the change of the screening when one inner electron is missing, the outer electron sees a field of charge $Z+1$, in the first approximation. Thus, for the transition $L_I \rightarrow L_{III}M_{IV}$, we can write, to a good approximation, $E_{kin} = L_I(Z) - L_{III}(Z) - M_{IV}(Z+1)$, where the level symbol denotes the energy of the corresponding level. For the transition to occur, we must have $[L_I(Z) - L_{III}(Z)] > M_{IV}(Z+1)$. CAUCHOIS [4.29] has constructed an energy-level diagram for the doubly ionized atoms from the data on the satellite lines. Table 4.2 shows $\Delta E(L_{III})$, the difference in the L_{III} ioniza-

Table 4.1. Possible Auger transitions of the Coster-Kronig type (X_iX_fY)

Process	Z	Process	Z	Process	Z	Process	Z
$L_IL_{II}N_I$	<70	$M_IM_{II}M_{IV}$	<34	$M_{III}M_VN_V$	<91	$N_IN_VN_V$	<87
N_{II}	<75	M_V	<34	N_{IV}	<89	$N_{II}N_{III}O_{III}$	>66 / <55
N_{III}	<81	$M_{III}N_V$	<87	N_{III}	<77		
N_{IV}	<92	N_1	<50	N_{II}	<70	O_{II}	<86 / >66 / <55
$L_{III}M_I$	<31	M_V	<34	N_I	<57		
M_{II}	<36	M_{IV}	<34	M_V	<37	N_{VII}	<80
M_{III}	<37	$M_{IV}M_V$	<44	$M_{IV}M_VO_{III}$	<85	N_{VI}	<80
M_{IV}	>73 / <50	M_{IV}	<43	O_{II}	<82	$N_{IV}N_V$	<60
M_V	>77 / <50	M_VM_V	<45	$N_IN_{II}O_{III}$	<88	N_{IV}	<60
		$M_{II}M_{III}N_{IV}$	>65 / <48	O_{II}	<85	N_VN_V	<62
$L_{II}L_{III}M_{IV}$	<30	N_V	>65 / <48			$N_{III}N_{IV}O_I$	<87
M_V	<30 / >90	$M_{IV}N_I$	<88	O_I	<79	N_{III}	<86
		M_V	<35	N_{VII}	<80	N_{VI}	<85
$M_IM_{II}N_{VII}$	<91	M_{IV}	<35				
N_{VI}	<91	$M_{III}M_{IV}N_V$	<85	N_{VI}	<80	N_V	<57
N_V	<72	N_{IV}	<84	N_V	<53	N_{IV}	<56
N_{IV}	<71	N_{III}	<71	N_{IV}	<53	N_VO_I	<89
N_{III}	<53	N_{II}	<65	$N_IN_{III}N_{VI}$	<92	N_{VIII}	<87
						N_{VI}	<86
N_{II}	<53	N_I	<55	N_V	<53	N_V	<57
N_I	<47	M_V	<36	N_{VI}	<53	$N_{IV}N_VO_{IV,V}$	<81
		M_{IV}	<36	$N_{IV}N_V$	<84	$N_VN_{VI}N_{VII}$	<91
				N_{IV}	<80	N_{VI}	<91
						$N_{VII}N_{VII}$	<91

tion energy for an atom already ionized in the $M_{IV,V}$ shell and a normal atom. We can write, in our notation

$$\Delta E(L_{III}) = (L_{III}M_{IV,V}-M_{IV,V}) - L_{III} = \left[L_{III}(Z)+M_{IV,V}(Z+1)-M_{IV,V}(Z)\right] - L_{III}(Z)$$
$$= M_{IV,V}(Z+1) - M_{IV,V}(Z) \quad . \qquad (4.44)$$

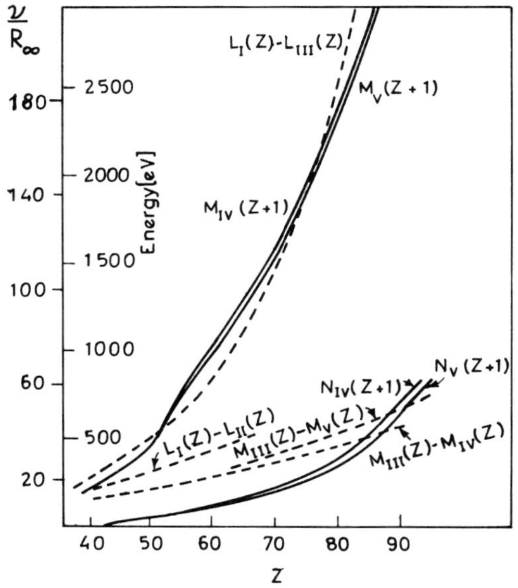

Fig. 4.8. Variation with Z of the ionization energies of the M_{IV}, M_V, N_{IV}, N_V levels (solid curves) and the differences in ionization energies $L_I - L_{II}$, $L_I - L_{III}$, $M_{III} - M_{IV}$, $M_{III} - M_V$ (dotted curves) to find for what ranges of Z the Coster-Kronig transitions can occur

Table 4.2. Difference of X-ray energy levels in eV between atoms doubly ionized and singly ionized in an inner shell [4.29]

Element	$\Delta E(L_{III})$	$\Delta E(M_V)$	$\Delta E(N_V)$	$\Delta E(O_V)$	$E_{M_{IV}}(Z+1) - E_{M_{IV}}(Z)$	$E_{M_V}(Z+1) - E_{M_V}(Z)$
78 Pt	89	52	19	0	94	89
79 Au	92	53	19	0.4	94	89
81 Tl	98	55	21	1.1	108	95
82 Pb	101	57	23	2.0	99	92
83 Bi	104	60	27	2.5	-	-
88 Ra	116	65	28	4.0	-	-
90 Th	125	70	36	6.3	122	109

This equation is approximately satisfied (Table 4.2). This table also shows $\Delta E(M_{IV,V})$, $\Delta E(N_{IV,V})$ and $\Delta E(O_{IV,V})$.

We have mentioned earlier that at the ionization of the inner shells by electrons or photons there is a finite probability for an outer-shell electron to be simultaneously excited to a bound state (shake-up) or to a continuum state (shake-off). In Auger-electron spectra these multiply excited initial states give rise to satellite peaks. Other satellite peaks occur because of the excitations in the final state.

The shake-off in the M shell at the ionization of the K shell in Ar gives a satellite on the low-energy side of the normal KLL Auger line. The energy of the normal Auger transition can be written as $E = K - LL$. For the case of shake-off in the M shell, we have $E^M = KM - LLM$. The shift is $\Delta = E - E^M = K - KM - (LL - LLM)$, where $KM - K$ = BE of the M electron in the ion with a K hole = ionization potential (I) of KII. Similarly, $LLM - LL$ = ionization potential of CaIII. Thus, $\Delta = I(\text{CaIII}) - I(\text{KII}) = 51.21 - 31.81 \simeq 19.4$ eV, which is in approximate agreement with the observed value 17.0 eV [4.30].

4.5.3 Auger Transitions and Widths of L and M Levels

When the probability of Auger transition from the L_I excited state is large, the lifetime of this state is reduced and consequently its width is increased. RICHTMYER et al. [4.31] found this to be so in their study of L_I, L_{II} and L_{III} absorption edges in ^{79}Au. A fit of the arctangent curve, (3.189), to the observed shape of the edge, yields the width of the initial state responsible for the absorption. The data on the widths of L levels are also available from precise measurements of the emission line widths. In Fig.4.9, we show the variation of the observed widths of the lines of the L series with Z [4.32,33].

We need more information on the widths of the final states to be able to correctly assess L-level widths from these data. However, the large increase of the widths of the group of L_I lines (γ_3, γ_2, β_3, β_4) for $Z > 70$ is obvious (Fig.4.9). The lines β_1, γ_1, β_2 originate from an initial L_{II} or L_{III} state.

This marked increase with Z of the widths of L_I lines is associated with an increase of width of the L_I level for just these elements ($Z > 73$), because of the possibility of the Auger transition. In the absence of a selective

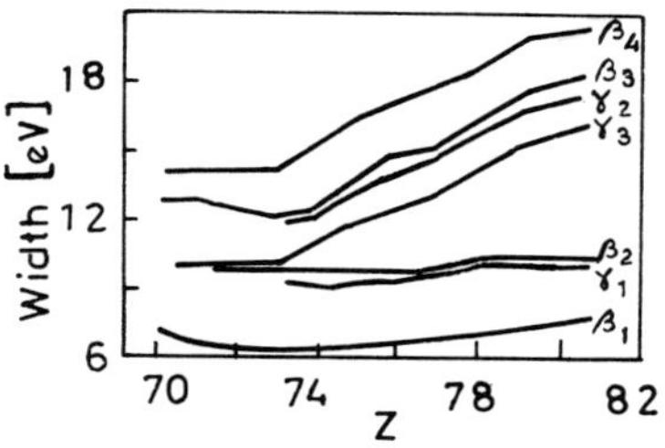

Fig. 4.9. Variation of the widths of the L series lines as a function of Z

Auger effect of this kind, the dependence of width on Z is determined [4.34] by the radiative transition probabilities, that are proportional to Z^4. The Auger width of the K level does not depend on Z [4.35]. Therefore, a general increase of the total widths with Z can be expressed as

$$\Gamma = A + RZ^4 \quad , \tag{4.45}$$

where $A = \mathrm{const}(Z)$ gives the Auger contribution, and RZ^4 the radiative contribution.

4.5.4 Auger Transitions and Relative Intensities of L-Series Lines

We have seen, (4.32), that the intensities of lines are influenced by the presence of the Auger process. The Auger width in (4.34) is increased for atoms in the L_1 state, as just discussed. Therefore, the intensity of lines, that have L_I as the initial state, is reduced, relative to the intensity of lines that have L_{II} and L_{III} as initial states. The physical reason is that because of the radiationless transitions, the holes in L_I are transferred to L_{II} or L_{III}. These levels consequently gain in holes at the expense of L_I, and so give relatively stronger emission lines when the radiative transitions finally occur.

In Fig.4.10, we show the variation with Z of the observed intensities [4.32,33,36] of the lines $L\beta_3$ ($L_I \rightarrow M_{III}$) and $L\beta_4$ ($L_I \rightarrow M_{II}$) relative to $L\beta_1$ ($L_{II} \rightarrow M_{IV}$). The variation can be understood on the basis of the Coster-Kronig theory.

Figure 4.8 shows that, for the transition $L_I \rightarrow L_{III}M_{IV,V}$, the kinetic energy of the Auger electron increases from zero for Z slightly above 50 to about 100 eV for Z = 41, and increases rapidly for smaller Z. On the basis of the nonrelativistic theory given earlier, COSTER and KRONIG estimated that

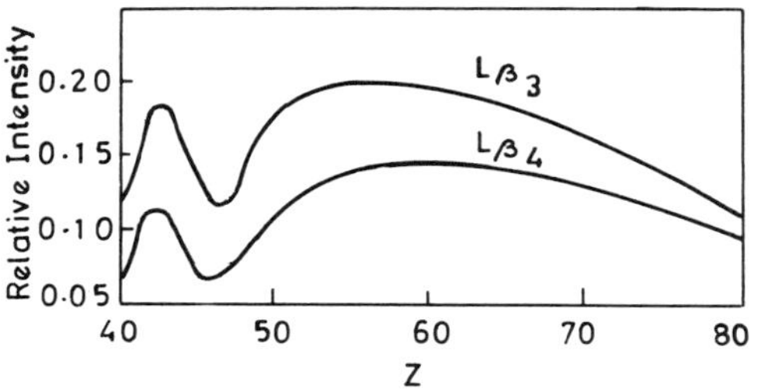

Fig. 4.10. Intensities of the lines $L\beta_3$ and $L\beta_4$, relative to that of $L\beta_1$, as a function of Z

the Auger transition for this transition becomes maximum for an ejected-electron energy of about 100 eV. A maximum for such a small value of E_{kin} implies that *each of the possible Auger transitions will occur with appreciable probability for Z values near the cut-off for the transition concerned.* With the help of Table 4.1, we can now use this rule of thumb to explain the variation shown in Fig.4.10.

For $Z = 40$, the Auger-transition probability for $L_I \to L_{II}M_{IV,V}$ is maximum by the above rule, and the relative intensities of $L\beta_3$ and $L\beta_4$ lines that originate in L_I show a decrease. For $Z = 42$ this Coster-Kronig transition is no longer possible (Fig.4.8); consequently the relative intensities increase. As Z increases further, the transition $L_I \to L_{III}M_{IV,V}$ becomes possible, with a large probability at $Z = 45$ or 46. Therefore, the curves of Fig.4.10 show a minimum around this Z value. This transition ceases to occur at $Z = 52$; therefore, the curves rise again. The final fall of the curves is because of the Coster-Kronig transition $L_I \to L_{III}M_{IV,V}$ that begins to occur at $Z > 73$.

Measurements [4.32] show that the intensities of $L\beta_2$, $L\beta_4$ relative to $L\beta_1$ and of $L\gamma_2$ ($L_I \to N_{II}$), $L\gamma_3$ ($L_I \to N_{III}$) relative to $L\gamma_1$ ($L_{II} \to N_{IV}$) decrease with increasing Z for $Z > 73$. On the other hand, the intensity of $L\beta_2$ ($L_{III} \to N_{VI}$) relative to $L\beta_1$ ($L_{II} \to M_{IV}$) remains constant for $Z > 73$, as would be expected on the basis of the foregoing arguments.

It is clear that the Auger process is very complex and is not yet fully understood. Even for free atoms, the calculations are difficult. Recently McGUIRE [4.37] has made calculations for heavy atoms where jj coupling is applicable.

4.6 Detection of Auger Electrons

Auger-electron spectroscopy (AES) has become an important tool in the study of atoms, molecules and solids. Therefore, several techniques have been developed for their detection and analysis.

In Robinson's magnetic-spectrograph analysis, the lines of Auger electrons could be easily separated from those of the photoelectrons, by using the fact that the kinetic energy of the former, unlike that of the latter, is independent of the incident-photon energy. Also, the relative intensities of photoelectrons for a given incident radiation do not remain constant for large changes of Z, whereas the Auger-electron intensities remain approximately constant. Auger electrons have also been detected in the magnetic spectra of electrons from radioactive sources [4.38,39].

The Wilson cloud-chamber method was used by AUGER [4.20], LOCHER [4.40], MARTIN et al. [4.41,42], and BOWER [4.43], to photograph the Auger tracks in gases and vapors. As the frequency of the narrow beam of incident X-rays is increased, the length of the tracks of the photoelectrons increases, but that of the Auger electrons remains constant. The length of the Auger-electron tracks increases with the atomic number Z of the gas. The typical approximate values of the Auger-electron energy in the transition K→LL for different gases are given in Table 4.3. For elements of low Z, the Auger tracks are very short. For 25 keV X-rays, ^{36}Kr ($E_K \simeq 14$ keV) gives photoelectrons and Auger electrons of almost the same range.

Auger used two cameras, mounted at right angles, to photograph the tracks. From the two photographs obtained for every track, the difference of the angular distribution of the photoelectrons and the Auger electrons could be easily deduced.

Auger electrons have small and characteristic energies (mean free path ~10 Å). They are useful in the study of solid surfaces. LANDER [4.44] has detected Auger electrons that come out of the metal surface. We have seen that, in general, three different levels are involved in an Auger process. The initial vacancy of a core level E_1 is filled by an electron from a higher level E_2, and the energy $E_2 - E_1$ set free by this transition is transferred to another electron in the valence band E_3, that comes out with a characteristic kinetic energy E_{kin} given by (see Fig.4.11),

$$E_2 - E_1 = E_F - E_3 + E_{kin} + \phi_{sp} \quad .$$

Note that the experiment does not distinguish between the transitions (I) $E_1 \to E_2E_3$ and (II) $E_1 \to E_3E_2$, because the initial and final electron configurations do not differ.

Table 4.3. Auger-electron energies for the Auger transition KLL for different gases

Element	$E_K - 2E_L$
18 A	3.14 keV
36 Kr	10.7
54 Xe	24.15
80 Hg	55.4

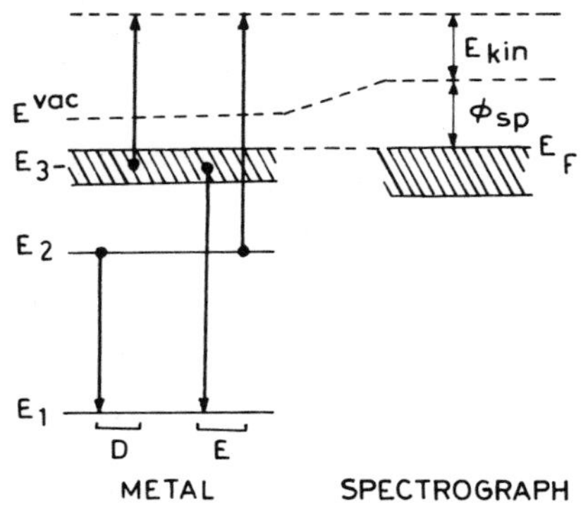

Fig. 4.11. Auger electrons come out of the surface with the same E_{kin} in the direct (D) and exchange (E) Auger processes I and II

The E_{kin} for the Auger electrons is a characteristic quantity of the element and so its determination can identify the presence of certain elements in the sample surface (ESCA). To detect these electrons, use can be made of the retarding-field analyzer (RFA), which collects all electrons with sufficient energy to surmount a variable retarding electric field in the detector Dispersive types of magnetic or electrostatic analyzers can also be used.

Recently McGILP et al. [4.45] measured the $N_{VI,VII}O_{IV,V}O_{IV,V}$ Auger spectr of Tl, Pb and Bi using MgKα X-ray excitation. The observed spectra are in good agreement with calculations if the initial and final states are treated in jj and jj - LS (intermediate) coupling, respectively. A broad background feature with a maximum at ~90 eV is attributed to the presence of NNO Coster Kronig and NNN super Coster-Kronig transitions that give rise to Auger electrons of kinetic energies between 20 eV and 120 eV. A super Coster-Kronig transition arises if the available energy is enough, not only to excite, but to eject an electron from the same shell. The available kinetic energy is shared by the super Coster-Kronig electron and the photoelectron, and so the latter shows a continuum instead of a sharp line in the spectrum.

4.7 Satellites

Close to prominent allowed diagram lines weak lines often occur in X-ray emi sion spectra. Because of their closeness to a strong parent line, they are called *satellites*. They are also called *nondiagram lines* because they cannot be understood as simple transitions on the usual X-ray energy-level diagram of a singly ionized atom (Fig.2.12). The reason is that the X-ray satellite lines arise from transitions in atoms multiply ionized in inner shells. For example, $K\alpha_3$, $K\alpha_4$, the satellite lines of the parent line $K\alpha_{1,2}$ ($K\rightarrow L$), are produced by the transitions $KL\rightarrow LL$, $KK\rightarrow KL$, in doubly ionized atoms. The satellite lines usually occur on the high-energy side of the diagram lines. They were first observed by SIEGBAHN and STENSTRÖM [4.46]. Atoms doubly ionized in inner shells may be produced 1) directly by the cathode-electron impact, or 2) by an Auger transition after a single ionization.

The lifetime of an X-ray state is very short, $\sim 10^{-16}$ s. Also, the density of electrons in the cathode beam is small compared to that of target atoms. Therefore, it is almost impossible for the same atom to be struck twice in succession, resulting in two electrons being knocked out. In fact, to produce the satellites on the high-energy side of a $K\alpha$ line, the cathode electrons in an X-ray tube had to be given a minimum energy $E_K + E_L$, by increasing the voltage, that was sufficient to knock out a K and an L electron simultaneously, by a single impact. This led WENTZEL [4.47] to postulate the existence of *an initial state of double ionization of the atom for the production of satellites*.

DRUYVESTEYN [4.48] used this postulate to assign double-ionization states for the observed satellites and to calculate their frequencies. In the Druyvesteyn-Wentzel theory, a satellite is produced when a *single* transition of the type $AY\rightarrow YB$ occurs among the doubly ionized states. The satellite is situated on the high-energy side of the parent line $A\rightarrow B$. The additional hole Y acts as a spectator.

The frequency $\nu_{K\alpha 3,4}$ of the satellite $K\alpha_{3,4}$ ($KL\rightarrow LL$) that belongs to the parent line $K\alpha_{1,2}$ ($K\rightarrow L$) of an element of atomic number Z, is given by

$$\left(h\nu_{K\alpha_{3,4}}\right)_Z = (KL)_Z - (LL)_Z \quad . \tag{4.46}$$

For the parent line

$$\left(h\nu_{K\alpha_{1,2}}\right)_Z = K_Z - L_Z \quad . \tag{4.47}$$

In the first approximation, we can write

$$(KL)_Z = K_Z + L_{Z+1} \quad , \tag{4.48}$$

$$(LL)_Z = L_Z + L_{Z'+1} \quad , \tag{4.49}$$

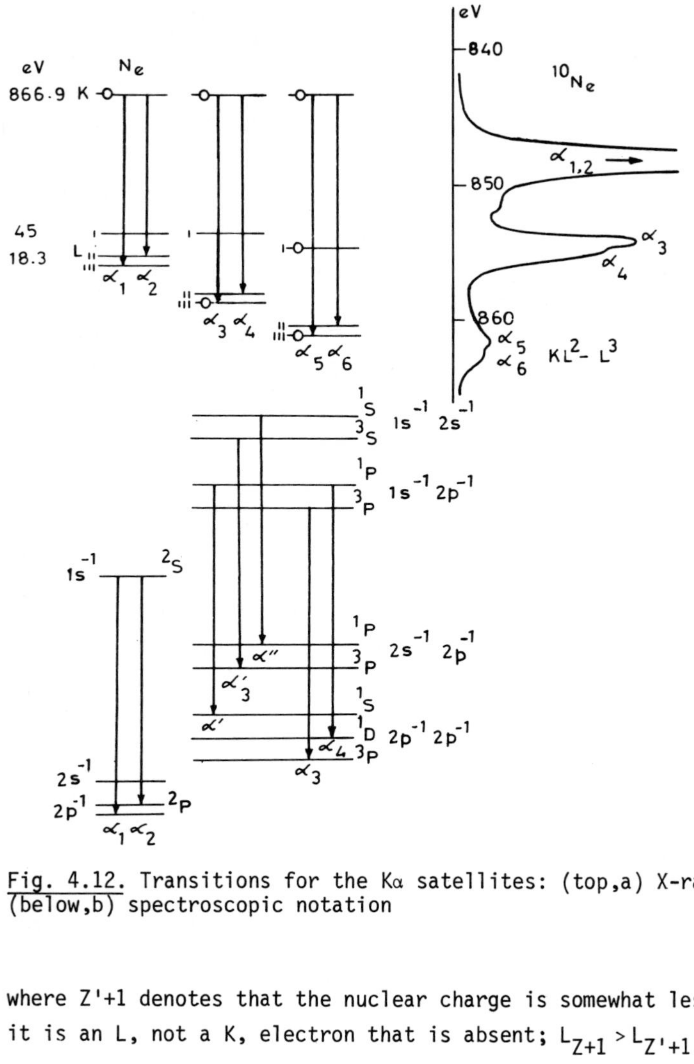

Fig. 4.12. Transitions for the Kα satellites: (top,a) X-ray notation, and (below,b) spectroscopic notation

where Z'+1 denotes that the nuclear charge is somewhat less screened, because it is an L, not a K, electron that is absent; $L_{Z+1} > L_{Z'+1}$. It follows that

$$\left(h\nu_{K\alpha_{3,4}}\right)_Z = \left(h\nu_{K\alpha_{1,2}}\right)_Z + \Delta \quad , \tag{4.50}$$

where $\Delta = (L_{Z+1} - L_{Z'+1}) > 0$ is a small *positive* quantity. Thus, more especially, the removal of the L_{III} (2p) electron increases the distance and hence the energy difference between L (2p) and K (1s) levels, with the resulting production of the $K\alpha_3$ ($KL_{III} \rightarrow L_{III}L_{III}$), $K\alpha_4$ ($KL_{III} \rightarrow L_{III}L_{II}$) satellites (Fig. 4.12). Note that KL_{II} would be the same as KL_{III}. With improved experiments [4.49] and quantum mechanical calculations, five Kα satellites have been identified [4.50-56]. The spectroscopic notations now generally accepted are shown in Fig.4.12b. The notation is valid for the light elements ($Z < 16$), where LS coupling is applicable.

The configuration ($1s^{-1}\ 2s^{-1}$) gives rise to two even terms, $^1S_0$ and $^3S_1$. The J values that arise from $2p^{-1}$ (or $2p^5$) are the same as for a single p electron. Four terms with $J = 0,1,1,2$ arise. In the LS coupling, the terms of the ($1s\ 2p^5$) and ($2s\ 2p^5$) configurations can be classified as a singlet $^1P_1$ and a triplet $^3P_{0,1,2}$, the singlet-triplet difference being large compared to the small spread of the triplet (shown as one level in Fig.4.12b). The number of terms that arise from $2p^4$ (or $2p^{-1}\ 2p^{-1}$) is limited by Pauli's principle, because we have equivalent holes, and only $^1D_2$, $^1S_0$, $^3P_{0,1,2}$ occur in the LS coupling. In the LS coupling, we get the following five KL→LL multiplets:

Initial configuration	Transition	Multiplet	Satellite
$KL_{II,III}$	$(1s^{-1}\ 2p^{-1})^1P \rightarrow (2p^{-1}\ 2p^{-1})^1S$	$^1P - {}^1S$	α'
KL_I	$(1s^{-1}\ 2s^{-1})^3S \rightarrow (2s^{-1}\ 2p^{-1})^3P$	$^3S - {}^3P$	α_3'
KL_I	$(1s^{-1}\ 2s^{-1})^1S \rightarrow (2s^{-1}\ 2p^{-1})^1P$	$^1S - {}^1P$	α''
$KL_{II,III}$	$(1s^{-1}\ 2p^{-1})^3P \rightarrow (2p^{-1}\ 2p^{-1})^3P$	$^3P - {}^3P$	α_3
$KL_{II,III}$	$(1s^{-1}\ 2p^{-1})^1P \rightarrow (2p^{-1}\ 2p^{-1})^1D$	$^1P - {}^1D$	α_4

The splitting of ^1S and ^1D states can be found from α' and α_4 that have the same initial state.

For $10 \leq Z \leq 40$, a more detailed study given by DEMEKHIN and SACHENKO [4.57] is suitable where both LS and jj (and intermediate js) couplings are considered. The $K\alpha''\alpha_3\alpha_4$ (KL→LL) and $K\alpha_5\alpha_6$ (triple hole KLL→LLL) satellites have been observed recently in ^{10}Ne, under both primary excitation [4.58] and secondary excitation.

In early days, it was believed that satellites are not found in the fluorescent radiation. They have now been observed when the energy of the incident photon is sufficient to remove two electrons from an atom. In secondary excitation, α satellites have been studied by BONNELLE and SENEMAUD [4.59], and ÅBERG et al. [4.60] in recent years. SAWADA et al. [4.61] observed the $K\alpha_{3,4}$ satellites when the incident-photon energy was enough to remove a K electron of atom Z and an L electron of atom Z+1. They found (Table 4.4), for example, that a Ni $K\alpha_1$ incident photon is able to excite only the $K\alpha_{1,2}$ lines of Fe, but a Cu $K\alpha_1$ photon is able to excite the $K\alpha_{3,4}$ satellites as well. DESLATTES

Table 4.4. Double-ionization energies and incident-photon energies for Z = 24, 25 and 26

Element	^{24}Cr [eV]	^{25}Mn [eV]	^{26}Fe [eV]	Lines excited
K_Z	5973	6526	7094	
L_{Z+1}	654	720	796	
$K_Z + L_{Z+1}$	6627	7246	7890	
Incident-photon energies used	FeKα_1:6386	CoKα_1:6912	NiKα_1:7458	K$\alpha_{1,2}$ lines only
	CoKα_1:6912	NiKα_1:7458	CuKα_1:8028	K$\alpha_{1,2}$ and K$\alpha_{3,4}$

[4.62], and CARLSON and KRAUSE [4.63] have shown that the probability of double ionization is appreciable, provided that the energy of the incident photon is great enough. This has led to a theory of X-ray satellites based on the sudden approximation calculation [4.64,65].

The transitions for the Kβ satellites are now also understood. For example, the satellite Kβ''' of the parent line K$\beta_{1,3}$ ($K \to M_{II,III}$) is assigned the transition $KL_{III} \to L_{III}M_{II,III}$. The relevant equations are

$$\left(h\nu_{K\beta_{1,3}}\right)_Z = K_Z - (M_{II,III})_Z \quad , \quad \text{(parent line)} \quad , \tag{4.51}$$

$$(h\nu_{K\beta'''})_Z = (KL_{III})_Z - (L_{III}M_{II,III})_Z \quad , \quad \text{(satellite)} \quad . \tag{4.52}$$

We can write

$$\begin{aligned}(h\nu_{K\beta'''})_Z &= \left[K_Z+(L_{III})_{Z+1}\right] - \left[(L_{III})_Z+(M_{II,III})_{Z+1}\right] \\ &= (K-L_{III})_Z + (L_{III}-M_{II,III})_{Z+1} \\ &= (K-M_{II,III})_Z + \left[(L_{III}-M_{II,III})_{Z+1}-(L_{III}-M_{II,III})_Z\right] \\ &= \left(h\nu_{K\beta_{1,3}}\right)_Z + \Delta \quad ,\end{aligned} \tag{4.53}$$

where

$$\Delta = (L_{III}-M_{II,III})_{Z+1} - (L_{III}-M_{II,III})_Z = \Delta_{Z+1} - \Delta_Z \quad . \tag{4.54}$$

Although $L_{III} \to M_{II,III}$ is a forbidden transition ($\Delta l=0$), both Δ_Z and Δ_{Z+1} can be computed in terms of the known energy levels of atoms of atomic number Z and Z+1. It clearly shows that the satellite photon energy cannot be determined by a transition on the energy-level diagram of the atom Z alone. By Moseley's law, $\Delta_{Z+1} > \Delta_Z$, $\Delta > 0$, and the satellite is displaced by Δ towards the high-energy side of the parent line. The Wentzel-Druyvesteyn theory shows a close approximation with the data for the K-series satellites.

Wavelengths of the $K\alpha$ satellites have been compiled by SANDSTRÖM [4.66]. We can use Moseley's law for the forbidden (diagram) line $L_{III} \rightarrow M_{II,III}$ to write

$$\Delta_{Z+1} = \tilde{\nu}_{Z+1}/R_\infty = a^2[(Z+1)-\sigma]^2 \quad ,$$
$$\Delta_Z = \tilde{\nu}_Z/R_\infty = a^2(Z-\sigma)^2 \quad ,$$
$$\Delta = (\Delta\tilde{\nu})/R_\infty = 2a^2\left[Z-\left(\sigma-\frac{1}{2}\right)\right] \quad . \tag{4.55}$$

Thus, $\Delta \propto Z$. It is difficult to verify this relation because the satellites appear as diffuse and faint lines at the foot of the parent line and the same satellite rarely occurs for a wide range of Z values. In fact, some workers have shown that $((\Delta\tilde{\nu})/R_\infty)^{\frac{1}{2}}$ versus Z plot (semi-Moseley plot) for the satellites gives straight lines over a limited range of elements and helps to identify single types of satellites for various elements.

RICHTMYER [4.67] suggested that high-energy satellites were caused not by a single transition of the Druyvesteyn type but as a result of an inner and an outer transition occurring simultaneously (double electron jumps), so that their energy changes are combined and come out as a single quantum. On this theory, one inner level (K or L) is ionized simultaneously with excitation of an outer-level electron to one of the optical levels of the atom. BLOCH [4.68] calculated the probability for such an occurrence. Such an excited state can undergo double transitions: 1) the inner vacancy is filled from a level of higher quantum number, giving the parent line $h\nu_i$, and 2) the outer excited electron falls to one of the outer atomic levels ε eV below. When both of these transitions are completed in a single act, a satellite line $h\nu_s$ is produced such that,

$$h\nu_s = h\nu_i + \varepsilon \quad .$$

Because ε will correspond to a soft X-ray line, RICHTMYER suggested that by Moseley's law

$$\varepsilon^{\frac{1}{2}} = \left[h(\nu_s-\nu_i)\right]^{\frac{1}{2}} = \text{linear function of } Z \quad . \tag{4.56}$$

There is, however, little experimental support for this relation. IDEI [4.69] showed, with the same data as were used by RICHTMYER to verify (4.56), that $(\nu_s-\nu_i)$ itself is a linear function of Z.

Other attempts to explain the high-energy satellites are also found in the literature. HULUBEI [4.70] explained the $K\beta_6$ satellite in ^{42}Mo to ^{45}Rh in terms of a double transition $K\rightarrow M_{III}$ and $M_I\rightarrow M_{II,III}$ in a doubly ionized (KM_I) atom. CANDLIN [4.53] explained the $K\alpha$ satellites for the elements $16<Z<42$ by considering atomic transitions between states of double ionization. His perturbation calculations included the effect of allowed l and s orientations with jj coupling. Some of his assignments are

$$\alpha' : 1s_{1/2}\ 2s_{1/2}\ (1) \quad 2s_{1/2}\ 2p_{3/2}\ (2) \quad ,$$

$$\alpha_3 : 1s_{1/2}\ 2p_{3/2}\ (2) \quad 2p^2_{3/2}\ (2) \quad ,$$

$$\alpha_4 : \begin{cases} 1s_{1/2}\ 2p_{1/2}\ (1) & 2p_{1/2}\ 2p_{3/2}\ (2) \\ 1s_{1/2}\ 2p_{1/2}\ (0) & 2p_{1/2}\ 2p_{3/2}\ (1) \end{cases} \quad ,$$

where J values are given in parentheses. Thus, $K\alpha_4$ is a narrow doublet.

High-energy satellites have also been observed in the L and M spectra. COSTER and KRONIG [4.26] were the first to consider the Auger process as responsible for double ionization of the atom and consequent emission of L series satellites. In fact, according to this view, the appearance and intensity of the L-satellite lines should directly reflect the probability of the associated Auger transition. There is ample evidence for this. Figure 4.13 shows the variation with Z of the ratio of satellite to parent intensities for the satellites of $L\alpha_1$ $(L_{III}\rightarrow M_V)$ and $L\beta_1$ $(L_{II}\rightarrow M_{IV})$. We have seen that the Coster-Kronig transition probability for $L_I\rightarrow L_{II}M_{IV,V}$ is about a maximum for $Z=40$ and zero for $Z=42$ (compare with Fig.4.13b). The Coster-Kronig transition $L_I\rightarrow L_{III}M_{IV,V}$ has maximum probability at $Z=45$ or 46 and zero at $Z=52$ (compare with Fig.4.13a). This process again becomes possible for $Z>73$, when $L_I\rightarrow L_{III}$ energy is enough to ionize the M_V sublevel. For $Z>77$, this energy can also ionize the M_{IV} sublevel. The relative intensities of $L\alpha_1$ satellites show sharp rises at $Z>73$ and $Z>77$ (Fig.4.14).

Thus, the Coster-Kronig [4.26] theory gives satisfactory correlations for the L-series satellites. This success led PARRATT [4.72] to classify the observed Ag L satellites, RICHTMYER and RAMBERG [4.73] to compute correctly the satellites of $L\alpha_{1,2}$ and $L\beta_2$ in ^{79}Au, PINCHERLE [4.74] to propose a general classification of the L satellites for $37<Z<56$, and CAUCHOIS [4.75] to es-

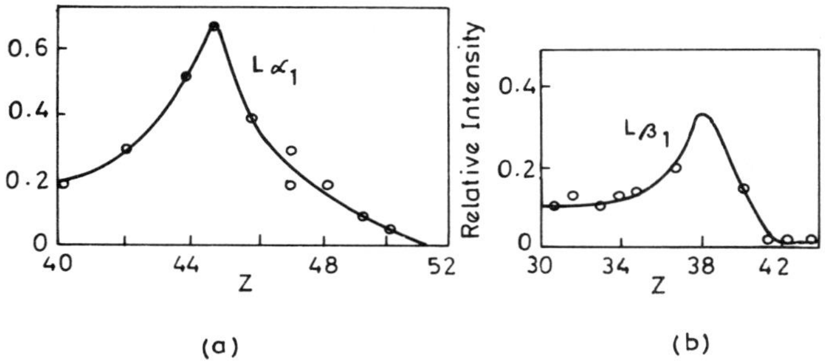

Fig. 4.13. Variation with Z of the intensity of L-series satellite relative to the parent line (a) $L\alpha_1$, (b) $L\beta_1$ [4.71]

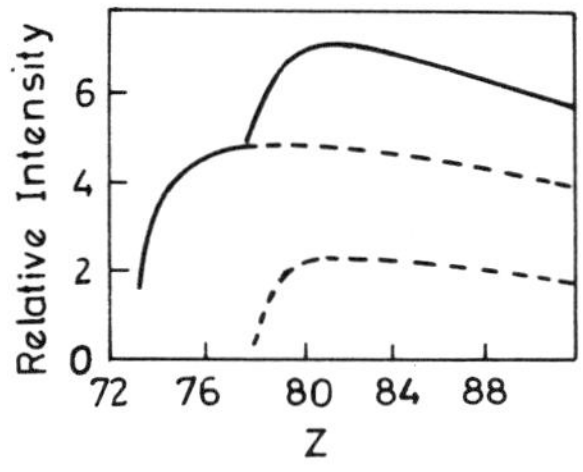

Fig. 4.14. Variation with Z (>72) of the intensity of satellites in percent of $L\alpha_1$. Solid curve: total relative intensity; dotted curve: breakup in the groups $L_{III}M_V$ and $L_{III}M_{IV}$

tablish systematics for the L satellites in heavy elements. KRAUSE et al. [4.76] studied in detail the ZrL satellites.

In the M series, the Coster-Kronig process is responsible for the high-energy satellite ($M_VN_{IV,V} \rightarrow N_{IV,V}N_{VII}$) of the $M\alpha_1$ ($M_V \rightarrow N_{VII}$) line. The relative intensity of this line increases from ^{78}Pt to ^{82}Pb and decreases sharply for ^{90}Th and ^{92}U. The reason is that the radiationless transition $M_{III} \rightarrow M_V$ cannot create a hole in the M_V level and simultaneously in the N_{IV} level (Auger electron emitted) for $Z > 89$ and N_V level for $Z > 91$ (see Table 4.1).

4.7.1 Low-Energy Satellites

X-ray K and L satellites have also been observed on the low-energy side of the parent lines. In the $K\beta$ series, BEUTHE [4.77] observed such a satellite in many elements from ^{23}V to ^{32}Ge and named it $K\eta$. He found it to be a broad line with a sharp edge on the high-energy side. FORD [4.78] detected $K\eta$ in six elements from ^{19}K to ^{24}Cr. VALASEK [4.79] observed it in ^{17}Cl. BLOKHIN [4.80] explained it by the forbidden transition $K \rightarrow M_I$ and simultaneous ejection of an $M_{II,III}$ electron to the Fermi level, for the elements ^{23}V to ^{29}Cu. SAWADA et al. [4.81] resolved $K\eta$ into two components from ^{24}Cr to ^{30}Zn and named the longer-wavelength one Kl. They explained them on the basis of two-electron jumps in a doubly ionized atom,

$$K\eta = KL_{II} - L_I M_I = K_Z + (L_{II})_{Z+1} - (L_I)_Z - (M_I)_{Z+1} \quad ,$$

$$Kl = KL_{III} - L_I M_I = K_Z + (L_{III})_{Z+1} - (L_I)_Z - (M_I)_{Z+1} \quad .$$

The $K\beta'$ satellite of ^{24}Cr is explained [4.80] by the transition $K \rightarrow M_{II,III}$ (parent line $K\beta_1$), with simultaneous ejection of an electron from the $M_{IV,V}$ level of the atom into the conduction band of the solid target. The short-wavelength end of $K\beta'$ corresponds to the Fermi energy and the line shape mirrors the density of states of the conduction band.

HULUBEI [4.82] observed a low-energy satellite of $K\alpha$ for $33 < Z < 44$. The energy separation from the $K\alpha_1$ line was very nearly equal to the energy of the M absorption edge of the element one higher in atomic number. A similar line was observed by HULUBEI et al. [4.83] on the low-energy side of $K\beta_1$. ÅBERG and UTRIAINEN [4.84,85] detected a very complex structure on the low-energy side of $K\alpha$ and $K\beta$. These observations are related to that of BLOCH and ROSS [4.86] who found a low-energy satellite of $K\beta_5$ (quadrupole) line in ^{42}Mo.

The best explanation seems to lie in terms of the simultaneous excitation of an outer-shell electron and emission of an X-ray photon. This is called the *radiative Auger effect*. Because an electron in the outer (say, M) shell absorbs a part of the radiation energy, the energy of the satellite is decreased relative to that of the parent line.

A radiative Auger transition is a double-electron transition, where an outer-shell electron $n_f l_f$ makes a transition to an inner-shell hole $n_i l_i$ and where another outer-shell electron $n_{f'} l_{f'}$ is simultaneously excited into a bound or continuous state εl. (If ε is a bound or optical state, the transition is often called a semi-Auger transition.) These transitions correspond to a discrete and continuous distribution of photon energies $h\nu$ such that

$$h\nu = E_i - E_{ff'} - \varepsilon \quad . \tag{4.57}$$

Here E_i is the $n_i l_i$ ionization energy of the neutral system, and $E_{ff'}$ the energy required for the simultaneous ionization of the two electrons $n_f l_f$ and $n_{f'} l_{f'}$ from the neutral system. The range $0 \leqq \varepsilon \leqq (E_i - E_{ff'})$ gives a distribution of photon energies (broad diffuse satellite). Note that $(E_i - E_{ff'})$ is just the energy of the Auger electron in the radiationless transition $i \rightarrow ff'$.

Radiative Auger processes have been interpreted recently by ÅBERG [4.87] in terms of shake-off and interaction between single- and double-hole configurations in the final state. ROOKE [4.88] suggested that the energy of the principal line is attenuated by simultaneous plasmon excitation. A molecular orbital interpretation has been given by URCH [4.89].

To conclude this section, we can say that there exist many processes (including crossover transitions in compounds) that contribute to the satellite structure. We have discussed some of them here. Recently BRIAND et al. [4.90] observed KK→KL X-rays in coincidence with KL→LL X-rays in a few elements ($Z \geq 31$). They called these new weak satellites *hypersatellites*. Heavy-ion collision can also give rise to K hypersatellites [4.91,92]. They also occur in L spectra [4.93].

Multihole states created by shake-off and Coster-Kronig processes give X-ray satellites that may or may not be distinguishable from the parent diagram lines. According to the analysis of ZrL satellites by KRAUSE et al. [4.76], the satellites caused by LM Coster-Kronig state appear on the high-energy side of the parent line, and those caused by LN shake-off state fall within the parent line. Thus, only a fraction of satellites is observable and a diagram line may be contaminated by hidden satellites.

The subject of satellites is not simple and we are far from any satisfactory understanding of this complex problem. Some useful review articles are by HIRSH [4.94], DEODHAR [4.95], EDWARDS [4.96] and ÅBERG [4.97].

4.8 Fluorescence

Fluorescence competes with the Auger-electron emission as a de-excitation process, following the creation of an inner-shell hole. The former dominates in the heavier atoms and the latter for $Z < 15$, Table 4.5. The fluorescent radiation is in the X-ray region and consists only of the characteristic line spectra.

Table 4.5. Probability (P) of the Auger-electron emission relative to the X-ray emission, as a function of Z

Z :	10	15	20	25	30	40	45	50	55	65	70
P :	0.99	0.95	0.86	0.71	0.55	0.31	0.22	0.16	0.11	0.09	0.07

Suppose X-rays incident on unit volume of an element eject q-type (q=K,L,M,...) photoelectrons from n_q atoms per second. In the steady state, $n_q = n_{q,A} + n_{q,R}$ will return to the un-ionized or normal configuration, where $n_{q,A}$ is the number of atoms that emit Auger electrons (radiationless transition) and $n_{q,R}$ is the number of atoms that emit fluorescent radiation (radiative transition). Of these $n_{q,R}$ atoms, a number $n^1_{q,R}$ will emit the line $q\alpha_1$ as a step in the process, $n^2_{q,R}$ will emit $q\alpha_2$, etc. Thus

$$n_{q,R} = n^1_{q,R} + n^2_{q,R} + \dots = \sum_i n^i_{q,R} \quad . \tag{4.58}$$

The sum is over all lines of the q series.

We define the *fluorescence yield* $\tilde{w}_q$ by

$$\tilde{w}_q = \frac{\sum_i n^i_{q,R}}{n_q} = \frac{n_{q,R}}{n_{q,A}+n_{q,R}} \quad . \tag{4.59}$$

Thus, for a sample of many atoms, $\tilde{w}_q$ is equal to the ratio of the number of photons emitted when vacancies in the shell are filled and the number of primary vacancies in the shell. In terms of the corresponding transition probabilities (P) and widths (Γ),

$$\tilde{w}_q = \frac{P_{q,R}}{P_{q,A} + P_{q,R}} = \frac{\Gamma^R_q}{\Gamma^A_q + \Gamma^R_q} \quad . \tag{4.60}$$

Great care is needed in the proper definitions of the quantities that are measured [4.22].

The dominant contribution to $P_{q,R}$ comes from the electric-dipole transitions,

$$P_{q,R}\, dt = \frac{64\pi^4\nu^3}{3hc^3} |\underline{M}|^2\, dt \quad , \quad \underline{M} = e\int \psi^*_f\, \underline{r}\, \psi_i\, d\tau \quad . \tag{4.61}$$

Theoretical calculations of $\tilde{w}_K$ have been made by BURHOP [4.98] and PINCHERLE [4.99] using a nonrelativistic theory, and by MASSEY and BURHOP [4.100] using a relativistic theory.

Integration of (4.36) gives complicated expressions for the Auger transition rate, but it shows that the rate is independent of the atomic number. On the other hand, the integration of (4.61) gives for the radiative transition rate a Z^4 dependence[1]. Therefore, we can approximate (4.60) as,

$$\tilde{w}_K = \frac{bZ^4}{a+bZ^4} = \frac{1}{1+a_K Z^{-4}} \quad , \tag{4.62}$$

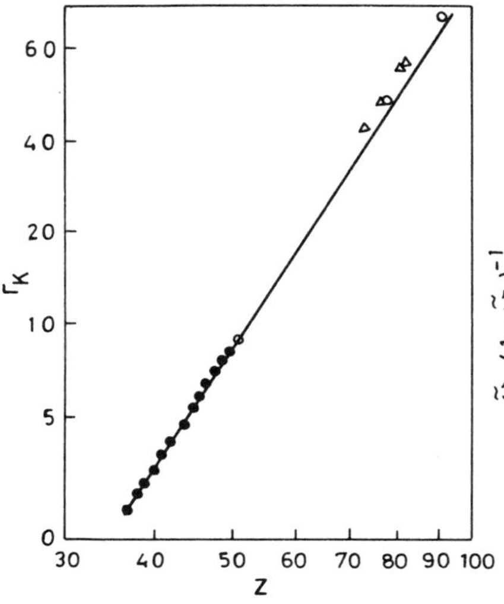

(a)

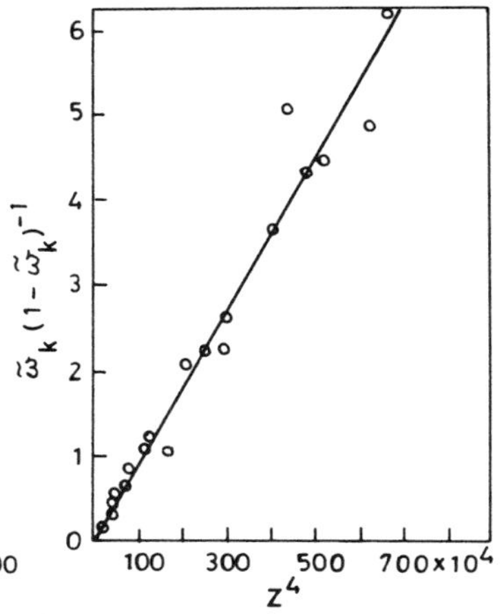

(b)

Fig. 4.15. (a) K-level width as a function of Z. The straight line represents $\Gamma_K = 1.73 \times Z^{3.93} \times 10^{-6}$ eV. The measurements are by MASSEY [4.27] o; GOKHALE [4.102] •; LASKAR [4.103] Δ. (b) Plot of $\tilde{w}_K/(1-\tilde{w}_K)$ versus Z^4. The points represent averaged values of various measurements

where $a_K = a/b$ is an adjustable constant. LEISI et al. [4.101] used Kα line-width information with measurements of L_{II} and L_{III} level widths to show that for $Z > 40$, K-level widths can be represented by the relation

$$\Gamma_K = 1.73 \times Z^{3.93} \times 10^{-6} \text{ eV.}$$

It is compared with experiments [4.27,102,103] in Fig.4.15a. In Fig.4.15b, we show a plot of $\tilde{w}_K/(1-\tilde{w}_K)$ versus Z^4 along with the experimental points. A straight line given by (4.62) with $a_K = 1.12 \times 10^6$ provides a good fit. The theoretical value is close to this [4.104].

[1]This can also be seen by regarding the radiating atom as a damped harmonic oscillator. From (3.3) $\langle -dU/dt \rangle = (2e^2/3c^3)\langle \ddot{x}^2 \rangle = (2e^2/3c^3)(\ddot{x}_0^2/2)$, where $\ddot{x}_0$ is the peak acceleration. For weak damping $\ddot{x} = 2\pi\nu\dot{x}_0$. With $U = m\dot{x}_0^2/2$, we have relaxation time = (energy stored)/(mean energy loss per s) = $U/\langle -dU/dt \rangle$ = $3mc^3/(8\pi^2 e^2 \nu^2)$. By Moseley's law, $\nu \propto Z^2$, relaxation time $\propto 1/Z^4$, or the transition probability $\propto Z^4$.

A relation similar to (4.62) is found to be applicable for the L-shell fluorescence, although the agreement with the experiments is not so good as in the K-shell case. K-shell radiative transition probabilities have been calculated by SCOFIELD [4.105] to give K_β/K_α X-ray intensity ratio in agreement with experiments.

4.9 Measurement of Fluorescence Yield

AUGER [4.20], BOWER [4.43], and MARTIN and EGGLESTON [4.42] used (4.59) to determine $\tilde{w}_q$ in gases filled in the cloud chamber. To find $n_q = n_{q,A} + n_{q,R}$ it is necessary only to detect *all* of the photoelectron tracks whose range corresponds to the ejection of electrons from the q level of atoms in the gas. This range is determined by $h\nu - E_q$, where $h\nu$ is the incident X-ray photon energy. To determine $n_{q,R} = n_q - n_{q,A}$, we need only locate all of the Auger electron tracks ($n_{q,A}$) that are emitted from the *same* atoms. For example, the Auger electrons will have energies $E_K - 2E_L$ and $E_L - 2E_M$ when the K-shell vacancy is filled by Auger transitions.

It is a statistical method. If in all, n tracks are observed, and m of these are recognized as Auger tracks, then the fluorescence yield is (n-m)/n. The probable error of m/n is $0.6475\ [m(n-m)/n^3]^{1/2}$.

In ^{36}Kr, Auger found 223 tracks of K photoelectrons. Of these, 109 were accompanied by an Auger electron. Therefore, $\tilde{w}_K = (223 - 109)/223 = 0.51$ (Table 4.6).

For solid specimens, the power can be directly measured in the incident X-ray beam and in the fluorescent radiation from the atoms. The rate of inner-shell ionization is related to the absorption coefficient. For this, we must know the spectral composition of the primary beam.

To ensure a definite composition of the incident beam, three methods are available: 1) crystalline diffraction, 2) suitable filters [4.106], and 3)

Table 4.6. Auger's results on fluorescence yield

Element	$\tilde{w}_K$	$\tilde{w}_L$
18 A	0.07	-
36 Kr	0.51	0.1
54 Xe	0.71	0.25

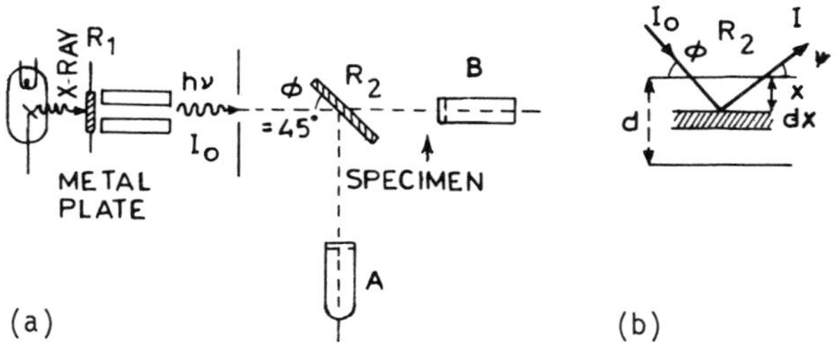

Fig. 4.16. (a) Compton's arrangement for the measurement of fluorescence yield, and (b) path of X-rays in the excitation of fluorescence in solids

pure characteristic fluorescent radiation from a metal plate exposed to primary X-rays.

A typical experimental arrangement used by COMPTON [4.107] and STEPHENSON [4.108] is shown in Fig.4.16a. X-rays fall on a metal plate R_1 and its characteristic K-series fluorescent radiation is used as the incident beam for the speciment R_2 under study. The ionization chamber A measures the ionization current I produced by the fluorescent radiation from R_2, and B measures the current I_0 produced by the direct beam with R_2 removed.

The intensity of incident X-rays at a depth x in R_2 is

$$I_x = I_0 \, e^{-\mu x/\sin\phi} \quad , \tag{4.63}$$

where $\mu = \sigma + \tau$ is the total linear attenuation coefficient, and ϕ is the glancing angle of incidence of X-rays from R_1 on R_2, Fig.4.16b. The decrease of the initial-beam intensity when it passes through a layer dx at x is

$$dI_x = -\frac{\tau + \sigma}{\sin\phi} I_0 \, e^{-\mu x/\sin\phi} \, dx \quad . \tag{4.64}$$

If I_0 is the energy of X-rays per cm^2 falling on R_2, then the radiation incident on a surface s of R_2 is $I_0 s \sin\phi$. Therefore, the part of this energy that undergoes photoelectric absorption in dx at x is

$$dE = \tau s I_0 \, e^{-\mu x/\sin\phi} \, dx \quad . \tag{4.65}$$

This is the energy absorbed by all of the q levels of the atoms in R_2 in the layer dx. Writing

$$dE = \sum_q dE_q \quad , \quad \tau = \sum_q \tau_q \quad , \tag{4.66}$$

and using (3.237), we have from (4.65),

$$dE_q = \frac{r_q - 1}{r_q} \tau s I_0 \; e^{-\mu x/\sin\phi} \, dx \quad . \tag{4.67}$$

Here r_q is the absorption-jump ratio at the q edge, and $\tau_q/\tau = (r_q - 1)/r_q$ gives the fraction of the total number of photoelectrons that come from the q level when X-rays are absorbed.

If $n_q sdx$ atoms in the volume sdx of R_2 are excited in the q level, we can write

$$dE_q = h\nu n_q sdx = h\nu (n_{q,R}/\tilde{w}_q) sdx \quad , \tag{4.68}$$

where we have used (4.59) and $n_{q,R} = \sum_i n^i_{q,r}$. If p_i is the probability that an atom excited to the q level will emit a line i of the q series, then $n^i_{q,R} = p_i \, n_{q,R}$ and we have, from (4.67,68),

$$n^i_{q,R} \; sdx = p_i \frac{\tilde{w}_q}{h\nu} \frac{r_q - 1}{r_q} \tau s I_0 \; e^{-\mu x/\sin\phi} \, dx \quad . \tag{4.69}$$

It follows that the energy of the secondary X-rays with a frequency ν_i of the line i emitted in *all* directions from the volume sdx is

$$(dE_q)_i = h\nu_i \; n^i_{q,R} \; sdx = \frac{\nu_i}{\nu} \frac{r_q - 1}{r_q} \tau \tilde{w}_q \; p_i \; s \, I_0 \; e^{-\mu x/\sin\phi} \, dx \quad . \tag{4.70}$$

This energy leaves R_2 at an angle ψ (Fig.4.16b) and undergoes absorption along the path $x/\sin\psi$. The intensity dI_i of the line i of the secondary spectra at a distance R from the volume sdx is,

$$dI_i = \frac{(dE_q)_i}{4\pi R^2} \; e^{-\mu_i x/\sin\psi}$$

$$= \frac{\nu_i}{\nu} \frac{I_0}{4\pi R^2} \tilde{w}_q \frac{r_q - 1}{r_q} \tau p_i s \; e^{-(\mu/\sin\phi + \mu_i/\sin\psi)x} \, dx \quad , \tag{4.71}$$

where μ_i is the total linear attenuation coefficient of the line i in R_2. Integration over x from 0 to d gives

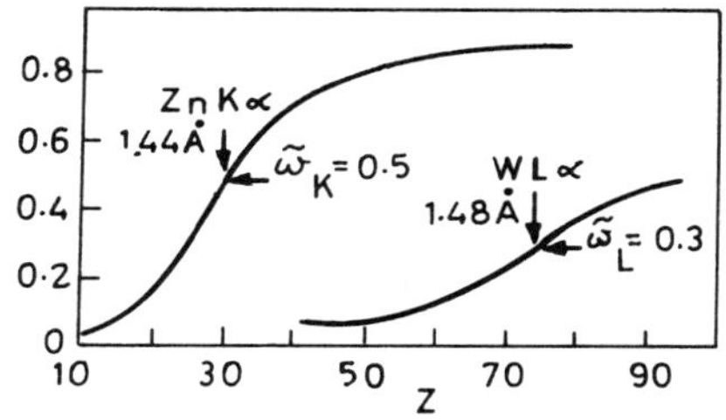

Fig. 4.17. Fluorescence yield as a function of Z is of form (4.62)

$$I_i = \frac{I_0}{4\pi R^2}\,\tilde{w}_q\,\frac{r_q-1}{r_q}\,s\,\sin\phi\,\frac{\tau}{\mu}\,p_i\,\frac{\nu_i/\nu}{1+\frac{\mu_i}{\mu}\frac{\sin\phi}{\sin\psi}}\times\left[1-e^{-(\mu/\sin\phi+\mu_i/\sin\psi)d}\right] \quad . \tag{4.72}$$

The usual choice is $\phi = \psi = 45^\circ$.

The total intensity of the q-series spectra at a distance R from R_2 is

$$I_q = \sum_i I_i \quad . \tag{4.73}$$

The probability p_i can be found from the relative intensities of the lines of the q-series from R_2,

$$p_i = I_i'/\sum_i I_i' \quad . \tag{4.74}$$

The measurement of I_q/I_0 allows us to determine $\tilde{w}_q$ from (4.72-74).

The results of $\tilde{w}_K$ and $\tilde{w}_L$ measurements as a function of Z are shown in Fig.4.17. For a given Z, the difference $\tilde{w}_K - \tilde{w}_L$ is large, 0.5 to 1.0, but for a given λ this difference is small, 0.15 to 0.2. A plot of $\tilde{\omega}_K/(1-\tilde{\omega}_K)$ versus Z^4 gives approximately a straight line (Fig.4.15). There is an increasing demand for accurate fluorescence-yield data [4.109], as it provides useful information for quantitative X-ray spectroscopy.

4.10 Autoionization and Internal Conversions

A secondary process that resembles Auger ionization is *autoionization*. It can occur when a sample is exposed to photons or electrons of suitable energies. In the initial state, an inner electron is promoted to an outer (optical) level. If the excitation energy exceeds the ionization energy of any of the other electrons present, the excited atom will eject an electron and reorgan-

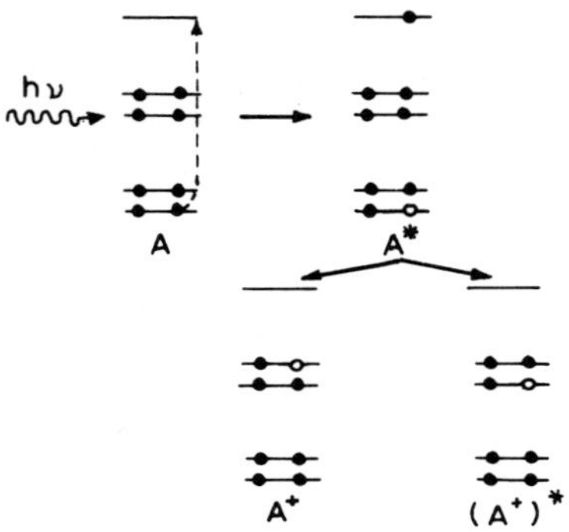

Fig. 4.18. Autoionization process possible in an atom A; following an initial excitation process $A \rightarrow A^*$; •: electron, ○: hole

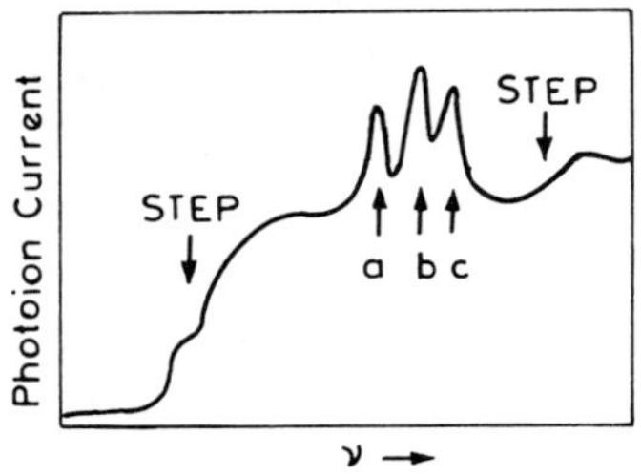

Fig. 4.19. The resonance autoionization structure: a,b,c, indicate autoionization levels

ize to an ion (Fig.4.18). The ejected electron is called the *autoionization electron*. In the Auger process, the initial step requires the ejection of an inner-shell electron, rather than a promotion to a higher level, and the final state is a doubly ionized ion, rather than a singly ionized ion.

The incident photon hν, that is absorbed in the process, must have an energy equal to the difference of energy between the ground-state atom A and the excited atom A^* (Fig.4.18). If hν is varied, autoionization levels will show up as resonance peaks superimposed on the expected steps for the various ionization potentials (Fig.4.19). Autoionization will be detected in the usual photoelectron spectroscopy only if hν happens to coincide with an autoionization level.

The possibility of the *internal conversion* process was discussed in Sect. 4.5. The β-ray spectra from a nucleus consist of electrons that have a continuous range of energy from zero up to a definite upper limit. Sometimes superimposed on this continuous distribution there occur several sharp lines. These lines correspond to electrons that are homogeneous in energy, and arise from the internal conversion of the γ-radiation, that accompany the β-decay, by the electrons of the atom of the origin of γ-radiation. It is analogous to the Auger process [4.18,110].

4.11 Muonic X-Rays

When a negatively charged particle (like μ^-, π^-, K^-) replaces an electron in an atom, a *mesonic atom* is formed. In this atom, the energy level of the mesons are analogous to those of electrons in a normal atom and characterized by the same quantum numbers n,l. When the half-spin μ^- replaces the electron, it is called a *muonic atom*. Muonic atoms have the same quantum number j, as normal atoms. Except for the mass ($m_\mu \simeq 207\ m_e$), the μ^- behaves like an electron. The simplest system is *muonic hydrogen* (p^+,μ^-).

The energy levels E(n,l) of a mesonic atom are related to $E_0(n,l)$ of an ordinary atom by

$$E(n,l) = (M^*/m^*)E_0(n,l) \quad ,$$

where $M^*(m^*)$ is the reduced mass of the meson (electron). The mean distance $\bar{r}_M$ of the meson from the nucleus in the ground state is smaller than the Bohr radius a_0 in the same ratio,

$$a_0/\bar{r}_0 = Z(M^*/m^*) \quad .$$

For muonic silver this gives $\bar{r}_0 \simeq 5 \times 10^{-13}$ cm $\simeq$ radius of the silver nucleus. Therefore, the muon spends a considerable part of its time in the interior of the nucleus.

When μ^- are brought to rest in a target, muonic atoms of the target element are formed, usually by the replacement of one of the valence electrons by the μ^-. This forms a highly excited state. The de-excitation occurs in about 10^{-3} s, first by means of radiationless Auger processes in which the μ^- falls successively to states of lower excitation, the excess energy being used to eject electrons from the atom. When the principal quantum number n reduces to less than about 5 in heavier atoms, the radiative transitions become more probable than Auger transitions, and *muonic X-rays* are emitted. The single muon in the muonic atom continues to make de-excitation transitions until it reaches the K shell, because the Pauli principle does not inhibit it.

The $K\alpha_{1,2}$ lines in the muonic X-rays were resolved by FITCH and RAINWATER [4.111]. KESSLER et al. [4.112] observed the muonic X-ray Lyman ($np \rightarrow 1s$) and Balmer ($nd \rightarrow 2p$) series of ^{22}Ti. The energies of the first four lines in the H Lyman series in eV are 10.15, 12.04, 12.69, 13.00. The electron is so far away from the nucleus in hydrogen that the latter behaves as a point charge. If we calculate the Lyman series of muonic Ti on this picture, they would be exactly $Z^2(m_\mu^*/m_e^*) \simeq 10^5$ times the energies for hydrogen; 1015, 1204, 1269, 1300 (in keV, X-ray region). The observed values are 935, 1124, 1190, 1221 keV, all with energies about 80 keV less than expected on the assumption of a point nucleus. The muonic Ti Balmer-series lines are found almost at the expected positions of 188, 254, 284, 301 keV. This shows that the 1s level is strongly affected by the nuclear size, whereas the 2p level (with larger value of $\bar{r}$) is little affected.

Nuclear-size effects become significant with increase of Z, because the nuclear radius increases, whereas the mean radius of the 1s orbit, $\bar{r}_0$, decreases. The measurements of muonic X-ray spectra in various elements suggest a nuclear-charge distribution of the form

$$\rho(r) = \rho_0\left(1 + e^{(r-R)/z}\right)^{-1} \quad ,$$

where z determines the thickness of the nuclear surface, and R is practically the radius at which the electric charge density has become one-half its value at the center. A useful review article is by BURHOP [4.113]. In contrast with μ^-, π^- have a strong interaction with nucleons. Therefore, π^--mesonic atoms are not useful for studying nuclear radii.

5. Scattering of X-Rays

An X-ray beam that passes through an absorber is attenuated. The degree of attenuation depends upon both photoabsorption and scattering processes. Scattering occurs when an X-ray photon interacts with one of the electrons of the absorbing element. If this collision is elastic (no energy is lost in the collision process), the scattering is said to be *coherent* (Rayleigh scattering). Because no change of energy is involved, the coherently scattered radiation remains *unmodified* (same wavelength as the incident beam), and there is a definite relationship between the phase of the scattered beam and that of the incident beam. The X-ray diffraction is a special case of coherent scattering. In X-ray spectroscopy, this diffraction process provides a method for the wavelength separation.

It can also happen that in the collision process a small fraction of the energy of an incident X-ray photon is transferred to a loosely bound electron of the target atom. Then the scattering is said to be *incoherent* (Compton scattering). The incoherently scattered radiation is *modified* (slightly longer wavelength than that of the incident beam) and has no fixed relation to the phase of the incident beam. It cannot produce interference effects and so unavoidably contributes to the uniform background of diffraction patterns.

In Fig.5.1 we show the more important of the processes responsible for the attenuation of the primary beam that passes through an absorber.

5.1 Classical Theory of Thomson and Rayleigh (Coherent) Scattering

The coherent scattering of radiation by a free electron was first discussed by J.J. THOMSON in terms of classical electrodynamics. Let an unpolarized X-ray beam propagate along the y axis and encounter an electron at 0 (Fig.5.2). Let $E = E_0 \exp(i\omega t)$ be the electric field of the incident wave at the point 0 at time t. If the electron has a velocity small compared to c, it would receive an acceleration $\underline{a}$, such that $\underline{a} = e\underline{E}/m$. This accelerated electron, in

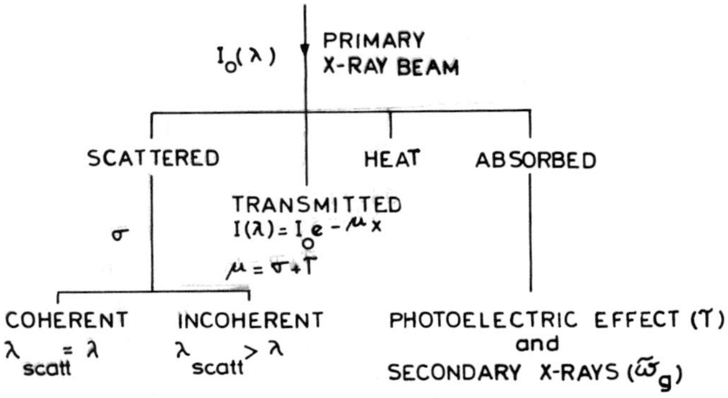

Fig. 5.1. Attenuation processes

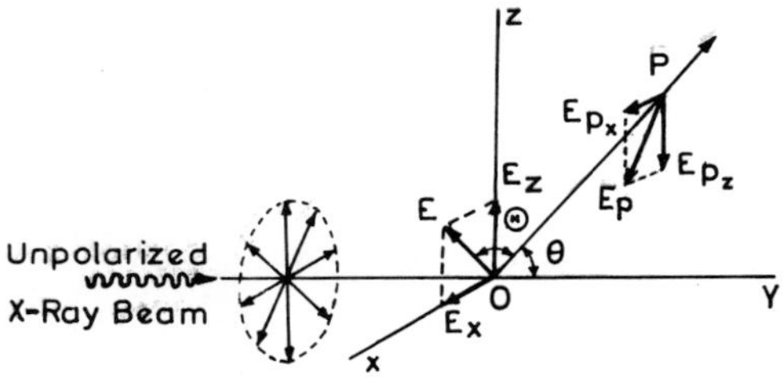

Fig. 5.2. Coherent scattering of X-rays by a single electron at 0

turn, becomes a source of radiation of the same frequency ω. This kind of interaction gives the scattered wave at a point P, at the same time t, of the form (1.73)

$$\underline{E}_P = -\underline{E}\frac{e^2}{mc^2R}\sin\Theta \quad , \quad \underline{E} = \underline{E}_0\, e^{i\omega\left(t-\frac{R}{c}\right)} \quad , \tag{5.1}$$

where R is the distance OP and Θ the angle between the scattering direction and the direction of acceleration of the electron. The negative sign simply means that the wave scattered by the free electron in the forward direction has a phase opposite to that of the primary wave[1].

[1]For an oscillator $E(t) = (\omega^2/cR)ex(t) = (\omega^2e^2E_0/mc^2R)(\omega_0^2-\omega^2+i\gamma\omega)^{-1}\exp(i\omega t)$, (Sect.3.3). For a free electron $\gamma=0$, $\omega_0=0$, and so $E(t)/E_{free}(t) = (-\omega^2)(\omega_0^2-\omega^2+i\gamma\omega)^{-1}$. The minus sign means that the wave scattered by the free electron has an opposite phase to that of the incident wave.

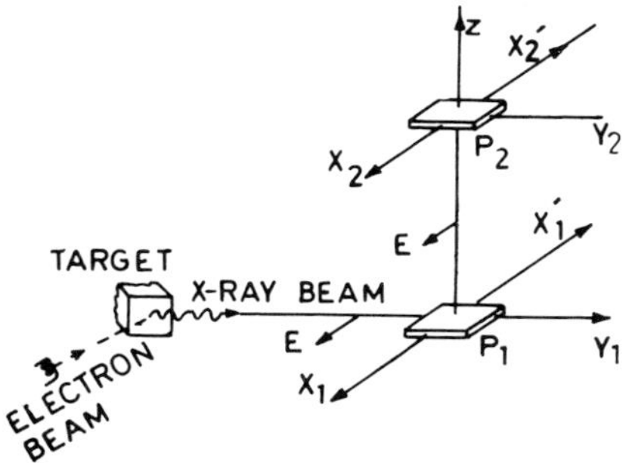

Fig. 5.3. Polarization of X-rays by scattering

For convenience, let the point of observation P lie in the yz plane. We consider the components of $\underline{E}$ along the x and z axes. From (5.1), in the scattered beam at P (Fig.5.2),

$$\underline{E}_{Px} = \frac{e^2\underline{E}_x}{mc^2R} \quad , \quad \underline{E}_{Pz} = \frac{e^2\underline{E}_z}{mc^2R}\cos\theta \quad . \tag{5.2}$$

The energy per unit volume that flows past the point P is proportional to $E_P^2 = E_{Px}^2 + E_{Pz}^2$. For an unpolarized primary beam, on the average, $<E_x>^2 = <E_z>^2 = <E>^2/2$. Therefore,

$$<E_P>^2 = \frac{e^4<E>^2}{m^2c^4R^2}\left(\frac{1+\cos^2\theta}{2}\right) \quad , \tag{5.3}$$

and the intensity ($I=c|E|^2/4\pi$) of coherent scattering by one electron is obtained by multiplying both the sides of (5.3) by $c/4\pi$,

$$I_e = I_0\frac{e^4}{m^2c^4R^2}\left(\frac{1+\cos^2\theta}{2}\right) = I_0\frac{r_e^2}{R^2}\left(\frac{1+\cos^2\theta}{2}\right) \quad . \tag{5.4}$$

This is the *Thomson equation*. The quantity in parentheses is called the *polarization factor*. Plane polarization occurs for $\theta = 90^\circ$, $\underline{E}_{Pz} = 0$.

COMPTON and HAGENOW [5.1] studied polarization of X-rays scattered by two thin foils P_1 and P_2 (Fig.5.3). The primary beam from an X-ray tube with its axis parallel to x is polarized so that its electric vector $\underline{E}$ is parallel to x. Consequently, the electrons in P_1 are accelerated only along x_1x_1' and a scattered beam is observed in the z direction but not parallel to x. Because the polarization of the primary beam is not complete, a much larger intensity is merely observed along z than along x, as found by BARKLA (Sect.1.13). The X-ray beam scattered along z can have $\underline{E}$ parallel to x only and so can accelerate the electrons in P_2 only along x_2x_2'. Thus, the observed intensity of the twice-scattered beam changes from a maximum along y to a minimum along x_2x_2'. COMPTON and HAGENOW found this to be so. It shows that the electromagnetic-wave treatment of Thomson is correct and that in coherent scattering by electrons the X-rays get polarized. The nucleus of an atom cannot participate in this kind of scattering because it cannot be accelerated by $\underline{E}$ because of its large mass.

The classical differential cross section per electron for Thomson scattering is given by

$$\left(\frac{d\sigma_e}{d\Omega}\right)_{\text{Thom}} = \frac{\text{Energy radiated/unit time/unit solid angle}}{\text{Incident-energy flux in energy/unit area/unit time}}$$

$$= \frac{I_e R^2}{I_0} = r_e^2 \left(\frac{1+\cos^2\theta}{2}\right) \quad . \tag{5.5}$$

The total power scattered is obtained by integration over the surface of a sphere of radius R,

$$P_e = \int_0^\pi I_e 2\pi R^2 \sin\theta \, d\theta = I_0 \pi r_e^2 \int_0^\pi (1+\cos^2\theta)\sin\theta \, d\theta = I_0 \frac{8\pi}{3} r_e^2 \quad . \tag{5.6}$$

The *classical scattering cross section per electron* is

$$\sigma_e = \int_0^\pi \left(\frac{d\sigma_e}{d\Omega}\right)_{\text{Thom}} 2\pi \sin\theta \, d\theta = \frac{P_e}{I_0} = 6.66 \times 10^{-25} \text{ cm}^2 \quad . \tag{5.7}$$

We now consider the cooperative effect of all of the electrons in an atom (*Rayleigh scattering*). The electrons are all forced into vibration at the frequency of the incident X-rays. If this frequency is much greater than the natural frequencies of the atomic electrons, the scattered radiation from all of the electrons will be in antiphase with the incident radiation [5.2]. The scattering is coherent and so the amplitudes are added before squaring to give the intensity. The cross section per electron is now nonadditive, so we speak of an atomic scattering cross section σ_R such that

$$\sigma_R = \frac{8\pi}{3} r_e^2 f^2(\theta) \quad ,$$

where $f(\theta)$ is the atomic scattering factor. It is defined with respect to σ_e for a free electron.

If the material contains n electrons per unit mass that scatter X-rays independently of each other according to (5.4), we can define the *mass scattering coefficient* as

$$\sigma_m = n\sigma_e = (N_A Z/A)\sigma_e \ \mathrm{cm}^2\ \mathrm{g}^{-1} \quad . \tag{5.8}$$

Setting $Z/A \simeq 0.5$, we get $\sigma_m \simeq 0.20$. Thus, σ_m is a universal constant and does not depend on the scattering material.

It may be noted that electrons do not scatter X-rays independently of each other, and Thomson's theory does not include the quantum effects. For light elements, however, (5.7) is in good agreement with experiments.

5.2 Incoherent (Compton) Scattering

The relativistic expression for the energy of a free particle is

$$E = \left[(pc)^2+(mc^2)^2\right]^{\frac{1}{2}} \quad , \tag{5.9}$$

where m is the rest mass and p the momentum of the particle. The velocity at this momentum is

$$v = \frac{dE}{dp} = \frac{pc^2}{E} = \frac{pc^2}{(p^2c^2+m^2c^4)^{\frac{1}{2}}} \quad .$$

For a photon $v = c$. Therefore, the *photon rest mass must be zero*, and $E = pc$, or

$$p = E/c = h\nu/c \quad , \quad (\textit{photon}) \quad . \tag{5.10}$$

Consider a photon with initial momentum $\underline{p}$ ($p=h\nu/c$) and energy $h\nu$, incident upon a stationary electron of rest mass m. After the collision, the photon has momentum $\underline{p}'$ ($p'=h\nu'/c$), and the electron recoils at an angle ϕ with momentum $\underline{P}$, see Fig.5.4. The paths of the incident and the scattered photon define the *scattering plane* and θ is the *scattering angle*. The momentum normal to

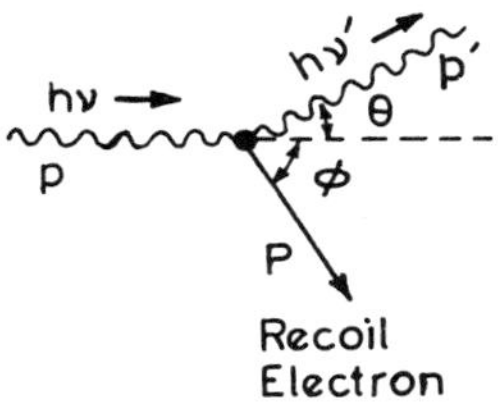

Fig. 5.4. Compton scattering

this plane is zero. Therefore, the path of the recoil electron must also lie in the scattering plane.

Conservation of momentum, in the direction of the incident photon, gives

$$(h\nu/c) = (h\nu'/c)\cos\theta + P\cos\phi \quad , \tag{5.11}$$

and normal to the direction of the incident photon and in the scattering plane, gives

$$0 = (h\nu'/c)\sin\theta - P\sin\phi \quad . \tag{5.12}$$

Conservation of energy can be expressed as

$$h\nu = h\nu' + E_{kin} \quad , \tag{5.13}$$

where E_{kin} is the kinetic energy of the recoil electron, or as

$$h\nu + mc^2 = h\nu' + (P^2c^2+m^2c^4)^{1/2} \quad . \tag{5.14}$$

From (5.11,12),

$$P^2c^2 = (h\nu)^2 + (h\nu')^2 - 2(h\nu)(h\nu')\cos\theta \quad . \tag{5.15}$$

From (5.14),

$$P^2c^2 + m^2c^4 = (h\nu-h\nu')^2 + 2mc^2(h\nu-h\nu') + m^2c^4 \quad . \tag{5.16}$$

This can be combined with (5.15), to give

$$mc^2(\nu-\nu') = h\nu\nu'(1-\cos\theta) \tag{5.17}$$

or equivalently, the equation for the *Compton effect* [5.3] is

$$\lambda' - \lambda = (h/mc)(1-\cos\theta) = 0.0243\,(1-\cos\theta)\ \text{Å} \quad . \tag{5.18}$$

The term h/mc is called the *Compton wavelength* of the electron. Note that $\lambda' > \lambda$. The increase of the wavelength of the incoherently scattered line ($h\nu'$) depends only on the scattering angle θ. This is confirmed by experiments (Fig.5.5a).

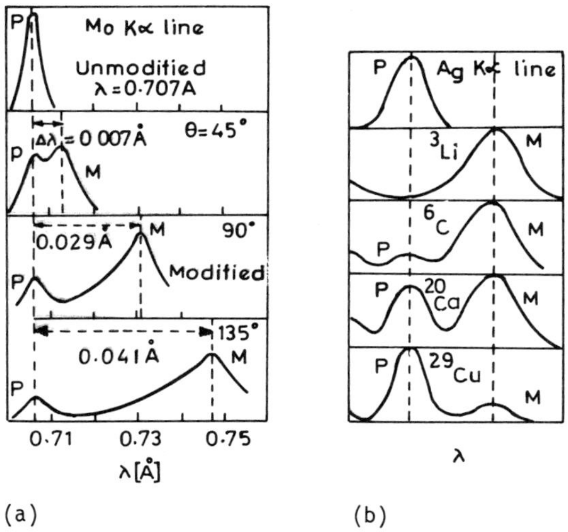

Fig. 5.5. Compton scattering (a) of MoKα line by graphite at various angles, and (b) of AgKα line by different atoms at 120°

The *modified line* (λ') is observed to be broader than the primary or *unmodified line* (λ). The reason is that the electrons in the target atom are not initially at rest, as assumed in the above derivation, but have a range of initial momenta and energies.

The unmodified line (no change of wavelength) is also observed along with the modified line. The reason is that for a part of the time the electron behaves essentially as a *free* particle at rest, as assumed above, and for a part of the time it behaves as *bound* to the rest of the atom. In the latter case, the collision is between the incident photon and the whole atom of mass $M \gg m$. Then h/mc is to be replaced by h/Mc in (5.18). Because $h/Mc \simeq 0$, we get $\lambda' = \lambda$. The observed fact, that the unmodified line is hardly detected for low Z atoms and becomes stronger for high Z atoms (Fig.5.5b), confirms this conjecture.

The Compton shift of wavelength, $\lambda' - \lambda$, for a given θ, is independent of the energy $h\nu$ of the incident photon. On the other hand, the associated shift of energy, $h\nu - h\nu'$, determined by (5.17), increases as $h\nu$ increases. This follows from the reciprocal relationship between frequency and wavelength.

From (5.11-13), we get

$$E_{kin} = h\nu \, \frac{2\alpha \cos^2\phi}{(1+\alpha)^2 - \alpha^2 \cos^2\phi} = h\nu \, \frac{\alpha(1-\cos\theta)}{1+\alpha(1-\cos\theta)} \quad , \tag{5.19}$$

$$\cos\phi = (1+\alpha)\,\frac{1-\cos\theta}{\sin\theta} = (1+\alpha)\,\tan\frac{1}{2}\theta \quad , \tag{5.20}$$

where $\alpha = h\nu/mc^2$. For $\alpha \ll 1$ ($h\nu \ll mc^2$, X-ray region) the recoil electron comes out with energy less than $h\nu$ by a factor of approximately $2\alpha\cos^2\phi$. Thus, the Compton recoil electrons possess much less energy than the photoelectrons and are confined to angles less than 90°. The maximum energy of the recoil electron is when $\phi = 0$,

$$E_{kin,max} = \frac{h\nu}{1+(2\alpha)^{-1}} = h\alpha\left[1-(2\alpha)^{-1}+(2\alpha)^{-2}-\dots\right] \quad . \tag{5.21}$$

The recoil electrons have been observed by WILSON [5.4] in a cloud chamber. BLESS [5.5] used a magnetic spectrograph to verify (5.19,20). The angles θ and ϕ were measured by HOFSTADTER and McINTYRE [5.6].

KLEIN and NISHINA [5.7] applied Dirac's relativistic theory of electron to the Compton-scattering problem. They obtained a general expression for the differential cross section for initially unbound and stationary electrons. A perturbation calculation in quantum electrodynamics gives (see, for example [5.8]) the unpolarized cross section as

$$d\sigma_e = \frac{1}{2}\, r_e^2\left(\frac{\nu'}{\nu}\right)^2\left(\frac{\nu}{\nu'}+\frac{\nu'}{\nu}-\sin^2\theta\right)d\Omega \;\frac{\mathrm{cm}^2}{\mathrm{electron}} \quad , \tag{5.22}$$

where the scattered photon $h\nu'$ goes into the solid angle $d\Omega = 2\pi\sin\theta\, d\theta$. Substituting $\nu/\nu' = 1+\alpha(1-\cos\theta)$ from (5.17), we get

$$d\sigma_e = r_e^2\left[\frac{1}{1+\alpha(1-\cos\theta)}\right]^2\left(\frac{1+\cos^2\theta}{2}\right)\left\{1+\frac{\alpha^2(1-\cos\theta)^2}{(1+\cos^2\theta)\,[1+\alpha(1-\cos\theta)]}\right\}d\Omega \; . \tag{5.23}$$

In the forward direction $\theta = 0$, it approaches the classical value (5.5) $(d\sigma_e/d\Omega)_{\theta=0} = r_e^2 = 79.41\times10^{-27}$ cm^2/(electron × steradian). The Klein-Nishina formula (5.23) is plotted in Fig.5.6, and is in excellent agreement with experiments. It reduces to (5.5) for $\lambda = \infty$. The integration of (5.23) over $d\Omega$ gives the total cross section

$$\sigma_e = 2\pi r_e^2\left\{\frac{1+\alpha}{\alpha^2}\left[\frac{2(1+\alpha)}{1+2\alpha}-\frac{\log(1+2\alpha)}{\alpha}\right]+\frac{\log(1+2\alpha)}{2\alpha}-\frac{1+3\alpha}{(1+2\alpha)^2}\right\}\frac{\mathrm{cm}^2}{\mathrm{electron}} \; . \tag{5.24}$$

For $\alpha \ll 1$ ($h\nu \ll mc^2$), this becomes

$$\sigma_e = \frac{8\pi}{3}\, r_e^2(1-2\alpha) \simeq \frac{8\pi}{3}\, r_e^2 \quad , \tag{5.25}$$

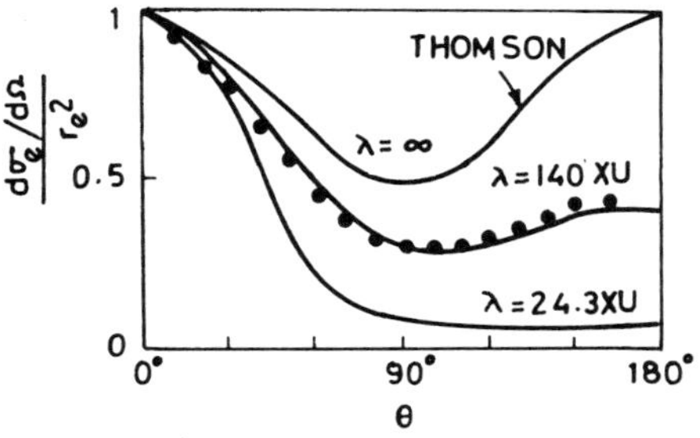

Fig. 5.6. $(d\sigma_e/d\Omega)/r_e^2$ versus θ plot. The experimental points are for carbon at λ = 140 XU

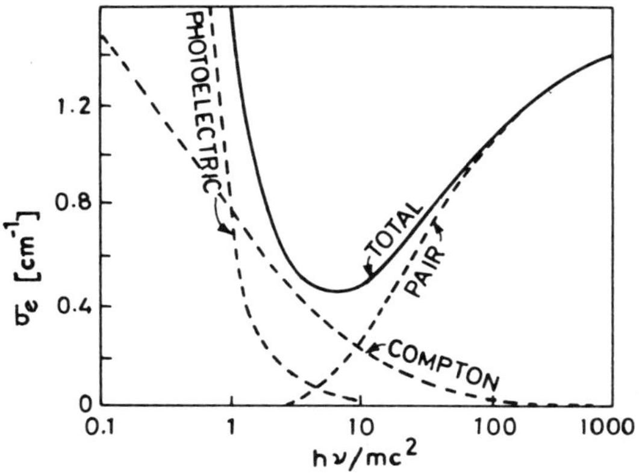

Fig. 5.7. Total linear attenuation coefficient for lead as a function of $h\nu/mc^2$

and for $\alpha >> 1$ $(h\nu >> mc^2)$,

$$\sigma_e = \pi r_e^2 \frac{1}{\alpha}\left(\log 2\alpha + \frac{1}{2}\right) \quad . \tag{5.26}$$

Thus, the Compton cross section drops off at high energies. At energies above a few MeV, the dominant absorption process is pair production (photon → electron + positron).

In a thin absorber of thickness dx, having N_a atoms per cm^3, there are N_aZ electrons per cm^3 and N_aZdx electrons per cm^2. Suppose a beam of n photons, each of energy hν, passes through the absorber. The number dn of these photons that experience Compton collisions and are thereby removed from the beam is

$$-\frac{dn}{n} = (N_aZdx)\sigma_e \quad . \tag{5.27}$$

We define the Compton total linear attenuation coefficient as

$$\sigma_1 = N_a Z\sigma_e \text{ cm}^{-1} = N_A \rho (Z/A)\sigma_e \text{ cm}^{-1} \quad , \tag{5.28}$$

where the number N_a is given by

$$N_a\left[\frac{\text{atoms}}{\text{cm}^3}\right] = N_A\left[\frac{\text{atoms}}{\text{mole}}\right] \frac{\rho[\text{g/cm}^3]}{A[\text{g/mole}]} \quad . \tag{5.29}$$

Figure 5.7 shows the total linear attenuation coefficient for ^{82}Pb as a function of $h\nu/mc^2$.

The polarization is the same for unmodified and modified X-rays. A detailed review on the Compton effect is by EVANS [5.9].

5.3 Scattering by Bound Electrons

We first discuss the classical theory under the following assumptions:

1) The electrons are distributed throughout a volume comparable with the atomic dimensions.

2) Each electron is loosely bound to the atom so that it scatters coherently according to the Thomson equation as a free electron.

3) The electrons move so slowly that there is no appreciable change of the configuration during a large number of complete alternations of the field-vector of the incident X-rays.

As shown in Sect.5.1, the atomic scattering cross section can be expressed as

$$\sigma_R = \frac{8\pi}{3} r_e^2 f^2(\theta) \quad .$$

In the forward direction there would be no path difference between the waves scattered by electrons situated in various parts of the atom. Because the electrons are all loosely bound, any change of phase on scattering is the same for each of them. Therefore, the amplitude of the scattered wave in the forward direction would be Z times that due to a single electron. For other directions, there would be path differences between the waves scattered from electrons in different parts of the atom. The resulting interference effects would produce a resultant amplitude less than that produced in the forward direction. This happens because the distances between electrons in an atom are comparable to the wavelengths of X-rays.

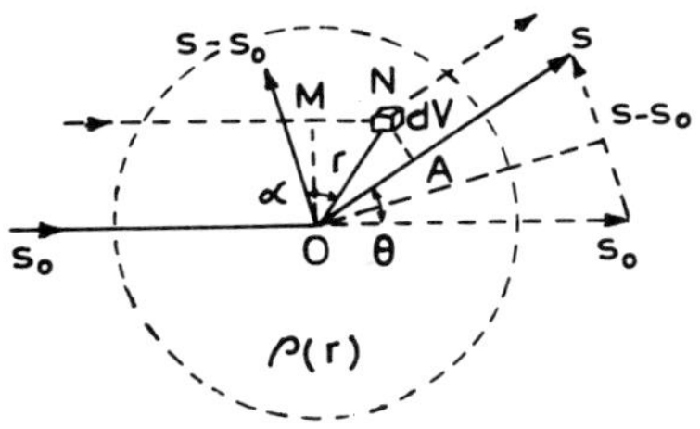

Fig. 5.8. Calculation of the atomic scattering factor f. The circle represents the charge density given by a radially symmetric function that gradually goes to zero as $\underline{r}$ increases

As in Sect.5.1, let $E = E_0 \exp(i\omega t)$ be the electric field of the incident wave at the center of the atom O, at time t. Then, the field of the radiation scattered by an arbitrary electron at O, is given by (5.1),

$$E_P = -\frac{e^2 \sin\theta}{mc^2 R} E_0 e^{i\omega\left(t - \frac{R}{c}\right)} \quad , \tag{5.30}$$

where the point of observation P is at a distance R from O, and θ is the angle between the scattering direction and the direction of the field E.

We also consider the radiation scattered by an electron that at the same time t is not at the center of the atom, but at the point N characterized by a radius vector $\underline{r}$ drawn from O (Fig.5.8). If the direction of incident X-rays is denoted by the unit vector $\underline{s}_0$ and the direction of the scattered radiation by the unit vector $\underline{s}$, then the path difference is (Fig.5.8)

$$OA - MN = \underline{s} \cdot \underline{r} - \underline{s}_0 \cdot \underline{r} = (\underline{s} - \underline{s}_0) \cdot \underline{r} \quad . \tag{5.31}$$

The corresponding phase difference is

$$\phi = \frac{2\pi}{\lambda} (\underline{s} - \underline{s}_0) \cdot \underline{r} = \frac{2\pi}{\lambda} \underline{S} \cdot \underline{r} \quad , \tag{5.32}$$

where $\underline{S} = \underline{s} - \underline{s}_0$, and from Fig.5.8,

$$S = 2 \sin\frac{1}{2}\theta \quad . \tag{5.33}$$

For R >> (size of the atom), the distance of the point of observation can be taken as R for any position of the electron in the atom. Therefore, the field at P, at the instant t, because of the scattering by the electron at N, is given by

$$(E_P)_N = -\frac{e^2 \sin\theta}{mc^2 R} E_0 e^{i\left[\omega\left(t - \frac{R}{c}\right) + \phi\right]} = E_P e^{i(2\pi/\lambda)\underline{S} \cdot \underline{r}} \quad , \tag{5.34}$$

where we have used (5.30) for E_P. It follows that the amplitude scattered by one electron in the actual atom referred to the amplitude that would be scattered by an electron at the center of the atom is obtained by the superposition of contributions from all possible points N in the atom,

$$f = \frac{\sum_N (E_P)_N}{E_P} = \int \rho(\underline{r})\, e^{i(2\pi/\lambda)\underline{S}\cdot\underline{r}}\, dV \quad , \tag{5.35}$$

where $\rho(\underline{r})\, dV$ is the probability of finding the electron in the element of volume dV at $\underline{r}$ (Fig.5.8).

If $\underline{r}$ makes an angle α with $\underline{S}$, then

$$(2\pi/\lambda)\underline{S}\cdot\underline{r} = (2\pi/\lambda)rS\cos\alpha = (4\pi/\lambda)r\sin\tfrac{1}{2}\theta\cos\alpha \quad . \tag{5.36}$$

Suppose the distribution of the charge density is spherically symmetric. Then the volume element can be chosen to be an annular ring of radius $r\sin\alpha$, width $rd\alpha$, and thickness dr, so that $dV = 2\pi r^2\, dr\sin\alpha\, d\alpha$, and (5.35) becomes

$$f = \int_0^\infty \int_0^\pi \rho(r)\, e^{ikr\cos\alpha}\, 2\pi r^2\, dr\sin\alpha\, d\alpha \quad , \tag{5.37}$$

where

$$k = 4\pi\left(\sin\tfrac{1}{2}\theta\right)/\lambda \quad . \tag{5.38}$$

We have extended the r integration to infinity, because only the charge density within the atom can contribute to the integral.

Integration over α gives,

$$\int_0^\pi e^{ikr\cos\alpha}\sin\alpha\, d\alpha = \left(\frac{e^{ikr\cos\alpha}}{ikr}\right)_0^\pi = \frac{2\sin kr}{kr} \quad . \tag{5.39}$$

Therefore, the *atomic scattering factor* f is given by

$$f = \int_0^\infty U(r)\, \frac{\sin kr}{kr}\, dr \quad . \tag{5.40}$$

where $U(r)dr = 4\pi r^2\, dr\, \rho(r)$ gives the probability that an electron lies between radii r and $r + dr$ in an atom. We can calculate f from (5.40), if we know $\rho(r)$. From (5.35) we see that f is the *ratio* of the radiation amplitude scattered by the charge distribution in an atom to that scattered by a point electron.

5.3.1 Quantum-Mechanical Approach

For the hydrogen atom in the ground state, the charge density is given by

$$\rho(r) = |\psi_{100}|^2 = \left|\left(\pi a_0^3\right)^{-\frac{1}{2}} e^{-r/a_0}\right|^2 = \left(\pi a_0^3\right) e^{-2r/a_0} \quad , \qquad (5.41)$$

where $a_0 = 0.53$ Å. Therefore, for the hydrogen atom

$$f_H = \frac{4}{a_0^3 k} \int_0^\infty r\, e^{-2r/a_0} \sin kr\, dr = \left[1+(a_0 k/2)^2\right]^{-2} \quad , \qquad (5.42)$$

because the integral is of the form $\int_0^\infty x\, e^{-ax} \sin bx\, dx = 2ab(a^2+b^2)^{-2}$. Thus, f_H depends only on $\lambda^{-1}\sin(\theta/2)$. For the forward direction ($\theta=0$), we have $k=0$ and $f_H = 1$, as expected. As θ or $\lambda^{-1}\sin(\theta/2)$ increases, f_H decreases. This is true for other atoms also, and is one of the reasons why higher-order Bragg reflections (which occur at higher angles) are weaker than low ones.

For a crystal $\lambda^{-1}\sin(\theta/2) = [2d_0(hkl)]^{-1}$, where hkl are integers. This shows that the scattering factor has a constant value for each crystal plane and is independent of wavelength.

In the quantum mechanical approach, Hartree wave functions are used to evaluate $\rho(r)$ for a many-electron atom. The scattering intensity is determined by f^2, and both coherent (Rayleigh) and incoherent (Compton) scattering are present.

5.4 Scattering by Crystals (Laue Equations)

We now determine the amplitude of the wave scattered in a given direction by all of the atoms in the unit cell. The *structure amplitude* F(hkl) for a given hkl Bragg reflection is the *ratio* of the reflection amplitude and the amplitude of the wave scattered by a single point electron for the same wavelength.

The value of F(hkl) is given by

$$F(hkl) = \sum_i f_i\, e^{i\phi_i} = \sum_i f_i\, e^{i(2\pi/\lambda)(\underline{S}\cdot\underline{r}_i)} \quad , \qquad (5.43)$$

where the sum is over all atoms in a unit cell, ϕ_i is the phase of the wave scattered by the ith atom referred to that of the origin, $\underline{r}_i$ is the vector from the origin of the ith atom, and f_i is the atomic structure factor of the ith atom.

Let the unit cell be defined by the translation vectors $\underline{a}$, $\underline{b}$, $\underline{c}$. Then,

$$\underline{r}_i = u_i\underline{a} + v_i\underline{b} + w_i\underline{c} \quad , \tag{5.44}$$

where u, v, w are integers. The successive atoms in the row $\underline{a}$ would scatter in phase if

$$\underline{a} \cdot \underline{s} - a \cdot \underline{s}_0 = \underline{a} \cdot \underline{S} = h\lambda \quad . \tag{5.45}$$

The incident-beam unit vector $\underline{s}_0$, the scattered beam unit vector $\underline{s}$, and $\underline{a}$ need not be coplanar. In general, the diffracted beam $\underline{s}$ can be regarded as forming a cone about the row of atoms for a given $\underline{s}_0$. For a triply periodic array, the condition that the entire array scatter in phase is that the following three *Laue equations*,

$$\underline{a} \cdot \underline{S} = h\lambda \quad , \quad \underline{b} \cdot \underline{S} = k\lambda \quad , \quad \underline{c} \cdot \underline{S} = l\lambda \quad , \tag{5.46}$$

are satisfied simultaneously. This permits only one direction of the diffracted beam, because three cones can mutually intersect along only one line.

From (5.44,46),

$$\underline{S} \cdot \underline{r}_i = \lambda(hu_i+kv_i+lw_i) \quad . \tag{5.47}$$

If α_0, β_0, γ_0 (α,β,γ) are the angles made by the incident (diffracted) beam with the coordinate axes a, b, c, the direction cosines satisfy the well-known relations

$$\cos^2\alpha_0 + \cos^2\beta_0 + \cos^2\gamma_0 = \cos^2\alpha + \cos^2\beta + \cos^2\gamma = 1 \quad ,$$

$$\cos\theta = \cos\alpha \cos\alpha_0 + \cos\beta \cos\beta_0 + \cos\gamma \cos\gamma_0 \quad ,$$

where θ is the scattering angle. Therefore, the Laue equations give

$$(\cos\alpha-\cos\alpha_0)^2 + (\cos\beta-\cos\beta_0)^2 + (\cos\gamma-\cos\gamma_0)^2$$
$$= 2 - 2\cos\theta = \left[(h/a)^2+(k/b)^2+(l/c)^2\right]\lambda^2 \quad ,$$

or

$$2d(hkl)\, \sin\frac{1}{2}\theta = \lambda \quad , \quad d(hkl) = \left[(h/a)^2+(k/b)^2+(l/c)^2\right]^{-\frac{1}{2}} \quad . \tag{5.48}$$

This is just the Bragg law. The h,k,l are related to the Miller indices and may contain a common integral factor n.

From (5.43,47),

$$F(hkl) = \sum_i f_i \, e^{i2\pi(hu_i+kv_i+lw_i)} \quad , \tag{5.49}$$

and

$$|F|^2 = \left[\sum_i f_i \cos 2\pi(hu_i+kv_i+lw_i)\right]^2 + \left[\sum_i f_i \sin 2\pi(hu_i+kv_i+lw_i)\right]^2 \quad . \tag{5.50}$$

If all of the atoms are identical, (5.49) becomes

$$F(hkl) = f \sum_i e^{i2\pi(hu_i+kv_i+lw_i)} \quad . \tag{5.51}$$

The body-centered cubic cell has two atoms of the same kind located at 0,0,0 and 1/2,1/2,1/2. We find

$$F = f\left[1 + e^{i\pi(h+k+l)}\right] \quad , \tag{5.52}$$

$$F = 2f \quad \text{if} \quad (h+k+l) \text{ is even} \quad ;$$

$$F = 0 \quad \text{if} \quad (h+k+l) \text{ is odd} \quad .$$

Therefore, the diffraction spectrum would not contain lines such as (100), (300), (111), etc., but lines such as (200), (110), (222), etc., would occur.

5.5 X-Ray Raman and Plasmon Scattering

Scattered radiation, whose frequency is changed corresponding to transitions between the bound states of the target atom, was discovered in the visible region by RAMAN [5.10], and is known as *Raman scattering*. The frequency ν'_R of the modified line of Raman type is given by [5.11]

$$h\nu'_R = h\nu - \delta E \quad , \tag{5.53}$$

where ν is the frequency of the incident radiation and δE is the energy difference required for the transition of the atom from the ground state to an optical (excited) state.

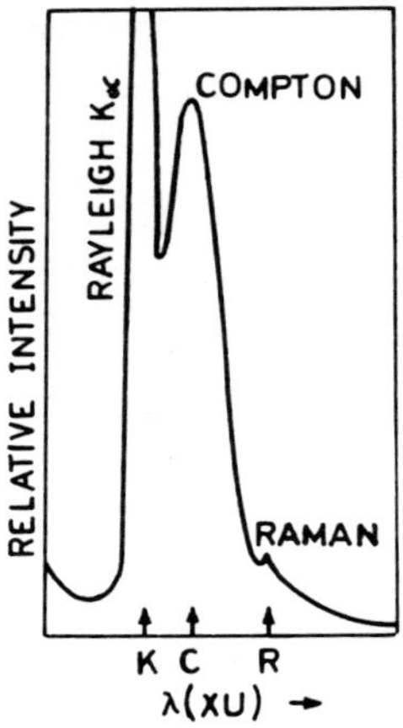

Fig. 5.9. Spectroscopic analysis of $CuK\alpha_{1,2}$ radiation scattered at 90° by polystyrene (carbon): K (Rayleigh $CuK\alpha_{1,2}$ 1540 XU); C (Compton peak at 90°); R (Raman line 1597.1 XU)

The existence of Raman lines in the X-ray region was predicted by DUMOND [5.12] and later by SOMMERFELD [5.13]. They have now been unambiguously observed by DAS GUPTA [5.14]. The Raman lines occur on the low-energy side of the Rayleigh lines, the energy difference is independent of the scattering angle and equal to the excitation energy of the scattering atom. The Raman lines are weak and overlap the tail of the Compton peak. The presence of *X-ray Raman scattering* has been confirmed by several workers (for example [5.15-17]).

DAS GUPTA [5.14] used a bent mica (d=2.554 Å) Cauchois spectrograph (diameter of focal circle = 9 inches) as analyzer, K X-rays of ^{29}Cu and ^{42}Mo as incident radiation, and carbon as scattering material. His main result for the $CuK\alpha_{1,2}$ radiation scattered at $\sim 90^\circ$ is shown in Fig.5.9. The observed shift of the Raman line relative to the primary (Rayleigh) wavelength is 57.1 XU. This corresponds to an energy shift of 282 eV, which is close to the excitation energy of the K electrons of carbon (284 eV).

The Raman peaks can come from core-electron excitation, as discussed above, and from plasmon excitation. The *plasmon* peak was predicted by PINES [5.18] and has been observed recently [5.19-21]. The frequency ν_P of the modified line of *plasmon scattering* is given by

$$h\nu_P' = h\nu - E_P \quad , \tag{5.54}$$

where E_P is the plasmon-excitation energy of the scattering material.

Consider the conduction-band electrons that form an electron gas. In the absence of collisions, the equation of motion of a free electron in an elec-

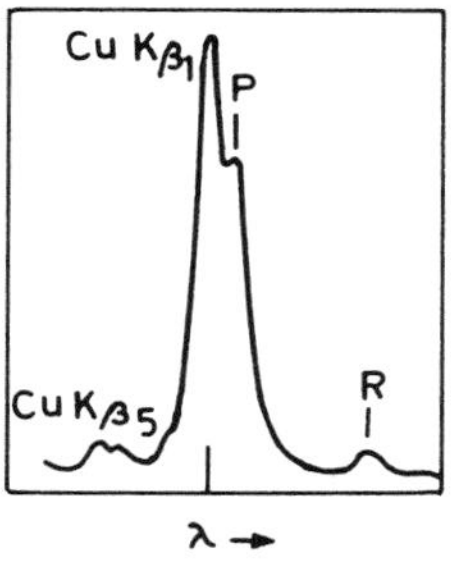

Fig. 5.10. The Rayleigh line ($CuK\beta_1$, $\lambda = 1.39222$ Å), plasmon scattering line P ($\lambda = 1.39528$ Å), the Raman scattering line R ($\lambda = 1.41087$ Å) from beryllium for the scattering angle 10°. Displacement of P is 19.6 eV, and of R is 117.8 eV

tric field is $md^2x/dt^2 = -eE$. If x and E have the time dependence $\exp(-i\omega t)$, then we get $x = eE/(m\omega^2)$. The dipole moment p of the electron is $p = -ex = -e^2E/(m\omega^2)$, and the polarization is $P = -nex = -(ne^2/m\omega^2)E$, where n is the electron concentration. The dielectric function at frequency ω is $\varepsilon = 1 + 4\pi P(\omega)/E(\omega)$, (3.37), or

$$\varepsilon(\omega) = 1 - \frac{4\pi ne^2}{m\omega^2} = 1 - \frac{\omega_p^2}{\omega^2} \quad , \tag{5.55}$$

where $\omega_p = 4\pi ne^2/m$ is called the *plasma frequency*. A *plasmon* is a quantized plasma oscillation. It can be excited by passing (or reflecting) an electron or an X-ray photon through a thin metallic film. The charge of the electron, or the electric field of the radiation, couples with the electrostatic field fluctuations of the plasma oscillations. In general, we can write $E_p = \hbar\omega_p = \hbar\, 4\pi ne^2/m_{eff}$, where m_{eff} is the effective mass of the electron. In Fig.5.10 we show a microphotometer record of the spectrogram of $CuK\beta_1$ line scattered by beryllium at scattering angle 10°.

6. Chemical Shifts and Fine Structure

The study of chemical bonding by X-ray and photoelectron spectroscopy has been a major area of research over the past two decades. The information obtained by these two methods is complementary. Besides this we shall also discuss here the correlation of factors, like effective nuclear charge and bond length, with X-ray energies and spectral shapes.

6.1 Solid-State Effects and Bonding

6.1.1 Metallic Bond

Each atom has a set of discrete energy levels. Consider an assembly of N such free atoms. The quantum-state distribution for this system is just that of an atom duplicated N times. As atoms are brought together to form a crystal, the wave functions of the outer electrons in the neighboring atoms begin to *overlap*. That is, the outer electrons, originally bound to one nucleus, begin to feel the potentials of the neighboring nuclei. This introduces a perturbation. We have seen in (2.46), that the general effect of a perturbation is to remove the degeneracy (to separate out the levels that had the same unperturbed energies). Therefore, N-fold degenerate outer levels would get separated and form a *band* of energy levels. The degree of the removal of degeneracy would depend on the strength and nature of the perturbation, and on the symmetry of the system. The perturbation felt by the core electrons is weak. Therefore, the inner levels remain sharp (localized orbitals).

Each atomic level can accomodate two electrons by the Pauli exclusion principle. Therefore, the band of certain energy width would consist of 2N levels. If each atom happens to have only one valence electron, then only the lower half of the levels in the energy band would be occupied, and the upper half would remain empty, at absolute-zero temperature.

The unoccupied part of the band is called the *conduction band* and the occupied one the *valence band*. The energy E_F of the highest occupied level is called the *Fermi energy*. The energy states in the bands are very closely

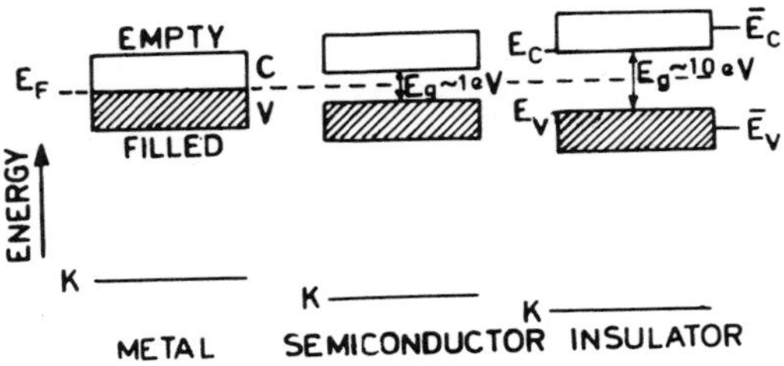

Fig. 6.1. Simplified energy-level diagram for a metal, a semiconductor, and an insulator. The shaded (blank) area of average energy $\bar{E}_V$ ($\bar{E}_C$) indicates the occupied (unoccupied) valence V (conduction C) band of edge E_V (E_C). K is an inner level. E_g is the forbidden gap. E_F is the Fermi energy

spaced ($\sim 10^{-22}$ eV). Therefore, the distribution of the energy levels in the entire band can be conveniently described by the *density of states* N(E), N(E) dE being the number of electronic states per unit volume with energies between E and E + dE. A band is a state of high delocalization.

In a metal (Fig.6.1) the states above E_F are empty and easily accessible. Therefore, metals (like copper, silver, gold, etc.) are good conductors of electricity. If a compound is formed (for example, an oxide of copper) or in some elements (like silicon and germanium) the conduction band gets separated from the valence band and a forbidden region of energy, called the *energy gap* (or *band gap*), E_g, appears between them (Fig.6.1). When $E_g \sim 1$ to 2 eV, light quantum, or even thermal energy at room temperature can easily excite a few electrons across the gap from the highest point in the valence band (*valence-band edge*, E_V) to the lowest point in the conduction band (*conduction-band edge*, $E_C = E_g + E_V$). This makes conduction possible on a limited scale and the material is a *semiconductor*.

When $E_g \sim 5$ to 10 eV, the conduction is not possible at room temperature and the material is an *insulator*.

We have also *semimetals*, like arsenic, antimony, and bismuth. In a semimetal, the conduction-band edge is only slightly lower in energy than the valence-band edge.

The function

$$f(E,T) = \frac{1}{\exp[(E-E_F)/kT]+1} \quad , \quad \text{(k=Boltzmann constant)} \quad , \qquad (6.1)$$

is called the *Fermi function*. It gives the fraction of states with energy E that are filled at the temperature T (Fig.6.2). All states with $E < E_F$ are filled at T = 0, and there is at least one state with energy $E = E_F$ occupied

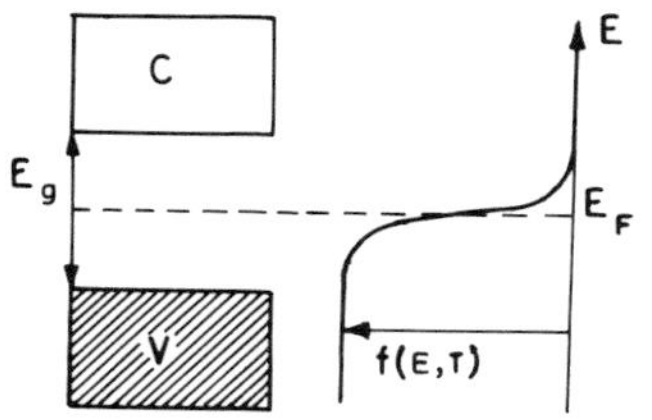

Fig. 6.2. Location of the Fermi level at T = 0. For most semiconductors, the Fermi level is at the *middle* of the energy gap.

It is usual, in solid-state physics, to take $E_F = 0$. Then the conduction-band states have positive energies and the valence-band and inner bound states have negative energies. The energy-level diagram is aligned with E_F above the core levels. Therefore, for solids it is usual to draw the inner K level at the bottom, rather than at the top, of the energy level diagram.

Some workers have regarded the crystal as a giant molecule (for example, see HULLINGER and MOOSER [6.1]) and used chemical-bonding language to discuss the X-ray spectra. We shall now briefly define the technical words used in such discussions.

6.1.2 Ionic and Covalent Bonds

KOSSEL suggested in 1916 that elements placed just before (after) a noble gas in the periodic table can attain a noble-gas structure by gaining (losing) electrons and forming negatively (positively) charged ions. Thus ^{17}Cl, with a structure KL $3s^2 3p^5 = 2.8.7$ can become a chloride ion Cl^- with a structure [Ar] = 2.8.8 by gaining an electron, and ^{11}Na, KL 3s = 2.8.1, can form a sodium ion Na^+, [Ne] = 2.8, by losing an electron. Both Na^+ and Cl^- have stable structures and form Na^+Cl^- under an *ionic bond*. When many ions are held together by such electrostatic forces, we get an *ionic crystal*.

Two chlorine atoms cannot form an ionic bond to give Cl_2 because neither can transfer an electron to the other. To explain the formation of such molecules, LEWIS suggested in 1916 that atoms can attain a noble-gas structure, not only by complete transference of electrons, as in ionic bonding, but also by sharing electrons. The shared pair of localized electrons constitute a *covalent bond*. The covalent bond is directed in space.

The force that binds a metal atom to a number of electrons within its sphere of incluence is called a *metallic bond*. We have seen that the valence electrons in a metallic bond are spread over the crystal, more or less uni-

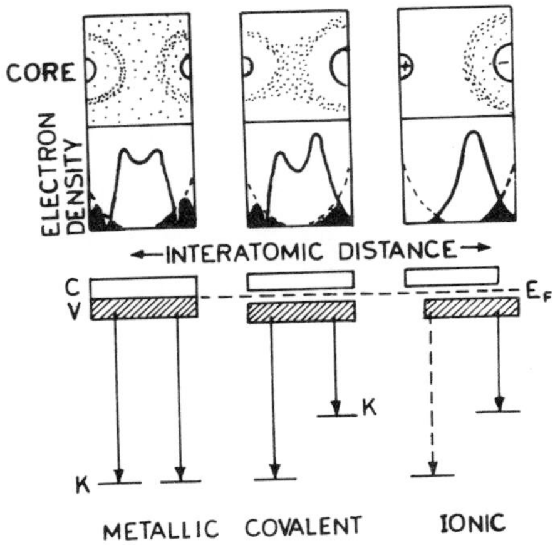

Fig. 6.3. Valence-electron distribution for metallic, covalent and ionic bondings (dots: valence electrons). Qualitative electron-density distributions for third-row cations (Al, Si, etc.) and second-row anions (C, N, O, etc.) are indicated (solid curve: 2p+3p; dashed curve: 1s electron density). The fully shaded regions show where the valence and 1s electron densities overlap. The K spectra shown below arise from these regions. In ionic solids, the valence electron that undergoes transition for the cation K spectra belongs to the anion (crossover transition, dashed line) (After Nagel, D.J.: Eighteenth ACAXA, Denver 1969)

formly. Therefore, metallic bonds are nondirectional and weaker than covalent bonds. Most of the actual bonds are intermediate in type (mixture of ionic and covalent, or of covalent and metallic forces).

The sharing of electrons in strong bonds among atoms essentially determines the electronic structure of interest in the X-ray spectroscopy. In Fig.6.3 we show how the valence-electron localization increases in going from the metallic through covalent to ionic bonding. In ionic compounds, the electrons are localized on one of the atoms, in covalent compounds between the atoms, and in metals throughout the lattice.

The dipole-matrix element for the X-ray K emission (and absorption) process involves the wave functions of the outer bonding states and the inner 1s state. Therefore, the K spectra arise from those regions where the valence and 1s electron densities overlap (Fig.6.3).

We shall use both the band and bond theories to discuss the X-ray spectra. For metallic and ionic solids reliable band calculations have become available in recent years.

6.1.3 Hybridized Orbitals

When two atoms are brought together, their atomic orbitals overlap. In a simple approximation, the *hybridized orbital* Φ is a linear combination (superposition) of two atomic orbitals ψ_{nl},

$$\Phi_{ll'} = \psi_{nl} \pm \psi_{nl'} \quad . \tag{6.2}$$

The associated probability density contains the interference effects given by $\pm(\psi^*_{nl}\psi_{nl'} + \psi^*_{nl'}\psi_{nl})$. The valence orbitals are usually the s and p ones, with occasionally a small admixture of d states. Thus, a hybridized orbital mixes the s and p, and possibly d, atomic orbitals. This leads to the formation of *bonding* and *antibonding* states (Fig.6.4) with an energy separation (gap) $E_{g,bond}$. We have, PHILLIPS [6.2,3],

$$E_{g,bond} = (\bar{E}_c - \bar{E}_v) > E_g \quad , \tag{6.3}$$

where $E_g = E_c - E_v$, and $\bar{E}_c$ $(\bar{E}_v)$ is the *average* conduction- (valence-) band energy. For example, in Si and ZnO, E_g is 1.1 and 3.6 eV, compared to $E_{g,bond}$ = 4.8 and 12 eV, respectively.

To see the origin of $E_{g,bond}$, consider the simplest case of a diatomic molecule AB. The molecular orbital Φ in the region of each atom resembles the atomic orbitals ψ_A and ψ_B. We can write, in the approximation of the linear combination of atomic orbitals (LCAO),

$$\Phi = C_A\psi_A + C_B\psi_B \quad , \tag{6.4}$$

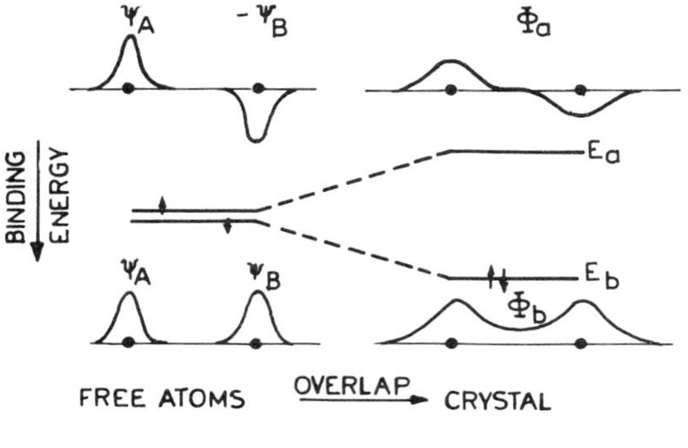

Fig. 6.4. Formation of bonding and antibonding states in the LCAO approximation. E_b (E_a) is the bonding (antibonding) energy level

where constants C_A and C_B are chosen to give lowest energy to the molecular orbital. Because Φ^2 is a measure of electron density at any point, it gets contributions from ψ_A^2 and ψ_B^2 in the ratio of $C_A^2:C_B^2$.

For homonuclear diatomic molecule (A=B), $C_A^2 = C_B^2$, or $C_A = \pm C_B$. Thus

$$\Phi_a = \psi_A - \psi_B \quad , \quad (\textit{antibonding}) \quad , \tag{6.5}$$

$$\Phi_b = \psi_A + \psi_B \quad , \quad (\textit{bonding}) \quad . \tag{6.6}$$

The energies of these molecular orbitals are

$$E_a = \left(\int \Phi_a^* H \Phi_a d\tau\right) / \int \Phi_a^* \Phi_a \, d\tau = \left[\int (\psi_A - \psi_B) H (\psi_A - \psi_B) d\tau\right] / \int (\psi_A - \psi_B)^2 \, d\tau$$

$$= \frac{1}{2} (H_{AA} + H_{BB} - 2H_{AB}) = \frac{1}{2} (E_A + E_B - 2E_{AB}) \quad , \tag{6.7}$$

where H is the Hamiltonian and we have taken $\int \psi_A \psi_B \, d\tau = \delta_{AB}$ for the denominator. Similarly,

$$E_b = \frac{1}{2} (E_A + E_B + 2E_{AB}) \quad . \tag{6.8}$$

Because E_A, E_B and E_{AB} are all negative quantities, Φ_b is more stable than the 1s orbital of the atom A (or B), whereas Φ_a is less stable (Fig.6.4). Clearly, $E_{g,bond} = 2E_{AB}$ in this case.

In addition to classifying orbitals as bonding and antibonding, it is useful to classify them in terms of their symmetry about the internuclear axis. In Fig.6.5, we illustrate the formation of molecular orbitals out of 2s and $2p_z$ atomic orbitals of two atoms A and B.

In order of increasing energy the molecular orbitals are: $1s\sigma < 1s\sigma^* < 2s\sigma < 2s\sigma^* < 2p\sigma < 2p_z\pi = 2p_y\pi < 2p_z\pi^* = 2p_y\pi^*$ ... etc. The asterisk indicates an antibonding level. The subscripts g (gerade=even) and u (ungerade=odd) indicate behavior under the operation of a twofold rotation. For example, an s orbital is rotated into itself, whereas a p_z orbital is turned into minus itself (the phases of its lobes change sign).

The overlapping orbitals can be combined to form molecular orbitals if they have similar energy and similar symmetry. The strength of such a covalent bond depends on the extent of overlap achieved. This is known as the *principle of maximum overlap*. For example, in a polar diagram (Fig.6.6) the p orbitals are concentrated in a particular direction and their lobes are longer than the radius of the corresponding spherical s orbital. Therefore, they can overlap with other s or p orbitals more effectively than two s or-

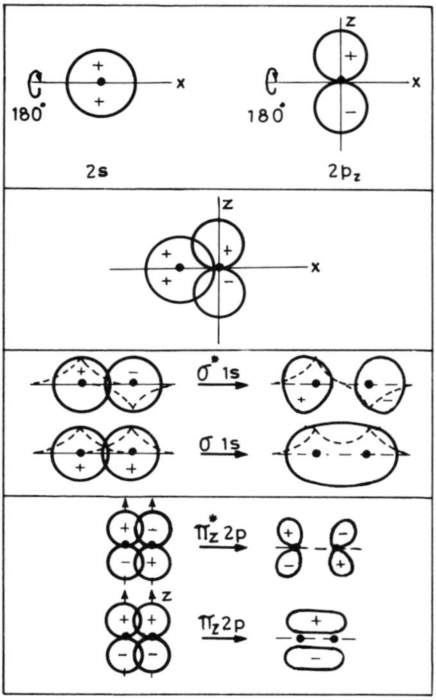

Fig. 6.5

Fig. 6.6

Fig. 6.5. Formation of molecular orbitals in simple cases. First row: atomic orbitals (2s is symmetric about x axis, and $2p_z$ is unsymmetrical about x axis because of θ, ϕ dependence, r cosθ). Second row: nearly zero overlap and so molecular orbital is not formed. Third row: σ* 1s (antibonding) and σ1s (bonding) molecular orbitals from two 1s atomic orbitals, with symmetry about the nuclear axis. Fourth row: π_z^* 2p and π_z 2p bonds that are unsymmetrical about the nuclear axis. The dashed curves represent the wave functions and the solid curves the boundary surface of an orbital

Fig. 6.6. Overlap between two s orbitals, and one p and one s orbital

bitals can overlap (Figs.6.5,6); s-p σ bonds are stronger than s-s bond, and p-p σ bonds are stronger still. The relative bond strengths are of the order s-s : s-p : p-p = 1 : 1.732 : 3.

Sometimes certain symbols of group theory are also used. Lower-case symbols (a_{1g}, a_{2u}, e_g, etc.) are used to indicate one-electron wave functions (orbitals). Upper-case symbols (A_{1g}, E_g, T_{1u}, etc.) are used to describe electronic energy levels. The symbols A (or a) and B (or b), with any suffixes, indicate single degeneracy, E (or e) indicates double degeneracy, and T (or t) indicates triple degeneracy. The symbol t_{2g}^3 indicates that these orbitals are occupied by three electrons. Other subscripts (for example, 1, 2, or primes) serve only to distinguish general symmetry properties. Thus, t_{1g} and t_{2g} represent two sets of triply degenerate orbitals, all centrosymmetric (g).

6.1.4 Coordination

In covalent structures, the directed orbitals are oriented towards the nearest-neighbor atoms, so that the energy of occupied bonding orbitals is decreased. Semiconductors usually have structures such that each atom is *tetrahedrally coordinated*, corresponding to the sp^3 orbitals directed towards the corner of a tetrahedron. A number of binary compounds $A^N B^{8-N}$ with 8 valence electrons per atom pair occur in this coordination. If $\bar{N}$ is the average number of electrons per atom, then for such structures the *coordination number* is $8-\bar{N}$.

Besides the tetrahedral sp^3 ($\bar{N}=4$) structures, we also have layer structures of hexagonal symmetry with sp^2 bonds lying in the xy plane (like graphite).

An exception to the rule that the sp^3 bonding requires $\bar{N}=4$, is provided by *fluorite structures* with formula RX_2. They contain eight electrons per formula unit ($A_2^{II}B^{IV}$: Mg_2Si), or sixteen ($A^{II}B_2^{VII}$: CdF_2 or $BaCl_2$; $A^{IV}B_2^{VI}$: GeO_2). Mg_2Si is covalent, CdF_2 is ionic, and GeO_2 has some covalent and some ionic properties.

The rule that s and p orbitals hybridize to form directed bonds is usually valid so long as $E_{g,bond} > E_{np} - E_{ns}$, where $E_{np} - E_{ns}$ is the energy difference between the np and ns valence electrons of the free atom. If $E_{np} - E_{ns}$ is large the energy gap between the bonding $(sp^3)^+$ states and antibonding $(sp^3)^-$ states is weakened or even destroyed, and the crystal becomes metallic (for example, Pb with large n).

The column VI elements (except oxygen) are called *chalcogenides*. The crystals of S, Se and Te are semiconductors with a structure of spiral chains parallel to the z axis.

A combination of one s-, three p- and two d-atomic orbitals leads to six hybrid orbitals that are directed *octahedrally* (away from the origin in the ±x, ±y and ±z directions). We can have d^2sp^3 or sp^3d^2 octahedral hybridization. An example is provided by sulphur hexafluoride. The ground-state arrangement of electrons in the sulphur atom is

	1s	2s	$2p_x$	$2p_y$	$2p_z$	3s	$3p_x$	$3p_y$	$3p_z$	3d
S :	↓↑	↓↑	↓↑	↓↑	↓↑	↓↑	↓↑	↓	↓	
S** :	↓↑	↓↑	↓↑	↓↑	↓↑	↓	↓	↓	↓	↓↓

Before sulphur can form octahedral bonds by sp^3d^2 hybridization, a double *promotion* is necessary to give the excited state S**. The overlap with 2p orbitals of fluorine will then yield the octahedral SF_6 molecule.

In PCl_5, we have sp^3d-hybrid orbitals directed towards the corners of a *trigonal* bi-pyramid.

The *spinels* are a group of complex oxides with a general formula $M^{2+}M_2^{3+}O_4$, where M may be Mn, Fe or Co, or a number of other transitional metals. The oxides Fe_3O_4 and Mn_3O_4 are spinels, with formulas $Fe^{2+}Fe_2^{3+}O_4$ and $Mn^{2+}Mn_2^{3+}O_4$. In the crystal structure of a spinel, one-third of the metallic ions are surrounded, tetrahedrally, by four O^{2-} ions, and two-thirds are surrounded, octahedrally, by six O^{2-} ions. In a *normal* spinel (Mn_3O_4), all of the M^{3+} ions occur at octahedral sites, and all of the M^{2+} ions at tetrahedral sites. In the *inverted* spinel (Fe_3O_4), half of the M^{3+} ions occur at tetrahedral sites and the other M^{3+} and M^{2+} ions occupy octahedral sites.

6.1.5 Ionic Character of Covalent Bonds

If A and B are unlike atoms, the pair of shared electrons may not be shared equally in a covalent bond. For example, if B has stronger attraction for electrons than A, the shared pair will be attracted towards B and away from A. Such displacement of electrons in a covalent bond gives the bond some ionic character. The actual bond then can be represented as a *resonance* hybrid between the covalent and ionic forms,

$$A - B \leftrightarrow A^+B^- \quad .$$

The *ionicity of a bond* is defined as the fraction f_i of ionic or heteropolar character in the bond compared to the fraction f_c of covalent or homopolar character, such that

$$f_i + f_c = 1 \quad . \tag{6.9}$$

a) *Coulson's Definition of Ionicity*. The valence wave function can be written as (6.4), or simply as

$$\Phi_{valence} = \psi_{sp^3}(A) + \lambda\psi_{sp^3}(B) \quad , \tag{6.10}$$

for the orbitals in tetrahedrally coordinated A^NB^{8-N} semiconductors, considered here for simplicity. The trial wave function (6.10) is put in the wave equation and the best value of λ is obtained by minimizing the energy. Because $\psi(A)$ and $\psi(B)$ are assumed to be orthogonal, following COULSON et al. [6.4], we can define ionicity as

$$f_i = \frac{P(A) - P(B)}{P(A) + P(B)} = \frac{1-\lambda^2}{1+\lambda^2} \quad , \tag{6.11}$$

where P(A) is the probability of finding a valence electron on A.

b) *Pauling's Definition of Ionicity*. PAULING [6.5] noticed that, in molecules the reaction $AA + BB \rightarrow 2AB$ is usually exothermic if the A - B bond is partially covalent. This means that the energy of an A - B bond is lower (more negative) than the average energies of A - A and B - B covalent (homopolar) bonds,

$$D_{AB} > \frac{1}{2}(D_{AA} + D_{BB}) \quad , \tag{6.12}$$

where D represents the magnitude of energy. The difference is called the extra-ionic energy Δ_{AB} because A - B bond is partially ionic, whereas A - A and B - B bonds are covalent. We have $\Delta_{AB} > 0$. This would be so if (say) atom B has greater power to attract electron than A. Let a number x_B, called the *electronegativity* of B, denote this power. Then, we can think that a certain number of electrons proportional to $|x_A - x_B|$ has been transferred from A to B. Clearly, the Coulomb interaction, between the ionic charge left behind and the valence charge transferred, would be proportional to $|x_A - x_B|^2$. This would give rise to the extra-ionic energy responsible for the inequality (6.12). We can write, following PAULING,

$$\Delta_{AB} = \text{const.} \times (x_A - x_B)^2 \quad . \tag{6.13}$$

Thus electronegativity is the power of an atom in a molecule to attract electrons to itself (Pauling). We can take $|x_A - x_B|$ to be a measure of ionicity. As $|x_A - x_B|$ increases, $f_i \rightarrow 1$, $0 \leqq f_i \leqq 1$. Also, the ionicity of an A - B bond should be the same as that of a B - A bond, $f_i(A,B) = f_i(B,A)$. These considerations led PAULING [6.5] to define the ionicity of a single bond as

$$f_i = 1 - e^{-0.25(x_A - x_B)^2} \quad , \tag{6.14}$$

where the constant in (6.13) is chosen so that x_A increases by 0.5 when Z changes by 1 in the first-row elements.

In molecules, the coordination number of an atom is less than or equal to its valence N. On the other hand, in crystals the coordination number usually exceeds N. In A^NB^{8-N} crystals it is usually 4 or 6. To accommodate this fact, PAULING introduced the concept of *resonating bonds*. The N valence bonds per atom are pictured as shared with all of the 4 (or 6) nearest neighbors of each atom. This implies that the degree of covalency $f_c = 1 - f_i$ is also shared and we must replace the single-bond ionicity f_i by a resonating-bond ionicity $f_i^{(res)}$,

$$1 - f_i^{(res)} = (N/n)(1-f_i) = (N/n)\, e^{-0.25(x_A-x_B)^2} \quad , \tag{6.15}$$

where n is the number of nearest neighbors. This definition is applicable to *crystals*. Pauling's values of x_A are given in Appendix J.

c) *Effective-Charge Approach*. An electron in a bond is assumed to be attracted by one of the two nuclei with a Coulomb force $Z_{eff}\, e^2/r^2$, where eZ_{eff} is the effective nuclear charge felt by the electron at a mean distance r from the nucleus. Slater's method can be used to estimate Z_{eff}. Pauling's value of x_A can be represented [6.6-9] by a formula of the form,

$$x_A = a\left(Z_{eff}/r^p\right) + b \quad . \tag{6.16}$$

ALLRED and ROCHOW [6.6] use $a = 0.359$, $p = 2$, $b = 0.744$. Values of partial charges in polyatomic ions (like NH_4^+, OH^-, CO_3^{2-}, etc.) have been given by SANDERS [6.10]. His two principles are: 1) When two or more atoms, initially different in electronegativity, combine chemically, they become adjusted to an equal intermediate electronegativity in the compound. 2) The electronegativity of all of the atoms in a compound is the geometric mean of the electronegativities of all of the atoms before combination. SANDERS further assumes that a group of atoms that make up a positive or negative ion can be treated as a single atom that has an electronegativity that corresponds to their geometric mean.

d) *Phillips' Definition of Ionicity*. For a crystal A^NB^{8-N}, the potential can be written as

$$V(\text{crystal}) = \sum_\alpha V_A(\underline{r}-\underline{r}_\alpha) + \sum_\beta V_B(\underline{r}-\underline{r}_\beta) \quad , \tag{6.17}$$

where $\underline{r}_\alpha$, $\underline{r}_\beta$ label the lattice sites of the sublattices of the A, B atoms, respectively, and V_A, V_B are the screened Coulomb potentials seen by valence electrons of atoms A, B. For each cell, we can separate V(crystal) into the parts that are even and odd with respect to the interchange of A and B. Following PHILLIPS [6.2,3], we associate these parts with covalent and ionic potentials,

$$V_{covalent} = V_A + V_B \quad , \tag{6.18}$$

$$V_{ionic} = V_A - V_B \quad . \tag{6.19}$$

We expect f_c and f_i to be related to certain average values of (6.18,19) calculated with respect to suitable wave functions of the crystal.

In crystallography, it is found useful to express the *bond length* r_{AB} between nearest neighbors, in a given state of valence hybridization, as the sum of *covalent radii* r_A and r_B,

$$r_{AB} = r_A + r_B \quad . \tag{6.20}$$

The tetrahedral covalent radii in Å have been given by PAULING: C(0.77), O(0.66), F(0.64), Cu(1.35), S(1.04), Cl(0.99), Pb(1.46), etc.

The effective potential is taken to be of the form $V = (Ze^2/r)\exp(-k_s r_{AB}/2$ where Ze^2/r is the Coulomb part and $\exp(-k_s r_{AB}/2)$ is the Thomas-Fermi exponential screening factor [6.11]. The screening wave number k_s is equal to $(\pi a_0)^{-1} 4k_F$, where k_F is the Fermi momentum of a free electron gas of density equal to that of valence electrons, and a_0 is the Bohr radius.

PHILLIPS [6.2] expresses the energy gap between the bonding and antibonding states as,

$$E_{g,bond} = E_h + iC \quad , \tag{6.21}$$

$$E_{g,bond}^2 = E_h^2 + C^2 \quad , \tag{6.22}$$

where C represents the ionic (charge transfer) and E_h the covalent contribution to $E_{g,bond}$. The complex energy gap (6.21) means that in the (E_h, C) plane we can define a phase angle ϕ by $\tan\phi = C/E_h$. The requirements $f_i(A,B) = f_i(B,A)$ $0 \leqq f_i \leqq 1$ and $f_i + f_c = 1$, suggest the definitions

$$f_i = \sin^2\phi = C^2/E_{g,bond}^2 \quad ; \quad f_c = \cos^2\phi = E_h^2/E_{g,bond}^2 \quad . \tag{6.23}$$

f_i defined by (6.23) is called the *spectroscopic ionicity*.

PHILLIPS uses the one-electron approximation of PENN [6.12] for diamond-like semiconductors to determine E_h and C. In this model, the static dielectric constant is given by

$$\varepsilon(0) = 1 + (\hbar\omega_p/E_{g,bond})^2(1-E_{g,bond}/4E_F) \quad , \tag{6.24}$$

where ω_p is the plasma frequency (5.55), $E_F = \hbar^2 k_F^2/2m$ is the Fermi energy, and $E^2_{g,bond} = E_h^2 + C^2$. PENN has found that $E_h \propto r_{AB}^{-2.48}$. Also, E_h can be found from

$$d \log E_h/d \log a = s \quad ,$$

where a is the cubic-lattice parameter and $s = -2.5$. In Penn's model, the contribution to $\varepsilon(0)$ from the polarization of the d (or f) shell has been neglected. For heavier atoms, there is s-p-d mixing [6.13] in the outer levels that results in an increase of the effective number of valence electrons per atom (n_{eff}). To account for this multi-electron effect, a polarization correction factor $D = (4n_{eff})^{-1}$ is introduced, so that

$$\varepsilon_D(0) = \varepsilon(0)D(A,B) \quad . \tag{6.25}$$

It is difficult to find n_{eff} accurately. VAN VECHTEN [6.7] has given an empirical relation

$$D(A,B) = \Delta_A\Delta_B - (\delta_A\delta_B-1)(z_A-z_B)^2 \quad ,$$

where $(\Delta_A,\delta_A) = (1,1)$ for the I- and II-row elements, (1.12, 1.0025) for the III row, (1.21, 1.0050) for the IV row, and z_A is the number of valence electrons on atom A.

From the knowledge of $\varepsilon(0)$ and E_h, we can calculate C. We expect C to depend on $V_A - V_B$. A suitable form turns out to be [6.3,7],

$$C(A,B) = b\left(z_A e^2/r_A - z_B e^2/r_B\right) \exp(-k_s r_{AB}/2) \quad , \tag{6.26}$$

where $b \simeq 1.5$. This would give C even if $\varepsilon(0)$ is not available. PHILLIPS has calculated C, E_h and f_i for many compounds (Table 6.1).

It is useful to compare (6.26) to Pauling's electronegativity difference $\bar{\Delta}_{AB} = x_A - x_B$. It turns out that approximately (Table 6.1),

$$C(A,B) = a_c\ \bar{\Delta}_{AB}(eV) + b_c \quad , \tag{6.27}$$

Table 6.1. Average energy gaps in some $A^N B^{8-N}$ compounds [6.2] $\Delta^{\pm}_{AB} = |x_A \pm x_B|$ are calculated on the Pauling scale. The forbidden energy gap is given for comparison

Crystal	$\Delta^-_{AB}(\Delta^+_{AB})$	$C(E_h)$ [eV]	$E_{g,bond}$ [eV]	f_i	E_g [eV]
Si	0 (3.6)	0 (4.77)	4.77	0	1.14
BP	0.1(4.1)	0.68(7.44)	7.47	0.006	-
SiC	0.7(4.3)	3.85(8.27)	9.12	0.177	3
AlSb	0.4(3.4)	2.07(3.53)	4.14	0.250	1.52
BN	1.0(5.0)	7.71(13.1)	15.2	0.256	-
GaSb	0.3(3.5)	2.10(3.55)	4.12	0.261	0.78
AlAs	0.5(3.5)	2.67(4.38)	5.14	0.274	2.16
BeS	1.0(4.0)	3.99(6.31)	7.47	0.286	-
GaAs	0.4(3.6)	2.90(4.32)	5.20	0.310	1.43
InSb	0.2(3.6)	2.10(3.08)	3.73	0.312	0.18
GaP	0.5(3.7)	3.30(4.73)	5.75	0.327	2.26
InAs	0.3(3.7)	2.74(3.67)	4.58	0.357	0.35
InP	0.4(3.8)	3.34(3.93)	5.16	0.421	1.35
GaN	1.4(4.6)	7.64(7.64)	10.8	0.500	-
ZnTe	0.5(3.7)	4.48(3.59)	5.74	0.609	2.26
ZnO	1.9(5.1)	9.30(7.33)	11.8	0.616	3.2
ZnS	0.9(4.1)	6.20(4.82)	7.85	0.623	3.6
CdS	0.8(4.2)	5.90(3.97)	7.11	0.685	2.42
CdTe	0.4(3.8)	4.90(3.08)	5.79	0.717	1.45

where $a_c = 5.75$ and $b_c = 0$. By the same reasoning, we expect E_h to depend on $V_A + V_B$, so that approximately (Table 6.1),

$$E_h(A,B) = a_h \Delta^+_{AB}(\mathrm{eV}) + b_h \quad , \tag{6.28}$$

where $\Delta^+_{AB} = x_A + x_B$, and our estimate is $a_h = 4.2$, $b_h = -11$. In this approximation, we can write

$$f_i = \left[1+(E_h/C)^2\right]^{-1} \simeq \left[1+(4.2\ \Delta^+_{AB}-11)^2/(5.75\ \Delta^-_{AB})^2\right]^{-1} \quad . \tag{6.29}$$

LEVINE [6.14] has extended Phillips' theory to more complex crystals (like Al_2O_3, $LiGaO_2$, and $NaClO_3$) where each bond is characterized by a polarizability, an ionicity, an average symmetric and antisymmetric potential, and an average number of electrons per bond (instead of the average number of electrons of Phillips' theory). The last point arises because the number of bonding electrons may differ in each type of bond. If μ denotes the different types of bonds in a crystal, LEVINE writes for individual bonds,

$$\left(E^{\mu}_{g,\mathrm{bond}}\right)^2 = \left(E^{\mu}_h\right)^2 + (C^{\mu})^2 \quad ,$$

$$f^{\mu}_i = \left(C^{\mu}/E^{\mu}_{g,\mathrm{bond}}\right)^2 \quad , \quad f^{\mu}_c = \left(E^{\mu}_h/E^{\mu}_{g,\mathrm{bond}}\right)^2 \quad ,$$

$$\varepsilon^{\mu} = 1 + 4\pi\chi^{\mu} \quad , \quad \chi = \sum_{\mu} F^{\mu}\chi^{\mu} = \sum N^{\mu}_b \chi^{\mu}_b \quad ,$$

$$\chi^{\mu} = A^{\mu}D^{\mu}(1/4\pi)\left(\hbar\omega^{\mu}_p\right)^2/\left(E^{\mu}_{g,\mathrm{bond}}\right)^2 \quad , \quad \left(\hbar\omega^{\mu}_p\right)^2 = \left(4\pi N^{\mu}_e e^2/m\right) \quad ,$$

$$A^{\mu} = 1 - E^{\mu}_{g,\mathrm{bond}}/\left(4E^{\mu}_F\right) \quad , \quad D^{\mu}(A,B) = \Delta^{\mu}_A\Delta^{\mu}_B - \left(\delta^{\mu}_{\alpha}\delta^{\mu}_{\beta} - 1\right)\left(Z^{\mu}_A - Z^{\mu}_B\right) \quad ,$$

$$E^{\mu}_F = \left(\hbar k^{\mu}_F\right)^2/2m \quad , \quad \left(k^{\mu}_F\right)^3 = 3\pi^2 N^{\mu}_e \quad ,$$

$$N^{\mu}_e = n^{\mu}_v/v^{\mu}_b \quad , \quad n^{\mu}_v = \left(Z^{\mu}_A/N_{cA}\right) + \left(Z^{\mu}_B/N_{cB}\right) \quad ,$$

$$v^{\mu}_b = \left(r^{\mu}_{AB}\right)^3/\left[\sum_{\nu}\left(r^{\nu}_{AB}\right)^3 N^{\nu}_b\right] \quad ,$$

$$E^{\mu}_h = 39.74\left(r^{\mu}_{AB}\right)^{-2.48}\ \mathrm{eV} \quad ,$$

$$C^{\mu} = b^{\mu}\exp\left(-k^{\mu}_s r^{\mu}_0\right)\left(\frac{z^{\mu}_A e^2}{r^{\mu}_A} - \frac{z^{\mu}_B e^2}{r^{\mu}_B}\right)\mathrm{eV} \quad ,$$

$$k^{\mu}_s = \left(4k^{\mu}_F/\pi a_0\right)^{1/2} \quad , \quad r^{\mu}_0 = \frac{1}{2}\, r^{\mu}_{AB} \simeq r^{\mu}_A \simeq r^{\mu}_B \quad .$$

Here χ^{μ} = susceptibility of μ-type bonds, F^{μ} = fraction of μ-type bonds, N^{μ}_b = number of μ-type bonds per unit volume, χ^{μ}_b = susceptibility of a single bond μ, N^{μ}_e = number of valence electrons of μ-type per unit volume, z^{μ}_A = number of valence electrons on the atom A of μ bond, n^{μ}_v = number of valence electrons per bond, v^{μ}_b = bond volume, N^{μ}_{cA} = coordination number of atom A, and r^{μ}_{AB} = bond length.

LEVINE has calculated b^{μ} values for many compounds and found that $b = 0.089\ \bar{N}^2_c$, where $\bar{N}_c$ is the average coordination number. For A_mB_n-type crystals LEVINE has suggested,

$$C^{\mu} = b^{\mu}\exp\left(-k^{\mu}_s r^{\mu}_0\right)\left(\frac{z^{\mu}_A e^2}{r^{\mu}_0} - \frac{n}{m}\,\frac{z^{\mu}_B e^2}{r^{\mu}_0}\right) \quad , \quad \text{for } n > m \quad ,$$

$$C^{\mu} = b^{\mu} \exp\left(-k_s^{\mu} r_0^{\mu}\right) \left(\frac{m}{n} \frac{z_A^{\mu} e^2}{r_0^{\mu}} - \frac{z_B^{\mu} e^2}{r_0^{\mu}} \right) \quad , \quad \text{for } n < m \quad ,$$

$$\bar{N}_c = \left(\frac{m}{m+n}\right) N_c(A) + \left(\frac{n}{m+n}\right) N_c(B) \quad .$$

LEVINE has used these relations to calculate the bond parameters for various binary and ternary compounds. The difficulty arises for the transition-metal compounds in which the d-electron effects predominate and the actual number of valence electrons is not always known. LEVINE has incorporated the d-electron effects by writing b as

$$b = b' \left(\frac{1+\Gamma}{1-f_c/f_i} \right)^{\frac{1}{2}} \quad , \quad b' = 0.089 \, \bar{N}_c^2 \quad ,$$

where Γ is the fraction of empty d levels.

e) *Concept of Effective Charge*. Mainly, three different approaches have been used to define the effective charge: 1) thermochemical theory of PAULING [6.5], 2) valence-bond molecular-orbital theory of COULSON [6.4], and 3) the dielectric theory of SZIGETI [6.15]. The formation of the chemical bond results in a rearrangement of the outer electrons of the participating atoms. In particular, the more-electronegative atom (anion B) receives electrons from the less-electronegative atom (cation A) (Fig.6.7). Therefore, the charge redistribution on bonding can be equated to alteration of the charge on an ion, in the first approximation. The effective nuclear charge is related to the electronegativity, (6.16). Following PAULING [6.5], and MOOSER and PEARSON [6.16], it has been suggested [6.17-19] that the effective charge q on a cation in a molecule can be expressed as

$$q = \sum_{\mu} f_{i,\mu} = \sum_{\mu} \left\{ 1 - \exp\left[-0.25(x_A - x_B)^2 \right] \right\}_{\mu} \quad , \tag{6.30}$$

where $f_{i,\mu}$ is the partial ionicity of the bond μ, as given by (6.14). The sum is over the bonds that connect the cation A with the different neighbors B.

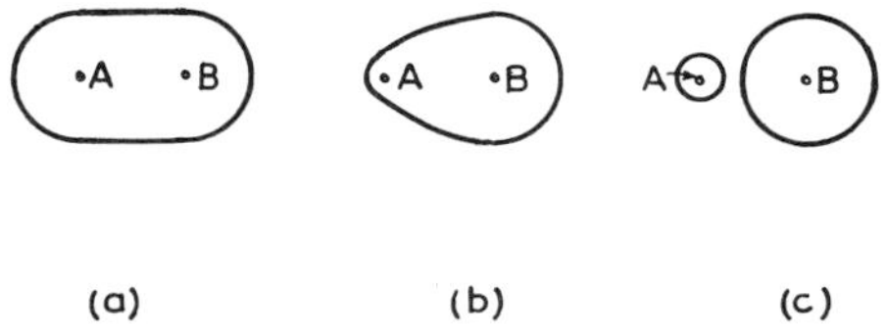

Fig. 6.7. Shape of orbital representing different bondings: (a) covalent (A-B), (b) covalent with ionic character ($A\text{-}B \leftrightarrow A^+ + B^-$), and (c) ionic ($A^+ + B^-$)

In estimating q methods with different degrees of sophistication and rigor can be used [6.20]. However, simple calculations of the Pauling type are in reasonable agreement with the sophisticated calculations based on self-consistent-field programs. We shall, therefore, consider only the simple free-ion models.

SUCHET [6.21] has expressed the effective charge as

$$q(\text{cation}) = n\left[1-0.01185(Z/r'+Z'/r)\right] = -(n/n')q(\text{anion}) \quad , \tag{6.31}$$

where Z(Z') and r(r') are the total number of electrons and the ionic radii (following Pauling) of the atoms present at the cation (anion) site, and n(n') gives the number of electrons transferred from the cation (anion).

SYRKIN and DYATKINA [6.22], and COULSON et al. [6.4], have given a concept of effective charge by considering localized molecular-orbital bonds in the A^NB^{8-N} compounds.

SZIGETI [6.15] calculated the macroscopic polarization by taking into account the effect of only near neighbors and defined effective charge in terms of dynamic charges based on the infrared spectra.

If the interatomic distance R between the atoms in a molecule AB is increased by x, then the static charge q(0) supported by A in the initial situation becomes

$$q(x) = q(0) + x(dq/dx) \quad . \tag{6.32}$$

The dipole moment P can be written as the sum of the relative shifts of two atoms with charges ±q(0) and the transfer of the charge dq/dx over R,

$$P = xq(0) + xR(dq/dx) \quad . \tag{6.33}$$

If e is the "dynamic charge" defined as $P = ex$, we get

$$e = q(0) + R(dq/dx) \quad . \tag{6.34}$$

BATSANOV and OVSYANNIKOVA [6.23] have expressed the coordination effective charge q as

$$q = N - nf_c \quad , \tag{6.35}$$

where N is the valency, n is the coordination number, and f_c the covalency.

6.2 Chemical Shifts of Emission Lines

LINDH and LUNDQUIST [6.24] were first to observe the effect of chemical combination on X-ray line spectra. RAY [6.25], BÄCKLIN [6.26], and DEODHAR [6.27] showed that the wavelength of $SK\alpha_{1,2}$ lines depends on the particular sulphate in which the sulphur is found.

Valence electrons penetrate the ion core slightly ($\sim$10 percent or less of their probability distribution) (Fig.2.14). When they happen to be inside the core, their charge repels the charge of the core electrons. In other words, the effect of the valence charge in the core region is to *screen* partially the core electron-nucleus attraction. This reduces the binding energies of the core levels. In the bond formation, some valence charge is removed (or transferred) from the atom. Consequently, the screening is reduced, the cation-core electrons (like K, L) are attracted more strongly to the nucleus, and their binding energies become more negative (larger in magnitude). This change of the binding energies would be more if the number of valence electrons that participate in the bond (oxidation number) were greater. Detailed calculations based on approximate many-electron functions [6.28,29] have shown that both of the core levels, K and $L_{II,III}$, that participate in the $SK\alpha_{1,2}$ transitions, are shifted in the same direction (towards higher binding energy) (Fig.6.8 and Table 6.2), as expected on the basis of the changes of the screening of core electrons.

Photoelectron studies [6.30] have directly confirmed this predicted increase of the binding energies of the inner electrons of sulphur with oxidation (Fig.6.9). They found that: 1) The K, L binding energies increase with oxidation number. 2) All of the three levels K, L_I, $L_{II,III}$ show an energy shift of about 1 eV per degree of oxidation. 3) The shifts are somewhat larger for the K level than for the L levels. These observations are supported by the X-ray-emission chemical-shift data.

FAESSLER and GOEHRING [6.31] have observed the shifts of the sulphur Kα doublet in a large number of different compounds. They find that the effective charge of the atom determines the chemical shifts of the lines with respect to their positions in the element: S^{2-} (-0.14 eV), S^{2+} (+0.31 eV), S^{4+} (+0.95 eV), S^{6+} (+1.19 eV) (Table 6.2). The maximum shift is 1.33 eV, or about 3 XU (between S^{6+} and S^{2-}). The observed shifts are consistently smaller than the calculated shifts, although the trend is the same.

It must be noted that, by photoelectron spectroscopy, the *total* shifts of each individual level (K, L_I, $L_{II,III}$, etc.) are measured separately, whereas in the X-ray spectra the small *differences* of shifts of the two core

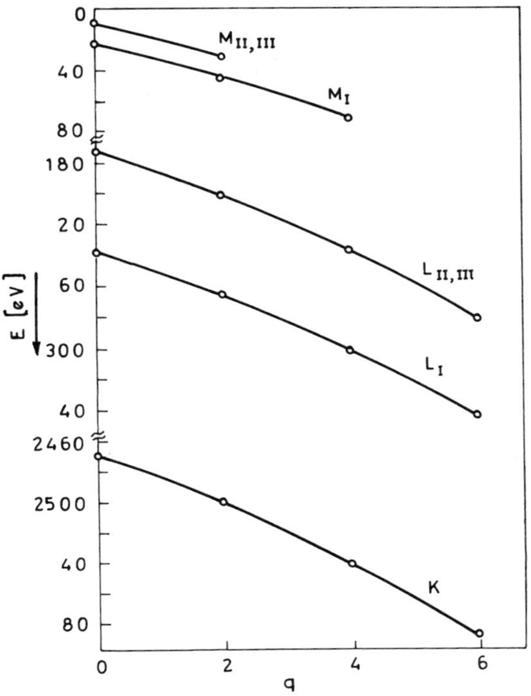

Fig. 6.8. Level energies for S, S^{2+}, S^{4+} and S^{6+} [6.29]

Table 6.2. Ionization and $K\alpha_{1,2}$ energies for S, S^{2+}, S^{4+} and S^{6+}, in eV [6.29]

Atom or ion		S	S^{2+}	S^{4+}	S^{6+}
M-shell		$s^2p_x^2p_y^2$	$s^2p_z^2$	s^2	-
$M_{II,III}$ (3p)		10.0	31.5		
M_I (3s)		23.0	44.7	70.6	
$L_{II,III}$ (2p)		172.4	200.5	235.9	280.2
L_I (2s)		237.0	265.4	300.9	342.9
K (1s)		2473.0	2502.7	2540.9	2587.4
$K\alpha_{1,2}$ (calc.)		2300.6	2302.2	2305.0	2307.2
Shift	(calc.)	0	1.6	4.4	6.6
	(obs.)	0	0.31	0.95	1.19

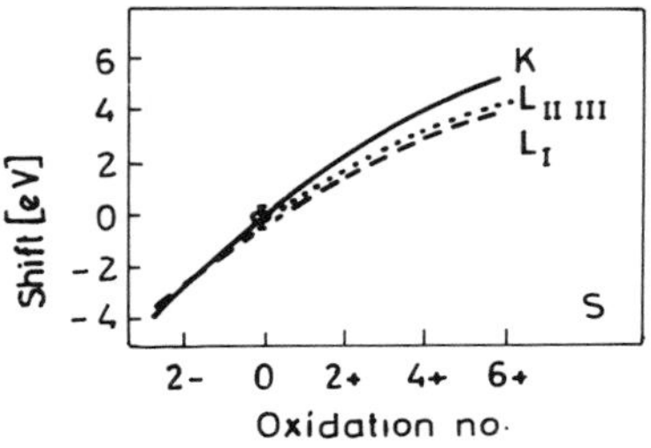

Fig. 6.9. Chemical shifts of the K, L_I, $L_{II,III}$ levels of S for different oxidation numbers, by photoelectron spectroscopy

levels involved in the transition are measured. Because the shifts of the K and L levels are about the same, and in the same direction, much smaller shifts (~1 eV) of the $K\alpha_{1,2}$ emission lines are obtained in the lighter elements (like ^{14}Si, ^{15}P, ^{16}S) and the theoretical interpretation is more difficult. The shifts become even smaller (~0.1 eV) as the atomic number increases. For example, the $CuK\alpha_1$ line shifts by only 0.11 eV [6.32] in going from Cu (metal) to CuO.

Sodium thiosulphate, $Na_2S_2O_3$, has the following structure:

$$2Na^+ \left[O - \overset{\overset{O}{|}}{\underset{\underset{O}{|}}{S}} - S \right]^{2-} ,$$

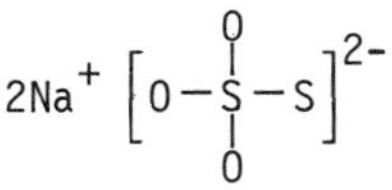

with two sulphur atoms in *different* chemical positions with oxidation number 6+ (central atom) and 2- (ligand atom). In the ground state, the electronic structure of S is [KL] $3s^2 3p^4$. In the thiosulphate ion, the s and p orbital hybridize in the central sulphur atom to give six $s^2 p^4$ hybrid orbitals and a resulting sixfold covalent bond. On the other hand, the ligand sulphur atom is bounded by receiving two hybrid orbital electrons from the central sulphur into its two half-filled 3p orbitals. Therefore, the outer orbital-electron density is different for the two sulphur atoms with a resulting difference of the binding energies of core electrons. Direct evidence has come for this from photoelectron spectroscopy [6.17] where two (2p level) lines are obtained (Fig.6.10a), one from the central S^{6+} atom and the other at lower energy (by about 6 eV) from the ligand S^{2-} atom. Similarly, two (K level) lines have been observed[1] by FAHLMAN et al. [6.30].

---

[1] Electron binding energies determined by the photoelectron-spectroscopic method have the Fermi level (E_F) as a reference level. One can argue that the level shifts are observed because the Fermi level is shifted to different positions in the forbidden energy gap E_g between the valence-band edge and the conduction-band edge. This explanation is not acceptable because different energies of an electron line are found from sulphur atoms in different chemical positions in the same substance.

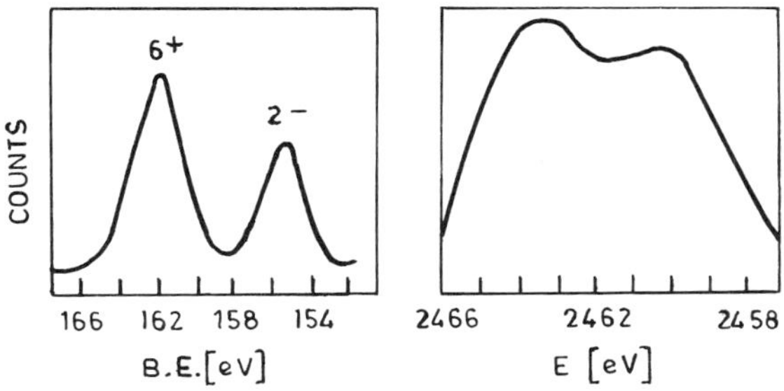

(a) (b)

Fig. 6.10. Sulphur spectrum from sodium thiosulphate by (a) photoelectron (2p level, MgKα line as source), and (b) X-ray (SKβ line) spectroscopy

In the X-ray emission spectra also, two doublets ($K\alpha_{1,2}$) are observed [6.33] corresponding to the two kinds of sulphur atoms: $K\alpha_1$ (5358.49 XU), $K\alpha_2$ (5361.35 XU) from S^{6+} central atom, and $K\alpha_1$ (5361.35 XU), $K\alpha_2$ (5363.95 XU) from S^{2-} ligand atom. (Because $S^{6+}K\alpha_2$ and $S^{-2}K\alpha_1$ overlap, only three lines are observed.) Further confirmation comes from the SKβ (3p→1s) line (Fig. 6.10b) where two peaks are observed.

The energy difference dE per angular increment dθ of a Bragg spectrometer is determined by

$$\lambda(\text{Å}) = (2d/n)\sin\theta = 12398.1/E[\text{eV}] \quad ,$$

$$dE/d\theta = -(12398.1/\lambda)\cot\theta \quad ,$$

$$|dE| = (4.32/\lambda)\cot\theta\ [\text{eV}] \quad , \qquad (6.36)$$

for $d\theta = 0.02^\circ$ (usual angular precision). This gives about 5 eV for $\lambda = 1$ Å and 0.5 eV for $\lambda = 10$ Å. Thus, the dispersion is not enough for the measurements of X-ray chemical shifts at shorter wavelengths that occur in heavier elements. For chemical shifts of the order of 0.1 eV to 1 eV, high-precision spectrometry becomes necessary [6.34].

GOKHALE et al. [6.35] have carefully measured the chemical shift of the $K\alpha_1$ line of tin (^{50}Sn) in its oxides [E(SnO)-E(Sn)=0.09 eV, E(SnO_2)-E(Sn) = 0.173 eV]. The chemical shifts of Kα and Kβ lines in heavy elements ($Z > 36$) have been observed by PETROVICH et al. [6.36] and theoretically analyzed by COULTHARD [6.37]. KOSTER and MENDEL [6.38] have extensively studied the X-ray Kβ emission spectra (Kβ''', β_5, β'', β_1, β') of 3d-transition metals and com-

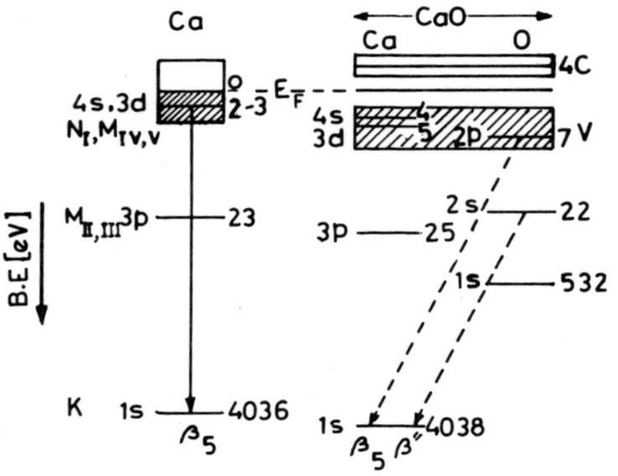

Fig. 6.11. Energy levels in eV of Ca and CaO. The crossover transitions are shown by dashed lines. E_F is Fermi level at zero energy, C is the empty conduction band, and V is the full valence band

pounds. The energies involved in electron transitions in these compounds can be used to construct a composite energy diagram, if use is made of the photoelectron data for which the Fermi energy provides the reference level (Fig. 6.11). In such a diagram, the levels belong to the constituent atoms or ions and represent the energy states of the compound.

The transitions for the X-ray emission are possible not only between levels of the same species but *crossover* transitions are also possible. The binding energy of 1s electron of Ca in CaO is 4038 eV [6.39]. The energy of oxygen $K\alpha$ is 525 eV [6.40]. The energy of $K\beta_5$ is found to be 4031 eV. The $K\beta_5$ of a pure metal results from the $4s \rightarrow 1s$ transition. In a compound all valence electrons are used in bond formation and so the $K\beta_5$ line arises between the level mainly characterized by 2p(O) and 1s(Ca). If this is a crossover transition, the oxygen $L_{II,III}$ would be at 7 eV, and the K binding energy in CaO would be $525 + 7 = 532$ eV. This is in good agreement with the value 532 eV, as measured in MgO and Al_2O_3 [6.39,41]. The residual charge on the oxygen is nearly equal for these three compounds (in electron units: -1.8 for MgO, -1.6 for Al_2O_3, -1.8 for CaO). The line $K\beta''$ at -15 eV with respect to $K\beta_5$ is explained by the crossover transition $2s(O) \rightarrow 1s(Ca)$; the L_I level of oxygen then is situated at 22 eV. From photoelectron spectroscopy, this value is 23.7 eV [6.41]. The position of the CaK-absorption edge at 4041 eV in CaO [6.42] places the conduction band centered at 4 eV above E_F. Two satellite peaks at 527 and 528 eV are found in the oxygen spectrum [6.40]. These can be explained by the valence-band transitions $3d(Ca) \rightarrow 1s(O)$ and $4s(Ca) \rightarrow 1s(O)$.

We expect larger shifts for the $K\beta_5$ emission band because the transition involves the valence band and an inner level. The shift of the valence band is large because of the appearance of the forbidden energy gap. KOSTER and MENDEL [6.38] observed a marked chemical shift only for the $K\beta_5$ band. In Ca, the $K\beta_5$ appears as a broad line, with shape dependent only on the bonding of the calcium atoms. In CaO, the band-like nature of this line arises because of the formation of molecular orbital of Ca and O. FISCHER [6.42a] has drawn similar conclusions about the $K\beta$ emission bands of Al and Al_2O_3. The X-ray emission spectra support the conclusions drawn on the basis of photoelectron spectroscopy.

The shift of energy of a photoelectron or an X-ray line on bonding is determined by self-atom and neighbor-atom effects. Many attempts have been made to correlate the observed shifts of lines in photoelectron and X-ray spectroscopy with parameters such as ionicity, effective charge, valence, and coordination number. We shall first discuss the self-atom (free-ion) effects. It would be wrong to assume that (ion) core-level shifts can be ignored while discussing X-ray emission and absorption spectra. As mentioned earlier, the change of the screening of the (ion) core electrons by the penetrating parts of the valence-electron wave functions on bond formation, results in a change of the core-level-energy position. This ionic model for core-level shifts is both simple and useful for finding correlations.

The energy of a core level (nl) is essentially given by

$$E_{nl} = (13.6/n^2)(Z-\sigma_{nl})^2 \quad , \tag{6.37}$$

where σ_{nl} is the screening constant. The derivative of this equation with respect to σ_{nl} gives

$$|\Delta E_{nl}| = (27.2/n^2)(Z-\sigma_{nl})\Delta\sigma_{nl} \quad . \tag{6.38}$$

The $\Delta\sigma_{nl}$ can be estimated from the assumed charge transfer and the known screening constants. This method has been used by VARLEY [6.43], SHUBAEV [6.44], and VEIGELE et al. [6.45].

Using Coulomb (C) and exchange (A) integrals over the occupied orbitals [6.46], NEFEDOV [6.47] has calculated the level shifts. For the loss of one valence ns electron on bonding, he finds that the core 1s level shifts by

$$\Delta E_{1s} = C(1s,ns) - \frac{1}{2} A(1s,ns) \quad . \tag{6.39}$$

SHUBAEV [6.48] has used differences in the radial wave functions $P_{nl}(r)$ and screening constants σ_{nl} to estimate ΔE. ALDER et al. [6.49] have calculated level shifts by considering the change of $|P_{nl}(r)|^2$ on bonding.

Reliable free-ion ΔE calculations [6.50] are based on the Hartree-Fock-Slater technique [6.51,52].

FAHLMAN et al. [6.30] related the change of binding energy of the core electron to oxidation number (Fig.6.9) by using the simple electrostatic model in which the atomic orbitals are concentric spherical shells, each with charge e_{eff}. The classical theory gives the potential energy of an electron inside such a shell as

$$U = \frac{1}{4\pi\varepsilon_0} \cdot \frac{e_{eff}^2}{r} \quad , \tag{6.40}$$

where r is the radius of the sphere. The contribution to the energy of any inner-shell electron from the valence electrons is then given by U, with r equal to the mean radius of the valence-electron orbital. Because of bonding, if a valence electron of effective charge e_{eff} is removed from the atom, the binding energy of an inner electron will increase by U. The same shift is obtained for all inner electrons (K, L, M, etc.). The shift is proportional to the number of valence electrons removed; if an electron is gained (as in an anion) the shift changes sign. The curves of Fig.6.9 can be explained on this model. For the sulphur 3p orbital, $r = 1.0$ Å. For the observed level shift of 1 eV per oxidation number, we need $e_{eff} = 0.07$ electron units. Refined calculations [6.50] give a somewhat larger shift for the K shell than for the L shells. For A^2B^6 semiconductors VESELEY and LANGER [6.53] found that ΔE of least-bound core levels is directly proportional to the spectroscopic ionicity f_i.

We now consider the free-ion relations for the X-ray line shifts ΔE_ν involving various aspects of the emitting substance. KARALNIK [6.54] related the shifts to the changes of electron screening. He equated the shift of X-ray energy per unit change of valence-electron screening to the known difference of X-ray energy for a unit change of nuclear charge $(Z \rightarrow Z+1)$,

$$\frac{\Delta E_\nu}{\Delta\sigma} = \frac{E(Z)-E(Z+1)}{1} \quad . \tag{6.41}$$

Many attempts have been made to relate ΔE_ν to effective charge q [6.55,56]. CLEMENTI [6.57] has related $\Delta E_{K\alpha}$ for III-row elements to the degree of ionization, from a self-consistent free-ion model. SHUBAEV [6.58] writes $\Delta E_\nu = b\Delta q$, where Δq is the charge transfer and b ranges from 0.33 for Mg to 1.32 for S in units of eV per charge unit. Line shifts for the 3d transition elements depend on the participation of 4s and 3d electrons in the bonding, and so [6.56]

$$\Delta E_{K\alpha} = b_{4s}\Delta q_{4s} + b_{3d}\Delta q_{3d} \quad . \tag{6.42}$$

SUMBAEV [6.59] suggested

$$\Delta E_{\nu} = \left(f_i^S - f_i^R\right) \sum_l N_l C_l \quad , \tag{6.43}$$

where S and R denote the sample and reference, f_i is the ionicity on the Pauling scale, N_l is valence decomposed according to the angular momentum of the contributing orbital, and C_l is in eV per l subshell electron. SUMBAEV [6.59] measured $K\alpha_1$ shifts for $32 \leqq Z \leqq 74$ using a Cauchois curved-crystal spectrograph and used (6.43). He found straight-line plots according to the equation

$$\Delta E_{K\alpha_1}/f_i = xC_{sp} - (N-x)C_{4d} \quad , \tag{6.44}$$

where N is taken as a single number for each element, $f_i^R = 0$ for the pure-element reference, and x is the number of sp electrons. For the IV, V, VI rows, he found $C_{sp} \simeq 80$ meV, $C_d = -115$ meV. Thus, the participation of s,p electrons, in the bonding (Cu-Se, Ag-Te) gives positive shifts, whereas d (and f) electron bonding gives negative shifts.

TILGNER et al. [6.60] studied the chemical shifts of Kα and Kβ doublets of Ga and As, and used them to calculate effective atomic charges on the basis of free-ion model.

SIEGBAHN et al. [6.17,61] used a simple electrostatic model to take into account both the self-atom and neighbor-atom effects for calculating the level shifts,

$$\Delta E_i = b(\Delta q_i/r) - \sum_j (\Delta q_j/R_{ij}) + l \quad , \tag{6.45}$$

where $q_i(q_j)$ is the charge on atom i(j), r is the valence-electron shell radius, R_{ij} is the interatomic distance and b, l are adjustable constants. The first term on the right is the free-ion contribution if the electron is assumed to be removed to infinity, and the second term gives the correction to account for the fact that electrons are transferred to near-neighboring atoms.

GOLDMAN et al. [6.62] wrote for the tetrahedral AB compounds,

$$\Delta E(\text{cation}) = \Delta q(2.6/r - \alpha/R)e \quad , \quad \Delta q = \left[Z - 4(1-f_i)\right]e \quad , \tag{6.46}$$

where Δq is the charge transfer, r(R) is the interatomic spacing in the metal (compound), and α is the Madelung constant. Such models cannot be used for the X-ray line shifts as no difference is made between different core levels.

6.3 Width and Fine Structure of Emission Lines

The shift of $K\alpha$ lines on bonding is a second-order effect that results from the change of screening of the inner 1s and 2p electron when a change occurs in the valence-electron shell. The direct effect of the chemical bond is to perturb the valence-electron states. Therefore, the line of shortest wavelength in each series is always strongly affected by the chemical state of the emitting atom. In this case, there is not only a wavelength shift but also a change of the width and shape of the line.

The kinetic energy E of an electron in the valence band of a metal of *width* $E_F - E_0$ is $E_{kin} = E - E_0$, where E_0 is the energy of the lowest state of the valence band and $E_0 < E < E_F$. The smallest cell in the phase space, as given by the uncertainty principle is

$$\Delta p_x \Delta p_y \Delta p_z \Delta x \Delta y \Delta z = (\Delta p)^3 \Delta V = h^3 \quad . \tag{6.47}$$

By the Pauli principle, it can contain two electrons. If $\Delta V = GV_a$, where G is a large whole number and V_a is the atomic volume, (6.47) becomes

$$(\Delta k)^3 = (GV_a)^{-1} \quad , \tag{6.48}$$

where $p = hk$, $\Delta p = h\Delta k$. Thus, each pair of electrons occupies a cell of volume $(GV_a)^{-1}$ in the wave-vector space (k_x, k_y, k_z). We can write $V_a = V_e/n$, where V_e is the volume of the unit cell and n the number of atoms in the cell.

If there are N valence electrons per atom, or GN electrons per volume GV_a, that behave as almost-free electrons in the valence band, then the corresponding states occupy a volume

$$\frac{1}{2} GN(\Delta k)^3 = N/(2V_a) \quad , \tag{6.49}$$

in k space. At absolute-zero temperature, these electrons will occupy a sphere in k space such that

$$N/(2V_a) = (4/3)\pi k_F^3 \quad , \tag{6.50}$$

where k_F is the radius of the *Fermi surface*. The maximum kinetic energy is given by $E_F - E_0 = (hk_F)^2/2m$, so that (6.50) gives

$$N = (8\pi/3)(2m)^{3/2}\left(V_a/h^3\right)(E_F - E_0)^{3/2} \quad . \tag{6.51}$$

If the observed emission-band width is ΔE, we can write

$$\Delta E = (E_F - E_0) + (\Delta E)_i \quad , \tag{6.52}$$

where $(\Delta E)_i$ is the width of the inner level involved. From (6.51,52),

$$N/V_a = 4.53 \times 10^{-3}\left[\Delta E - (\Delta E)_i\right]^{3/2} \quad , \tag{6.53}$$

where V_a is in Å^3 and energy in eV.

In brass, with Cu 70% + Zn 30% by atoms, each Zn atom gives two valence electrons to one from each Cu atom. The unit cell of brass is fcc with $a = 3.67$ Å and $n = 4$. Therefore, for the $K\beta_{2,5}$ lines of Cu and Zn that arise from a transition between the composite valence band of the brass to the K levels of Cu and Zn [6.63],

$$N = (30\times2+70\times1)/100 = 1.3 \quad , \quad V_a = a^3/4 = 12.4 \text{ Å}$$

$$\Delta E - (\Delta E)_K = 8 \text{ eV.}$$

From the absorption spectrum, we get $(\Delta E)_K$ as 1.3 ± 0.5 eV for Cu and 1.5 ± 0.5 eV for Zn [6.64]. Therefore, the width of the calculated emission band is 9.3 eV for Cu and 9.5 eV for Zn. The observed width of $CuK\beta_{2,5}$ is 8.8 eV and of $ZnK\beta_2$ is 9.0 eV. The agreement is within experimental error. The observed width increases with the increase of N as the Zn content of the alloy is increased, in agreement with (6.53). The theory works because the α brass behaves as a metal. The widths of bands in Cu and Zn are found to be nearly the same. This shows that the valence band is common to both of the components of the alloy.

The measured and calculated values [6.65-68] of emission-band widths associated with the outer electrons are given in Table 6.3. For metals and metallic alloys, the two are close to each other, but for the nonmetals (B, C, Si, P and S) the two differ considerably. This shows that the theory of free electrons is not applicable to nonmetals.

It has been found [6.69] that in many cases (like Ti, Mn, Cu) the observed lines are *asymmetrical*. The excess width on the low-energy side can arise

Table 6.3. Observed and calculated values of bandwidths

Band	ΔE [eV] Obs. (Calc.)	Band	ΔE [eV] Obs. (Calc.)
LiK	4.2 (4.7)	NaL_{III}	3.7 (3.2)
BeK	14.7(14.3)	MgL_{III}	7.2 (7.2)
BK	31 (25.2)	SiL_{III}	18.2(12.7)
CK	33 (21.9)	PL_{III}	43 (11.5)
MgK	7.4 (7.2)	SL_{III}	49 (14.3)
AlK	12.7(11.7)	CuL_{III}	6.8 (7.1)
SiK	18 (12.7)	CuM_{III}	7.5 (7.1)

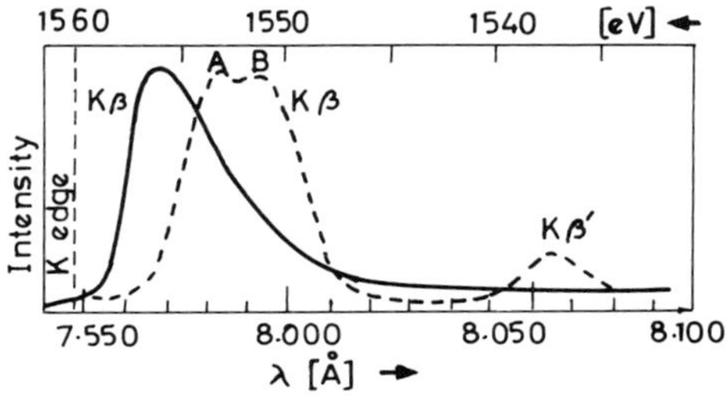

Fig. 6.12. The AlK β-band emission spectra of Al metal (solid curve) and sixfold coordinated corundum (α-Al_2O_3, dotted curve). The Kβ' is a satellite

from transitions between excitation states of the valence-electron configuration type [6.70], from unidentified satellites, and from Compton scattering.

The $L\beta_2$ and $L\beta_5$ bands have been studied by MANDE [6.71] in Au-Pd alloys. DODD and GLEN [6.72] have studied the K-emission spectra of Mg, Al, and Si in elements and in oxides. Figure 6.12 shows how both the shape and wavelength of the AlKβ-emission band [6.42a,72] change on oxidation. The observed change from line to band shape on oxidation is common for such light elements. The electronic configuration for ^{13}Al atom is $[KL]3s^2 3p^1$.

O'BRIEN and SKINNER [6.73] considered the usual band model for the weak, broad $K\beta_{1,3}$ line ($3p \rightarrow 1s$, $M \rightarrow K$) in metal. The approximate atomic-orbital ionization energies in eV are: Al 3s(10.6) 3p(6.0), O 2s(28.4) 2p(13.6). The band-like nature of the line in Al_2O_3 arises because of the formation of the common energy band of the molecular orbital of Al 3s3p and O 2p. The cross-over transitions (Fig.6.13a) for the band Kβ and the satellite Kβ' give the energy difference between them as equal to Al 3p(6.0) minus O 2s(28.4), that

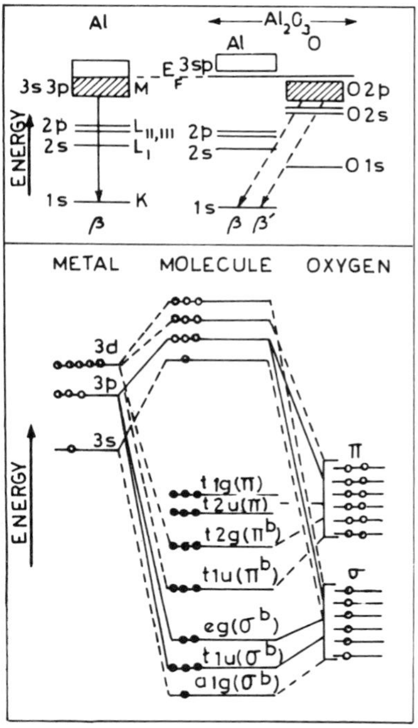

Fig. 6.13. (a,top) The Kβ emission transitions in Al and α-Al_2O_3 in the band picture. (b,bottom) The molecular-orbital energy-level diagram for octahedrally (sixfold) coordinated α-Al_2O_3. Solid dot is occupied (↑↓)

is, ~22 eV. The observed value is ~16 eV, which is of the same order. Thus, the band picture can account for the high-energy shift of the Kα line (discussed earlier), the low-energy shift of the Kβ line, together with a change of the shape from line to band, and the appearance of the low-energy satellite Kβ'.

DODD and GLEN [6.72] and URCH [6.74] found an interpretation based on the molecular-orbital picture more rewarding. In particular, URCH has shown that the observed relative intensities of low-energy satellites are in better agreement with this approach.

Following the work of BALLHAUSEN and GRAY [6.75] on the transition-metal complexes, the molecular-orbital energy-level diagram can be drawn for octahedrally coordinated oxides of Mg, Al and Si (Fig.6.13b). The relative energy levels are approximate. The 3p(π) bonding orbitals, with group t_{1u} symmetry, are occupied and lie at a higher energy than the 3p(σ) bonding orbitals also with t_{1u} symmetry. Assuming that 3p→1s transition can occur for both 3p bonding orbitals (π or σ), DODD and GLEN suggest that the higher-energy peak A in the α-Al_2O_3 Kβ band arises from a 3p(π^b)→1s transition, and the lower-energy peak B from a 3p(σ^b)→1s transition, Figs.6.12,13b.

6.4 Absorption Spectroscopy

Lambert's law (3.198) is

$$I = I_0 e^{-\tau_m m} , \qquad (6.54)$$

where $m = \rho x$ is the mass-per-unit area of the absorbing film. Defining I and τ_m as the intensity and the mass absorption coefficient on the low-energy side of the absorption edge, and I' and τ_m' the values on the high-energy side SANDSTRÖM [6.76] suggested that the optimum thickness (m_{op}) of the absorber for the maximum contrast (I-I' maximum) is obtained by setting $d(I-I')/dm = 0$. This gives

$$\left[d(I-I')/dm\right]_{m=m_{op}} = I_0\left[-\tau_m e^{-\tau_m m_{op}} + \tau_m' e^{-\tau_m' m_{op}}\right] = 0 ,$$

or

$$m_{op} = \frac{\ln \tau_m' - \ln \tau_m}{\tau_m' - \tau_m} = 2.30 \frac{\log\tau_m' - \log\tau_m}{\tau_m' - \tau_m} , \quad [\text{in g cm}^{-2}] . \qquad (6.55)$$

Values of τ_m and τ_m' can be found from (3.206,236), or similar formulas.

KURYLENKO [6.77] has shown that for the photography of absorption fine structure on the high-energy side of the edge, a suitable practical formula is obtained if we maximize the small change of intensity $\delta I' = \delta[I_0 \exp(-\tau_m' m)]$ $= -mI_0 \exp(-\tau_m' m)\delta\tau_m'$ on the high-energy side of the edge. Let us equate the derivative of $\delta I'$ with respect to m to zero,

$$I_0 \delta\tau_m'(\tau_m' m_{op} - 1) e^{-\tau_m' m_{op}} = 0 ,$$

or

$$m_{op} = \rho x_{op} = 1/\tau_m' , \qquad (6.56)$$

where τ_m' is the average mass absorption coefficient in the fine-structure region. KRISHNAN and NIGAM [6.78] showed that the smaller thickness given by (6.56) gives more-pronounced fine structure in Ni foils.

When the effects on absorption introduced by a counter used as a detector are included, PARRATT et al. [6.79] suggested that the thickness should be such that $I/I_0 \simeq 0.10$, which leads to

$$m_{op} = -\ln(0.10)/\tau_m' \quad . \tag{6.57}$$

These results are applicable if the readings of the detector (such as an ionization chamber or a counter) are proportional to intensity I. In the photographic method, the microphotometer reading is a measure of optical density (blackening) and is not proportional to I. Therefore, we should find a relation between the optical density D and the exposure time t.

Optical density is defined in terms of the film's ability to transmit visible light,

$$D = \log_{10}(J_0/J) \quad , \tag{6.58}$$

where $J_0(J)$ is the light incident upon (transmitted through) the X-ray film under study. The deflection l on a microphotometer record is proportional to J. Therefore,

$$D = \log_{10}(l_0/l) \quad , \tag{6.59}$$

where $l_0(l)$ is the deflection from the unexposed (exposed) portion of the X-ray film. The deflections are measured from the line of no transmission taken as the zero line; $l_0 > l$.

The quantity D has been shown to be proportional to the total number of X-ray quanta that fall on unit area of the film (containing a given amount of silver per unit area) provided that D is less than 0.7. Therefore, for $D < 0.7$,

$$D = \log_{10}(l_0/l) = bIt \quad , \tag{6.60}$$

where b is a constant. If D(D') is the optical density on the low- (high-) energy side of the edge ($D > D'$), we get

$$D = \log(l_0/l) = bIt \quad , \quad D' = \log(l_0/l') = bI't \quad ,$$

$$\Delta l = l - l' = l_0(e^{-2.30bIt} - e^{-2.30bI't}) \quad . \tag{6.61}$$

The optimum value of t is given by

$$(dI/dt)_{t=t_{op}} = 2.30bI_0\left(I\ e^{-2.30bIt_{op}} - I'\ e^{-2.30bI't_{op}}\right) = 0 \quad ,$$

or

$$I/I' = \exp\left[2.30b(I-I')t_{op}\right] \quad ,$$

$$t_{op} = \frac{\log(I/I')}{b(I-I')} \quad . \tag{6.62}$$

Using $I/I' = D/D'$, and $b(I-I') = (D-D')/t$, we finally get

$$t_{op} = \frac{\log(D/D')}{D-D'}\ t \quad . \tag{6.63}$$

The right-hand side can be estimated from an exposure for any reasonable time t.

For small variations $\delta D'$ of the optical density D' on the high-energy side of the absorption edge (fine-structure region), we can write $D - D' = \delta D'$, and

$$t_{op} = \frac{\log(1+\delta D'/D')}{\delta D'}\ t = \frac{1}{\delta D'}\left[0.434\ \ln(1+\delta D'/D')\right]t \approx (0.434/D')t \quad . \tag{6.64}$$

For $t = t_{op}$, we get

$$(D')_{op} = 0.434 \quad . \tag{6.65}$$

A simple rule is just to select *half-value thickness*, that is, the thickness that reduces the intensity on the high-energy side of the edge to about half its initial value. On the basis of the statistical errors in X-ray absorption measurements, NORDFORS [6.80] recommended that it is preferable to choose an absorber that is too thick than one that is too thin.

For absorption experiments in gases, the knowledge of the gas pressure that would give the maximum intensity difference at the edge, is needed. Jönson's formula,

$$\tau_m = (Z/A)(\tau_e)_K N_A(E_\alpha/E_K) \quad ,$$

where E_α is the value of the energy level of the long-wavelength edge, gives τ_m between two absorption edges. Therefore, (6.55) gives [6.81]

$$\rho x = \frac{A E_K \left(\log E_K - \log E_{L_I}\right)}{Z(\tau_e)_K N_A \left(E_K - E_{L_I}\right)} \quad , \tag{6.66}$$

where x is the path length and ρ is the density of the absorbing gas. This formula overestimates the pressure. Using (3.233), AGARWAL [6.82] suggested the form

$$\rho x = \frac{A\, aZ^b \log(aZ^b)}{Z(\tau_e)_K N_A (aZ^b - 1)} \quad , \quad [\mathrm{g\ cm^{-2}}] \quad , \tag{6.67}$$

where $a = 1.8053$ and $b = -0.6207$. For oxygen, this formula gives $\rho x = 0.8 \times 10^{-4}$ g cm^{-2}. The density and pressure (in mm) are related by $\rho = (A/22400)(p/760)$. For $x = 2$ cm, the pressure p comes out to be less than 3 mm Hg.

In Fig.6.14a, we show the transitions responsible for the last emission line of a series and for the corresponding absorption edge and accompanying absorption band or structure on the high-energy side, in a metal. The line has the structure of an emission band with a high-energy limit that corresponds to the valence-band edge and the main absorption edge corresponds to the conduction-band edge. Therefore, in a metal, the observed edge of the emission band coincides with that of the main-absorption edge [6.64] (Fig. 6.14b). The shape of the line and of the absorption edge depends on the width

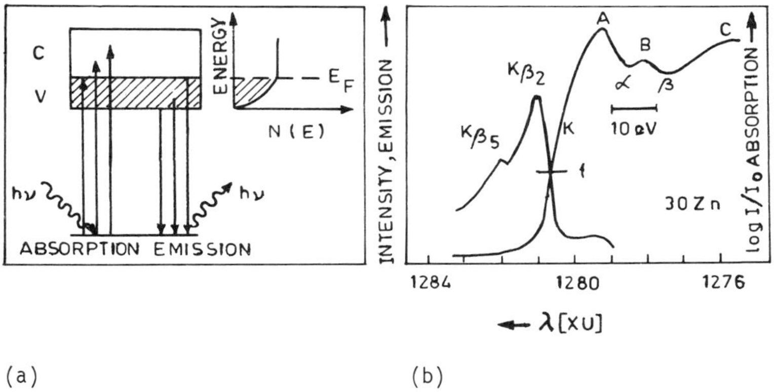

Fig. 6.14. (a) Transitions for the emission and absorption bands in metals. (b) The relation between the K emission and the K absorption bands in Zn. The absorption maxima (A, B, C,...) and minima (α, β, γ,...) occur on the high-energy side of the main absorption edge ($\mathrm{Zn}\lambda_K = 1280.7$ XU). The inflection point is marked f

of the inner level involved and the transition probabilities. Let us now try to indicate the senses in which various terms are used, sometimes vaguely, in discussing absorption spectra.

Absorption discontinuity is the general region of an X-ray absorption spectra where the transmitted intensity, or absorption coefficient, varies suddenly (almost discontinuously) with λ or ν.

Absorption edge refers to the portion of absorption discontinuity that extends from the maximum of transmitted intensity (minimum absorption coefficient) to the minimum of transmitted intensity (maximum absorption coefficient).

Absorption limit refers to the inflection point of the longest wavelength in any curve (absorption coefficient or microphotometer curve) that represents an absorption edge.

Main edge refers to the entire jump of the absorption edge, including any fine structure within the edge.

Extended fine structure is the undulation of the transmitted intensity (or absorption coefficient) found on the high-energy side of the main edge and extending about 300 eV from the edge, or more.

White line is a sharp line-like absorption often observed in the initial absorption rise at the edge of a semiconductor, an insulator, or a semimetal. It is called a white line because of its appearance on the X-ray film. Its presence masks the true position of the absorption edge.

There are several prescriptions in use for locating the position of an edge. We give here the two methods that are usually followed.

1) The true absorption coefficient curve is easily drawn if the ionization chamber (or counter) is used, because then the intensity is directly proportional to the reading of the detector. The arctangent curve, (3.189), is fitt to the observed edge shape and the inflection point is located with sufficien accuracy. Even without this fit, the inflection point is sufficiently well defined on the absorption-coefficient curve. The wavelength of this point gives the absorption limit [6.83].

2) If the photographic method is used, the microphotometer curve itself is directly used for locating the absorption limit, because it is not easy to convert it into the absorption-coefficient curve, in view of the nonlinearity of the intensity versus blackening curve. This method does not involve statistical counting errors; moreover, it simultaneously records all

wavelengths of interest. The inflection point is usually found near the middle of the absorption edge. When not clearly marked, the usual practice is to identify the middle of the absorption jump with the absorption limit.

Microphotometer curves are not true curves of relative transmitted intensity, mainly becuase the relation between incident intensity and film blackening varies with λ. In the absorption-limit measurements, we are concerned with a very small portion (a few XU) of the curve, so this photographic effect is negligible. The thickness of the absorber also has a minor effect on the location of the inflection point [6.79,84]. In most cases, these effects are small compared to the experimental errors.

MATTHEWS [6.85] compared the absorption-limit values determined from the absorption-coefficient curves with those obtained from microphotometer curves. The arctangent equation (3.189) and Lambert's law can be written as

$$\tau(\nu) = C\left\{\frac{1}{2} - \frac{1}{\pi}\tan^{-1}\left[a(\nu_0-\nu)\right]\right\} \quad , \tag{6.68}$$

$$I/I_0 \equiv I_r = e^{-\tau x} \quad , \tag{6.69}$$

where C and $a^{-1} = \Gamma_i/2$ are constants, and ν_0 is the frequency of the absorption limit. These equations give

$$d\tau/d\nu = (Ca/\pi)\left[1+a^2(\nu_0-\nu)^2\right]^{-1} \quad ,$$

$$d^2\tau/d\nu^2 = (2Ca^3/\pi)(\nu_0-\nu)\left[1+a^2(\nu_0-\nu)^2\right]^{-2} \quad ,$$

$$dI_r/d\nu = -x\, e^{-\tau x}(d\tau/d\nu) \quad ,$$

$$d^2I_r/d\nu^2 = x^2\, e^{-\tau x}(d\tau/d\nu)^2 - x\, e^{-\tau x}(d^2\tau/d\nu^2) \quad .$$

Consider the absorption limit of frequency ν_1 on a curve that represents the actual transmitted intensity, such that $\nu_0 - \nu_1 = \delta\nu$. At $\nu = \nu_1$, $d^2I_r/d\nu^2 = 0$, or

$$x(d\tau/d\nu)^2 = d^2\tau/d\nu^2 \quad ,$$

$$x(Ca/\pi)^2\left[1+a^2(\delta\nu)^2\right]^{-2} = (2Ca^3/\pi)(\delta\nu)\left[1+a^2(\delta\nu)^2\right]^{-2} \quad ,$$

whence

$$\delta\nu = Cx(2a\pi)^{-1} \quad . \tag{6.70}$$

The asymptotic behavior of (6.68) gives $C = \tau_h - \tau_l$, where $\tau_h(\tau_l)$ is the limiting value at the high- (low-) frequency side of the main edge. Also, $a^{-1} = \Gamma_i/2$, where Γ_i is the width of the initial atomic level. Therefore, (6.70) becomes

$$\begin{aligned} \delta\nu &= \nu_0 - \nu_1 = \left[x\Gamma_i(\nu)/4\pi\right](\tau_h - \tau_l) > 0 \quad , \\ \delta\lambda &= \lambda_0 - \lambda_1 = -\left[x\Gamma_i(\lambda)/4\pi\right](\tau_h - \tau_l) < 0 \quad , \end{aligned} \tag{6.71}$$

where $\Gamma_i(\nu) = \nu_{3/4} - \nu_{1/2}$, (3.191), and $\delta\nu/\Gamma_i(\nu) = -\delta\lambda/\Gamma_i(\lambda)$. The difference $\delta\nu = \nu_0 - \nu_1$ being positive, this analysis shows that the absorption limit obtained from an intensity curve would have lower frequency.

BECKMAN et al. [6.86] found that, for the K edge of gold, the inflection point of the intensity curve occurs at 0.026 ± 0.010 XU longer wavelength than the inflection on the corresponding absorption-coefficient curve. They found $\tau_h = 16.2\ \mathrm{mm}^{-1}$, $\tau_l = 5.49\ \mathrm{mm}^{-1}$, and $\Gamma_i(\lambda) = 0.14$ XU as read directly from the curve. When these values are put in (6.71), we get $\delta\lambda \simeq -0.011$ XU, which agree rather well with the observed value, within the experimental errors.

The difference $\delta\lambda$ given by (6.71) is opposite in direction to the shift of the absorption limit caused by the finite thickness of the absorber. Therefore, it is possible [6.85] that the absorption-limit values obtained from microphotometer curves may be nearer the true absorption limit than most of those obtained from absorption-coefficient curves.

6.5 Nature of the Main Absorption Edge and the White Line

KOSSEL [6.86a] suggested that the fine structure of the main X-ray absorption edge of an atom is connected with the transition of ejected K electron to discrete optical levels of an atom. PARRATT [6.87] and SCHNOPPER [6.88] have carefully recorded the K absorption edge of argon (Fig.6.15).

In the free atoms in a *gas*, the final states in the hydrogenic approximation should be the sharp atomic states for resonance lines that are broadened to about 0.58 eV, which is the finite width (Γ_i) of the inner K level in ^{18}Ar (the width of the 4p level is negligible). PARRATT [6.70] has called them the

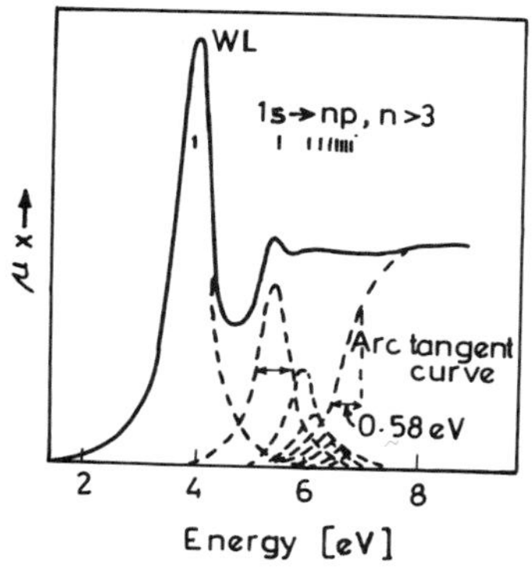

Fig. 6.15. The solid curve represents the K absorption spectrum (uncorrected) of gaseous ^{18}Ar. The resonance lines and the main absorption edge (arctangent curve) at the series limit are shown by dashed curves

bound-ejected-electron excitation states. Thus, the spectrum should exhibit a few allowed absorption lines of a Rydberg series ($1s \to np$, $n > 3$) converging to the absorption limit ($1s \to \infty$), that are broadened by Auger-effect decay of the corresponding discrete lines. This broadening makes most of the lines of the series, except the first few (Fig.6.15), merge smoothly into one another and into the adjoining continuum (positive energy states), which marks the series limit of the atomic levels. The actual photoabsorption limit, is, therefore, not observed, because it is masked by the absorption lines. The first strong line appears as a *white line* (WL) on the photographic plate. According to FONDA and NEWTON [6.200] the X-ray absorption edge occurs at the energy corresponding to raising the atom to its first vacant excited state, and should not be identified with the onset of the photoeffect. The reason is that the sudden onset of the K-shell photo cross section is compensated by a nearly equal and opposite discontinuity in the elastic X-ray cross section.

PARRATT calculated the outer atomic states by assuming that the ^{18}Ar atom with a hole in the K level behaves approximately as a ^{19}K atom. The first allowed transition in argon is $1s \to 4p$, followed by transitions to the subsequent np states whose energy separation is the same as that of the optical levels in potassium. The p-level separations for potassium are $4p\text{-}5p = 1.44$ eV, $5p\text{-}6p = 0.53$ eV, $6p\text{-}7p = 0.25$ eV, etc. The arctangent curve with $\Gamma_i = 0.58$ eV is first drawn at the Rydberg series limit to merge with the observed continuous curve. The first two resolved peaks are then drawn as emission lines with the Lorentz shape (3.177) and equal width (0.58 eV). Very little freedom is left, so the remaining lines are drawn at Rydberg positions with heights adjusted to give the observed curve when added to the arctangent curve (Fig.6.15). The relative intensities of these lines are found to be 100 : 34 : 18 : 8.5, etc.

The inflection point on the arctangent curve gives the true position of the photoabsorption edge. Similar attempts to analyze the K edge in argon and other inert gases have been made by WATANABE [6.89], BAGUS [6.90], and MITCHELL [6.91].

VAINSHTEIN and NARBUTT [6.92] used hydrogenic wave functions for the initial and final states to calculate the square of the dipole transition matrix for the transition $1s \to np$. Remembering that the Coulomb attraction of the 1s electron is much stronger than of the highly screened np electron, they found the area of the absorption line θ_n for the nth level to be

$$\theta_n = f(Z,\eta)\,\frac{n^2-1}{n^5} \quad , \tag{6.72}$$

where n is the effective principal quantum number, η is the charge of the K-ionized atom, and Z is its atomic number.

From (3.173), we can write the probability of dipole transition per unit time from the level 1s to the level np as

$$P_n = \frac{4\pi^2 e^2}{\hbar}\,\rho(E_n) \sum |(r_e)_{1s\to np}|^2 \quad , \tag{6.73}$$

where $\rho(E_n)$ is the radiation density per unit interval of energy, $(r_e)_{1s\to np}$ is a matrix element of component r_e of the radius vector in the direction of polarization for a transition from the state 1s to one of the states np, and the summation is over all np states and is doubled for the two electrons in the K shell. If $P_n(E)dE$ denotes the differential probability of the same transition, but with the absorption of a quantum of definite energy between E and $E+dE$, and if we relate the atomic absorption coefficient $\tau_n(E)$ with the ordinate of the absorption line for the energy E, then

$$\theta_n = \int_{-\infty}^{+\infty} \tau_n(E)\,d(E-E_n) = \frac{E_n}{c\rho(E_n)} \int_{-\infty}^{+\infty} P_n(E)\,d(E-E_n)$$

$$= \frac{E_n}{c\rho(E_n)}\,P_n = \frac{4\pi^2 e^2 E_n}{c\hbar} \sum |(r_e)_{1s\to np}|^2 \quad . \tag{6.74}$$

We also know that the atomic coefficient of absorption τ (E) in the K shell with a transition to the contiuum is given by

$$\tau(E) = \frac{4\pi^2 e^2 E}{c\hbar} \sum |(r_e)_{1s\to E}|^2 \quad , \tag{6.75}$$

where the dipole-matrix element corresponds to the transition to one of the allowed states of the continuum with energy $E - E_\infty$. The summation is over all E (l=1) components and is doubled. If, in (6.74,75), integration is performed over the angular coordinates and the summation is carried out, we get [6.93],

$$\sum |(r_e)_{1s\to np}|^2 = g\Big[\int_0^\infty r^2 R_{1s}(r) R_{np}(r) d^3r\Big]^2 \quad ,$$

$$\sum |(r_e)_{1s\to E}|^2 = g\Big[\int_0^\infty r^3 R_{1s}(r) R_{E(l=1)}(r) d^3r\Big]^2 \quad , \qquad (6.76)$$

where g is a constant and R's are the appropriate radial wave functions.

In (6.72), θ_n was found from (6.74,76). We can find τ_∞ in the same manner from (6.75,76). Another way is to note that at the boundary of the discrete and continuous spectra [6.94],

$$\tau_\infty \equiv \lim_{E\to E_\infty} \tau(E) = \lim_{n\to\infty} \frac{\theta_n}{E_{n+1}-E_n} \quad . \qquad (6.77)$$

The Rydberg equation for the position of the lines can be written as

$$E_n = E_\infty - \frac{\eta^2}{n^2} Ry \quad . \qquad (6.78)$$

Substituting θ_n from (6.72) and $E_{n+1} - E_n$ from (6.78), we can find τ_∞ from (6.77). Also, assuming a dispersion form for the absorption lines with half-width Γ (total width at half height), we have $\theta_n = \pi(\Gamma/2)\tau_n$. Therefore, the ratio of heights of the nth absorption line τ_n to that of the continuous absorption curve τ_∞ is [6.93]

$$\frac{\tau_n}{\tau_\infty} = \frac{4\eta^2}{\pi\Gamma} \frac{n^2-1}{n^5} \quad . \qquad (6.79)$$

BARINSKII and NADZHAKOV [6.95] used the set of equations

$$E_n = E_\infty - \frac{\eta^2}{n^2} Ry \quad ,$$

$$E_{n+1} = E_\infty - \frac{\eta^2}{(n+1)^2} Ry \quad ,$$

$$\cdots\cdots\cdots\cdots\cdots \quad ,$$

$$\frac{\pi\Gamma\tau_n}{\tau_\infty} = 4\eta^2 \frac{n^2-1}{n^5} \quad ,$$

$$\frac{\pi\Gamma\tau_{n+1}}{\tau_\infty} = 4\eta^2 \frac{(n+1)^2-1}{(n+1)^5} \quad ,$$

$$\cdots\cdots\cdots\cdots\cdots \quad , \qquad (6.80)$$

to reproduce the entire shape of the main K absorption edge for a number of gas atoms and molecules, with Γ, n and η as adjustable parameters. The lines were assumed to have the Lorentzian shape; the true edge was assumed to have the arctangent shape, as in Parratt's analysis. The arbitrariness of the choice of the three parameters was removed, to a large extent, by requiring self-consistency, and using the observed position of the second absorption line.

BARINSKII [6.96] extended the foregoing treatment to the L_{III} absorption spectra, for which the initial state has p symmetry and the symmetry of the final states is s and d. The final expressions are of the form,

$$\frac{\tau_n^s}{\tau_\infty^s} = \frac{4\eta^2}{\pi\Gamma_s}\frac{1}{n_s^2} \quad , \quad \text{(s final states)} \quad ,$$

$$\frac{\tau_n^d}{\tau_\infty^d} = \frac{4\eta^2}{\pi\Gamma_d}\frac{\left(n_d^2-1\right)\left(n_d^2-4\right)}{n_d^7} \quad , \quad \text{(d final states)} \quad . \qquad (6.81)$$

There is very little difference in the results of PARRATT and of BARINSKII and NADZHAKOV for the fit to the observed absorption curve. The latter method however, permits inference of the effective charge η from the relative height of the lines.

In *solids* of heavier atoms, the width of the inner level increases and the dispersion on the high-energy side of the edge becomes poor. Consequently, in the L-absorption spectrum of atoms bound in compounds, the lines are not resolved; they are merged to give an asymmetrical WL structure [6.96]. It has been possible to analyze, as in the case of argon, the L_{III}-absorption edges of transition atoms bound in compounds that exhibit strong WL structure [6.96]. AGARWAL and AGARWAL [6.97] have analyzed the WL structure observed by them at the ErL_{II} edge in Er_2O_3 and the HoL_{III} edge in Ho_2O_3, by the method of PARRATT.

According to CAUCHOIS and MOTT [6.98] a WL structure arises if there are unoccupied electronic states of the required symmetry with a high density of states at the absorption edge. The well-known cases are the $L_{II,III}$ edges ($2p \rightarrow nd$) of transition elements where there is a high density of empty d states. The white line is observed in the $L_{II,III}$ edge of transition elements in the metallic state [6.76,99-106] but not for the L_I edge, as expected on this view. However, in most of these cases the WL is not very sharp, the arctan structure of the main edge is retained (not replaced by a sharp line shape, as in Ar), and only a pronounced maximum (usually marked A) immediately

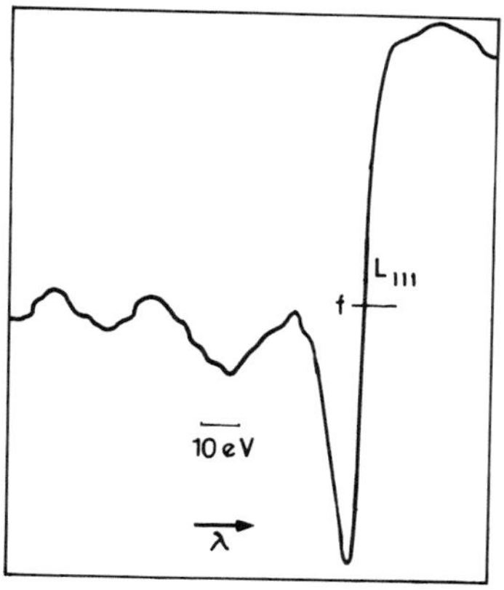

(a)

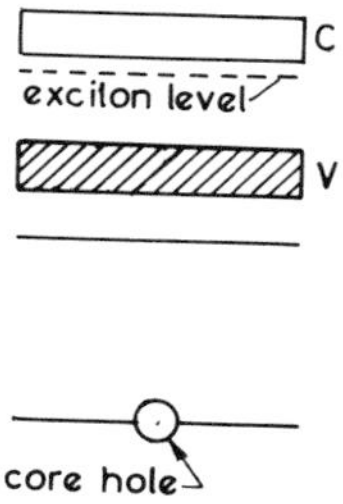

(b)

Fig. 6.16. (a) Microphotometer record of the WL in the Gd L_{III} edge. Spectrograph: 40 cm Cauchois curved crystal. Crystal: mica (100). Absorber thickness: Gd_2O_3 12 mg cm^{-2}. Target: Pt. Exposure: 4 hour at 32 kV, 5 mA. Reference lines: Fe (3 min.), Ni (3 min.), Co (9 min.), [6.117]. (b) Formation of exciton with a vacancy in the core

on the high-energy side of the edge is observed. The situation is further complicated by the self-absorption lines [6.107]. A similar pronounced maximum is also observed in ^{33}As metal edge [6.98,108]; note that As is a semimetal.

A strong WL (Fig.6.16a) is usually observed in the K-absorption spectra of transition-metal compounds [6.98,109-112], and $L_{II,III}$ absorption spectra of rare-earth compounds [6.97,113-116]. Because of scattering effects, the WL gets broadened as the absorber becomes thick [6.79,112,118].

If a sharp WL is observed, it is possible that the screened hole in the inner-core level and the ejected electron form hydrogenlike bound states, called the *exciton states*, just below the conduction-band edge and within the forbidden energy gap (Fig.6.16b). The initial transition being to this sharp exciton level [6.119], an absorption line is observed that merges with the main edge because the exciton levels converge to a series limit. These are not the usual exciton states, in which a hole in the valence band and a corresponding electron in the conduction band are attracted to each other to form a bound system with energy levels given by the Bohr formula; such an exciton is free to move through the crystal. The kind of exciton postulated here (Fig.6.16b) cannot move through the crystal because the hole occurs in an inner-core level and the hole-electron separation is of the order of the atomic radius; the energy levels are similar to those of a free atom of atomic number $Z+1$ rather than hydrogenlike.

SEKA and HANSON [6.111] prefer to give a molecular-orbital picture to explain the white-line structure observed in the transition-metal compounds.

BEEMAN and BEARDEN [6.120], VAINSHTEIN [6.121], BHIDE and BHAT [6.122], SINGH and AGARWAL [6.123], and AGARWAL and JOHRI [6.124], have investigated the K-absorption spectra of ions in *solutions*. In general, it is found that, compared to the spectra of gaseous molecules, the selective-absorption lines show smaller intensity, greater width and a high energy shift that is independent of the nature of the solvent shell around the ion. The greater width is because external levels of ions are split under the perturbation caused by the solvent molecules that form a shell around the ion.

The theory of the shape and width of the main edge given by RICHTMYER, BARNES and RAMBERG is applicable both to *metals* and *semiconductors*, provided that the transitions of the inner electrons beyond the Fermi levels involve a wide band of allowed states. Only then the assumption of the conduction band as a region of continuous distribution of available energy values, made in the theory, is satisfied. The inflection point of the arctangent curve characterizes the position of the Fermi level, and its shape yields the width Γ_i of the inner level [6.64].

The K-absorption edge of ^{30}Zn provides a good example. The electronic structure of the Zn atom is $[KL]3s^2 3p^6 3d^{10} 4s^2$. Thus, the relatively narrow 3d band is full. In the solid state, we can expect some hybridization of 4s-4p bands that makes the upper part of the 4s band free; except for the beginning, the entire 4p band is also made free. The allowed transition from 1s to the very wide 4p band gives the arctan form to the main edge [6.125], see Fig.6.17.

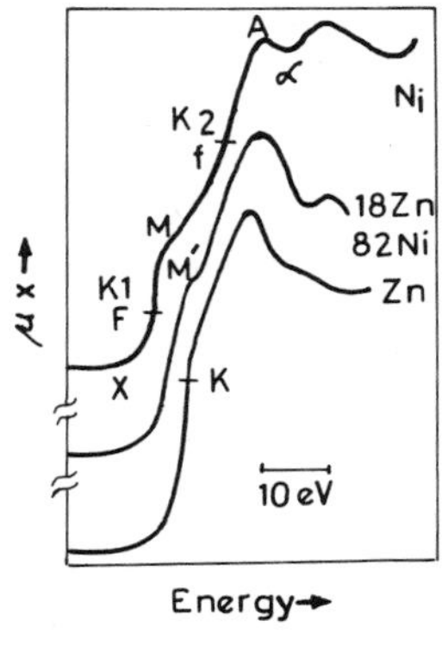

Fig. 6.17. The K-absorption edge of ^{30}Zn, ^{28}Ni in metals and of Zn in alloy (18Zn 82Ni)

The interpretation of the main edge becomes complicated in the case of transition metals with incomplete d shells. For example, the electronic structure of the ^{28}Ni atom is [KL]$3s^2 3p^6 3d^8 4s^2$. In the metal, 4s-3d hybridization occurs. We have 60% atoms with $3d^9 4s^1$ and 40% with $3d^{10} 4s^0$ electronic structure. In fact, the next highest and normally unoccupied 4p orbitals overlap both of the 4s and 3d orbitals. We thus have s-p-d hybridization of the orbitals. It is convenient to describe this situation in terms of the density-of-states curve, N(E), which gives the number of orbitals available at each admissible energy value E. We can write

$$N(E) = N_{oc}(E) + N_{unoc}(E) \quad , \tag{6.82}$$

$$\tau(E) \propto P_{ab}(E) \cdot N_{unoc}(E) \quad , \quad \text{(absorption)} \quad , \tag{6.83}$$

$$I(E) \propto P_{em}(E) \cdot N_{oc}(E) \quad , \quad \text{(emission)} \quad , \tag{6.84}$$

where P is the transition probability, and N_{oc} (N_{unoc}) denotes the density of occupied (unoccupied) states.

BEEMAN and FRIEDMAN [6.64] assumed that $P_{ab}(E)$ in (6.83) is either constant or a very slowly varying function of energy E in the region studied. If this is true, the absorption spectrum should mirror the density-of-states curve of the metal or semiconductor.

The transition probability cannot be calculated exactly. The $1s \rightarrow ns$ and $1s \rightarrow nd$ transitions are Laporte forbidden. However, according to the band calculations [6.126], the final states have sufficient admixtures of s, p and d orbitals to permit us to relax the selection rule and to allow transitions from the inner 1s level to the admixed high density of d states. IRKHIN [6.127] concluded that the calculated relative probabilities for dipole transitions to 3d + 4s + 4p admixed states in Ni were a hundred times greater than the corresponding probabilities for the quadrupole transitions. YEH and AZAROFF [6.125] have studied the ^{30}Zn and ^{28}Ni K edges in metals and solid solutions of Zn in fcc Ni (Fig.6.17).

The metallic ^{28}Ni K-absorption edge, XMA in Fig.6.17, shows structure (K1, K2). The initial absorption rise XM (called K1) arises from transitions to the states of s-p-d admixed symmetry, and the second absorption rise MA (called K2) from transitions to the Laporte-allowed states of nearly pure 4p symmetry. The inflection point F on K1 locates the Fermi level E_F. Above E_F, the s-p-d admixed states first acquire pure s symmetry just around M and finally pure 4p symmetry around the inflection point f of K2. The kink at M arises because the transition probability suddenly decreases for the pure s

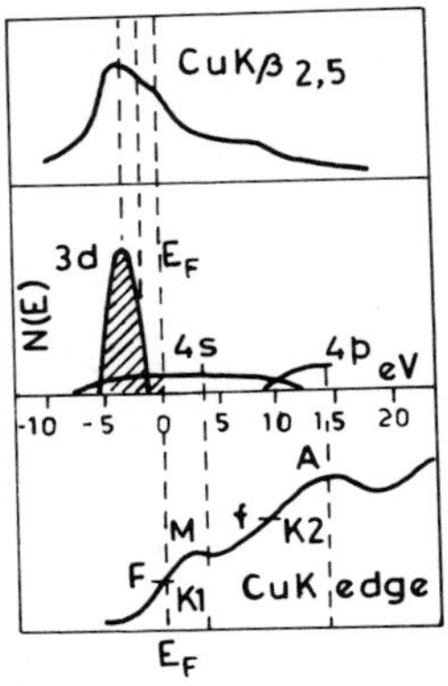

Fig. 6.18. (Top) $CuK\beta_{2,5}$ emission, (middle) density-of-states curve, and (bottom) K-absorption spectrum of ^{29}Cu metal [6.128]

character of states at M and this causes a decrease of the absorption. The structure K2 that ends in peak A is assigned $1s \rightarrow 4p$ transitions, and the region beyond A shows maxima of absorption, corresponding to the transitions $1s \rightarrow np$, $n = 5,6,7$, etc.

In the solid state, the solubility of Zn in Ni is less than 25 at.%, so that 3d holes of Ni atoms are not saturated completely. However, the area below the curve K1 of the Ni edge decreases monotonically as the Zn content increases. On the other hand, the simple Zn metal edge develops a kink M' (Fig.6.17) as it donates s and d electrons to Ni and develops empty high-density d states. This kink becomes more peaked as the Zn content decreases because there can then be more s-d vacancies per atom.

It is instructive to compare the theoretical density-of-states curve for ^{29}Cu metal [6.128] with the observed $CuK\beta_{2,5}$ emission and CuK-absorption spectra (Fig.6.18). According to BURDICK [6.128] the filled (shaded) states have largely 3d symmetry, but there is an increasing admixture of 4s and 4p symmetry as E_F is approached. About 4 eV above E_F, the 4s symmetry becomes more predominant, whereas at higher energies it becomes purely 4p.

Just *below* E_F, $N_{oc}(E)$ is not high, but relatively large emission is found because the states have strongly admixed p symmetry. At lower energies, $N_{oc}(E)$ increases but the admixture of p symmetry decreases, so that the emission curve does not continue to rise.

Just *above* E_F, $N_{unoc}(E)$ states of s-p-d hybridization exist, so that the absorption rises rapidly (K1). At about 4 eV above E_F, the states have predominantly s symmetry, so that the transition probability decreases and gives less absorption at M (Fig.6.18 bottom). At slightly higher energies, the states assume pure 4p character, and the absorption rises rapidly again. The subsequent maxima arise from transitions between 1s→np, n≥5, states.

Both ^{28}Ni and ^{29}Cu have identical crystal structures; they differ only in that Ni has one less electron. Therefore, they show nearly similar K1, K2 structures in the main absorption edge. Although the P(E) term is not calculated in (6.83,84), the positions of the observed structures in the spectra near E_F can be correlated with the corresponding regions of the calculated energy band [6.129,130].

The K-absorption spectra of the 4d transition metals (^{37}Rb, ^{38}Sr, ^{39}Y, ..., ^{47}Ag) have been recorded, for example, by BHIDE and BHAT [6.131], AGARWAL and SINGH [6.132], BHIDE and KAICKER [6.133], SINGH and AGARWAL [6.123], KOSTROUN et al. [6.134], AGARWAL and JOHRI [6.124]. The Rb K edge in the metal (in vacuum) was first recorded by AGARWAL and SINGH [6.132]. The measurements of KOSTROUN et al. [6.134], and AGARWAL and JOHRI [6.124] clearly revealed the existence of structures K1 and K2 with the kink M between them, as in the case of the 3d transition metals. The kink M disappears for ^{47}Ag which has a filled d band [6.134].

The observed split structure (K1, K2) can also be looked upon as arising from the effect of edge singularity in the absorption spectra of core electrons in metals with an empty or incomplete shell, as discussed by KOTANI and TOYOZAWA [6.135]. They treat the empty or incomplete shell as a nondegenerate localized state that interacts with the conduction band through a mixing interaction. In the initial state of the transition, the empty state of each atom is high enough above E_F for the mixing of the two states to be negligible. However, in the final state of the transition, the empty level of the excited atom is lowered because of the potential of a core hole left behind. When the lowered empty state remains close to, but above E_F, a singular absorption edge K1 appears at the Fermi energy, as expected, with an asymmetric antiresonance dip K2 around the lowered empty state. The theoretical shape of the absorption spectra, given by KOTANI and TOYOZAWA, shows a striking resemblance to the Rb metal edge observed by AGARWAL and JOHRI [6.124].

6.6 Chemical Shifts of Absorption Edges

Even in early days of X-ray spectroscopy, BERGENGREN [6.136], and LINDH [6.137] showed that chemical combination affects X-ray emission and absorption spectra. Since then, considerable work has been reported on chemical shifts of

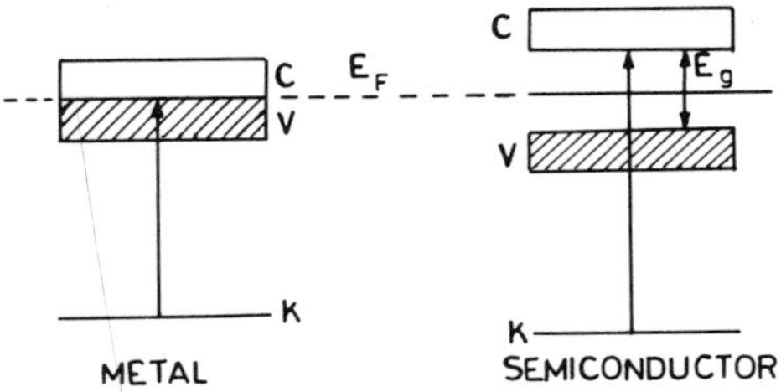

Fig. 6.19. Relative-energy diagram of a metal and the corresponding semiconductor

X-ray absorption edges. These have been reviewed by CAUCHOIS [6.138], MEISEL [6.139], AZAROFF and PEASE [6.130], and NAGEL and BAUN [6.140].

As shown in Fig.6.19, two main factors contribute to the observed high-energy shifts of X-ray absorption edges: 1) the tighter binding of the core level because of the change of the effective charge (or screening) of the nucleus caused by the participation of the valence electrons in the chemical-bond formation, and 2) appearance of the energy gap E_g when we go from a metal to a compound (semiconductor or insulator).

Direct evidence for the first factor is provided by photoelectron spectroscopy. The second factor, E_g, is known to be related to phenomena such as covalency, effective charge, coordination number, crystal structure, etc. In Phillips' approach, the *bond parameters* C, $E_{g,bond}$, f_i are important. Because the various factors contribute simultaneously, it is very difficult to give a satisfactory and general theory for the shift of the edges.

By combining the inner-level shift data from photoelectron spectroscopy and X-ray absorption shift data, some valuable information may be obtained on the changes that occur in the conduction band alone, when bonding occurs [6.141]. Ionic bonds being stronger than covalent bonds give larger shifts.

Because of the selection rules, the absorption transitions for the different edges probe different regions of the conduction band. A combination of such information can tell us about the nature of the redistribution of electrons in the conduction band when chemical bonds are formed. For emission spectra, WIECH [6.142] found that the sum of different valence-band spectra (that is, K and $L_{II,III}$ band spectra) approximates the total density of states $N_{oc}(E) = N_{oc,s}(E) + N_{oc,p}(E)$, where the suffixes s, p denote the symmetry involved. The same would hold for absorption spectra, where information about $N_{unoc}(E)$ would be obtained. For compounds few N(E) calculations exist.

The shift of the X-ray absorption edge i (i=K,L,M,...) of an element in a compound, with respect to that of the pure element is written as

$$\Delta E_{ab} = E_i(\text{compound}) - E_i(\text{element}) \quad .$$

In general, the shift ΔE is positive (towards high energy) and ranges usually from ~1 eV to ~15 eV.

AGARWAL and VERMA [6.143] suggested an empirical rule for ΔE_{ab}: *In general, the chemical shift is towards the high-energy side of the metal edge; it increases progressively with increase of the valence of the cation, unless the shift is either suppressed by the covalent character of the bond or enhanced by the formation of a metal-metal bonding.*

The first part of the rule, namely the valence dependence of the shift, is well known [6.138,144]. BOEHM et al. [6.145], MANDE and CHETAL [6.146], and MILLER [6.147] used this dependence to determine the valences of the absorbing ions in compounds. SAPRE and MANDE [6.148] tested this rule and found that the covalence suppresses the shift, not only for the cations but also for the anions. Moreover, anions show a negative (low-energy) shift. The evidence for the large shift in the presence of the metal-metal bond is not conclusive [6.149]. This bond is known to be present in Cl-Hg-Hg-Cl (mercurous chloride). Other examples are the Pb-Pb bond in $Pb_2(CH_3)_6$ and Cu-Cu bond in $[Cu(Acetate)_2H_2O]_2$.

OVSYANNIKOVA et al. [6.150], and DEY and AGARWAL [6.112,151] suggested the following empirical formulas, respectively,

$$\Delta E_{ab} = \text{const.}[N-(1-f_i)n] \quad , \tag{6.85}$$

$$\Delta E_{ab} = N^2 f_i \quad , \tag{6.86}$$

where N is the valence, f_i is the Pauling's ionicity for a crystal (6.15), and n is the coordination number.

Comparison of (6.85) with (6.35) shows that $\Delta E_{ab} \propto$ effective charge q. Quite often, Suchet's method of evaluating q has been employed to relate ΔE with q [6.148,152]. ADHYAPAK et al. [6.153] found a correlation between the chemical shift of the edge and Phillips' electronegativity difference C. The chemical shifts of the K edges in spinels, chalcogenides and intercalation compounds [6.154-156] yield valuable information about the nature of bondings in such substances.

LEVY and VAN WAZER [6.157], GLEN and DODD [6.110], and SEKA and HANSON [6.111] initiated the application of molecular-orbital theory to explain the nature and shifts of the X-ray absorption edges in transition-metal octahedral and tetrahedral complexes. GLEN and DODD [6.110] observed that the main peak A in the Mn K edge shifts towards the high-energy side in the manganese oxide spectra as the oxidation state increases: MnO (oxidation 2; peak A above edge

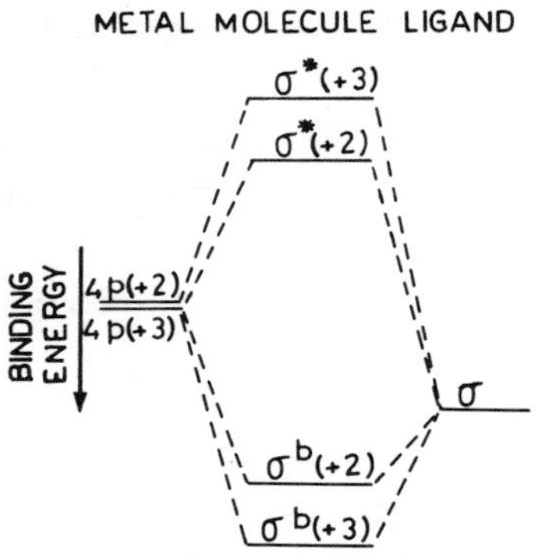

Fig. 6.20. Simplified molecular-orbital diagram for the changes of energy levels as a function of oxidation state [6.110]

16.0 eV), Mn_3O_4 (2,3; 20.3 eV), MnO_2 (4; 23.6 eV). This can be explained with the help of Fig.6.20. The more-oxidized metal forms a stronger bond with the ligand (oxygen); therefore, there is a greater overlap of the metal and ligand orbitals. This gives more-stable occupied bonding orbitals. On the other hand the empty antibonding orbitals, which are obtained by changing the signs of the ligand wave functions and the overlap integral, are less stable because of the stronger repulsive forces. The absorption transitions are to the unoccupied antibonding orbitals. Therefore, the spectral shifts are to higher energies. In this approach also, quantitative calculations are not available.

6.7 Extended Fine Structure of Absorption Edges

If a small region in Fig.3.12, containing (say) the K edge and the smooth-looking absorption curve on the short wavelength side of the K edge, is enlarged, it reveals fluctuations or *fine structure* that extends about 1000 eV from the edge. This absorption spectrum is usually separated, for the purpose of calculation, into three groups: main-edge structure (such as K1, K2), near edge (Kossel) structure (such as WL or excitonic peaks) to within 20 eV on the high-energy side of the main edge, and extended fine (Kronig) structure up to about 500 eV or more on the high-energy side of the Kossel structure. We have already considered the first two groups. Here we discuss the *extended fine structure* (EFS); it is also called the *extended X-ray-absorption fine structure* (EXAFS). The early work has been reviewed by CAUCHOIS [6.138] and AZAROFF [6.158]. Fluctuations in $\tau(E)$, (6.83), are assumed to be related to rapid changes either in $P_{ab}(E)$ or in $N_{unoc}(E)$.

The EFS was first theoretically examined by KRONIG, both in diatomic molecules [6.159] and in crystalline absorbers [6.160]. The former initiated calculations for cases in which the ejected-electron final states are those that follow scattering by neighboring atoms: *short-range-order* (sro) theories, $\tau(E) \sim P_{ab}(E)$. The latter has led to *long-range-order* (lro) theories, in which plane waves (Bloch functions) are used to describe the ejected electron, and the allowed final-state energies in a crystal are calculated, $\tau(E) \sim N_{unoc}(E)$.

6.7.1 Kronig's Theory for Diatomic Molecules

KRONIG [6.159] considered absorption τ by an atom which is part of a diatomic molecule. Usually, τ is taken as a function of λ or the photon energy $h\nu$. An electron of binding energy E_i in an isolated atom is ejected with kinetic energy,

$$E = h\nu - E_i \quad , \tag{6.87}$$

in the absorption process. Let $\tau_a(E)$ represent the absorption coefficient for a single isolated atom A, as a function of E. In a diatomic molecule AB, the ejected photoelectron from A moves in the field of both A and B. The bonding perturbs the core levels of A and also the continuum of A. Consequently, the probability P of ejection of a photoelectron with kinetic energy E is affected, causing a change of the absorption coefficient.

Let $\tau_M(E)$ denote the absorption coefficient for a single atom A in a molecule. According to KRONIG, the change of the absorption by a diatomic molecule is determined by the ratio,

$$\chi(E) = \frac{\tau_M(E)}{\tau_a(E)} = \frac{P_M(E)}{P_a(E)} \quad , \tag{6.88}$$

where P_a and P_M are the corresponding probabilities of the transitions.

The ejected positive-energy electron can be represented by a plane wave,

$$\psi_k(x,y,z) = N\, e^{ikr} \quad , \tag{6.89}$$

where N is the normalization constant and $k = (2mE)^{\frac{1}{2}}/\hbar$. If $\psi_i(x,y,z)$ is the bound-state wave function of the initial core state, the probability P is proportional to the square of the matrix element of the oscillator amplitude, (3.142),

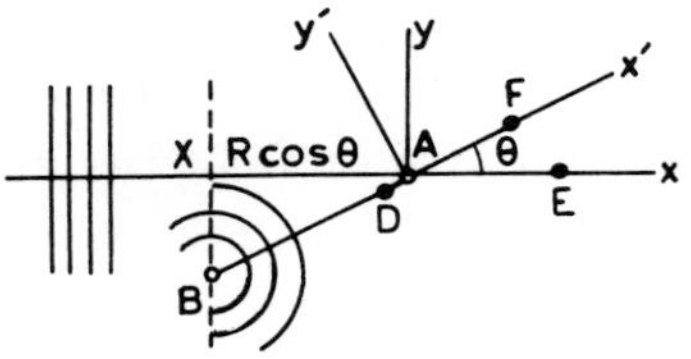

Fig. 6.21. Calculation for absorption by a diatomic molecule. $AB = R$, $AD = R(1-\cos\theta)$, $DB = R\cos\theta$, $AE = x$, $AF = x'$

$$P \propto \left(|x_{i\to k}|^2 + |y_{i\to k}|^2 + |z_{i\to k}|^2\right) \quad ,$$

$$x_{i\to k} = \iiint \psi_k^*(x,y,z)\, x\, \psi_i(x,y,z)\, dV \quad ,$$

where $dV = dx\,dy\,dz$.

If we designate the matrix elements $x_{i\to k}$, $y_{i\to k}$, $z_{i\to k}$ for an isolated atom by x_a, y_a, z_a, and for a bound atom by x_M, y_M, z_M, then

$$\chi(E) = \frac{x_M^2 + y_M^2 + z_M^2}{x_a^2 + y_a^2 + z_a^2} \quad . \tag{6.90}$$

The wave function for the inner state $\psi_i(x,y,z)$ is the same in both cases. Let the electron in a state k move along the x axis that passes through the atom A. The plane wave that represents the electron has the form $\psi_k(x) = N\exp(ikx)$. Then, for a single atom,

$$x_a = \iiint \psi_k^*(x)\, x\, \psi_i(x,y,z)\, dx\, dy\, dz \quad , \quad y_a = z_a = 0 \quad . \tag{6.91}$$

The plane-wave front from A is elastically scattered by B if it reaches the point X on the x axis (Fig.6.21). In elastic scattering, the magnitude of the wave vector, k, is conserved. The amplitude of the scattered spherical wave near the atom A can be expressed as (see, for example, [6.160a]) $f(\theta)/R$ where R is the distance between the atoms, θ is the scattering angle, and $f(\theta)$ is the angle-distribution function. If $R \gg \lambda$, the spherical wave near A can be approximated by a plane wave $(f(\theta)/R)\exp(ikx')$, that propagates along the x' axis,

$$\psi_{sc}^B(x') = N(f(\theta)/R)\exp(ikx') \quad . \tag{6.92}$$

Let us investigate the scattered wave at the point F on the interatomic axis x' (Fig.6.21). We can find the time and path lengths traversed by the incident wave and the scattered wave from the instant when the former reached the atom A. In this time interval, the scattered wave traversed the distance $BD = XA = R\cos\theta$ along the x' axis. At some subsequent time t, when the incident wave is at the point E, with coordinate x, the scattered wave will be at the point F with coordinate x'. Clearly,

$$x = AE = DF = DA + x' = R(1-\cos\theta) + x' \quad .$$

It follows that we can write, at F,

$$\psi_{sc}^{B}(x') = N\,\frac{f(\theta)}{R}\,e^{ik[R(1-\cos\theta)+x']} = q\psi_k(x') \quad , \tag{6.93}$$

where q does not depend on x', and we can take $N = 1$,

$$q = \frac{f(\theta)}{R}\,e^{ik\,R(1-\cos\theta)} \quad . \tag{6.94}$$

We now have two waves (incident + scattered from B) that act simultaneously on the atom A. The total wave function $\Psi_{tot}(x,y,z)$ can be expressed as

$$\Psi_{tot}(x,y,z) = \psi_k(x) + \psi_{sc}^{B}(x') = \psi_k(x) + q\psi_k(x') \quad . \tag{6.95}$$

The matrix element x_M then is

$$\begin{aligned} x_M &= \iiint \Psi_{tot}^{*}(x,y,z)\,x\,\psi_i(x,y,z)\,dV \\ &= x_a + q^{*}\iiint \psi_k^{*}(x')\,x\,\psi_i(x,y,z)\,dV \quad , \end{aligned} \tag{6.96}$$

where x_a is given by (6.91). If ψ_i has spherical symmetry, we get

$$\psi_i(x,y,z) = \psi_i(x',y',z') \quad ,$$

where the frame x', y', z' obtained by rotation through an angle θ around the z axis is related to x,y,z by,

$$x = x'\cos\theta - y'\sin\theta \quad , \quad y = x'\sin\theta + y'\cos\theta \quad , \quad z = z' \quad ,$$

$$dx\,dy\,dz = dV = dx'\,dy'\,dz' = dV' \quad .$$

It follows that (6.96) can be written as

$$x_M = x_a + q^* \iiint \psi_k^*(x')(x' \cos\theta - y' \sin\theta)\psi_i(x',y',z')dV'$$
$$= x_a + q^*(x_a \cos\theta - y_a \sin\theta) = x_a(1+q^* \cos\theta) \quad ,$$

because $y_a = 0$. In the same way, we get

$$y_M = q^* x_a \sin\theta \quad , \quad z_M = 0 \quad .$$

We can now evaluate (6.90),

$$|x_M|^2 = x_a^2|1+q^* \cos\theta|^2 = x_a^2(1+q^* \cos\theta)(1+q \cos\theta)$$
$$= x_a^2\left[1+(q+q^*)\cos\theta+|q|^2 \cos^2\theta\right] \quad ,$$
$$|y_M|^2 = x_a^2|q|^2 \sin^2\theta \quad ,$$
$$\chi(E) = \frac{x_M^2+y_M^2}{x_a^2} = 1 + (q+q^*)\cos\theta + |q|^2 \quad . \tag{6.97}$$

We can average this expression over $\cos\theta$,

$$\langle \chi(E)-1 \rangle = \frac{\int_{-1}^{+1}\left[(q+q^*)\cos\theta+|q|^2\right]d(\cos\theta)}{\int_{-1}^{+1} d(\cos\theta)} = \frac{1}{2}\int_0^{\pi}\left[(q+q^*)\cos\theta+|q|^2\right]\sin\theta \, d\theta \quad . \tag{6.98}$$

To find q, we have to evaluate $f(\theta)$ by taking some potential form for the scattering by the atom B [6.161]. For a symmetrical multi-atomic molecule [6.162]

$$\langle \chi(E)-1 \rangle = \sum_j \langle \chi_j - 1 \rangle \quad . \tag{6.99}$$

An analogous theory was given by CORSON [6.163], who applied it to the GeK-absorption spectrum in $GeCl_4$.

6.7.2 Kronig's (lro) Theory for Crystalline Matter

We shall follow a slightly different approach than that used originally by KRONIG [6.160] to derive the basic result. We assume that the ejected photoelectron moves out of the absorbing atom A almost as a free electron, with kinetic energy E, momentum $\hbar k = (2mE)^{\frac{1}{2}}$ and de Broglie wavelength $\lambda_B = 2\pi/k$. If the direction of motion is the x axis, it forms an incident plane wave,

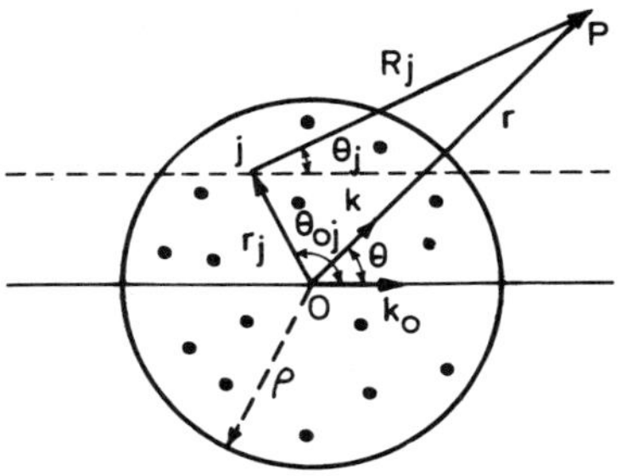

Fig. 6.22. Scattering of plane wave by an assembly of scatterers

$$\psi_i(x) = \exp(ikx) = \exp(ikr\cos\theta) \quad , \tag{6.100}$$

where r, θ define the polar coordinates. Let this electron be scattered by an assembly of N scatterers, all distributed inside a sphere of radius ρ, the jth scatterer being located with respect to the center of the sphere by the vector $\underline{r}_j$ (Fig.6.22).

Each scatterer is assumed to scatter independently. Then each scatterer creates a scattered wave radially outward, having an angle-distribution function f(θ). Therefore, the scattered wave at a point P is

$$\psi_{sc} = \sum_{j=1}^{N} \frac{f(\theta_j)}{R_j} e^{i(\underline{k}_0\cdot\underline{r}_j + kR_j)} \quad , \tag{6.101}$$

where we have assumed the phase to be zero at the origin, and $|\underline{k}_0| = k$. For $r \gg \rho$ (Fig.6.22), we get, $1/R_j \to 1/r$, $R_j \to r - r_j\cos(\underline{r}_j,\underline{k})$,

$$\psi_{sc} \simeq \frac{e^{ikr}}{r} f(\theta) \sum_{j=1}^{N} e^{i\underline{K}\cdot\underline{r}_j} \quad , \tag{6.102}$$

where $\underline{K} = \underline{k}_0 - \underline{k}$, $|\underline{k}_0| = |\underline{k}| = k$, and $K = 2k\sin(\theta/2)$.

To obtain Kronig's formula, let us assume that all of the scatterers are arranged in regularly spaced planes (long-range order), perpendicular to the vector $\underline{k} - \underline{k}_0$. We can now expect *coherent* scattering. Suppose that there are M such planes, a distance d apart, each containing L scatterers, so that $ML = N$ and $d < 2\rho$. We set the x axis parallel to $\underline{K}$. The x coordinate of the jth plane is then $x = x_1 + (j-1)d$, and (6.102) becomes[2]

[2] With $y = \exp(iKd)$, use $S_M = 1 + y + y^2 + \ldots + y^{M-1}$ and $S_M - S_M y = 1 - y^M$ to get $S_M = (y^M-1)(y-1)^{-1} = (y^{M/2}/y^{1/2})(y^{M/2}-y^{-M/2})(y^{1/2}-y^{-1/2})^{-1}$.

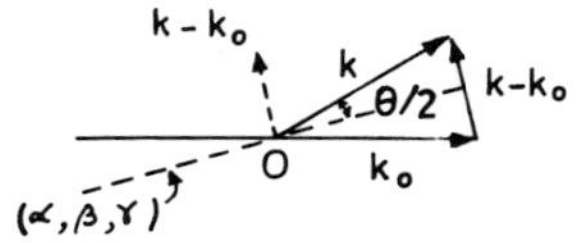

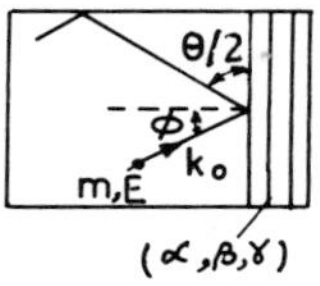

Fig. 6.23. (a) ($\underline{k}-\underline{k}_0$) is perpendicular to the reflecting plane (α,β,γ) and remains so as θ (or ϕ) takes on various values starting from 0° (or 90°). (b) The ejected photoelectron cannot travel as a free particle in the lattice because it is Bragg reflected if (6.104) is satisfied

$$\psi_{sc} = f(\theta)\,\frac{e^{ikr}}{r}\,L\,e^{iKx_1}\sum_{j=1}^{M} e^{iK(j-1)d}$$

$$= N\,f(\theta)\,\frac{e^{ikr}}{r}\,e^{iK(x_1+Nd/2)}\left\{\frac{\sin[kMd\,\sin(\theta/2)]}{M\,\sin[kd\,\sin(\theta/2)]}\right\} \quad .$$

Thus, a strong scattered wave is produced if

$$kd\,\sin(\theta/2) = n\pi \quad , \tag{6.103}$$

where n is an integer and $k = 2\pi/\lambda_B$. This is just the Bragg condition ($\underline{K}$ is kept perpendicular to the planes of scatterers), (5.48), and could have been written directly for the de Broglie wave [$2d\,\sin(\theta/2)=n\lambda_B$]. From (6.103), we obtain, by using $\lambda_B = h/p = h/(2mE)^{1/2}$ and $d = a/(\alpha^2+\beta^2+\gamma^2)^{1/2}$,

$$E = \frac{h^2}{8m}\,\frac{(\alpha^2+\beta^2+\gamma^2)}{a^2\,\cos^2\phi} \quad , \tag{6.104}$$

where α, β, γ are integers, a is the periodic spacing length, and ϕ is the angle between the ejected electron's propagation direction and the normal to a reflecting plane in the crystal (complement of the Bragg angle $\theta/2$), Fig.6.23a. We have absorbed n in the α,β,γ indices.

If the ejected photoelectron happens to have such a combination of kinetic energy E and angle ϕ inside a crystal that (6.104) is satisfied, then it is Bragg reflected by the planes (Fig.6.23). KRONIG approached this interpretation from the Brillouin-zone point of view. The electron with E, ϕ satisfying (6.104) cannot propagate as a free electron in the lattice because of the Bragg reflection (Fig.6.23b) by the (α,β,γ) planes. For an unpolarized X-ray beam (or a polycrystalline absorber), ϕ can take on all possible values and we should average over all possible propagation directions. This reduces (6.104) to[3]

$$E = \frac{h^2}{8m} \frac{(\alpha^2+\beta^2+\gamma^2)}{a^2} \quad , \tag{6.105}$$

where $h^2/8m = 37.5$ eV Å^2. This corresponds to the situation where $\phi = 0^\circ$ in (6.104) and the photoelectron is ejected in the direction of the normal itself with the minimum energy $E = E_{min}$. The three integers (α,β,γ) play the role of the Miller indices. The photoabsorption process is then like electron diffraction by certain critical planes that form the Brillouin-zone boundaries. Here a is the cell edge, if the lattice is cubic.

The Brillouin-zone boundaries given by (6.105) separate the allowed energy-value states from those in the forbidden region. Therefore, KRONIG suggested that the boundary energies given by (6.105) should correspond to *zero absorption* or *absorption minima* (the photoelectron cannot be ejected because it cannot propagate, in view of the Bragg reflection). KRONIG used (6.105) to calculate the forbidden energies (zero absorption) for several metals and found that, because of the large number of alternative (α,β,γ) values, they were much too closely spaced for experimental resolution (AZAROFF, SANDSTRÖM [6.158,83]). He grouped them into regions of higher density and associated their midpoints with the absorption minima $(\alpha,\beta,\gamma,\ldots)$ in the observed absorption curves. This introduces a certain amount of arbitrariness in the analysis. The admissible values of (α,β,γ) would depend on the type of the unit cell [6.164]. It is a long-range-order (lro) theory because the changes of the transition probability are ignored and the long-range order of the lattice implied in (6.105) is used in the argument. WEBER [6.165] recently compared Kronig's theory with experimental data.

In the absorption process, the ejected photoelectron can go to one of the possible free states of the electron in the lattice, beyond E_F given by (6.51)

$$E_F - E_0 = \frac{h^2}{8m}\left(\frac{3N}{\pi V_a}\right)^{2/3} = \frac{h^2}{8ma^2}\left(\frac{3n_c}{\pi}\right)^{2/3} \quad , \tag{6.106}$$

where $N/V_a = nN/a^3 = n_c/a^3$, n_c being the number of electrons in the unit cell. Comparison of (6.105,106) shows that the absorption coefficient can exhibit

---

[3] $\left| \int (1/\cos^2\phi)d\Omega / \int d\Omega \right| = \frac{1}{4\pi} \left| \int_0^\pi (2\pi \sin\phi/\cos^2\phi)d\phi \right| = \frac{1}{2} \left| \int_{-1}^{+1} d(\cos\phi)/\cos^2\phi \right| = 1.$

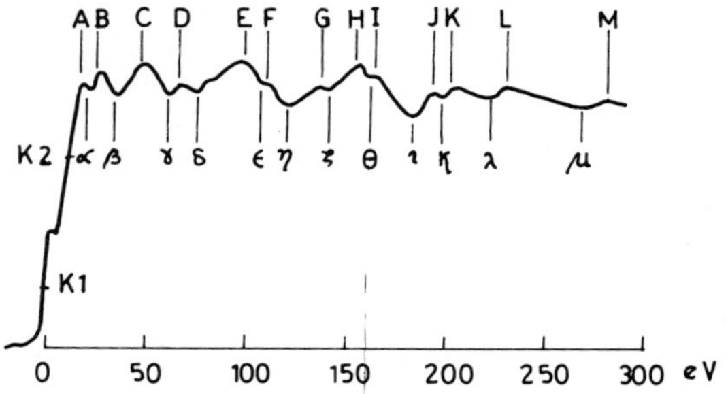

Fig. 6.24. Extended fine structure at CuK edge [6.141]

spectral fluctuations only at energy discontinuities on the surfaces of the Brillouin zone for which, with E_0 set equal to 0,

$$(\alpha^2+\beta^2+\gamma^2) \geqq (3n_c/\pi)^{2/3} \quad . \tag{6.107}$$

Thus, for Cu the *Kronig formula* (6.105) is valid beyond 67 eV; for Fe it is valid beyond 11 eV. KRONIG takes $P_{ab}(E) \simeq \text{const}$, $\tau(E) \simeq N_{unoc}(E)$, in (6.83).

Kronig's formula (6.105) implies that substances that have same crystalline structure should show similar fluctuations of fine structure. The energy distances from the main edge to the minima α, β, γ, ... (Fig.6.24) of the fluctuations should be proportional to $1/a^2$. This has been observed.

6.7.3 Hayasi's Modification of Kronig's (lro) Theory

HAYASI [6.166] gave a different interpretation to (6.105). He emphasized that it corresponds to the case $\phi = 0$, that is, the case where *the initial photoelectron* (ψ_{k_0}) *is reflected back* (ψ_k) *to the absorbing atom* by crystallographic planes formed by its nearest-neighbor atoms at right angles to the possible propagation directions. This creates a localized-wave pattern (a *quasi-stationary state*), where there is an increased overlap of the two wave functions. This leads to an increase of the transition probability. For cubic crystals, the energy values are still determined by (6.105), but now they correspond to the absorption *maxima* A, B, C, ... (instead of the minima α, β, γ,... in Kronig's interpretation). Because it is difficult to locate the true zero exactly along the energy scale, in an absorption spectrum, and because a relative displacement could take us from positions of minima to maxima, it is difficult to decide which interpretation is correct [6.112]. Note that only the positions of absorption maxima (A, B, C,...) are predicted by Hayasi's theory.

6.7.4 Short-Range-Order (sro) Theories

For molecules, KRONIG showed that 1) the density of allowed states N(E) varies slowly, 2) the absorption process is primarily determined by the transition probabilities, and 3) the ejected photoelectron can be described by a plane wave that is scattered by the surrounding atoms whose potentials (or scattering amplitudes) enter into the expression for the absorption coefficient. This approach was used by KOSTAREV [6.167] to calculate the EFS of solids (infinite molecules). He used the WKB (Wentzel-Kramers-Brillouin) method to calculate the scattering amplitudes (or phase shifts) caused by the surrounding atoms. SHIRAIWA et al. [6.168] used the Born approximation to obtain the scattering amplitude. We give here their calculations; $\tau(E) \sim P_{ab}(E)$.

The need for a short-range-order (sro) theory arose mainly because of the observation of EFS in amorphous substances [6.169,170,118].

Consider Fig.6.22, where the absorbing atom is at the origin O and the ejected photoelectron is represented by a plane wave

$$\psi_i = e^{i\underline{k}_0\cdot\underline{r}} \quad , \tag{6.108}$$

where $|\underline{k}_0| = k$. The scattered wave is given by

$$\psi_{sc,j} = \frac{f(\theta_j)}{R_j}\, e^{i(\underline{k}_0\cdot\underline{r}_j + kR_j)} \quad . \tag{6.109}$$

In order to find the transition probability from a 1s state, we need the wave function in the space near the nucleus (at O), that is, where the wave function of the inner (1s) state exists. Then the spherical wave (6.109) is replaced by the plane wave

$$\psi_{sc,j} = \frac{e^{ikr_j}}{r_j}\, f(\pi-\theta_{0j})\, e^{i(\underline{k}_0\cdot\underline{r}_j + kr_j)} \quad , \tag{6.110}$$

where θ_{0j} is the angle between $\underline{k}_0$ and $\underline{r}_j$ (Fig.6.22). In obtaining (6.110) we have merely moved the point P to O in Fig.6.22.

SHIRAIWA et al. [6.168] write the final-state wave function as

$$\Psi_{tot} = \psi_i + \sum_j \psi_{sc,j} + \sum_{j,l} \psi_{sc,jl} + \sum_{j,l,n} \psi_{sc,jln} + \dots \quad , \tag{6.111}$$

where ψ_i is the emitted (or incident) wave, $\psi_{sc,j}$ is the emitted wave scattered by the jth atom, $\psi_{sc,jl}$ is this wave rescattered by the lth atom, and so on.

If we neglect multiple scattering, then

$$\Psi_{tot} = \psi_i + \sum_j q_j \, e^{i\underline{k}_{0j}\cdot\underline{r}} \quad , \tag{6.112}$$

$$q_j = \frac{f(\pi-\theta_{0j})}{r_j} \, e^{i(\underline{k}_0\cdot\underline{r}_j + kr_j)} \quad , \tag{6.113}$$

and the transition probability (6.90) of Kronig's (sro) theory is proportional to,

$$\sum_{x,y,z} |\int \Psi^*_{tot} x \psi_{1s} d\tau|^2 = \left[1+\sum (q_j+q_j^*) \cos\theta_{0j} + \sum_j |q_j|^2 | \int \psi_{1s} x \, e^{ikx} \, d\tau|^2\right] \quad , \tag{6.114}$$

where ψ_{1s} is the wave function of the inner 1s state, and

$$|\int \psi_{1s} x \, e^{ikx} \, d\tau|^2$$

corresponds to the transition probability in the case of the isolated atom, as in (6.97).

To calculate q_j, the potential of each neighboring atom is assumed to be of square-well type of depth V and radius a. In the Born approximation,

$$f(\theta) = -\frac{1}{4\pi} \int U(r) \, e^{i\underline{K}\cdot\underline{r}} \, d\tau = -\int \frac{\sin(Kr)}{Kr} U(r) r^2 dr \quad ,$$

where $d\tau$ is the volume element, $U(\underline{r}) = 2mV(r)/\hbar^2$ and $\underline{K}$ is the momentum transfer. In our case (see, for example [6.160a])

$$f(\pi-\theta_{0j}) = -\frac{2m}{\hbar^2} Va^3 \left[\frac{\cos(Ka)}{(Ka)^2} - \frac{\sin(Ka)}{(Ka)^3}\right] \quad ,$$

where $K = 2k \sin[(\pi-\theta_{0j})/2]$. This can be substituted in (6.113) to obtain q_j.

For unpolarized X-rays, we have to average (6.114) over all directions of the vector $\underline{k}_0$, that is, over a solid angle $d\Omega$. Thus, the required result that gives the difference of the transition probability from that of the isolated atom is

$$\chi_{solid} = \frac{1}{4\pi} \int d\Omega \left[\sum (q_j+q_j^*) \cos\theta_{0j} + \sum |q_j|^2\right] \quad . \tag{6.115}$$

The first term on the right-hand side is of the form

$$\int q_j \cos\theta_{0j}\, d\Omega = \frac{2\pi}{r_j} e^{ikr_j} \int_{-1}^{+1} e^{ikr_j \cos\theta_{0j}} \cos\theta_{0j}\, f(\pi-\theta_{0j}) d(\cos\theta_{0j}) \quad . \tag{6.116}$$

For $ka < kr_j$, we can expand it in a series in the powers of $1/kr_j$ by partial integration. Thus,

$$\int_{-1}^{+1} e^{ikr_j\alpha} \alpha F(\alpha) d\alpha = \frac{1}{ikr_j} e^{ikr_j\alpha} \alpha F(\alpha) \Big|_{-1}^{+1} - \dots \quad ,$$

where $\alpha = \cos\theta_{0j}$ and $F(\alpha) = f(\pi-\theta_{0j})$. We now get

$$\int q_j \cos\theta_{0j}\, d\Omega = 2\pi\left(-\frac{2m}{\hbar^2} Va^3\right) \frac{1}{ikr_j^2} \left\{e^{2ikr_j}\left[M(K_0) - \frac{1}{3}\right]\right\} + \dots$$

where $M(K_0) = K_0^{-2} \cos K_0 - K_0^{-3} \sin K_0$, with $K_0 = 2ka$. Retaining only the first term here, we can finally write

$$\frac{1}{4\pi} \int d\Omega (q_j + q_j^*) \cos\theta_{0j} \simeq \left(-\frac{2m}{\hbar^2} Va^3\right) \frac{\sin 2kr_j}{kr_j^2} M(K_0) \quad . \tag{6.117}$$

A similar calculation shows that the last term in (6.115) changes slowly with k and so may be ignored. Therefore, the absorption coefficient fluctuation is essentially given by

$$\tau_{solid}(k) \propto - \sum_j \left(-\frac{2m}{\hbar^2} Va^3\right) \frac{N_j}{r_j^2} \sin 2kr_j \left[\frac{\cos(2ka)}{(2ka)^2} - \frac{\sin(2ka)}{(2ka)^3}\right] \quad , \tag{6.118}$$

where N_j is the number of atoms that are at equal distances r_j from the absorbing atom, and the summation extends over the different r_j (different shells of neighboring atoms). The form of this equation is a sinelike scattering from each shell of atoms at r_j with the EFS signal proportional to the number of atoms N_j and inversely proportional to r_j^2. Because $a \ll r_j$, the term in square brackets is a modulating factor for $\sin 2kr_j$. SAWADA et al. [6.118] showed that (6.118) gives a reasonable fit to the observed absorption curve (Fig.6.24) for Cu, and also for other metals, such as Ni and Fe. Usually, it is enough to sum for the first few neighbor-atom shells.

In the above theory, the contribution from one atom at a distance r_j has been simply multiplied by N_j to obtain the total contribution from the shell of radius r_j. VISHNOI and AGARWAL [6.171] improved this approximation by performing an integration corresponding to a uniform distribution of all of the atoms in a given shell over a spherical surface of radius r_j.

If $U(\underline{r}-\underline{r}_j)$ is the effective potential energy of an electron that arises from the atom located at the lattice point j of the crystal, we can write,

$$f(\pi-\theta_{0j}) = -\frac{1}{4\pi}\int U(\underline{r}-\underline{r}_j)\, e^{i\underline{K}\cdot\underline{r}}\, d\tau$$

$$= -\frac{1}{4\pi}\int U(\underline{r}-\underline{r}_j)\, e^{i\underline{K}\cdot(\underline{r}-\underline{r}_j)}\, e^{iK\cdot\underline{r}_j}\, d\tau$$

$$= -\frac{2m}{\hbar^2} Va^3\, M(Ka)\, e^{i\underline{K}\cdot\underline{r}_j} \quad ,$$

$$q_{j(p)} = \sum q_j \cos\theta_{0j} = -\frac{2m}{\hbar^2} Va^3 \sum \frac{e^{i(\underline{k}_0\cdot\underline{r}_j+kr_j)}}{r_j} M(Ka)\, e^{i\underline{K}\cdot\underline{r}_j} \cos\theta_{0j} \quad ,$$

$$M(Ka) = (Ka)^{-2}\cos(Ka) - (Ka)^{-3}\sin(Ka) \quad .$$

In $q_{j(p)}$, the summation over the pth nearest neighbors (all at a distance r_j) is to be evaluated by taking into account the dependence of the orientation of the scattering event with respect to the crystallographic axes. The almost exact method for evaluating $q_{j(p)}$, by averaging the scattering amplitudes for various directions having the same K, is not always easy to manage.

To a first order of approximation, we can replace the sum in $q_{j(p)}$ by an integral corresponding to a uniform distribution of all the pth nearest neighbors over a spherical surface of radius r_j. Then [6.171]

$$q_{j(p)} \simeq -\frac{2m}{\hbar^2} Va^3 \frac{N_j}{4\pi}\int \frac{e^{i(\underline{k}_0\cdot\underline{r}_j+kr_j)}}{r_j} M(Ka)\cos\theta_{0j}\, e^{iKr_j\cos\alpha}\, 2\pi\sin\alpha\, d\alpha$$

$$= -\frac{2m}{\hbar^2} Va^3 \frac{e^{i(\underline{k}_0\cdot\underline{r}_j+kr_j)}}{r_j} M(Ka)\cos\theta_{0j}\frac{\sin Kr_j}{Kr_j} \quad ,$$

$$\sum_j (q_j+q_j^*)\cos\theta_{0j} = -\frac{2m}{\hbar^2} Va^3 \sum_{r_j} M(Ka)\frac{2\cos(\underline{k}_0\cdot\underline{r}_j+kr_j)}{r_j}(\cos\theta_{0j})N_j\frac{\sin Kr_j}{Kr_j}$$

$$\tau(k) \propto \frac{2m}{\hbar^2} Va^3 \int\sum \frac{N_j}{r_j^2} M(Ka)\sin\left(2kr_j\cos\tfrac{1}{2}\theta_{0j}\right)\cos\left[kr_j(1+\cos\theta_{0j})\right]$$

$$\cdot\left(2\cos\tfrac{1}{2}\theta_{0j}\right)^{-1}(\cos\theta_{0j})d(\cos\theta_{0j}) \quad .$$

The integration over $\cos\theta_{0j}$ is to be done numerically. For copper we can take a = 0.42 Å and for the first seven shells (N_j, r_j) = (12, 2.5509 Å), (6, 3.6080

(24, 4.4198), (12, 5.1017), (24, 5.7006), 8, 6.2491), (48, 6.7470). The results of calculations for the absorption maxima (Fig.6.24) are given below:

	C	D	D'	E	E'	F	G	H	I	J	K
(a)	60	74	85	95	107	118	132	154	167	192	208
(b)	69	78	-	103	-	122	-	149	166	196	208
(c)	45	64	-	96	-	107	135	155	162	192	202
(d)	52	72	-	95	-	115	134	156	167	195	203

Here (a) (VISHNOI and AGARWAL [6.171]), (b) (KRONIG [6.160]) refer to the theoretical values in eV, and (c) (AGARWAL et al. [6.141]), (d) (PARRATT [6.70]) refer to the observed values for copper.

In general, $f(\theta)$ can be determined in terms of the phase shift δ_l,

$$f(\theta) = \frac{1}{k}\sum_l (2l+1)\, e^{i\delta_l} \sin\delta_l \, P_l(\cos\theta).$$

KOZLENKOV [6.172] compared the theories of PETERSEN [6.173], KOSTAREV [6.167], and SHIRAIWA et al. [6.174] and found that the sro theories for solids are more general and acceptable than the lro theories of KRONIG. He also demonstrated that the assumption of a simple square-well potential for an atom, for the purpose of calculating the scattering phase shifts, is reasonably good and that, essentially,

$$\tau(k) \propto -\sum_j \frac{N_j}{r_j^2} \sin\left[2kr_j+2\delta_j(k)\right] \quad , \tag{6.119}$$

where the subscript j denotes the jth shell, containing N_j atoms, at a distance r_j from the absorbing atom. Using phases given by MOTT and MASSEY [6.175], KOZLENKOV calculated curves for Cu, Fe, etc., which were in agreement with experiments. The $1/r_j^2$ factor in (6.119) reproduces the $1/a^2$ dependence of the spacing between maxima and minima of Kronig's formula (6.105). The temperature dependence of $\tau(k)$ has been discussed by SCHMIDT [6.176], and JOPE [6.177].

The approach of SHIRAIWA et al. has been further refined in recent years by STERN [6.178], LYTLE et al. [6.179], and STERN et al. [6.180]. They have shown that EFS can be accounted for by a "particle in a box" model to include scattering from the atoms of several coordination shells. It is a result of partial scattering of the ejected photoelectron by atoms in the neighborhood of the absorbing atom (sro theory). The contribution of the scattering to the final photoelectron wave function (6.111) can be expressed as a sum over the scattering contributions of all atoms. The fluctuations of $\tau(k)$ arise as a

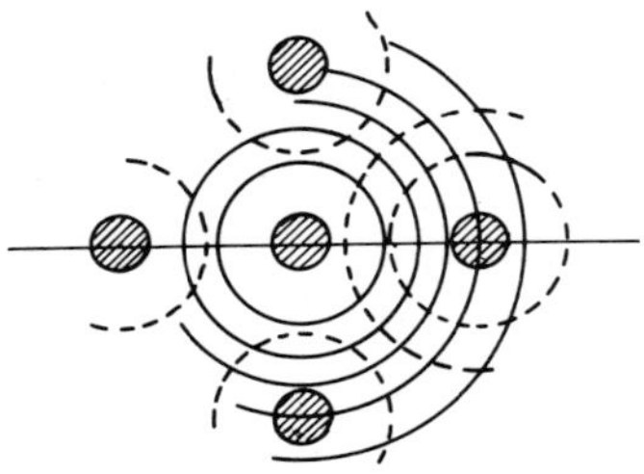

Fig. 6.25. The outgoing original wave fronts from the absorbing central atom are shown as full circles and the backscattered wave fronts from near atoms as dashed circles. The hatched circles are the atoms

result of *interference* between photoexcited outgoing electron waves and those same waves that have been backscattered from nearby neighboring atoms (Fig. 6.25) to near the origin where overlap with the initial state occurs. This means that $\tau(k)$ is to be calculated by putting the scattering contribution (6.111) in the square of the amplitude of the dipole transition matrix between the initial and final states involved in the absorption process and it is enough to obtain ψ_{sc} at the position of the excited atom. The interference is either constructive or destructive, depending on the wave number of the photoelectron, affects the dipole matrix element, and thus the $\tau(k)$. When the waves add constructively, there is an absorption maximum; when they add destructively, there is a minimum. LYTLE et al. [6.179] include the effects of the thermal motion of the lattice, the attenuation of the photoexcited electron as it moves away from the parent atom, and the change of the wave vector that is produced by multiple scattering. According to them, the normalized oscillatory part of the absorption rate is

$$\chi(k) = \frac{m}{4\pi\hbar^2 k}\sum_j \frac{N_j}{r_j^2} S_j(k)\sin\left[2kr_j+2\delta_j(k)\right] e^{-2r_j/\lambda}\, e^{-2k^2\delta_j^2} \quad , \qquad (6.120)$$

where $S_j(k)$ is the back-scattering amplitude from each of the N_j neighboring atoms in the jth shell, λ is the mean free path of the electron, and $\exp\left(-2k^2\delta_j^2\right)$ is a Debye-Waller factor that takes into account thermal vibrations of the positions of the atoms. The factor $\exp(-2r_j/\lambda)$ describes the decay of the photoelectron. We have sinelike scattering from each shell of atoms at r_j; $\delta(k)$ is the phase shift caused by the potential of the absorbing atom which is counted twice because both the outgoing and the backscattered electron waves experience this shift.

In (6.120), k is related to the kinetic energy E of the K-shell photoelectron (energy measured from the X-ray edge) by

$$\hbar^2k^2/2m = h\nu - E_K - E_0(k) = E - E_0(k) \quad , \qquad (6.121)$$

where $E_0(k)$ is the inner potential and locates the point below the Fermi energy where the electron is assumed to be at rest. We can identify $E_0(k)$ with the Fermi energy in metals. Because we are interested in an energy range of 50 to 1000 eV above threshold, k is not very sensitive to the choice of E_0 and we can take $E_0 = 0$.

According to STERN et al. [6.180] $\chi(k)$ is a superposition of terms of the form $\sin(2kr_j)$, where r_j measures the distance to the neighbors around the absorbing atom. The Fourier transform,

$$F(r) = \frac{1}{2\pi} \int e^{-i2kr} \chi(k) dk \quad , \tag{6.122}$$

should peak at the shell distances $r = r_j$ and should provide information on the structure of the solid. It is helpful to recall (3.19) with $\gamma = 0$.

LYTLE et al. [6.179] showed that the Fourier inversion of EFS data is a general and powerful method to determine many physical parameters. It requires a fast-Fourier-transform algorithm on the data, with about 2000 points, and analyzes them in terms of the real and imaginary components of 1000 frequencies. The lowest frequency has a wavelength of twice the range of the data; the remaining frequencies are the next 999 harmonics of this fundamental. Instead of following this method, it is also possible to determine the nearest-neighbor distances and phase shifts by use of a simple graphical technique.

The basic idea is that, usually, scattering from the first shell dominates the EFS curve; if the prominent peaks are used, the period is that of the first coordination shell.

The phase shift can be calculated only if we know the exact potential of the excited atom with a hole in the K shell. Available tables of atomic potentials [6.51] do not list such potentials. Also, the outer part of the potential will be modified compared to the atomic Hartree-Fock potential because metal electrons can shield the long tails. Therefore, we have to use $\delta(k)$ as a parameter.

The $\sin(2kr_j + 2\delta_j)$ factor arises from the phase change felt by the wave function on its outward and return trips [6.181], plus the phase shift $\delta_j(k)$ caused by the central atom. LYTLE et al. assume that δ_j is linear in k,

$$\delta_j(k) = -\alpha_j k + \beta_j \quad , \tag{6.123}$$

where α_j and β_j are constants. Substituting this into the argument of the sine, we can define an n_j by [6.179],

$$2kr_j - 2\alpha_j k + 2\beta_j = \frac{1}{2} n_j \pi \quad , \qquad (6.124)$$

where $n_j = 0,2,4,\dots$ for maxima and $n_j = 1,3,5,\dots$ for minima. For the first coordination shell

$$2k(r_1 - \alpha_1) + 2\beta_1 = \frac{1}{2} n_1 \pi \quad . \qquad (6.125)$$

Neglecting E_0, we have from (6.121), $k = (0.263\ E)^{1/2}$, where E is the energy of the peak in the EFS. A straight-line plot of n_1 against k (in $Å^{-1}$) for the observed maxima and minima determines $r_1 - \alpha_1$ from the slope.

Once α_1 is evaluated for a standard (known r_1), r_1 for the unknown compounds can be determined. For example, for Fe metal $\alpha_1 = 0.40$ Å, and for Fe in ferrocene $\alpha_1 = 0.44$ Å. LYTLE et al. associate this change of the value of α with a change of chemical bonding. They found that this graphical method is adequate for finding r_1, although the full Fourier analysis is richer in information.

6.7.5 Other Theories

Levy's method: If the argument of the sine scattering term is essentially taken as $2kr_1$ for the first coordination sphere and the first maximum at about 50 eV from the X-ray edge is identified with a kind of ionization or escape energy for the 1s electron, then the Bragg relation [6.182],

$$r_1(Å) = (151/\Delta E)^{1/2} \quad , \qquad (6.126)$$

where ΔE is the distance between the ionization maximum and the next minimum in eV, gives approximately the average bond length. The results obtained by use of this formula are in close agreement with those obtained from crystallographic data or with the approximate formula of Schomaker-Stevenson,

$$r_{AB} = r_A + r_B - 0.09|r_A - r_B| \quad , \qquad (6.127)$$

where r_A is the covalent radius of the atom A (for example, [6.112,124,183])

Single-potential model of Lytle: A pragmatic, and easily applied theory of the EFS was developed by LYTLE [6.184]. A nearly spherical atomic polyhedron (Wigner-Seitz unit cell) is constructed in the lattice and approximated by a sphere of equivalent volume with radius r_s. The energy states available to t ejected photoelectron are calculated by solving the Schrödinger equation,

$$\nabla^2\psi(r) + \frac{2m}{\hbar^2}[E-V(r)]\psi(r) = 0 \quad ,$$

where the potential V(r) seen by the electron is assumed to be the spherically symmetric potential well,

$$V(r) = \infty \quad , \quad r = r_s \quad ,$$
$$= 0 \quad , \quad r < r_s \quad . \tag{6.128}$$

Because of the spherical symmetry, we can write $\psi = R(r)Y(\theta,\phi)$. The radial equation for R is

$$\frac{1}{r^2}\frac{d}{dr}\left(r^2\frac{dR}{dr}\right) + \frac{2m}{\hbar^2}\left[E-V-\frac{l(l+1)}{r^2}\right]R = 0 \quad .$$

With $\beta^2 = (2m/\hbar^2)(E-V)$ and $\rho = \beta r$, we get for $r < r_s$ (V=0), the Bessel equation form

$$\frac{d^2R}{d\rho^2} + \frac{2}{\rho}\frac{dR}{d\rho} + \left[1-\frac{l(l+1)}{\rho^2}\right]R = 0 \quad .$$

For $r < r_s$, the electron moves freely, so that the states of motion with a well-defined value of the orbital angular momentum are given by the wave function,

$$\psi_{klm} = Aj_l(kr)Y_{l,m}(\theta,\phi) \quad , \tag{6.129}$$

where j_l are the spherical Bessel functions, $Y_{l,m}$ are the spherical harmonics, and k determines the energy of the electron,

$$E = \hbar^2k^2/2m \quad . \tag{6.130}$$

Because the electron cannot penetrate the infinite potential at the cell boundary, we must have

$$j_l(kr_s) = 0 \quad , \quad \text{or} \quad k = (1/r_s)X_{nl} \quad , \tag{6.131}$$

where X_{nl} are the zero roots of the lth-order spherical Bessel function with $n = 1,2,\ldots$ as the principal quantum number. The corresponding energies of the stationary states are,

Table 6.4. Energy levels Q_{nl} for a particle in a spherical well in units of $h^2/8mr_s^2$

n	l = 0	l = 1	l = 2	n	l = 0	l = 1	l = 2
1	(1s) 1.0	(2p) 2.04	(3d) 3.37	9	(9s) 81.0	(10p) 90.0	(11d) 9
2	(2s) 4.0	(3p) 6.04	(4d) 8.40	10	(10s)100	(11p)110	(12d)12
3	(3s) 9.0	(4p)12.0	(5d)15.4	11	(11s)121	(12p)132	(13d)14
4	(4s)16.0	(5p)20.0	(6d)24.4	12	(12s)144	(13p)156	(14d)16
5	(5s)25.0	(6p)30.0	(7d)35.5	13	(13s)169	(14p)182	(15d)19
6	(6s)36.0	(7p)42.0	(8d)48.6	14	(14s)196	(15p)210	(16d)22
7	(7s)49.0	(8p)56.0	(9d)63.8	15	(15s)225	(16p)240	(17d)25
8	(8s)64.0	(9p)72.0	(10d)80.5	16	(16s)256	(17p)272	(18d)28

Table 6.5. Various atomic polyhedra. The dimensions are in units of the orthogonal unit cells for each lattice with unit cell dimension a, which contain 8, 2, and 4 atomic volumes for the diamond-cubic, bcc and fcc lattices, respectively

	Diamond-cubic	bcc	fcc
Nearest-neighbor distance, r_1	0.4330 a	0.866 a	0.7071 a
Volume of polyhedron, atoms	2	4	4
r_s Radius of equivalent sphere	0.3908 a	0.7814 a	0.6203 a
a =	2.559 r_s	1.280 r_s	1.612 r_s
r_1 =	1.108 r_s	1.108 r_s	1.140 r_s

$$E = \frac{\hbar^2}{2mr_s^2} X_{nl}^2 = \frac{h^2}{8mr_s^2} Q_{nl} \quad , \tag{6.132}$$

with $Q = (X_{nl}/\pi)^2$. Few values of Q_{nl} are given in Table 6.4.

LYTLE argued that if the electron wave front has a node in the neighborhood of the surrounding atoms, the electron wave satisfies the conditions fo resonance in a way analogous to resonance in any cavity (6.131). This condition will increase the amount of energy absorbed by the system (an allowed transition). Therefore, we can identify E with the energies of absorption maxima in the EFS. For example, the K absorption would be to levels of p symmetry, and the $L_{II,III}$ transitions would be to s and d levels. If the observ E values are plotted against the appropriate Q values, a straight line of slope M is obtained, such that

$$r_s(\text{Å}) = (37.60/M)^{\frac{1}{2}} \quad . \tag{6.133}$$

This value of r_s can be related to the nearest-neighbor distance r_1 with the help of Table 6.5 [6.185]. The quantitative agreement with experimental data is good [6.184].

Double-potential model: It is found that the linear E vs. Q plot passes through the origin as expected (6.132) in metals, but not in compounds (for example, [6.187,112]). We know, from (6.16), that in a compound an electron in the outer region experiences a force Z_{eff}/r^2. The effect of this force on the ejected photoelectron in a compound can be included by modifying the Lytle potential, (6.128), as shown in Fig.6.26a [6.186].

The potential well of finite depth V_c and range a_c at the center takes care of the attractive force because of the screened Coulomb potential experienced by the escaping electron while it is near the one of the two nuclei in a bond. Outside of this potential well, the electron moves, as before, in the Lytle potential well.

From $r = a_c$ to $r = r_s$ we have sine or cosine wave functions, with E given by (6.132). From $r = 0$ to $r = a_c$, we again have sine or cosine functions, but with energy eigenvalues determined by equations of the form [6.188],

$$\left.\begin{array}{l} l = 0 \quad \xi \cot\xi = -\eta \\ l = 1 \quad \xi^{-1}\cot\xi - \xi^{-2} = \eta^{-1} + \eta^{-2} \end{array}\right\} \xi^2 + \eta^2 = \frac{2mV_c a_c^2}{\hbar^2} \quad , \tag{6.134}$$

where

$$\xi = \left[\frac{2m}{\hbar^2}\left(V_c - |E|\right)\right]^{\frac{1}{2}} a_c \quad , \quad \eta = \left[\frac{2m}{\hbar^2}|E|\right]^{\frac{1}{2}} a_c \quad . \tag{6.135}$$

At $r = \pm a_c$, the two sets of solutions must join so that ψ and $d\psi/dr$ are continuous (Fig.6.26b). These conditions, together with normalization, specify each ψ_{nlm} and the associated E_n. The low amplitude of ψ_{nlm} inside the central well indicates that the electron is unlikely to be found there.

Equation (6.134) can be solved numerically or graphically. However, in our simple model, it would be enough to consider only the first bound state. This

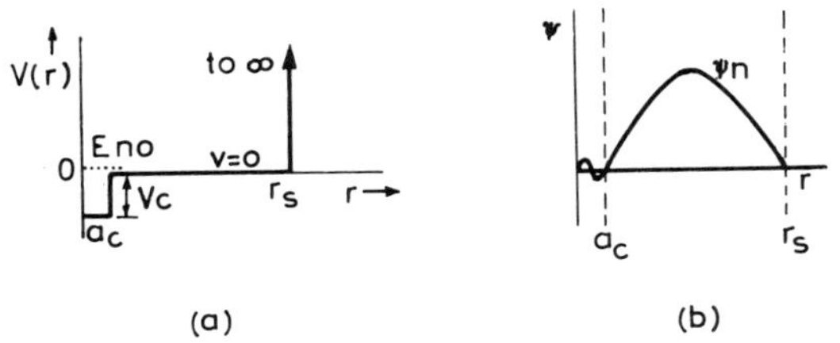

Fig. 6.26. (a) The double potential, and (b) the wave function [6.186]

occurs when $\xi = \pi/2$, $\eta = 0$, or for an electron in an enclosure of size comparable to covalent radius ~1 Å,

$$V_c = \frac{\pi^2\hbar^2}{8ma_c^2} \simeq \pi^2 \text{ eV} \quad , \quad (m \simeq 0.5 \text{ MeV}, a_c \simeq 1 \text{ Å}) \quad . \tag{6.136}$$

This value is comparable with the observed intercepts [6.112,183,189] on the E axis of the Lytle plot for compounds. Thus, for compounds, Lytle's result (6.132) can be written in the modified form as

$$E = M\, Q_{nl} + V_c \quad , \tag{6.137}$$

where $M = h^2/(8mr_s^2)$ and V_c is the effective Coulomb potential (Fig.6.26) in which the escaping electron moves in a compounds for $r \leqq a_c$. It is found that V_c is a measure of Z_{eff} or electronegativity [6.186], $V_c \propto (x_B - x_A)^2$. It appears that we have neither well-settled data on EFS nor a fully acceptable theory. Much work remains to be done in this area for full evaluation of (6.83).

6.8 Isochromats

A spectrometer is set to reflect a certain definite wavelength λ_0 into the detector and the voltage is varied from a value below the threshold V_0 necessary to produce λ_0, $eV_0 = h\nu_0 = hc/\lambda_0$, to a value above V_0. For the voltage V_0, the wavelength λ_0 is the Duane-Hunt limit of bremsstrahlung. The curve of intensity as a function of the voltage in such an experiment is called an *isochromat*, because it represents the intensity variation at a *single* wavelength λ_0. The intensity suddenly rises at $V = V_0$. For $V > V_0$, DuMOND and BOLLMAN [6.190] observed distinct structures within about 15 volts above the threshold V_0.

The spectrometer can also be set to record (say) the $K\alpha_1$ line of the target element. As the voltage on the X-ray tube is slowly raised from just below to just above the excitation voltage V_K for the $K\alpha_1$ line, the observed intensity of the $K\alpha_1$ line begins to rise and then shows fine structure, as shown in Fig.6.27 [6.191,192]. JOHANSSON [6.193] found a similar fine structure near the Duane-Hunt limit of the continuous spectrum. SPIJKERMAN and BEARDEN [6.194] used this method to determine the value of h/e as $(1.37949 \pm 30 \text{ ppm}) \times 10^{-17}$ erg-s/esu.

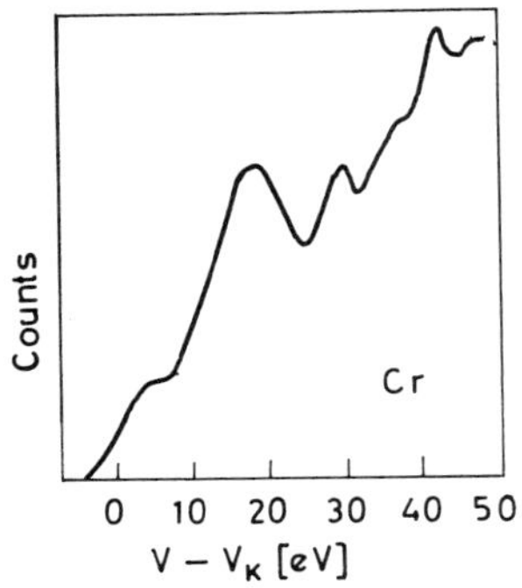

Fig. 6.27. A typical isochromat [6.191]

NIJBOER [6.195], ULMER [6.196], and BÖHM and ULMER [6.197] explained the fine structure near the initial rise in the isochromat by considering the solid-state effects, such as the allowed and forbidden energy bands for electrons moving in solids, and the density of states. This is similar to the theory of the EFS in X-ray absorption.

BERGWOLL and TYAGI [6.198] studied the isochromats of several transition metals. They attributed the first maximum in the initial rise ("foot" structure) to transitions to the empty d states and suggested that the subsequent structures are the "echoes" of the first two maxima, a phenomenon that is consistent with plasmon theory.

Experimental methods, like de Haas-van Alphen effect, cyclotron resonance, magnetic-acoustic effect, and electronic specific heat, give information only about the Fermi surfaces and electronic states in their vicinity. On the other hand, the X-ray emission and absorption spectra can explore the entire band structure. We know (p.163) that for inner-level widths $\Delta E_K > \Delta E_L > \Delta E_M$, etc. Also, energy measurements are more accurate in the long wavelength region, (6.36). These two considerations suggest that favourable conditions for the study of valence and conduction bands occur when shells near the valence band participate in the transitions. This takes us to the soft X-ray emission and absorption spectroscopy.

7. Soft X-Ray Spectroscopy

Soft X-ray spectroscopy has developed into a major area of research for the study of solids [7.1-6]. We shall give here only a brief account of this topic.

Air shows high absorption in the wavelength region 2 to 2000 Å. It becomes necessary to evacuate the spectrograph for study of such radiation. Therefore, it is called the *vacuum ultraviolet region*. SCHUMANN [7.7] built the first vacuum spectrograph, with fluorite prism as analyzer. The region from 1000 Å to 2000 Å is known as the *Schumann region*. The extreme ultraviolet region extends from 2 Å to 1000 Å, and the part from 20 Å to 500 Å is usually called the *soft X-ray region*.

We have seen that the levels of a valence electron in a solid form a continuum (valence band) that can be specified by a density function N(E)dE. The main aim of the soft X-ray spectroscopy (SXS) is to investigate this function experimentally.

The presence of the valence and conduction bands in solids drives us into the region of SXS to enable us to obtain the required information. The reason is that it would be impossible to try to disentangle the structure of the N(E) curve from the radiation emitted or absorbed in the transitions between pairs of levels both of which are members of the continuum. On the other hand, if one of the participating levels is a discrete level, the structure of the continuum is rather directly reflected in the emitted (or absorbed) radiation. The discrete (sharp) levels correspond to electrons bound more or less completely to individual atoms. The actual width of the discrete level used sets a limit to the resolution with which the continuum structure of allowed symmetry can be probed ($\Delta E_K > \Delta E_L > \Delta E_M$ etc. for core levels).

The resolution of the spectrometer, $E/|dE| = [n\lambda/(4d^2-n^2\lambda^2)^{\frac{1}{2}}](1/d\theta)$, increases with λ; (6.36) $dE \propto 1/\lambda$. For ^{8}O K emission ($\sim$525 eV, $\sim$24 Å), dE/E is 2/525 or 0.4% because two maxima only 2 eV apart are easily separated by the spectrometer. The longer wavelength L emission spectrum from ^{16}S would give far more detail than the ^{16}S K spectrum. Thus, for both theoretical and technical reasons, this resolution is maximum for the soft X-ray region ($\sim$20 to $\sim$500 Å). In this region, the inner X-ray levels are often sharp to within

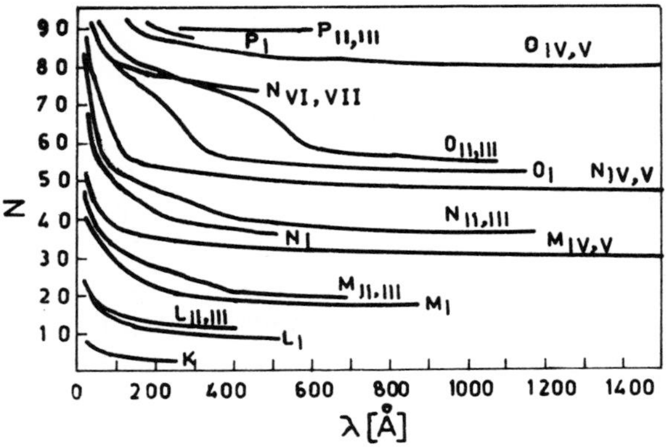

Fig. 7.1. Variation of the binding energy of atomic levels with Z

0.1 eV; this is enough for resolution of all the main features in the N(E) curve. We are, therefore, led to SXS for the study of bands in solids. As Z increases, the soft X-ray wavelength region shifts from K spectra to L, M, N... spectra (Fig.7.1).

7.1 Conventional Sources

For *emission* work, a glass soft X-ray tube is preferred because it can be outgassed more thoroughly. Cleanliness of target, filament and other accessories is necessary. A general diagram of the X-ray tube used by SKINNER is shown in Fig.7.2. It was run at 3500 volts and 250 mA. This voltage is more than enough to excite the soft X-ray wavelengths. For long exposures (many hours), the purity of the target material was ensured by use of a rotatable target; fresh layers of the material under study were periodically deposited

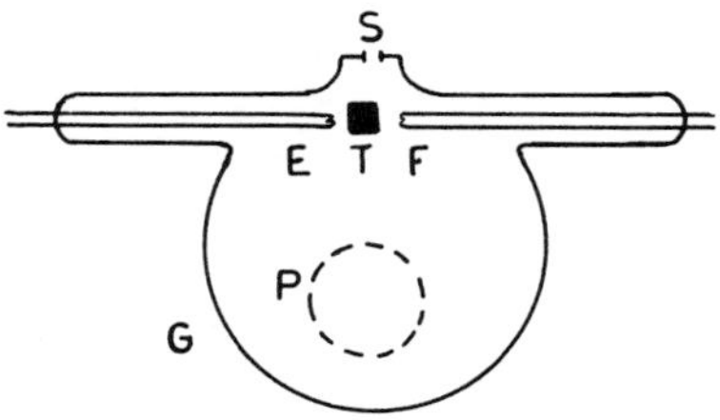

Fig. 7.2. Soft X-ray tube. G: glass bulb, P: pump tube, E: evaporating filament, T: rotatable target, f: X-ray filament, S: spectrograph slit

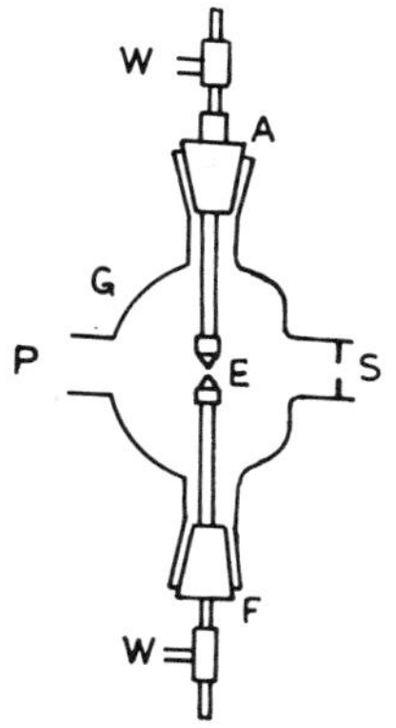

Fig. 7.3. Vacuum spark source. A: adjustable holder, E: electrodes, S: slit to spectrometer, G: glass bulb, F: fixed holder, P: pipe to pump, W: water end for cooling

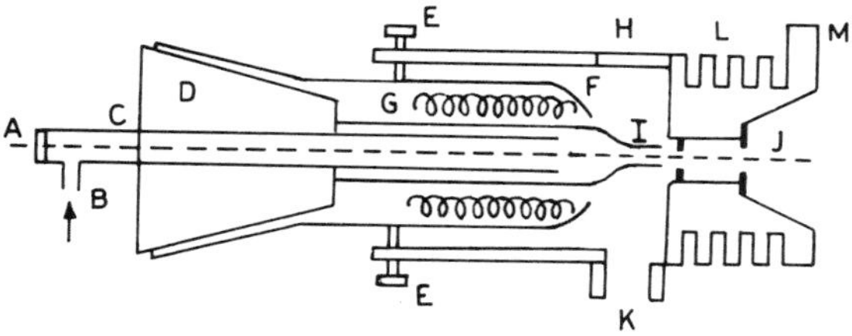

Fig. 7.4. Lyman discharge tube. A: glass window, B: gas inlet, C: brass cathode, D: pyrex taper joint, E: centering screws, F: glass tube, G: glass wool, H: glass window, I: quartz capillary, J: brass baffle plates, K: to pump, L: cooling fins, M: Al anode, below E: glass to metal seal

on it. High vacuum was maintained and a slit was used in the place of a window, because soft X-rays are highly absorbable.

For *absorption* work, the continuous radiation emitted by the Skinner tube at $V < 5$ kV is very weak and easily absorbed, even by the substrate of the absorbing material. Therefore, it becomes necessary to use line spectra emitted by highly ionized atoms. Such spectra are easily produced in condensed-spark discharges between metallic electrodes of Ag, Al, Cu, Fe and Mo (Fig.7.3). They give fairly dense line spectra in the region of interest. Another useful source is a Lyman discharge tube (Fig.7.4). In this source, a condensed spark is sent through a quartz capillary in which helium or argon is maintained at a pressure of about 1 mm Hg. The capillary is directed toward the slit of the spectrograph. It gives very strong lines in the soft X-ray region, for the absorption work. The source also acts as a filter. If the gas pressure is ~ 1 mm Hg, it will absorb radiation of wavelengths shorter than its absorption edge (He: 505 Å, Ar: 786 Å). This reduces the problem of overlapping orders

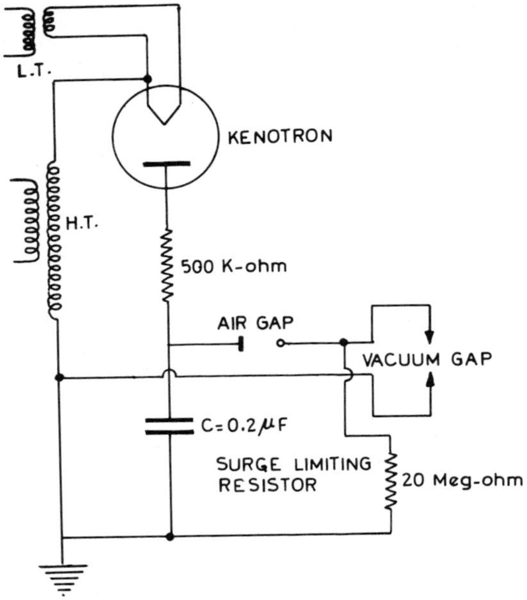

Fig. 7.5. Circuit diagram for condensed-spark source

and simplifies the analysis at longer wavelengths. If the gas pressure is 0.2 mm Hg, the shorter wavelengths are not strongly absorbed.

The periodic spark is obtained by discharging a 0.2 μF condenser that is charged to about 30 kV by transformers and rectifiers (Fig.7.5). Other laboratory line sources are also known. For example, radiofrequency oscillators can excite electrodeless ring discharges [7.8] and linear discharges [7.9] to produce soft X-rays. Use of a microwave source and cavity is also possible for this purpose [7.10-12].

7.2 Synchrotron as a Source

TOMBOULIAN and HARTMAN [7.13] showed that the magnetic bremsstrahlung (orbital radiation from a synchrotron) can be used as a strong continuous source for soft X-ray absorption spectroscopy. In recent years, much literature has grown on the extreme-ultraviolet spectroscopy of gases and solids by use of synchrotron radiation [7.6,14-17].

Qualitative radiation patterns from electrons in a circular orbit are shown in Fig.7.6. An electron moving through a perpendicular magnetic field

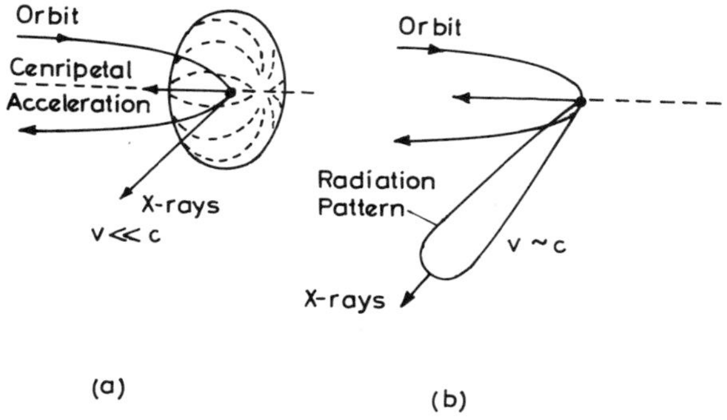

Fig. 7.6. Radiation pattern from electrons in a circular orbit (a) at low energy, and (b) at high energy

has an acceleration toward the orbit center. In a frame of reference moving with the electron, a Larmor radiation pattern is emitted (Sect.1.11). This pattern resembles a torus centered on the electron with the plane of maximum intensity tangent to the electron orbit (Fig.7.6a). For highly relativistic electrons, as viewed in the laboratory frame, the radiation is confined to a narrow cone tangent to the orbit (Fig.7.6b). The radiation is highly polarized with the electric vector $\underline{E}$ in the plane of the orbit [7.18].

The frequency of the radiation changes along with the pattern. At $v \ll c$ it is equal to the rotation frequency; at $v \sim c$ the radiant energy is distributed among many higher harmonics of the rotation frequency and the spectrum is essentially a continuum. Several synchrotrons are in use for SXS. The German Electron Synchrotron (DESY) produces an intense continuous spectrum in the hard X-ray region ($\sim$1 Å), whereas Tantalus I (University of Wisconsin) radiates most copiously in the soft X-ray region ($\sim$100 Å).

7.3 Vacuum Spectrograph

From (3.52,73), we can express the critical angle of reflection for radiation of wavelength λ as

$$\sin\theta_c = \left(\frac{e^2 n}{\pi m c^2}\right)^{\frac{1}{2}} \lambda \quad , \tag{7.1}$$

where θ_c is the grazing angle of incidence, and n is the number of electrons per unit volume of the medium. Thus, for a given λ, total reflection occurs at all grazing angles less than and up to θ_c, at which point the reflectance decreases rapidly. We can also say that, for a given grazing angle θ, all wavelengths longer than

$$\lambda_{min}[\text{Å}] = (3.33\times10^{14})n^{-1/2}\sin\theta \quad , \tag{7.2}$$

are totally reflected. For a glass (or Al) grating $n^{1/2} = 8.8\times10^{11}$, and so $\lambda_{min}[\text{Å}] = 379\sin\theta$, or $\lambda_{min}[\text{Å}] = 6.6\ \theta^{\circ}$ for small θ. For $\theta = 5^{\circ}$ this gives $\lambda_{min} = 39.6$ Å, as the order of magnitude to be expected in the experiment. For Pt, $n^{1/2} = 22.7\times10^{11}$ and $\lambda_{min}[\text{Å}] = 147\sin\theta$.

The decrease of reflectance with decreasing wavelength, of all grating materials (glass; Al, Au, Pt coatings) necessitates the use of grazing-incidence ($\theta\sim5^{\circ}$) spectrographs for SXS ($\sim$300 Å). If a concave grating is used, it serves as a focusing as well as a dispersing element.

Figure 7.7 is a simplified diagram of the *concave-grating vacuum spectrograph*. The simple grating equation is

$$d(\sin\phi - \sin\phi_n') = n\lambda \quad , \quad (\phi = 90^{\circ} - \theta) \quad , \tag{7.3}$$

where n is the spectral order, and $\phi(\phi_n')$ is the angle of incidence (diffraction). The zero order or the direct image is formed at $\phi = \phi_n'$. If ρ is the diameter of the Rowland circle, the angle $\phi - \phi_n'$ is simply related to the distance x along the plate from the zero order to the spectral line as

$$\phi_n' = \phi - (x/\rho) \quad . \tag{7.4}$$

The radius of the grating is ρ. We can use (7.3) with (7.4) for the numerical calculation of λ from the measured values of x.

A convenient choice for the region ~300 Å is: $\rho = 2$ m, $d = 8.47\times10^{-5}$ cm (the reciprocal of 30,000 lines per inch), $\phi = 85^{\circ}$ (grazing angle $\theta = 5^{\circ}$), n = The Al-coated grating becomes more transparent (less reflecting) as λ decreases below ~100 Å. The efficiency of concave gratings for SXS has been discussed by SPRAGUE et al. [7.19], and MORSE and WEISSLER [7.20].

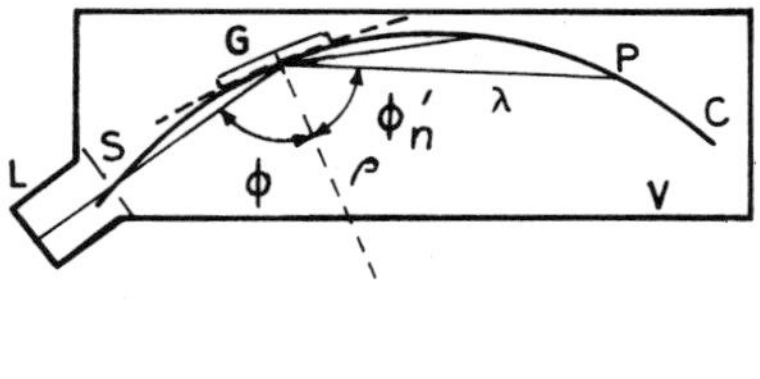

Fig. 7.7. Concave grating grazing incidence vacuum spectrograph. The slit S, the grating G and the plate P, all lie on the Rowland circle C of radius $\rho/2$. L is the source and V the vacuum chamber

For $\lambda > 500$ Å, it is possible to use the normal-incidence vacuum spectrograph with ϕ close to zero. Many other mountings are also used [7.3]. Adjustments of grating spectrographs have been discussed by HARRISON et al. [7.21]. MacADAM [7.22], and SAWYER [7.23] have discussed the optics of small displacements from the Rowland circle.

Synthetic crystals (organic esters such as octadecyl hydrogen maleate, $2d \sim 64$ Å; dioctadecyl adipate, $2d \sim 94$ Å; octadecyl hydrogen succinate, $2d \sim 97$ Å) with large interplanar spacings are available, which are useful for SXS in the region $25 < \lambda < 100$ Å. The synthetic crystal of potassium acid phthalate ($2d \sim 26$ Å) is convenient for $\lambda \sim 25$ Å [7.24].

7.4 Detectors

In the soft X-ray region, the energy of the photons (25 to 250 eV) is so low that they are easily absorbed by the protective gelatin layer of ordinary photographic emulsions. Special emulsions, on thin backing plates that may be bent along the Rowland circle, are therefore needed. In early days, Schumann plates were used. Eastman Short-Wave-Range (SWR) plates and Ilford Q-emulsion plates are found to be very suitable. In these emulsions, the density of silver bromide grains is high and the gelatin layer is very thin (less than 1000 Å).

In the range 100-800 Å, PIORE et al. [7.25] used a Be-Cu *photoelectron multiplier* as a detector that moves along the Rowland circle. It is 1000 fold more sensitive than the photographic plate. ROGERS and CHALKIN [7.26] used a *Geiger counter*, with 500 Å thick collodion window, in the range of about 20 to 200 Å. EDERER and TOMBOULIAN [7.27] used a Geiger counter between 100 and 300 Å to measure absolute intensities. A *silicon photodiode* can also be used [7.28] in the range 580 to ~1600 Å.

7.5 Emission Spectra

A large body of data exists on the soft X-ray emission and absorption spectra. Soft X-ray emission bands are produced when an electron is ejected from an inner shell of an atom and the resulting hole is filled by a transition from the band of valence electrons. In view of (2.121), the intensity is given by

$$I(E) = \nu^3 P(E) N(E) \quad , \quad P(E) \propto \sum_i \left| \int \psi_f^* x_i \psi_k \, d\tau \right|^2 \quad , \tag{7.5}$$

for $E \leq E_F$ and $I(E) = 0$ for $E > E_F$. In terms of a one-electron model the emission spectra give direct information about the occupied valence band $N_{oc}(E)$.

From (6.51), we can write for the free-electron model,

$$E_F = \frac{\hbar^2}{2m} \left(\frac{3\pi^2 N}{V_a} \right)^{2/3} \quad . \tag{7.6}$$

The corresponding density of orbitals at the Fermi energy is

$$N(E_F) = \frac{dN}{dE_F} = \frac{V_a}{2\pi^2} \left(\frac{2m}{\hbar^2} \right)^{3/2} E_F^{\frac{1}{2}} = \frac{3N}{2E_F} \quad . \tag{7.7}$$

Thus, the number of orbitals per unit energy range at the Fermi energy is determined by the total number of conduction electrons divided by the Fermi energy.

We can use (7.7) for the number of orbitals per unit energy range, $N(E)$, or the density of states, for $E < E_F$,

$$N(E) = \frac{dN}{dE} = \frac{V_a}{2\pi^2} \left(\frac{2m}{\hbar^2} \right)^{3/2} E^{\frac{1}{2}} \quad . \tag{7.8}$$

The density of single-particle states $f(E,T)N(E)$ is plotted in Fig.7.8 as a function of E, where $f(E,T)$ is given by (6.1). The average energy is increased when the temperature is increased from 0 to T, because electrons are thermally excited from region 1 to region 2. If spin is included, the electron density, that is the number of electrons per unit volume per unit energy, becomes

$$N_e(E) = 2N(E)f(E,T) \quad . \tag{7.9}$$

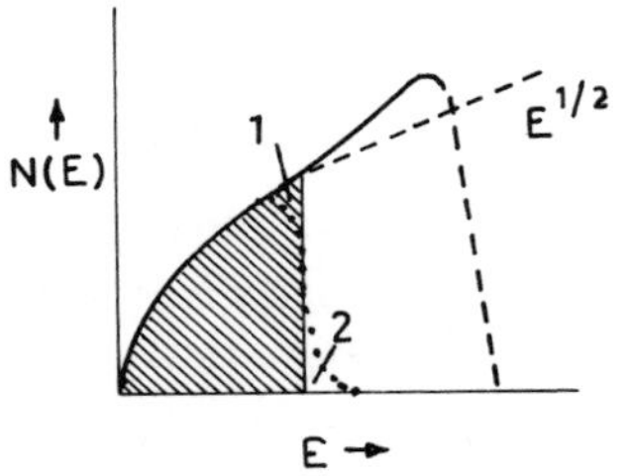

Fig. 7.8. N(E) vs E curve for a free-electron gas in three dimensions. The dotted curve represents the density $f(E,T)N(E)$ of occupied orbitals at a finite temperature T. The hatched area represents the occupied orbitals at $T = 0$. N(E) drops to zero at the Brillouin-zone boundary (dashed curve)

We can find a general expression for N(E). From (6.47), the number of allowed values of k for which the energy is between E and E + dE, $E = \hbar^2k^2/2m$, is

$$N(E)dE = \frac{V_a}{(2\pi)^3}\int_{\text{shell}} d^3k \quad , \tag{7.10}$$

where the integration is over the volume of the shell in $\underline{k}$ space lying between two surfaces, one of energy E and the other of energy E + dE. Consider a volume element $dS_E dk_\perp$, where dS_E is the cross section area and $dk_\perp$ is the perpendicular distance between these two surfaces, so that

$$\int_{\text{shell}} d^3k = \int dS_E dk_\perp \quad .$$

The gradient of E, $\nabla_{\underline{k}}E$, is also normal to the surface on which E is constant; also, obviously the quantity $|\nabla_{\underline{k}}E|dk_\perp = dE$. Thus, $dS_E dk_\perp = dS_E(dE/\nabla_{\underline{k}}E)$, and we have

$$N(E) = \frac{V_a}{(2\pi)^3}\int \frac{dS_E}{|\nabla_{\underline{k}}E|} \quad , \tag{7.11}$$

where the integration is on the surface S_E of constant energy E. The *van Hove singularities* are associated with points in the $\underline{k}$ space for which $\nabla_{\underline{k}}E$ vanishes.

When the free motion of the electron in the valence band is assumed to be modulated by the periodic potential provided by the lattice, energy discontinuities occur for certain values of k (see, for example, [7.29]). For the simple case of a cubic lattice, they occur at

$$\pi(\alpha^2+\beta^2+\gamma^2)d^{-1} + \alpha k_x + \beta k_y + \gamma k_z = 0 \quad . \tag{7.12}$$

Because $E = \hbar^2k^2/2m$, these discontinuities occur precisely where Kronig's formula for the Bragg reflection by (α,β,γ) planes, (6.105), holds. At these discontinuities, N(E) drops to zero (Fig.7.8) and ceases to be parabolic. This defines the boundary of the first Brillouin zone. Half of this zone would be filled for a monovalent metal, because $N_e = 2N(E)$ at absolute zero (Fig.7.8). Thus, the electrons in the first zone are those of lower kinetic energy and their de Broglie wavelengths are greater than the values required for the Bragg diffraction at the lattice planes. In an actual crystal, the evaluation of N(E) is quite complicated.

The zone boundaries are determined by (6.105), or by $n\lambda_B = 2\pi n/k = 2d\sin 90^\circ$, that is

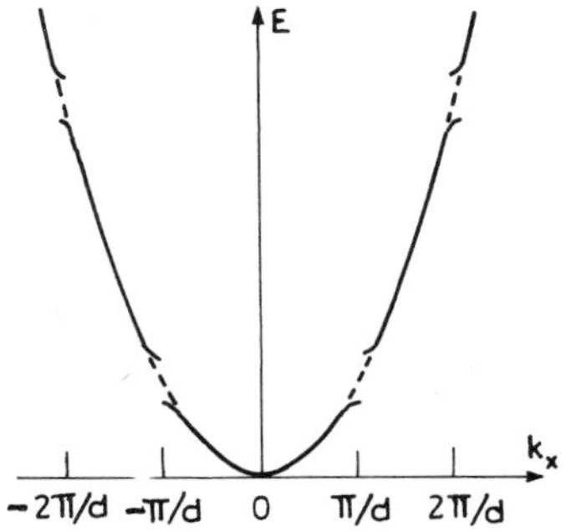

Fig. 7.9. E vs k plot for almost-free electrons in a simple lattice. The discontinuiti occur at values of k for which Bragg's law is satisfied

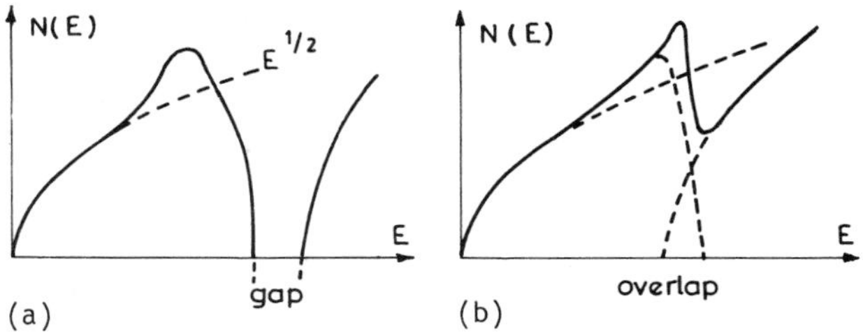

Fig. 7.10. N(E) curves for (a) nonoverlapping bands, and (b) overlapping ban

$$k = (n\pi/d) \quad , \quad n = 1,2,\ldots \quad . \tag{7.13}$$

In the simplest case, the energy discontinuities occur for these values of k (Fig.7.9). Qualitatively, we can say that, as the zone boundaries are approached, N(E) increases more rapidly than the free-electron $E^{\frac{1}{2}}$ behavior (as k approaches π/d, a small increase of E leads to a large increase of the num ber of k states, Fig.7.9), leading to a cusp (Fig.7.10a). N(E) then drops to zero if there is no energy overlap between the first and second zones. The level density is zero over the region of forbidden energies, beyond which it begins to rise as energies that correspond to the second zone are reached (Fig.7.10a). In case of overlap (like s-p hybridization), the levels in the second zone begin to contribute before the end of the first zone is reached (Fig.7.10b). A theoretical N(E) curve for overlapping bands was drawn in Fig.6.18 for ^{29}Cu metal.

From (7.5) we see that I(E) depends both on P(E) and N(E). In general,

$$I(E) \propto \nu^3 \int \frac{|M_{f,i}|^2}{\nabla_{\underline{k}}(E)} \, dS_E \quad .$$

If the matrix element $|M_{f,i}|$ were independent of $\underline{k}$,

$$I(E) \propto \nu^3 |M_{f,i}|^2 N(E) \quad , \quad \text{for } E \leq E_F \quad .$$

For K spectra, the final state is a 1s atomic level, and for $L_{II,III}$ spectra it is a 2p level. The ψ_k may be represented by $\exp(i\underline{k}\cdot\underline{r})$. The matrix element can then be evaluated for small k ($E\sim 0$). It turns out[1] that for K spectra $P(E)\sim E$, and for L spectra $P(E)\sim$constant. Therefore, with $N(E)\propto E^{\frac{1}{2}}$ for small E, we have for K bands

$$I(E) \propto \nu^3 |M_{K,k}|^2 N_p(E) \propto \nu^3 E^{3/2} \quad , \tag{7.14}$$

where $N_p(E)$ is the partial density of state of p symmetry, and for L bands

$$I(E) \propto \nu^3 |M_{L,k}|^2 N_{s,d}(E) \propto \nu^3 E^{\frac{1}{2}} \quad , \tag{7.15}$$

where $|M_{f,i}|$ is the matrix element between the initial and final states.

To obtain the intensity curve, an emission band is photographed twice, using two different exposure times. A calibration curve of the emulsion is thus obtained. By its use and the known instrumental dispersion, the spectrogram can be reduced so as to yield relative intensities within the band, as a function of E. For the Li K band, the result is shown in Fig.7.11a; the result for the Na $L_{II,III}$ band is shown in Fig.7.11b.

Let us now compare the simple theory, based on the one-electron approximation, with experiments [7.1,30-32]. The electronic configuration of Li atom

[1]The spontaneous-emission matrix element has the form

$$\langle\psi_{core}|\underline{p}\cdot\underline{A}|\psi_k\rangle = -i\hbar\,\underline{A}\cdot\left(\langle\psi_{core}|\nabla|\underline{k}\rangle - \sum_\alpha \langle\psi_{core}|\nabla|\alpha\rangle\langle\alpha|\underline{k}\rangle\right) \quad ,$$

because there are two terms in the single orthogonalized plane wave, the plane wave $|\underline{k}\rangle$ and the orthogonalization term. We can write $\langle\psi_{core}|\nabla|\underline{k}\rangle = i\underline{k}\langle\psi_{core}|\underline{k}\rangle$. For a core s state (K spectra), the overlap $\langle\psi_{core}|\underline{k}\rangle$ is nearly independent of $\underline{k}$; therefore, the first term on the right is proportional to k. In the second term, there are matrix elements $\langle\psi_{core}|\nabla|\alpha\rangle$, only for α corresponding to a p state. This matrix element is constant, but the overlap $\langle\alpha|\underline{k}\rangle\propto k$ for a p state. Thus, both of the terms are proportional to k, or $P(E)\sim E$, for the K spectra. For the L spectra, $\langle\psi_{core}|\nabla|\underline{k}\rangle$ contains a factor k from the gradient and a second factor of k from the overlap of the core p state with the plane wave. If this was the only term (a free-electron approximation), we would get $P(E)\propto E^2$ and $I(E)\sim E^{5/2}$. However, the second term gives matrix elements with α corresponding to a core s state; the corresponding overlap is constant; therefore, this term gives nearly constant contribution. This leads to $I(E)\sim E^{\frac{1}{2}}$ for L bands.

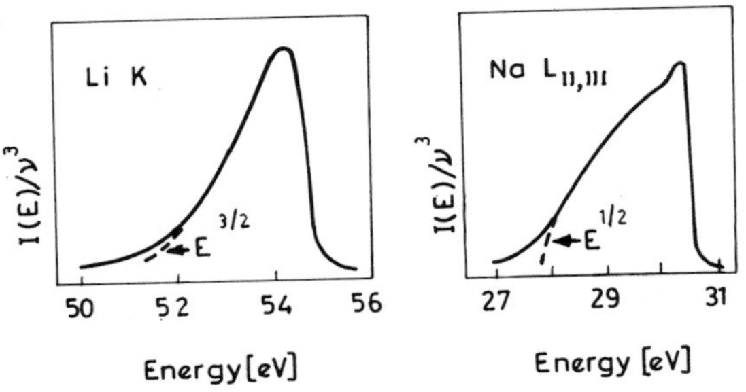

(a) (b)

Fig. 7.11. (a) The K emission band of Li, and (b) the $L_{II,III}$ band of Na

is $1s^2 2s^1$. Because the inner level can be only a K state, solid Li has one of the simplest energy level diagrams. The distribution within the observed band is quite smooth (Fig.7.11a) and should roughly represent the $N_p(E)$ curve or the $\nu^3 E^{3/2}$ law, (7.14). A relation of this form fits the observations quite well. It has a *low-energy tail* because of the Auger broadening. At E_F, the intensity decreases to only about 0.75 of its peak; we expect an abrupt drop, not a broad edge, in the simple model. This anomaly has attracted considerable attention. The band calculations [7.33] show that there is no contact of the Fermi surface with the Brillouin zone and that the N(E) curve increases near the Fermi surface. GOODINGS [7.34] ascribed this decrease of intensity to a change of N(E) in the presence of the 1s hole. He showed that the wave functions at the perturbed atom acquire more localized character at the expense of the band wave functions, resulting in a decrease of N(E) near E_F. A detailed discussion has been given by AUSMAN and GLICK [7.35]. The situation is not yet clear, because recent optical data support a connected Fermi surface [7.36].

The $L_{II,III}$ emission band of Na is shown in Fig.7.11b. The L_{II} band is considerably weakened by Auger decay, so the spectrum is almost pure L_{III}. The L_{III} emission spectra of Na, Mg, Al show a peak at the threshold. These peaks are probably caused by the X-ray edge singularity, given by the threshold converging factor $(\omega-\omega_T)^{-\alpha_0}$. A p-state hole excites electrons into the conduction band with mostly s-wave symmetry. The exponent α_0 is positive and gives a power-law divergence at threshold [7.37]. The earlier explanations depend on density-of-states arguments. The band structure of Na is similar to Li (alkali metal), and has the same density-of-states curve (Fig.7.12). If the Fermi energy were at the peak E_1 of the N(E) curve, or just below it,

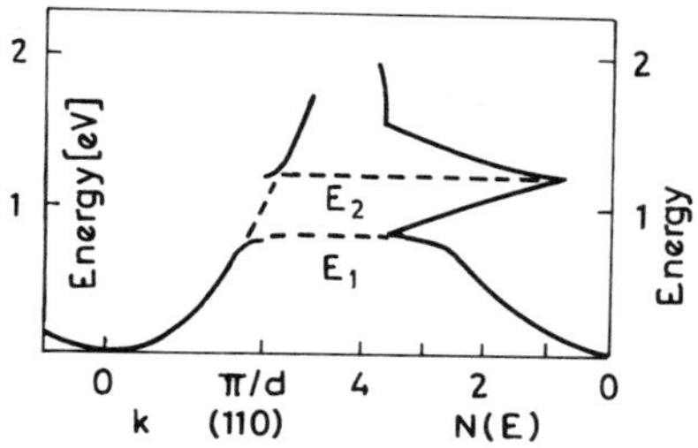

Fig. 7.12. Band structure of an alkali metal in the (110) direction

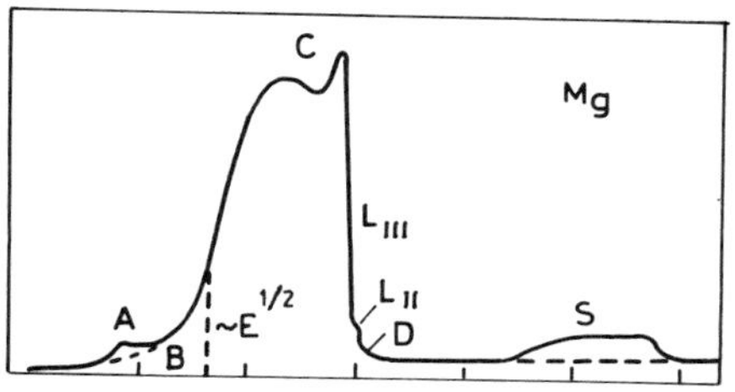

Fig. 7.13. A qualitative picture of Mg $L_{II,III}$ emission spectrum. A: line $L_I \to L_{III}$, B: low-energy tail, C: main band, D: high-energy tail, S: satellite band $L_{III}L_{II} \to L_{II}V$

the observed peak (Fig.7.11b) in Na near the X-ray threshold could be explained on this simple theory. However, the optical measurements show that E_F is considerably below E_1 [7.38]. Therefore, the idea that the Na L_{III} emission-band peak is caused by the band-structure effect must be discarded.

Thus, the explanation based on the one-electron model gives only limited information about the valence band. It has now been recognized that the transition probability, electron interactions and many-particle effects play an important role [7.39-42]. The proper treatment of these effects remains an open question.

K bands of ^{11}Na, ^{12}Mg, ^{13}Al, have been carefully observed [7.1,43-45] and compared with the corresponding N(E) curves. For recent reviews, on the K and L emission soft X-ray spectra, see TOMBOULIAN [7.2], APPLETON [7.46], and ROOKE [7.47].

Following TOMBOULIAN [7.2], we summarize in Fig.7.13 the various typical features of the observed Mg $L_{II,III}$ soft X-ray emission spectra. From such features, we can understand several aspects about the behavior of electrons in solids in the one-electron model. We briefly mention them here.

1) The absence of a sharp edge on the high-energy side of the band in nonmetals (like S, Si) enables us to distinguish them from metals (like Na, Al) (Fig.7.14).

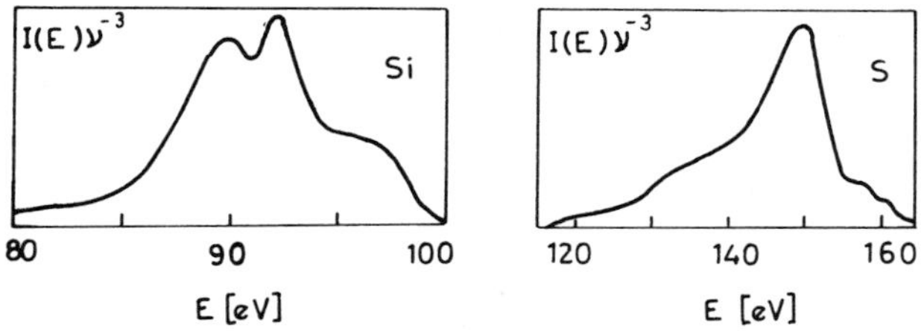

(a) (b)

Fig. 7.14. $L_{II,III}$ emission bands of (a) Si, and (b) S

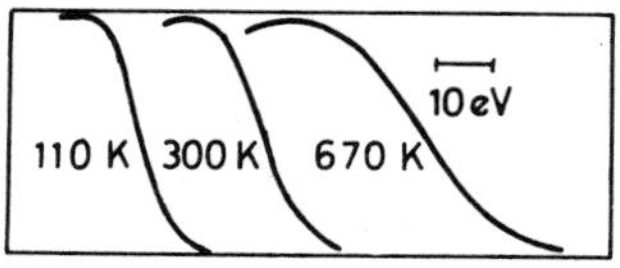

Fig. 7.15

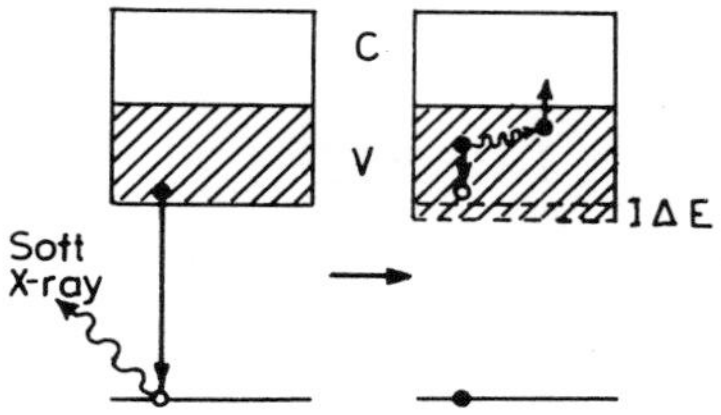

Fig. 7.16

Fig. 7.15. Temperature broadening of emission edges on the high-energy side of the Al L_{III} band

Fig. 7.16. Intra-band Auger transitions for the low-energy tail

2) The same material in two different crystalline forms (such as graphite and diamond) gives rise to emission bands of varying shapes.

3) The effect of the Fermi function f(E,T) on (7.9), and therefore on (7.5) can be directly seen by changing the temperature, T. As shown in Fig.7.8, the high-energy edge of the observed band should get broadened as the temperature is increased from absolute zero. This effect has been observed [7.1] (Fig.7.15) and is in agreement with the distribution given by (6.1). Thus, SXS provides a direct verification of the Fermi distribution that plays an important role in electronic processes in solids.

4) The low-energy tail (Fig.7.13) is associated with the radiationless transitions of electrons within the valence band. The transition of a valence electron to an L level creates a hole in the valence band. An electron from a level higher in the valence band makes a radiationless transition into this vacant level; as a result another electron is ejected into an empty conduction band level (Fig.7.16). The effect of the radiationless transition is to reduc

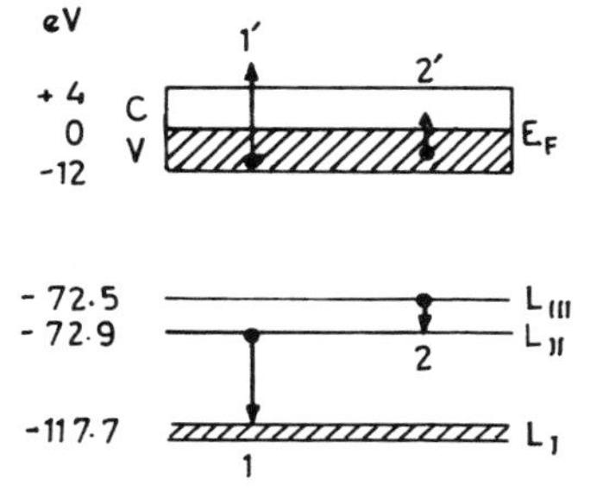

Fig. 7.17. Auger process for the intensity anomaly in Al L-emission spectra

the lifetime Δt of the initial valence-band state that participated in the soft X-ray radiative process. Therefore, this state acquires a width $\Delta E \sim \hbar/\Delta t$. This argument holds for all the levels of the valence band, except that Δt for a state near the bottom of the valence band is shorter than that for a state near E_F. Thus, the levels near the bottom of the valence band become broader and N(E) no longer drops to zero at the bottom of the band, but extends toward lower energies. This gives a gradual decline in I(E) toward the low-energy region of the emission band, that is, a *low-energy tail*.

5) *Intensity anomalies* caused by other Auger processes are observed in SXS (Chap.4). For example, in light elements the L_I (j=1/2) and M_I (j=1/2) emission bands have not been observed. Also, L_{II} (j=1/2) is much weaker and L_{III} (j=3/2) is much stronger, than would be expected from the statistical weight ratios determined by (2j+1). Thus, theoretically L_I:L_{II}:L_{III} = 1 : 1 : 2, whereas we observe approximately 0 : 1 : 10 for Na. Consider the partial energy-level diagram for Al (Fig.7.17). Let a hole be created in L_I. A radiationless transition transfers the hole to L_{II} and simultaneously ejects an electron from the valence band with energy (117.7-72.9) - 12 = 32.8 eV, which is sufficient to eject the electron from the metal. Auger electrons of this type have been detected [7.48]. The energy distribution of such Auger electrons, like the soft X-ray emission band, gives information about the density of states in the valence band. The second electron that participates in the Auger process (marked 2 in Fig.7.17) merely excites a valence-band electron near E_F to the conduction band (marked 2'). Auger transitions of the type $L_I \rightarrow L_{II,III}$ shorten the lifetime of the L_I excited state and give it a width of about 2 eV. In addition, this reduces the probability of a radiative transition $V \rightarrow L_I$, as confirmed by experiments. Similarly, the Auger transfer $L_{III} \rightarrow L_{II}$ reduces the probability of the radiative transition $V \rightarrow L_{II}$ and increases that of $V \rightarrow L_{III}$. This explains why the L_{III} band occurs with an enhanced intensity, L_{II} with reduced intensity, and L_I with near-zero intensity.

6) SKINNER [7.1] assumes that the high-energy tail (Fig.7.13) is a satellite of the type $L_{II}V \rightarrow VV$. He attributed the satellite band to the transition $L_{III}L_{II} \rightarrow L_{II}V$.

The L and M emission bands of heavier elements and compounds have also been studied [7.49-52].

7.6 Absorption-Spectra Recording

The absorber is prepared by evaporating the metal (~100 Å thick) on to a substrate. The substrate is usually a Zapon (cellulose acetate, collodion), Mylar (polyethylene tetraphthalate), or Formvar (polyvinyl formal) film of about 300 Å thickness, to avoid excessive absorption [7.53] of soft X-rays. Cellulose compounds can be dissolved in amyl acetate and Mylar in ethylene bromide. To prepare a collodion film, a drop of 10% solution is allowed to spread on the surface of distilled water in a tray. When dry, this surface film can be lifted on a small metal frame. Two absorbers of thicknesses x_1 and $x_2 > x_1$ are prepared.

The Lyman discharge tube gives a line spectrum. When the absorber is present, the lines in the wavelength region of absorption are reduced in intensity. Four exposures of the photographic plate are made, two through x_1 differing by a factor of two in the number of sparks (exposure), two through x_2 also differing from each other by a factor of two in exposure, but increased by a factor R so that the radiation reaching the plate through thin (x_1) and thick (x_2) absorbers is nearly the same.

A microphotometer trace is made of each spectrum. The deflections d_E for one exposure and twice that exposure, d_{2E}, are read and plotted as in Fig. 7.18a. This is done for both thin and thick absorbers and the average of all the points is taken. This curve is used to plot a relative-exposure versus deflection curve (Fig.7.18b). First, some arbitrary point d_a is taken on the first curve in the region of small deflection. This gives the relative exposure 1 and is plotted. For twice that exposure, at the corresponding point on the second curve, the deflection is d_{2a}. This is a relative exposure of 2. For deflection $d_b = d_{2a}$ on the first curve there is a corresponding d_{2b} on the second curve for which the exposure is 2×2 or 4. This process, carried out for the available values of d, yields a response curve. Intermediate points can be determined by taking other starting points on the first relative-exposure curve (Fig.7.18a) and repeating the process. This curve (Fig.7.18b) is

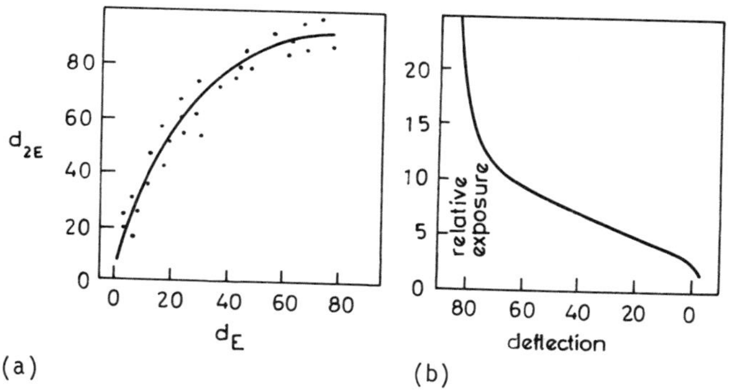

Fig. 7.18. (a) d_E vs d_{2E} plot. The solid curve is the average of the points from x_1 and x_2 absorbers. (b) Response curve

the *response curve* of the photographic emulsion used. It is based on the assumption that for the wavelength region covered on a plate, the response curve is independent of wavelength [7.54].

The response curve is used to determine the exposure for the deflection corresponding to each spectral line λ_i through both thick and thin samples. At any λ_i, the ratio of these exposures, thick to thin, divided by R, the ratio of the number of sparks through thick and thin absorbers, gives the transmittance T_i for the thickness $\Delta x = x_2 - x_1$. Then the absorption coefficient $\mu(\lambda_i)$ is found from the equation $T_i = \exp[-\mu(\lambda_i)\Delta x]$, for each λ_i in the region of interest. The points so obtained are plotted on a μ vs λ graph. These points are connected with a smooth curve to give the required absorption curve. When synchrotron continuous soft X-rays are used as source radiation the method of analysis is the same as for the hard X-ray region described earlier.

7.7 Interpretation of Absorption Spectra

The absorption coefficient $\mu(E)$ is given by [compare with (7.5)]

$$\begin{aligned} \mu(E) &= 0 \quad , \quad E \leqq E_F \quad , \\ &\propto P(E)N(E) \quad , \quad E > E_F \quad . \end{aligned}$$

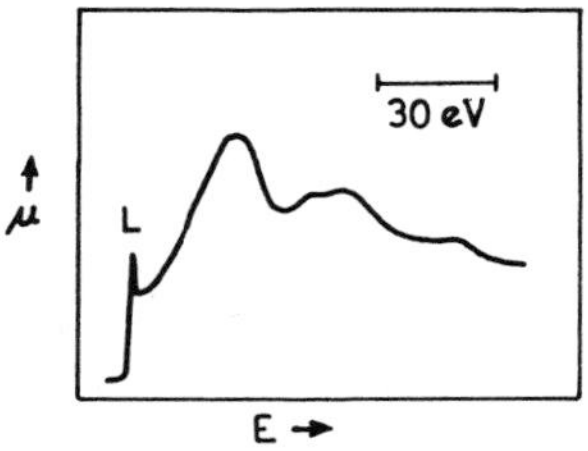

Fig. 7.19. Qualitative $L_{II,III}$-absorption shape for metals such as Na, Mg, Al. The spike is marked L

The K and $L_{II,III}$ absorption spectra of the lighter elements have been measured with the synchrotron source and reviewed by BROWN [7.6]. The earlier data from conventional sources has been reviewed by TOMBOULIAN [7.2]. In Fig.7.19, we show the general structure of $L_{II,III}$-absorption curves of Na, Mg, and Al. The absorption rises from a spike or line at the edge to a broad maximum 30 eV or more above threshold. The general rise from threshold is a matrix-element effect associated with transitions from the initial p state to the continuum d states [7.55], the transitions to continuum s states being very weak.

The characteristic sharp spikes or X-ray singularities close to the threshold for metals have been explained as a final many-body-state effect. The formation of a "core hole" in a metal during an X-ray absorption process gives rise to the sudden creation of a long-range hole (Coulomb) potential. This potential is screened in some way by the conduction electrons in the final state of the process. The deep hole being localized on the core level ψ_i of a definite atom, the screening resembles that around an excess nuclear charge and leads to the formation of an exciton [7.56]. There is not a one-to-one correspondence between the Wannier excitons (hole in the valence band and electron in the conduction band that forms a bound state) and the X-ray excitons (hole in the core state and electron in the bound state), but both lead to an enhanced transition probability, because the electron prefers to spend some time near the hole before moving away. MAHAN [7.39] originally derived this enhancement of threshold of the absorption (X-ray edge singularity) from the exciton theory. NOZIERES and DeDOMINICIS [7.41] assumed that in the final state, all of the conduction states ψ'_k are scattered by the screened potential V_j (hole) of the core hole and therefore differ from the corresponding unperturbed states ψ_k. Consequently, the transition probability $\langle \psi_i|\nabla|\psi'_k\rangle$ is modified.

As a reaction to the sudden creation of a hole potential, the conduction electrons rush in, so as to shield this potential. MAHAN [7.39] regarded this

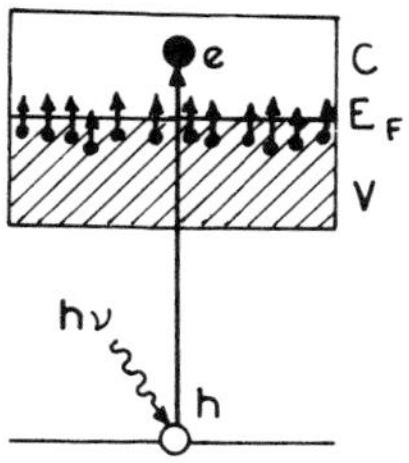

Fig. 7.20. Shielding of a suddenly created core hole in an energy picture for the absorption process. h: core hole; e: extra electron in the conduction band C due to the absorption transition

many-body effect as the plasmon excitation. NOZIERES and DeDOMINICIS [7.41] think of it as a scattering effect characterized by phase shifts δ_I in a partial wave analysis.

The theory of X-ray edge singularity is very complicated [7.37]. In a pedestrian way, we can say that in the ordinary picture of scattering in space, the conduction electrons are scattered by a spherically symmetric potential

$$V_j^{hole}(r) = \int \frac{\rho_j(r')}{|\underline{r}-\underline{r}'|} d^3r' \quad , \quad \rho_j(r') = e^2|\psi_j(r')|^2 \quad ,$$

where ψ_j is the hole wave function. A theorem by ANDERSON [7.57] states that the many-body ground state in the presence of a localized-potential well is orthogonal to the many-body state in the absence of the potential well, because an infinite number of electrons undergo an infinitesimal change of their states to produce the adjustment to the localized potential. This is the starting point of the X-ray edge theory. In the energy picture, this change of conduction-electron states can be thought of as an infinite number of transitions near the Fermi surface excited by a V_j^{hole} (Fig.7.20). Thus, the final-state effect is an assembly of many Auger processes, which gives rise to a large number of electron-hole pairs at E_F that accompany core excitation. We expect that the transition rate for these Auger-like processes would be maximum for the smallest excitation energy E measured above threshold E_T,

$$w(E) \propto \frac{1}{(E-E_T)^\alpha}\,\theta(E-E_T) \quad .$$

The critical exponent α (related to phase shift δ_1) for $p \rightarrow s$ transition is positive, so that an enhancement occurs for L shell absorption (Fig.7.19) at the threshold in Na, Mg, and Al. These spikes have been called excitons in a metal, although they differ from the usual excitons in an insulator (Fig. 6.16b). COMBESCOT and NOZIERES [7.58] have given a theory for excitons in

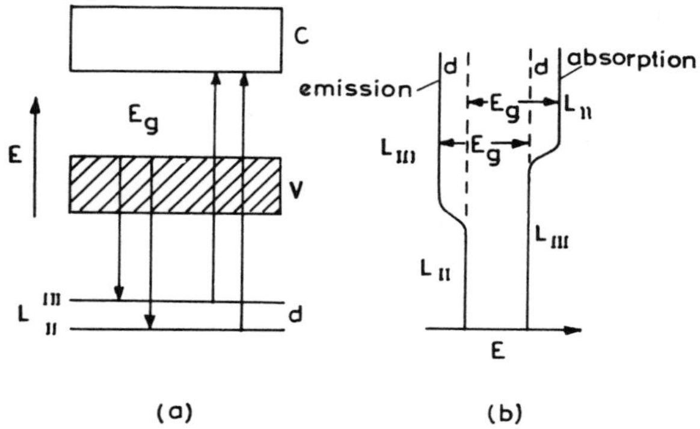

Fig. 7.21. (a) L_{II} and L_{III} emission and absorption transitions. (b) Estimation of E_g and L_{II}-L_{III} level separation in an idealized situation

which the transition from insulator to metal behavior can be followed. For a clear discussion of many-body effects in metals and Anderson orthogonality theorem, see KOTANI and TOYOZAWA [7.67].

The soft X-ray M-absorption spectra of transition metals and oxides have been measured by several workers (for example [7.6,59-64]). Comparison with the theoretical N(E) curve has led AGARWAL and GIVENS [7.61] to conclude that the antiferromagnetic nature of Cr can be directly demonstrated by study of soft X-ray absorption and emission spectra. This kind of interpretation has been further elaborated by MOTT and STEVENS [7.65]. The $M_{II,III}$-absorption spectra of Cr differ in high-energy details when studied with a conventional soft X-ray source [7.61] and by the synchrotron source [7.62], although the position and slope of the X-ray edge is nearly the same in both cases. SONNTAG and BROWN [7.66] have recently studied the soft X-ray absorption spectra of transition-metal layer compounds.

In a light element, such as ^{15}P, the L emission spectrum consists of two superimposed bands whose edges are separated by the energy difference between the L_{II} and L_{III} levels. The L absorption spectrum also gives two edges separated by the same amount (Fig.7.21a). The separation of L_{III} emission and L_{III} absorption should correspond to the energy gap E_g between the valence-band edge and the conduction-band edge. The idealized situation is shown in Fig.7.21b. Although the observed edges are not so sharp as shown, they can give a reasonable estimate of E_g and L_{II}-L_{III} level separation. The SXS thus gives information about inner-level separations, band gaps, the entire valence band (in emission), and the conduction-band (in absorption) density of states in the first approximation. In contrast, most other solid-state studies give information only about states near E_F.

8. Experimental Methods

The essential instrumentation required for X-ray spectroscopy consists of a) the primary X-ray source unit, b) the spectral analyzer, and c) the detector. We describe the elements of these units in this chapter.

8.1 X-Ray Tubes

X-ray tubes are mainly of two kinds: 1) Gas tubes, and 2) filament tubes. In gas tubes the electrons that strike the target are produced by ionization of the gas atoms; in filament tubes they are produced by heating the filament by passing an electric current through it. Filament tubes are further divided in two types: 1) sealed or Coolidge tubes, and 2) demountable tubes.

Gas tubes. The original tubes used by Röntgen consisted of a glass bulb evacuated to a pressure of 10^{-4} mm Hg. When a high voltage is applied between the cathode and anode, the residual air becomes partially ionized. The positive ions and free electrons are accelerated to the cathode and target (anode), respectively. The positive ions knock out some electrons on striking the cathode. (Metals of low Z disintegrate less under bombardment by positive ions, so Al is usually used as the cathode material). The electrons, or cathode rays, tend to collide with the gas ions on their way to anode. In this way, the ionization builds up. The final tube current is determined by the gas pressure, the applied voltage and the tube geometry. In Fig.8.1 we show the general structure of a gas tube with a glass envelope, and with a metal body [8.1]. The concave cathode produces an electrostatic focusing effect that gives a sharp focal spot on the target. In the Hadding tube (Fig.8.1b), gas is continuously leaked in through a needle valve and evacuated by a diffusion pump, to give the desired low pressure.

Coolidge tubes. W.D. Coolidge replaced the Al cathode with a tungsten filament in a sealed glass tube evacuated to less than 10^{-5} mm Hg, see Fig.8.2a. When the filament is heated to a temperature $T^{o}K$ by passing a current through it,

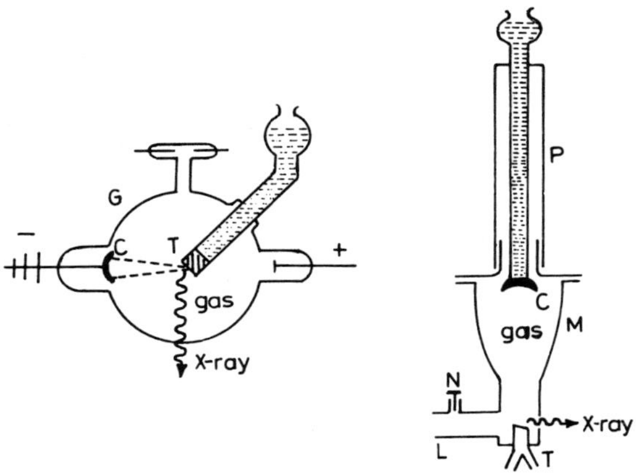

(a) (b)

Fig. 8.1. (a) Gas X-ray tube with glass envelope. (b) Hadding tube. C: Al cathode, T: water-cooled target, G: glass bulb, M: metal body, P: porcelain cathode assembly, N: needle valve to introduce air, L: to diffusion pump backed by rotary pump

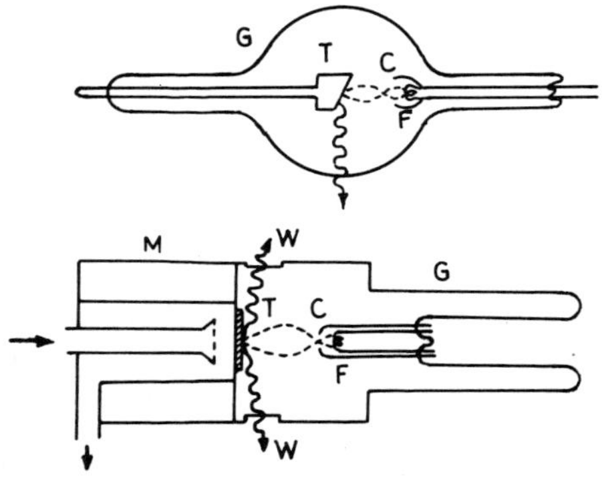

Fig. 8.2. Schematic cross section, of (top,a) Coolidge sealed-off tube, (bottom,b) commercial X-ray tube. T: water-cooled target, M: metal housing, W: beryllium window, F: filament, C: focusing hood, G: glass envelope

it emits electrons according to the Richardson-Dushman equation (see, for example [8.2]),

$$I = ze = AT^2 \exp(-B/T) \quad ,$$

where A, B are constants and the current I is the product of the number z of electrons that leave unit area of the surface per second by the electronic charge e. The thermionically emitted electrons are accelerated toward the target to produce X-rays. For a given voltage V on the tube, the tube current depends on the temperature of the filament. The advantage of Coolidge tubes, over gas tubes, is that the tube current i and the voltage V can be varied independently. This advantage would be lost, and the efficiency reduced, if the electrons from the filament can collide with gas atoms and produce extra electrons. Therefore, high vacuum is maintained inside the tube.

In Fig.8.2b, we show a typical sealed tube, of the Coolidge type, that is now commercially available. A metal focusing hood, at the same potential as the filament, helps to focus the thermionically emitted electrons to a small area (*focal spot*) of the target. Nonporous beryllium is used as window material. Because of its low atomic number, it is transparent to hard X-rays and does not produce secondary X-rays in this region.

Demountable (filament) tubes. For X-ray spectroscopic work, we often need long exposures and different target materials, either for emission studies or for suitable reference lines. With a demountable tube, we can dismantle the (porcelain) cathode and (metal) anode assemblies to replace the burnt-out filament and to change the target layer. An oil-diffusion vacuum pump, backed by a rotary pump, is used to produce the necessary vacuum (10^{-5} mm Hg) in the tube. The demountable tube has virtually an infinite lifetime.

The conventional X-ray tubes have been replaced, wherever facilities exist, by synchrotrons and storage rings as sources. The high-brilliance X-ray sources have been described by YOSHIMATSU and KOZAKI [8.30].

8.2 Line-Focus Filament

Large energy input in a tube with a small focal spot is not possible because it leads to excessive heating and melting of the target. Increase of the area of the focal spot in all directions causes bad geometry and loss of sharpness in photographs. The focus area should be increased in such a way that the cross section through the emitted X-ray beam at the focal spot is as small as possible.

The *line-focus filament* of Goetze provides a satisfactory solution. A long cylindrical spiral (obtained, say, by winding tungsten wire on a pin) of very small diameter gives a line focal spot (area ~1 mm × 10 mm) on the target that by virtue of its length can dissipate considerable energy without damage to the target. In the sealed-off tubes, the target is normal to the incident electron beam. When the line-shaped focal spot is viewed from one of its ends, the focal spot becomes foreshortened. By choosing the takeoff angle properly

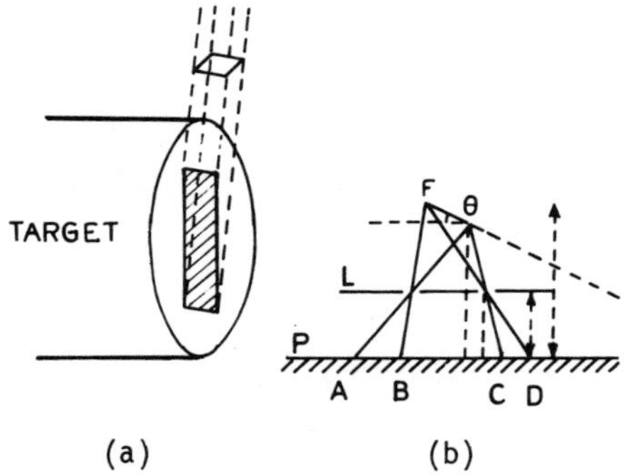

Fig. 8.3. (a) Line-focus of a filament. (b) Two-pinhole method for determining the angle of the focal spo F: focal spot, L: lead sheet, P: fil

($\sim 6^\circ$, $\sin 6^\circ \simeq 0.1$), the effective focal spot can be made to look like a smal square (1 mm × 1 mm) (Fig.8.3a). Thus, the intensity of the X-ray beam is increased whereas the effective size of the focal spot remains small, to give sharp images. The angle of the line-focus spot can be determined by photographing it through two pinholes in a lead sheet as shown in Fig.8.3b.

Even with a line-focus spot, the heating of the target is great. For example, a tube run at 50 kV, 20 mA, or at 10^3 W, all confined to a focal spot of area 1×10 mm^2, has a specific loading of 100 W/mm$^2 \simeq 25$ cal s^{-1} mm^{-2}. The efficiency of an X-ray tube is about 1 percent or less at 50 kV (Chap.2). Therefore, virtually all of this energy is used up in heating the target. It is enough to heat 100 cc of water from 25°C to the boiling point in about 5 min; it is essential to remove this heat as rapidly as possible, by coolin the target. If the target is earthed, it can be connected directly with the water mains and a flow of one or more liters per minute is enough to cool th target, under normal conditions. A water-circulation pump with a reservoir can also be used to supply this flow rate.

Copper is a good conductor of heat. Therefore, the anode assembly in X-ra tubes is usually made of copper, and the desired target metals are either plated onto its surface or mechanically fitted into it. After long use, the focal spot gets broadened and stray cathode electrons begin to strike the copper studs or the frame that holds the target material. This gives rise to Cu emission lines in old Mo or W target X-ray tubes.

8.3 High-Tension Circuits

A filament X-ray tube has the structure of a diode rectifier or a kenotron. Therefore, it gives half-wave *self-rectification* when alternating high vol-

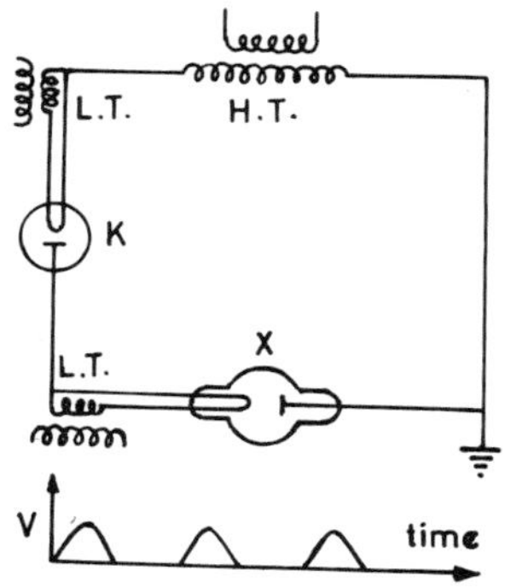

Fig. 8.4

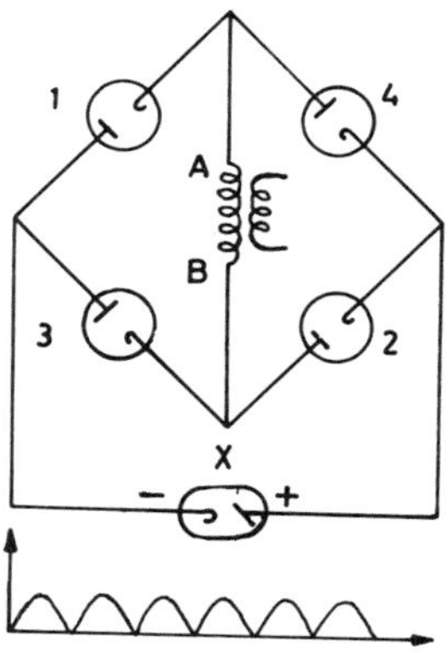

Fig. 8.5

Fig. 8.4. Half-wave-rectifier circuit. H.T.: high-tension transformer, L.T.: low-tension transformer for the filament current. K: kenotron, X: X-ray tube, V: voltage

Fig. 8.5. Full-wave-rectifier circuit

tage is directly applied from a transformer. For continuous operation, this is not recommended, because the target may get heated enough to produce thermionic electrons and therefore a reverse current. This can seriously damage the filament.

The recommended circuit for *half-wave rectification* involves a kenotron (Fig.8.4). To avoid production of X-rays by the kenotron, its voltage-ampere characteristics are kept different from that of the X-ray tube. Both of them have diode-like characteristic curves. For a given tube current, the voltage drop across the kenotron is much smaller (about 200 volts) than across the X-ray tube (about 20,000 volts).

The *Gratz circuit* uses four kenotrons for the *full-wave rectification* (Fig.8.5). When B, one end of the secondary winding, is negative and A is positive, electrons would flow from B through kenotron 3, the X-ray tube, and then through kenotron 4 to A. When A is negative the flow is $1 \rightarrow X \rightarrow 2 \rightarrow B$. Thus, in the X-ray tube the current always flows in one direction.

To minimize the voltage changes on the anode, a condenser C is introduced in the circuit (Fig.8.6). It draws current from the high-tension transformers when the transfer voltage happens to be higher than the condenser voltage; this may happen only over a small fraction of the total period of the cycle. Its capacity should be such that the decrease of the condenser voltage is small during the discharge part of the cycle.

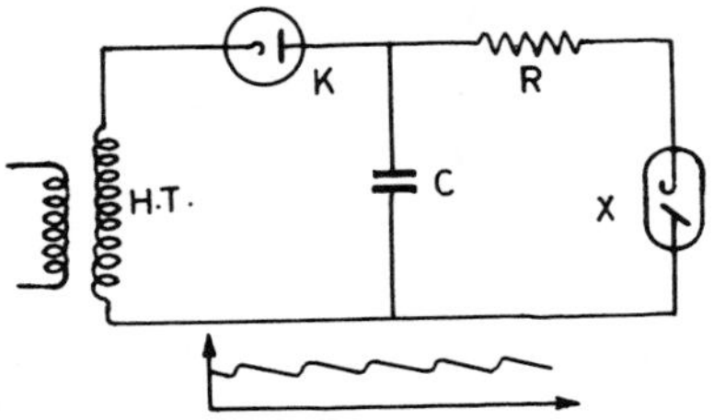

Fig. 8.6. Constant-voltage circuit. C: condenser, K: kenotron, X: X-ray tube, R: surge-limiting resistor

8.4 Wavelength Units

The *unit cell* of rock salt (NaCl) is shown in Fig.8.7. The location of the atoms can be formally described as follows:

$$\text{4 Na atoms at}\quad 0\,0\,0\;\Big|\;\tfrac{1}{2}\,\tfrac{1}{2}\,0\;\Big|\;0\,\tfrac{1}{2}\,\tfrac{1}{2}\;\Big|\;\tfrac{1}{2}\,0\,\tfrac{1}{2}$$

$$\text{4 Cl atoms at}\quad \tfrac{1}{2}\,\tfrac{1}{2}\,\tfrac{1}{2}\;\Big|\;\tfrac{1}{2}\,0\,0\;\Big|\;0\,\tfrac{1}{2}\,0\;\Big|\;0\,0\,\tfrac{1}{2}$$

We can determine the interplanar distance d (distance between planes of identical atomic population) as follows:

Molecular weight of NaCl = $M = 23.00 + 35.46 = 58.46$,

No. of atoms in 58.46 g NaCl = $2N_A = 2 \times 6.064 \times 10^{23}$,

Density of rock salt = $\rho = 2.163\ \mathrm{g\ cm^{-3}}$,

No. of atoms in 1 $\mathrm{cm^3}$ = $n = 2N_A\ \rho/M$,

Distance between Na and Cl along cube edge = d ,

No. of atoms in a row 1 cm long = $1/d$,

No. of atoms in 1 $\mathrm{cm^3}$ = $n = 1/d^3$.

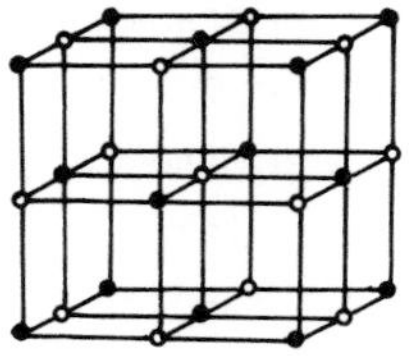

Fig. 8.7. Unit cell of rock salt: ● Na, ○ Cl

Equating the two expressions for n, we obtain

$$d^{-3} = 2 \times 6.064 \times 10^{23} \times 2.163/58.46 \quad ,$$

$$d = 2.81400 \times 10^{-8}\ \text{cm} \quad .$$

The crystal-grating spacing of the calcite ($CaCO_3$) crystal can be evaluated from the Bragg equation,

$$n\lambda = 2d_{NaCl}\ \sin\theta_{NaCl} = 2d_{CaCO_3}\ \sin\theta_{CaCO_3} \quad .$$

SIEGBAHN in this way found the calcite-cleavage spacing to be 3.02945×10^{-8} cm. SIEGBAHN recognized the uncertainty of the values of N_A and ρ used in his calculation and proposed a new unit called the X unit (XU) in which the d values were arbitrarily set as (for n=1)

$$d_{NaCl} = 2814.00\ \text{XU} = 2.81400\ \text{kXU} \quad , \quad (\text{at } 18^{\circ}\text{C}) \quad ,$$

$$d_{CaCO_3} = 3.02904\ \text{kXU} \quad , \quad (\text{at } 18^{\circ}\text{C}) \quad .$$

Thus, 1 kXU is approximately equal to 10^{-8} cm or 1 Å.

When ruled gratings became available for X-ray measurements, it was found that the wavelength values of the well-known lines were always about 0.2 percent higher than those obtained by use of the Siegbahn value of d for calcite. The ruled-grating measurements of λ of well-known lines made in 1935 by Bearden were free from the uncertainty involved in the values of ρ and N_A, and gave the corresponding value of d for calcite as 3.03560 Å. Bragg's conversion factor [8.3] is 1.00202 kXU = 1 Å. BEARDEN [8.4] took the W $K\alpha_1$ to define the X-ray wavelength standard as

$$\lambda_{WK\alpha_1} = 0.2090100\ A^* \quad .$$

This relation defines a new unit A^*. It approximates more closely to Å than to XU. To indicate the similarity to Å and yet distinguish the new unit as an independent entity, BEARDEN denoted it by A^*. The above choice makes Å/A^* very close to unity, but the conversion factor remains an experimentally determined quantity, subject to revision by future measurements.

In precision work, it is necessary to consider thermal expansion of the crystal by using

Table 8.1. Useful data on crystals

Crystal	d at 18°C [kXU]	d at 18°C [Å]	α
Rock salt, NaCl (200)	2.81400	2.81971	4.0×10^{-5}
Calcite, $CaCO_3$ (211)	3.02945	3.03560	1.02
Quartz, SiO_2 (1010)	4.24602	4.25465	1.04
Gypsum, $CaSO_4 \cdot 2H_2O$ (020)	7.56470	7.6001	3.78
Mica, $KAl_2[AlSi_3O_{10}] \cdot [OH]_2$ (001)	9.94272	9.9629	0.15

$$d_\infty = d_{0\infty}(1+\alpha\Delta t) \quad ,$$

where d_∞ is determined by (3.66), α is the linear coefficient of expansion, $d_{0\infty}$ is the grating constant at a reference temperature (18°C), and Δt is the temperature rise above the reference value. In Table 8.1 we give the data on the crystals normally used in X-ray spectroscopy.

Extensive lists of crystals and their properties are now available [8.5]. The atomic positions in the unit cell of calcite are [8.6]

2Ca	$\frac{1}{4}\frac{1}{4}\frac{1}{4}$	$\frac{3}{4}\frac{3}{4}\frac{3}{4}$				
2C	$0\,0\,0$	$\frac{1}{2}\frac{1}{2}\frac{1}{2}$				
6O	$\frac{1}{4}\frac{3}{4}0$	$\frac{3}{4}0\frac{1}{4}$	$0\frac{1}{4}\frac{3}{4}$	$\frac{1}{4}\frac{3}{4}\frac{1}{2}$	$\frac{3}{4}\frac{1}{2}\frac{1}{4}$	$\frac{1}{2}\frac{1}{4}\frac{3}{4}$

Thus, the unit cell does not have faces parallel to the cleavage planes of calcite; therefore, the indices of these planes are not (100) in terms of this unit cell, but are (211). The smallest *rhombohedron* of calcite that has faces parallel to the cleavage planes, and that contains an integral number of $CaCO_3$ molecules, has four molecules. It is not the true unit cell of calcite. The true unit has two molecules and does not have faces parallel to the cleavage planes.

8.5 Plane-Crystal Spectrograph

A simple arrangement [8.7] for absolute measurement (without using reference lines) of an unknown wavelength by use of a plane-crystal spectrograph is

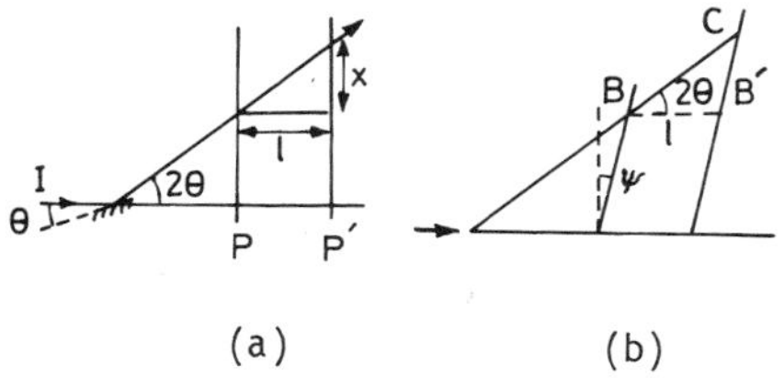

Fig. 8.8. (a) Absolute measurement of λ by a plane-crystal spectrograph. (b) Correction for the tilt of the plate holder

shown in Fig.8.8a. It is called the *method of displacement* [8.31-33]. The line of wavelength λ under study is photographed on the same film at two positions, P and P', of the plate holder held normal to the incident ray I. If the displacement of the plate holder is l and the distance between the recorded line images is x, then λ can be estimated under normal setting of the spectrograph by using the relations

$$2d\sin\theta = n\lambda \quad , \tag{8.1}$$

$$\tan 2\theta = x/l \quad . \tag{8.2}$$

It can happen that the plate holder is not normal to I. Suppose it is tilted by a small angle ψ (Fig.8.8b). From the triangle BB'C,

$$\frac{x}{\sin 2\theta} = \frac{l}{\cos(2\theta \pm \psi)} \quad . \tag{8.3}$$

From (8.1,3),

$$\psi = \cos^{-1}\left[\frac{ln\lambda}{xd}\left(1 - \frac{n^2\lambda^2}{4d^2}\right)^{\frac{1}{2}}\right] \pm 2\sin^{-1}(n\lambda/2d) \quad . \tag{8.4}$$

By recording a line of known wavelength λ, the angle ψ, a constant of the instrument can be determined, and then (8.3) can be used to find θ for the unknown λ [8.8]. Another method is to record the line λ on both sides of the incident beam. If the image is displaced by x on one side and by y on the other side, then

$$\tan 2\theta = \frac{2xy}{l(x+y)}\cos\psi \quad , \quad \sin\psi = \frac{l(x-y)}{2xy} \quad .$$

The *angular dispersion*, $D = d\theta/d\lambda$, defines the ability of the crystal to separate two wavelengths, λ_1 and λ_2. It represents the angular difference dθ (in radians) corresponding to the wavelength difference dλ (in Å). From (8.1)

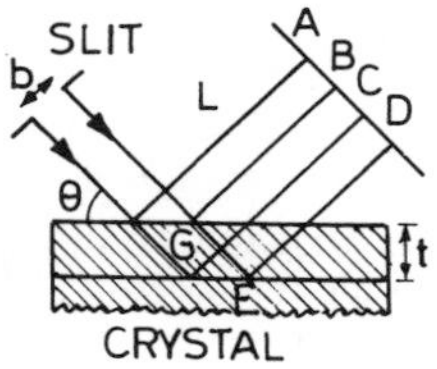

Fig. 8.9. Resolving power

$$D = \frac{d\theta}{d\lambda} = \frac{n}{2d}\,\frac{1}{\cos\theta} = \frac{\tan\theta}{\lambda} \quad . \tag{8.5}$$

D increases with order, n, and becomes large near 90°. The angular dispersion is different from the *resolving power*, $\lambda/d\lambda$, or *resolution*. The definition of resolution is somewhat arbitrary. ALLISON [8.9] assumed that two lines are just resolved if their separation is equal to the full width of the instrumental window. The expression for the resolution then follows by putting $d\theta =$ where b is the effective width of the instrumental window,

$$\frac{\lambda}{d\lambda} = \frac{\tan\theta}{d\theta} = \frac{\tan\theta}{b} \quad . \tag{8.6}$$

Thus, the resolving power is large if b is small. Note that $\lambda/d\lambda = E/|dE|$.

To see the effect of other parameters, consider the situation where the lines are recorded at a distance L from the crystal, and the X-rays penetrate a small distance t in the crystal (Fig.8.9). From the diagram,

$$AD = AB + BD = b + EG \sin 2\theta = b + 2t \cos\theta \quad .$$

Suppose that we wish to resolve λ and $\lambda + \Delta\lambda$. When the crystal is rotated $\Delta\theta$, the beam is rotated $2\Delta\theta$. From (8.1), we have $2d \cos\theta\ \Delta\theta = n\Delta\lambda$. The two images will not overlap; they will be just resolved, if $L \cdot 2\Delta\theta > b + 2t \cos\theta$, or

$$\Delta\lambda > (d/nL) \cos\theta (b+2t \cos\theta) \quad .$$

Clearly, the resolving power is large when $\Delta\lambda$ is small, or when L is large, b is small, and t is small.

The plane crystal is made to oscillate so that several wavelengths of interest present in the divergent X-ray beam can be simultaneously recorded. Another advantage is that the effect of lattice defects is partly obliterated because the whole crystal surface participates in the reflection of a line. The rays of wavelength λ that enter through the slit in a divergent beam stri

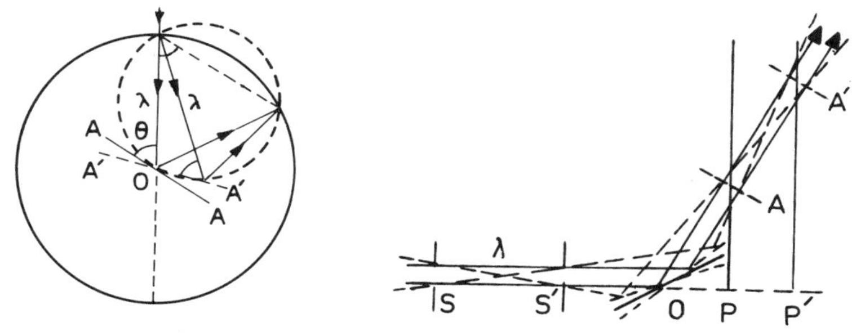

Fig. 8.10. (a) Parafocusing. (b) Parafocusing in the method of displacement. S and S' are collimating slits which are given the same width and the crystal oscillates about O. Distances are such that OA = OS', OA' = OS, SS' = AA', OP = OS' cos 2θ, and PP' = AA' cos 2θ, where θ is the Bragg angle for λ. The two lines of a given λ are recorded with identical widths

the rotating crystal at the Bragg angle θ. The reflected rays will cross the circle through the slit at one and the same point regardless of the position of the crystal at the moment of reflection (Fig.8.10a). This is called the *Bragg-Brentano parafocusing*. The term parafocusing was used by J.C.M. Brentano to distinguish it from true optical focusing, which can be obtained only by bending the crystal.

Parafocusing can be used with advantage in the method of displacement (Fig.8.10b).

Two slits, each of width 0.1 mm, separated by 1000 mm, give an angular width of about 20". This provides enough resolution for $\lambda \sim 1.5$ Å. The slits S and S' must be made vertical and the centers aligned on a straight line passing through the spectrometer axis. The crystal surface must also be made vertical and brought on the axis [8.20].

8.6 Curved-Crystal Spectrograph

Consider a thin crystal section of mica (or quartz) bent so that it becomes part of a cylindrical surface MN of radius or curvature R (Fig.8.11). The figure shows a cross section of this cylinder perpendicular to its axis. A divergent beam of X-rays from a broad focal spot (for example, a line focus

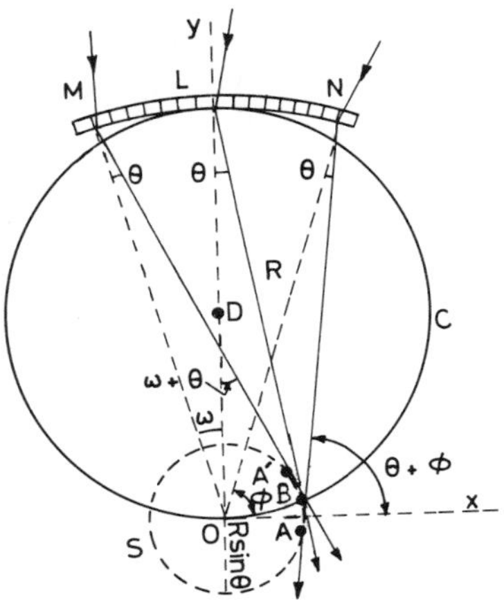

Fig. 8.11. Transmission-type curved-crystal Cauchois spectrograph. C: focal circle, S: caustic circle, MLN: crystal bent to radius R

looked at from a side rather than from an end) is incident from above. The radiation is transmitted through the thin bent crystal and diffracted by prominent transverse crystal planes. For simplicity, let us assume that these crystal planes are normal to the surface. This gives the basic geometry of the Cauchois spectrograph. The curved-crystal spectrograph acquired importance and maturity in the research of Professor Mlle Cauchois [8.10].

Consider a wavelength λ in the incident beam, for which the Bragg angle is θ. A ray of this wavelength is incident at point N $(R\cos\phi, R\sin\phi)$ with polar angle ϕ, and the diffracted ray is NA. Then the equation of the straight line NA that makes an angle $\theta+\phi$ with the x axis is given by

$$y = \tan(\theta+\phi)x + c \quad . \tag{8.7}$$

The constant c is determined by the fact that the line passes through the point N $(R\cos\phi, R\sin\phi)$. The resulting equation of NA is

$$y - \tan(\theta+\phi)(x-R\cos\phi) - R\sin\phi = 0 \quad . \tag{8.8}$$

As the point of incidence N moves along the arc NM, the angle ϕ varies; the envelope of the diffracted rays is obtained by differentiating (8.8) with respect to ϕ,

$$\cos^{-2}(\theta+\phi)(x-R\cos\phi) + \tan(\theta+\phi)R\sin\phi + R\cos\phi = 0 \quad . \tag{8.9}$$

This can be solved for x to give

$$x = (R\sin\theta)\sin(\theta+\phi) \quad . \tag{8.10}$$

From (8.10,8),

$$y = -(R\sin\theta)\cos(\theta+\phi) \quad . \tag{8.11}$$

We can combine (8.10,11) to get

$$x^2 + y^2 = (R\sin\theta)^2 \quad . \tag{8.12}$$

This represents a circle S with center O and radius $R\sin\theta$. It is called the caustic circle.

In practice, the segment MN is small (~ 1 cm) and $R \sim 40$ cm. Therefore, $\phi \sim 90^\circ$ and all of the diffracted rays that correspond to the incident angle θ pass in the immediate neighborhood (A to A') of point B (Fig.8.11),

$$x = R\sin\theta\cos\theta \quad , \quad y = R\sin^2\theta \quad . \tag{8.13}$$

There is no sharp focus at B, but the finite angular width of the diffracted rays that overlap here will produce a very great intensity (about hundred times more than in a plane-crystal spectrograph). This is a great advantage.

We would now like to find the locus of points, defined by (8.13), as the wavelength varies, that is, as θ varies. From (8.13), this locus is found by eliminating θ. It is a circle of radius R/2,

$$x^2 + \left(y - \frac{1}{2}R\right)^2 = \frac{1}{4}R^2 \quad , \tag{8.14}$$

whose center D is given by $x = 0$, $y = R/2$. This is the *focal circle* C. If a photographic film is wrapped along this circle, the various wavelengths will be simultaneously focused upon it in an inexact sense. Thus, a wide range of spectrum is recorded without any oscillation or rotation of the crystal. The line would be sharp on the edge at B and somewhat diffuse on the far edge.

Let 2ω be the angular width of the crystal at O. Coordinates of A are

$$x = R\sin\theta\cos(\theta-\omega) \quad , \quad y = R\sin\theta\sin(\theta-\omega) \quad ,$$

and of A' are

$$x' = R \sin\theta \cos(\theta+\omega) \quad , \quad y' = R \sin\theta \sin(\theta+\omega) \quad .$$

The concentration of rays occurs along the arc AA'. The distance over which we can displace the photographic plate (*depth of focus*) parallel to the y axis is

$$y' - y = R \sin 2\theta \sin\omega \quad . \tag{8.15}$$

Two spectral lines that correspond to the Bragg angles θ_1 and θ_2 subtend an angle $\theta_2 - \theta_1$ at L. The linear separation s between them is $s = R(\theta_2 - \theta_1)$. Using (8.6), $d\lambda/\lambda = \cot\theta \, d\theta$, we can write

$$ds = R d\theta \quad , \quad d\lambda/\lambda = (ds/R)\cot\theta \quad ,$$
$$ds/d\lambda = (R/\lambda)\tan\theta = nR/(2d \cos\theta) \quad . \tag{8.16}$$

For $ds = 1$ mm, $d\lambda$ can be calculated in the XU region from this formula.

A rough measure of the width of the line is $|x'-x| = 2R \sin^2\theta \sin\omega$. Therefore, an observed line is narrow if the aperture of the crystal is small.

A sheet of muscovite mica, $(OH)_2KAl_2(Si_3Al)O_{10}$, is easily bent. Its structure has been discussed by CLARK [8.11]. The cleavage plane of mica is (001). Planes normally used in curved-crystal spectrographs are listed in Table 8.2. The planes that belong to a zone can be selected by taking a Laue photograph of the mica sheet, aligned with the photographic film (Fig.8.12). The mica sheet can then be cut with edges perpendicular to the zone axis drawn with the help of Laue spots. These edges are bent along the cylinder of radius R to form the curved crystal. The zone axis is then parallel to the generators of the cylinder defined by the curved crystal. The various planes (h,k,l) that belong to the zone axis with indices [u,v,w], are related by the equation

$$uh + vk + wl = 0 \quad . \tag{8.17}$$

Table 8.2. Data on muscovite (monoclinic) crystal. F is the structure factor

Zone	h,k,l		d [Å]	F
[010]	$\bar{2}01$	$-5^\circ 30'$	2.578	182
[010]	100	$9^\circ 17'$	2.556	155
[110]	$\bar{3}3\bar{1}$	$-0^\circ 29'$	1.498	276

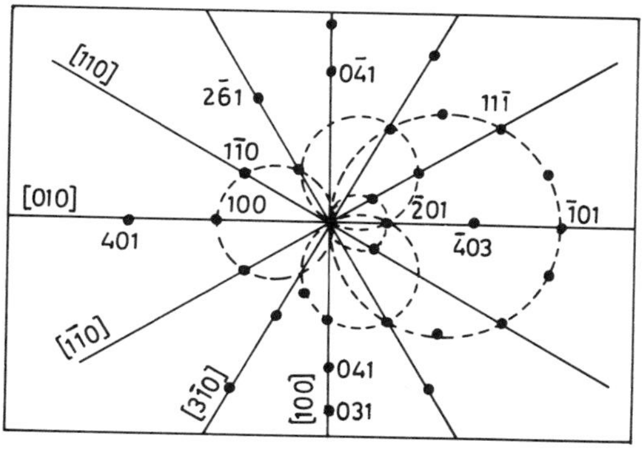

Fig. 8.12. Laue spots of mica obtained by placing the sheet normal to the incident beam

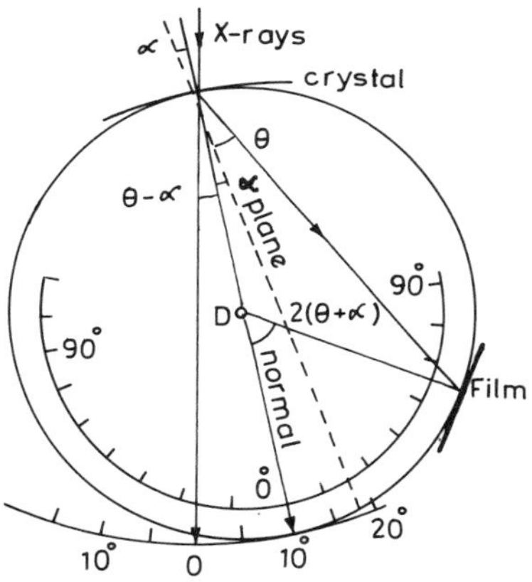

Fig. 8.13. The (θ-α) setting of the Cauchois spectrograph for the case $\theta > \alpha$

All of these various (h,k,l) planes intersect each other on lines parallel to the zone axis and are inclined at different angles α_{hkl} to the planes normal to the surface. The theory given here holds for these planes as well, except that θ must be replaced by $\theta \pm \alpha_{hkl}$. In practice, the crystal holder is rotated an angle $\theta \pm \alpha_{hkl}$ so that the (h,k,l) plane at L (Fig.8.11) makes Bragg angle θ with the incident beam. One such setting is shown in Fig.8.13.

DuMOND [8.12] used the Cauchois spectrograph in reverse, with a point source placed on the focal circle. The beam falls on the concave side of the crystal and the transmitted resolved radiation is detected by a Geiger counter, after it passes through a multiplate diaphragm.

In the Johann-type curved-crystal spectrograph [8.13] the divergent X-ray beam is reflected by a system of atomic planes parallel to the curved surface of radius R. The X-rays from a line-focus spot placed on the circle of

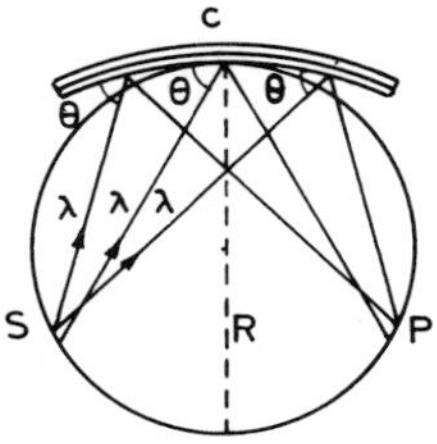

Fig. 8.14. Johann spectrograph. S: source, C: crystal, P: film

radius R/2 are reflected so as to give a fairly sharp line on the same circle (Fig.8.14). The principle of this spectrograph easily follows from that of the transmission-type instrument. To improve the focusing, JOHANSSON [8.14], the crystal of Johann, bent to a radius of R, is further ground to the radius of R/2.

8.7 Double-Crystal Spectrometer

The double-crystal spectrometer is an ionization spectrometer in which the incident beam first falls on the crystal A, and then on the crystal B. The theory has been developed by SCHWARZSCHILD [8.15], SPENCER [8.16], LAUE [8.17] and SMITH [8.18]. It has been reviewed by COMPTON and ALLISON [8.19], and THOMSEN [8.20]. We shall present here only a simple treatment of the basic results. We shall assume that the effects of tilt of the slit and vertical divergence are negligible.

The two possible settings, *plus position* (1,1) and *minus position* (1,-1) are shown in Fig.8.15, for the first-order reflections from A and B. In the (1,1) setting the incident ray and the ray reflected from B lie on the same side of the ray between A and B, whereas in the (1,-1) setting they lie on opposite sides and A is parallel to B for $\beta = 0$.

Consider a characteristic wavelength λ_0 corresponding to the peak position of a line in the incident radiation. Let $J(\lambda-\lambda_0)d\lambda$ be proportional to the intensity of X-rays between λ and $\lambda + d\lambda$. The function J thus gives the distribution of energy in the incident spectrum; for a line it can be approximated by the classic Lorentz form

$$J(\lambda-\lambda_0) = \frac{A}{1+[2(\lambda-\lambda_0)/\Gamma]^2} \quad , \tag{8.18}$$

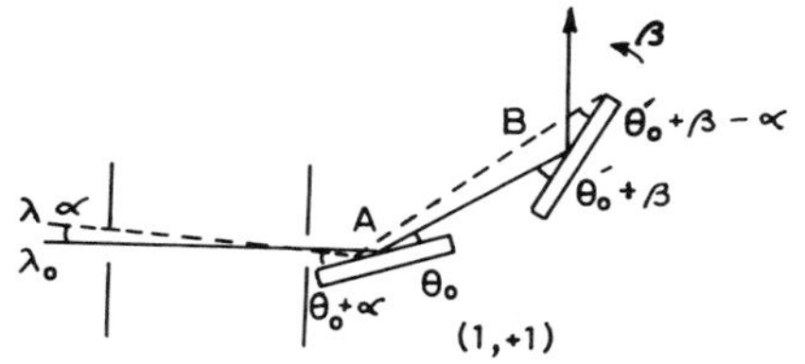

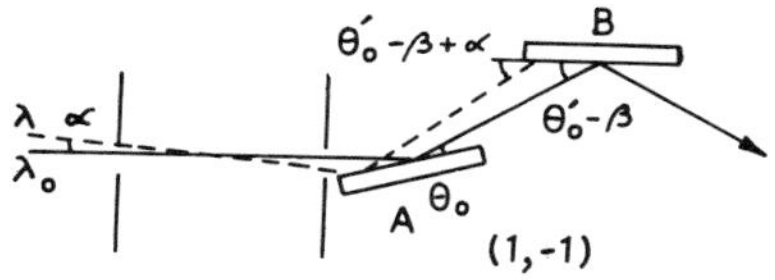

Fig. 8.15. The double-crystal spectrometer settings

where Γ is the linewidth. J is great only in the neighborhood of λ_0. Let a function $G(\alpha)$ describe the geometry of the instrument, that is, the slit size, the distribution of intensity in the focal spot, etc. Then,

$$\iint G(\alpha)J(\lambda-\lambda_0)d\alpha\ d\lambda \tag{8.19}$$

gives the total power in the incident beam.

Let a central ray λ_0 make the correct Bragg angle θ_0 with the first crystal A and the neighboring wavelength λ make an angle $\theta_0+\alpha$ with the face of the crystal A. The correct Bragg angle for λ according to the Bragg law $2d\ \sin\theta_0=\lambda_0$, is $\theta_0+\Delta\theta_0$ where

$$\Delta\theta_0 = (\lambda-\lambda_0)d\theta_0/d\lambda_0 = \left[(\lambda-\lambda_0)/\lambda_0\right]\tan\theta_0 \quad . \tag{8.20}$$

The power reflected from the crystal A will be given by a function $R(\delta\theta)$ of the deviation $\delta\theta$ of the actual angle $\theta_0+\alpha$ from the correct Bragg angle $\theta_0+\Delta\theta_0$, for the wavelength λ; $\delta\theta=(\theta_0+\alpha)-(\theta_0+\Delta\theta_0)=\alpha-\Delta\theta_0$. The *single-crystal diffraction-pattern function* $R(\delta\theta)$ would have its maximum value for $\delta\theta=0$. The total power in the beam after diffraction from the first crystal A can be expressed as

$$\iint G(\alpha)J(\lambda-\lambda_0)R(\alpha-\Lambda_0\ \tan\theta_0)d\alpha\ d\lambda \quad , \tag{8.21}$$

where $\Lambda_0=(\lambda-\lambda_0)/\lambda_0$.

Let the ray λ_0 make the correct Bragg angle θ_0' with the second crystal B in the original position (1,1) or (1,-1), where the prime indicates the angles at the second crystal B; $\theta_0' = \theta_0$. The deviation $\delta\theta'$ of the ray λ for the two settings can be calculated when the second crystal B has been rotated by an angle β (taken positive when counterclockwise) (Fig.8.15):

(1,1) *position*

glancing angle of ray $\lambda_0 = \theta_0' + \beta$,

glancing angle or ray $\lambda = \theta_0' + \beta - \alpha$,

$$\delta\theta_+' = (\theta_0'+\beta-\alpha) - (\theta_0'+\Delta\theta_0') = \beta - \alpha - \Lambda_0 \tan\theta_0' \quad . \tag{8.22}$$

(1,-1) *position*

glancing angle of ray $\lambda_0 = \theta_0' - \beta$,

glancing angle of ray $\lambda = \theta_0' - \beta + \alpha$,

$$\delta\theta_-' = -\beta + \alpha - \Lambda_0 \tan\theta_0' \quad . \tag{8.23}$$

We can combine the two results as

$$\delta\theta_\pm' = \pm\beta \mp \alpha - \Lambda_0 \tan\theta_0' \quad , \tag{8.24}$$

where the upper sign is for the puls position and the lower sign is for the negative position.

The power reflected by the second crystal B can now be written as

$$P(\beta) = \int_{\lambda_{min}}^{\lambda_{max}} \int_{-\alpha_m}^{\alpha_m} G(\alpha)J(\lambda-\lambda_0)R(\alpha-\Lambda_0 \tan\theta_0)R'(\pm\beta\mp\alpha-\Lambda_0 \tan\theta_0')d\alpha\, d\lambda \quad , \tag{8.25}$$

where the limits of λ are determined by the limits of wavelengths reflected by A, which depend on the horizontal divergence of the incident beam. If the limiting slits are rectangular apertures, of equal width w separated by the distance L, then $\alpha_m = w/L$. Equation (8.25) is the general equation of the instrument. We can deduce from it the properties and shapes of different types of *rocking curves* under suitable assumptions. A rocking curve is an ionization-versus-position curve obtained by rotating the second crystal B about its own axis. We shall assume the reflection coefficient R to be the same function of the angle of reflection for each crystal, $R = R'$.

An expression for the *dispersion* can be obtained by assuming that a perfectly discrete Bragg angle θ exists for any incident wavelength λ. Then the contribution to $P'(\beta)$ would occur from only that portion of the beam that has zero deviation from the correct Bragg angle, $\delta\theta = 0$. Thus,

$$\alpha - \Lambda_0 \tan\theta_0 = 0 \quad , \quad \text{(for A)} \quad , \tag{8.26}$$

$$\pm\beta \mp \alpha - \Lambda_0 \tan\theta_0' = 0 \quad , \quad \text{(for B)} \quad . \tag{8.27}$$

Eliminating α, we have

$$\beta - \Lambda_0(\tan\theta_0 \pm \tan\theta_0') = 0 \quad . \tag{8.28}$$

The dispersion D of the double-crystal spectrometer when crystal B alone is rotated is defined as $d\beta/d\lambda$,

$$D_\pm = d\beta/d\lambda = \lambda_0^{-1}(\tan\theta_0 \pm \tan\theta_0') \quad . \tag{8.29}$$

Obviously, $D_+ \neq 0$ and $D_- = 0$, and therefore the minus position is a case of zero dispersion. For the $MoK\alpha_1$ line $D_+ \simeq 68.57$ (angle in s/XU), which is twice the value for a single-crystal spectrometer. $D_- = 0$ is a useful property.

8.7.1 The Case of Zero Dispersion (Minus Position)

In (8.25), we note that $R(\alpha-\Lambda_0 \tan\theta_0)$ is large when the deviation from the Bragg angle is negligible, $\alpha - \Lambda_0 \tan\theta_0 \simeq 0$. Thus, the action of A is to separate the beam into monochromatic parallel bundles (Fig.8.16a). Similarly, $R(-\beta+\alpha-\Lambda_0 \tan\theta_0')$ is large when $\beta \simeq \alpha - \Lambda_0 \tan\theta_0'$ for the crystal B. From the condition on A, we conclude that $\beta \simeq 0$, for the (1,-1) setting, for large value of $P(\beta)$. Thus, the effective range of β is very small. The limits of λ integration are determined by $\lambda_\pm = \lambda \pm \alpha_m \, d\lambda_0/d\theta_0 = \lambda_0 \pm \alpha_m \lambda_0 \cot\theta_0$.

From the condition on A, we can write $G(\alpha) = G(\Lambda_0 \tan\theta_0)$. Also, because the important values of α lie very close to $\Lambda_0 \tan\theta_0$, that is, in a range that is too small compared to the range $2\alpha_m = 2w/L$, we can extend the limit of α inte-

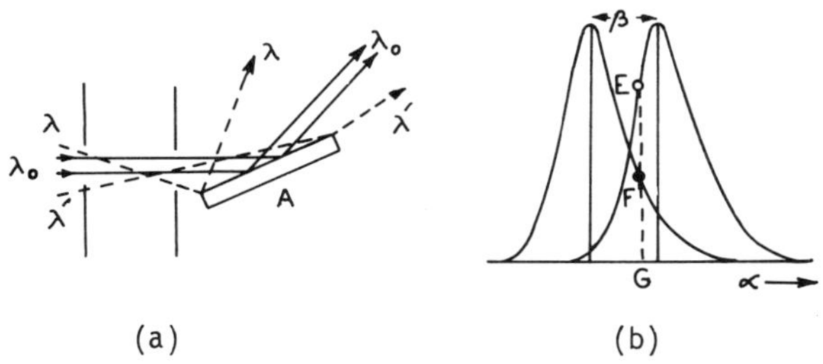

Fig. 8.16. (a) Action of crystal A. (b) Overlap of R(1) and R(1-β)

gration to $\pm\infty$ without affecting the value of the integral. For the minus (parallel) setting $R = R'$ and $\theta_0 = \theta_0' = \theta$. Therefore, (8.25) becomes

$$P(\beta) = \int_{\lambda_-}^{\lambda_+} G(\Lambda_0 \tan\theta)J(\lambda-\lambda_0)d\lambda \int_{-\infty}^{+\infty} R(l)R(l-\beta)dl$$

$$= K \int_{-\infty}^{+\infty} R(l)R(l-\beta)dl \quad , \tag{8.30}$$

where K is a number, $\lambda_\pm = \lambda_0 \pm \alpha_m\lambda_0 \cot\theta$, and we have replaced the variable α by the new variable $l = \alpha - \Lambda_0 \tan\theta$, $dl = d\alpha$.

The properties of the overlap integral (8.30) can be studied graphically. In Fig.8.16 we show plots of $R(l)$ and $R(l-\beta)$ as functions of l for a given β. The two curves have the same shape but $R(l-\beta)$ is displaced relative to $R(l)$. The integrand for any given l is given by the product of the ordinates of the two curves, for example, GF and GE. If a series a of these products is plotted against the corresponding values of α, the area under the resulting curve would give the value of the integral for the given β. Clearly, it has a peak at $\beta = 0$ (the two curves coincide) and falls off *symmetrically* as $|\beta|$ increases $P(\beta) = P(-\beta)$. This is an important property of the rocking curve and can be verified from the consideration of Fig.8.16 or by the use of the theorem

$$\int_{-\infty}^{+\infty} F(x)dx = \int_{-\infty}^{+\infty} F(x-a)dx \quad , \quad a = \text{constant} \quad , \tag{8.31}$$

for F finite and continuous everywhere. Thus,

$$P(\beta) = K \int_{-\infty}^{+\infty} R(l)R(l-\beta)dl = K \int_{-\infty}^{+\infty} R(l+\beta)R(l)dl = P(-\beta) \quad . \tag{8.32}$$

The power incident on the second crystal B can be written as

$$K \int_{-\infty}^{\infty} R(l)dl \quad .$$

We use this as a normalization factor, and write

$$P(\beta) = \int_{-\infty}^{\infty} R(l)R(l-\beta)dl / \int_{-\infty}^{\infty} R(l)dl \quad . \tag{8.33}$$

The intensity of an unpolarized X-ray beam can be considered as the sum of two equal parts for the independent components in the two directions of polarization. For unpolarized incident radiation, therefore, (8.33) becomes

$$P(\beta) = \frac{\int_{-\infty}^{\infty}\left[R_\sigma(1)R_\sigma(1-\beta)+R_\pi(1)R_\pi(1-\beta)\right]d1}{\int_{-\infty}^{\infty}\left[R_\sigma(1)+R_\pi(1)\right]d1} \quad , \tag{8.34}$$

where the suffix $\sigma(\pi)$ indicates that the electric vector is perpendicular (parallel) to the plane of incidence.

For a perfect crystal with negligible absorption, Darwin found that the angular widths of R curves, and so areas under them, are related as [8.19]

$$\int_{-\infty}^{\infty} R_\pi(1)d1 = \cos 2\theta \int_{-\infty}^{\infty} R_\sigma(1)d1 \quad . \tag{8.35}$$

We shall now consider how the area of the double-reflection curve, the integral of $P(\beta)$ with respect to β, written as $P_D = \int P(\beta)d\beta$, is related to the integrated reflection from the single crystal. Let us introduce a new variable γ such that $\beta = 1 - \gamma$, $d\beta = -d\gamma$. Then for either integral in the numerator of (8.34),

$$\int_{-\infty}^{\infty}\int_{-\infty}^{\infty} R(1)R(1-\beta)d1\ d\beta = \int_{-\infty}^{\infty}\int_{\infty}^{-\infty} R(1)R(\gamma)d1(-d\gamma)$$

$$= \int_{-\infty}^{\infty}\int_{-\infty}^{\infty} R(1)R(\gamma)d1\ d\gamma = \left[\int_{-\infty}^{\infty} R(1)\,d1\right]^2 \quad .$$

Thus,

$$P_D = \int_{-\infty}^{\infty} P(\beta)d\beta = \frac{\rho_\sigma^2+\rho_\pi^2}{\rho_\sigma+\rho_\pi} = \frac{1+\cos^2 2\theta}{1+\cos 2\theta}\,\rho_\sigma \quad ,$$

where

$$\rho_\sigma = \int_{-\infty}^{\infty} R_\sigma(1)d1 \quad , \quad \rho_\pi = \int_{-\infty}^{+\infty} R_\pi(1)d1 = \cos 2\theta\ \rho_\sigma \quad ,$$

are the integrated reflections from the single crystal for the respective polarizations σ, π. For the unpolarized radiation, we can define $\rho_D = (\rho_\sigma+\rho_\pi)/2 = (1+\cos 2\theta)\,\rho_\sigma/2$, so that

$$P_D = \frac{2(1+\cos^2 2\theta)}{(1+\cos 2\theta)^2}\,\rho_D \quad , \tag{8.36}$$

in Darwin's approximation. From (8.36), ρ_D can be calculated from the area of the observed rocking curve, and compared with Darwin's theory [8.21-23].

For $\beta = 0$, the numerator in (8.34) becomes $(1+\cos 2\theta)\int[R_\sigma(1)]^2 d1$ and

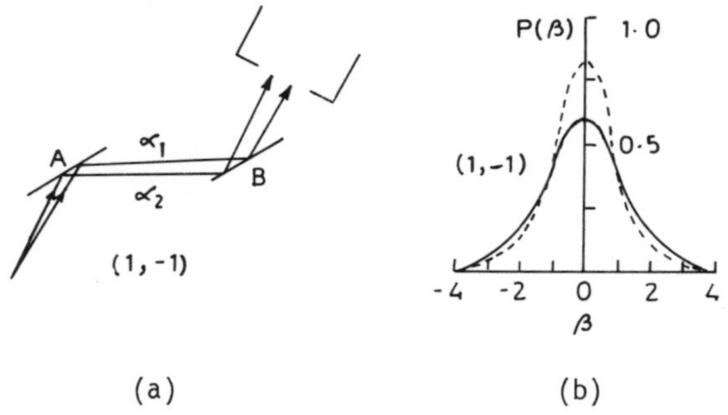

Fig. 8.17. (a) The (1,-1) position, D = 0; (b) A typical rocking curve (dotted: calculated, solid: observed)

$$P(\beta=0) = \frac{\int_{-\infty}^{\infty}\left[R_\sigma(l)\right]^2 dl}{\int_{-\infty}^{\infty} R_\sigma(l)dl} = \frac{32/15}{8/3} = \frac{4}{5} \quad , \tag{8.37}$$

in Darwin's theory. Because $P(\beta=0)$ is the peak value of the symmetrical rocking curve, it can be compared with the theoretical value 4/5.

The (1,-1) setting is characterized by zero dispersion, and so by economy of X-ray energy. For illustration, we consider $CuK\alpha_{1,2}$ lines. $K\alpha_2$ is at a slightly longer λ and has half the intensity of $K\alpha_1$. If α_1 has found a spot at which to reflect from A and therefore also from B (parallel to A), there is also a spot at which α_2 can reflect from A and B and enter the detector (Fig.8.17a). Thus, α_1 and α_2 and all nearby wavelengths are simultaneously recorded by the detector. Consequently, a strong but single peak is obtained in the rocking curve (Fig.8.17b). The observed peak shape and height can be made to approach the theoretical curve by polishing and etching the crystals A and B.

8.7.2 The Case of Nonzero Dispersion (Plus Setting)

For the (1,1) setting, we get from (8.25), with $\theta_0 = \theta_0' = \theta$ and $R = R'$,

$$P(\beta) = \int_{\lambda_{min}}^{\lambda_{max}} \int_{-\alpha_m}^{\alpha_m} G(\alpha)J(\lambda-\lambda_0)R(\alpha-\Lambda_0 \tan\theta)R(\beta-\alpha-\Lambda_0 \tan\theta)d\alpha\, d\lambda \quad .$$

For the reflection to be strong, we must have

$$\alpha \simeq \Lambda_0 \tan\theta \quad , \quad \text{(for A)} \quad ,$$

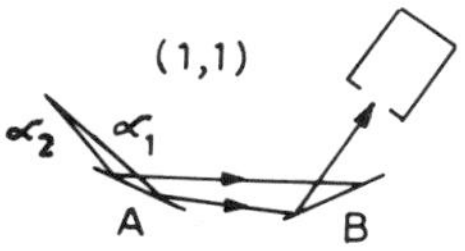

Fig. 8.18. The (1,1) position, $D \neq 0$

$$\beta \simeq \alpha + \Lambda_0 \tan\theta \simeq 2\Lambda_0 \tan\theta \quad , \quad \text{(for B)}.$$

Thus, unlike the (1,-1) case, in this case β has a large effective range.

The assumption $\theta_0 = \theta_0' = \theta$ means that the crystal B is positioned so that $K\alpha_1$ from the crystal A strikes B at the proper Bragg angle for reflection. However, $K\alpha_2$ requires a larger angle than does $K\alpha_1$; it leaves A at a large angle, and arrives at B at a smaller angle than the correct Bragg angle for $K\alpha_2$ (Fig.8.18). Consequently, we must rotate B from the $K\alpha_1$-reflection position to find the $K\alpha_2$-reflection position. (In the minus setting the conditions for α_1 and α_2 reflections were simultaneously satisfied.) The resulting high resolution of the double-crystal spectrometer makes it a very useful instrument. For Bragg-Laue and Laue-Laue schemes see PINSKER [8.34].

8.8 Use of Ruled Gratings

The discovery of total reflection of X-rays made it possible to measure wavelengths by use of ruled gratings provided that the glancing angle $\theta < \theta_c$. Such measurements are free from the uncertainties of determination of interplanar spacings in crystals. COMPTON and DOAN [8.24] were the first to obtain a spectrum with a metal grating; it had 50 lines mm^{-1}. Gratings with 200 lines mm^{-1} give improved results. THIBAUD [8.25] obtained the K spectrum of Cu with a ruled grating.

If d is the grating constant (Fig.8.19a), the grating equation is

$$n\lambda = d\left[\cos\theta - \cos(\theta + \alpha_n)\right] \quad . \tag{8.38}$$

Here n can be negative, with the corresponding glancing angles of diffraction lying between the plane of the grating and the totally reflected ray. However, the negative-order spectra are weak. ECKART [8.26] has given a general derivation of the grating equation.

Recently blazed replica gratings have been widely used in the region $20\ \text{Å} < \lambda < 100\ \text{Å}$. Figure 8.19b shows the cross section of a blazed grating.

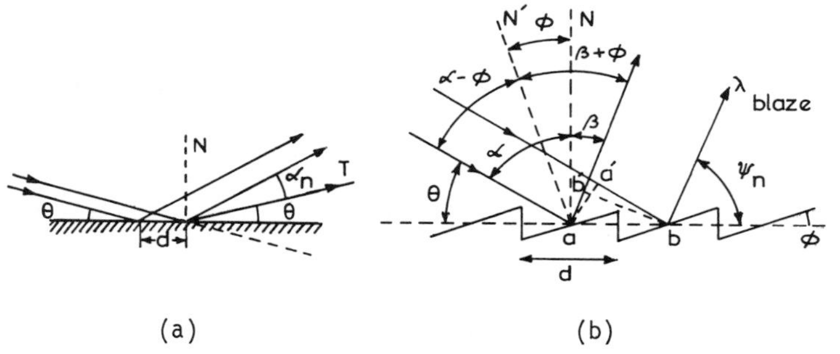

Fig. 8.19. (a) Diffraction from a ruled grating. T is the totally reflected ray. (b) Blazed grating: $n\lambda = ba' - ab' = d(\cos\theta - \cos\psi_n)$, $\psi_n = \theta + 2\phi$, ϕ = blaze angle

Let N be the normal to the macroscopic surface of the grating and N' to the groove face of blaze angle ϕ. The blaze angle is selected so that the wavelength of interest is diffracted in a direction that coincides with the direction of the specularly reflected beam from the surface of the facet, for greater intensity. The condition for this specular reflection is $\alpha - \phi = \beta + \phi$ (Fig.8.19b), where $\alpha(\beta)$ is the angle of incidence (diffraction). Therefore, $\beta = \alpha - 2\phi$ and $n\lambda = d(\sin\alpha - \sin\beta) = d[\sin\alpha - \sin(\alpha - 2\phi)] = d(\cos\theta - \cos\psi_n)$, where $\psi_n = \theta + 2\phi$. The use of gratings in the X-ray spectra has been discussed in detail by CUTHILL [8.27].

8.9 Detectors

Röntgen detected the X-rays for the first time with a fluorescent screen coated with barium platinocyanide. He also was the first to use photographic films for recording X-rays.

The X-ray detectors are mainly of three kinds: 1) photographic films, 2) ionization detectors, and 3) solid-state detectors. We shall briefly discuss them here.

8.9.1 Photographic film

Double-coated films, or single-emulsion films, can be used. When an X-ray beam is incident at an oblique angle on a double-coated film, two images, which are slightly displaced with respect to each other, are formed. Therefore, it is usual to remove the far-side emulsion before measuring lines.

The photographic emulsion is a colloidal suspension of small grains of silver bromide in gelatin. The emulsion layer is about 15 μm thick with a protective gelatin layer, about 1 μm thick, coated over it. The X-ray film base is about 200 μm thick.

The action of an X-ray photon is to free electrons from the bromide ions. These electrons tend to collect in regions of the crystals known as "electron traps". The excess negative charge in the traps neutralizes adjacent silver ions, which are thereby converted to uncharged silver atoms. Those atoms form a "latent image", which acts as a catalyst for development. The developer increases the sizes of such metallic-silver particles many millions of times. Overdevelopment produces "fog", because the developer tends to reduce unexposed silver bromide as well.

8.9.2 Ionization Detectors

Usually, an argon-filled detector is used to detect X-rays. It takes about 30 eV to create an electron-ion pair in Ar, $Ar + h\nu \rightarrow Ar^{+} + e^{-}$. An X-ray photon of $\lambda = 1$ Å has an energy of 12,400 eV. Therefore, for each photon absorbed by the gas, about 400 electron-ion pairs are formed. If there is no applied voltage on the enclosed gas, the liberated electrons and ions tend to recombine. On the other hand, if a metal wire, centrally mounted in a metal cylinder (Fig.8.20), is made about 100 V positive relative to the cylinder, the electrons are rapidly swept toward the positive wire anode, whereas the heavy ions move slowly toward the negative envelope. This prevents recombinations, and an electric current, proportional to the number of photons absorbed, is produced in the recording circuit. The detector used in this way is said to be an *ionization chamber*. Argon gas at a pressure of about 75 cm Hg is used for $\lambda > 1$ Å. Other useful gases are CH_3I, CH_3Br and krypton at about 20 cm Hg or less for $\lambda < 1$ Å. When the photon energy is greater than the absorption-edge energy of the counter gas, characteristic radiation will be produced from the gas, which will produce an additional "escape" peak that may interfere with measurements.

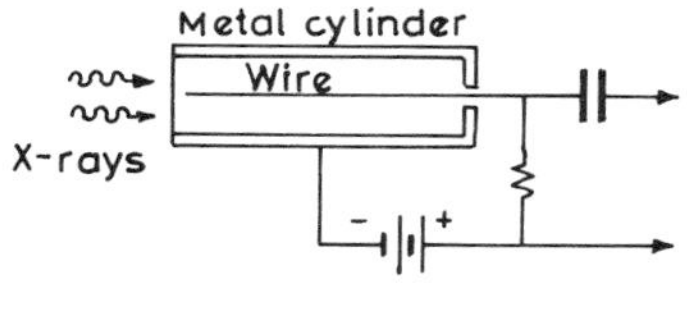

Fig. 8.20. Ionization detector

The number of electron-ion pairs can be increased by increasing the voltage on the wire. When the voltage exceeds approximately 300 V, each electron liberated by the absorbed photon acquires enough kinetic energy by acceleration so as to be able to produce, in turn, new electron-ion pairs, by collision with gas atoms. This amplification effect A is proportional to the applied voltage V in the range $300 < V < 900$ V. If n is the number of electron-ion pairs formed by each photon absorbed, then An electrons reach the wire. Individual electrical pulses can be measured if the photons come at reasonable intervals. The ionization detector (Fig.8.20) used in this voltage range is called a *proportional counter*, because the magnitude of the electrical pulses, detected for each X-ray photon absorbed, is proportional to the energy $h\nu$ of the X-ray photon.

If the voltage is increased to about 1 kV or more, an "avalanche effect" is triggered which gives a much increased value of A. This avalanche spreads throughout the counter so rapidly (about 1 μs) that almost the same-size pulse is formed by any absorbed X-ray photon, regardless of its energy. It now works as a *Geiger-Müller counter*.

8.9.3 Solid-State Detectors

Phosphor-coated screens have been used since early days to detect X-rays. The X-ray energy is converted into visible light by the process of fluorescence. Zinc calcium sulphide fluoresces at about 5300 Å for which the eye has maximum sensitivity.

Suppose pulses of visible light (*scintillations*) are produced by X-rays in a *single crystal* (such as thallium-activated NaI crystal) that is almost transparent to such light. Then we can amplify the light pulses with the help of a *photomultiplier*. Such an assembly is called a *scintillation counter*. Its sensitivity range is 0.2 to 2 Å.

Crystals can be used to detect X-rays in yet another way. If photons are absorbed by certain semiconductors or insulators, such as cadmium sulphide or silicon, the electrons are excited to the conduction band. The photoinduced current produced in this way can be used to detect the radiation. Such a *photoconducting detector* has been developed for the X-ray region. It is called a Si(Li) *detector*. It is a piece of p-type silicon that has been "doped" with lithium atoms in order to increase the sensitive, or "intrinsic" layer. To keep the lithium atoms from diffusing away, the crystal (3 to 5 mm thick) is kept at the liquid-nitrogen temperature [8.28].

8.9.4 Pulse-Height Discriminator

The Geiger-Müller counter is unable to distinguish the relative energies of X-rays and so is not of much use in X-ray spectroscopy. Proportional counters or solid-state detectors are commonly employed, because the magnitude of the pulse produced in them is proportional to the energy of the incident photon. This property enables us to use an electronic circuit that transmits only pulses greater than a certain magnitude or "height" and blocks lower-height pulses. Such a *pulse-height discriminator* can be combined with another circuit that transmits only those pulses that are smaller than a given height to form a *single-channel pulse-height analyzer*. In this way, it is possible to detect pulses that are formed by X-rays in a narrow range of energies. This is the principle of *energy-dispersion spectrometry*.

8.9.5 Intensity Measurement

The intensity I is the energy of X-rays that cross 1 cm^2 area/s perpendicular to the beam direction. If n photons of energy $h\nu$ in a monochromatic beam cross 1 cm^2 area/s, then $I = nh\nu$. For the continuous spectrum of X-rays,

$$I = \int_0^{\nu_{max}} \frac{dn}{d\nu} h\nu \, d\nu = \int_0^{\nu_{max}} I_\nu \, d\nu \quad ,$$

where $dn/d\nu$ is the numerical density of photons per unit of frequency that pass through 1 cm^2 in 1 s. If I_j is the intensity of the separate lines present, we have

$$I = \int_0^{\nu_{max}} I_\nu \, d\nu + \sum_j I_j \quad .$$

RUMP [8.29] directly measured the energy of X-rays by use of a calorimeter.

An ionization chamber can be used to measure I. A layer of dry air 1 cm thick at NTP with a density of 0.001293 g cm^{-3} has been chosen as a standard. The ionization coefficient of the gas is defined as

$$\gamma = \Delta I/I \quad ,$$

where ΔI is the fraction of the intensity I absorbed by a 1 cm layer of air.

The average energy required to ionize one molecule of air is 33.73 ± 0.15 eV. X-rays crossing normally a cube of air impart a fraction D of its energy in time t, so that the physical dose is

$$D = t\Delta I = \gamma I t \quad .$$

It has the dimension of energy per unit volume. The (old) *Röntgen unit* P is the dose that creates 1 esu of charge per 0.001293 g of air. Thus, $P = D/t = \gamma I$ or $I = P/\gamma = D/(\gamma t)$.

Appendix A
Rutherford Scattering for Attractive Field

A.1 Equation of Hyperbola

In Fig.A.1 establish at O the origin (inner focus) of polar coordinates r, ϕ. Now establish a point (external focus) O' at $\phi = \pi$, $r = 2a\varepsilon$. The perihelion $OB = a\varepsilon - a$, with $2a$ as the distance between the vertices of the two branches of the hyperbola that can be drawn between O and O' with *eccentricity* ε.

The hyperbola is the locus of all points whose distances r and r', from the foci O and O', have a constant difference $2a$, that is

$$r' - r = 2a \quad , \quad r'^2 = r^2 + 4ar + 4a^2 \quad . \tag{A.1}$$

From the triangle OO'A, we get

$$r'^2 = r^2 + (2a\varepsilon)^2 - 2r(2a\varepsilon)\cos(180-\phi) \quad .$$

Equating these values of r'^2, we get

$$r = \frac{a(\varepsilon^2-1)}{1-\varepsilon\cos\phi} \quad . \tag{A.2}$$

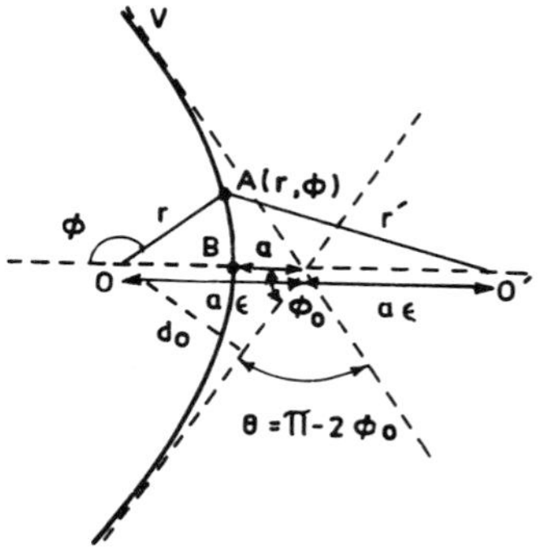

Fig. A.1. Hyperbolic orbit, which is characteristic of Coulomb attractive force

The *asymptotes* correspond to $r=\infty$ and make angles $\pm\phi_0$, such that

$$1 = \varepsilon \cos\phi_0 \quad . \tag{A.3}$$

If θ is the angle of deflection, then

$$\theta = \pi - 2\phi_0 \quad , \quad \cos\phi_0 = \sin\frac{1}{2}\theta = \frac{1}{\varepsilon} \quad , \quad \varepsilon^2 - 1 = \cot^2\frac{1}{2}\theta \quad . \tag{A.4}$$

The minimum positive value of r (vertex) is for $\phi = 180^\circ$, $\cos\phi = \cos\pi = -1$, and locates the vertex at

$$r_{OB} = a(\varepsilon-1) \quad , \quad \phi = \pi \quad . \tag{A.5}$$

Because the impact parameter is given by $\sin\phi_0 = d_0/a\varepsilon$ (Fig.A.1) we can eliminate a from (A.2), to get

$$r = \frac{d_0 \tan\phi_0}{1-\varepsilon \cos\phi} \quad . \tag{A.6}$$

A.2 Rutherford Scattering

An electron of mass m, charge -e and velocity v moves along the hyperbola if a heavy nucleus of charge +Ze is at the focus O. We have, at the vertex B, by conservation of energy

$$\frac{1}{2} mv^2 = \frac{1}{2} mv_B^2 - \frac{Ze^2}{a(\varepsilon-1)} \quad , \tag{A.7}$$

or, because $a = d_0/(\varepsilon \sin\phi_0) = d_0/\tan\phi_0$,

$$\frac{v_B^2}{v^2} = 1 + \frac{2k}{d_0}\frac{\sin\phi_0}{1-\cos\phi_0} \quad , \quad k = \frac{Ze^2}{mv^2} \quad . \tag{A.8}$$

The conservation of angular momentum gives

$$mvd_0 = mv_B\, a(\varepsilon-1) \quad , \tag{A.9}$$

or

$$\frac{v_B}{v} = \frac{d_0}{a(\varepsilon-1)} = \frac{\sin\phi_0}{1-\cos\phi_0} \quad . \tag{A.10}$$

From (A.8,10), we get

$$1 + \frac{2k}{d_0}\frac{\sin\phi_0}{1-\cos\phi_0} = \frac{1+\cos\phi_0}{1-\cos\phi_0} \quad ,$$

or

$$\tan\phi_0 = \frac{d_0}{k} = \frac{md_0v^2}{Ze^2} \quad . \tag{A.11}$$

Appendix B
Bohr's Formula for Energy Loss

In 1913 BOHR [1.36] derived a classical expression for the space rate of energy loss for a heavy charged particle of charge ze, mass m_z and velocity v, passing an atomic electron of mass m at a distance d_0 (impact parameter), see Fig.B.1. The impulse $\int_{-\infty}^{+\infty} F_x \, dt$ parallel to the path is zero by symmetry (for each position of the incident particle in the -x direction there is a corresponding position in the +x direction). However, there is always a force in the y direction. The associated impulse I_y is

$$I_y = \int_{-\infty}^{+\infty} F_y \, dt = \int_{-\infty}^{+\infty} \frac{ze^2 \sin\theta}{r^2} \frac{dx}{v} = \int_0^{\pi} \frac{ze^2 \sin\theta \, d\theta}{d_0 v}$$

because $dt = dx/v$, $r = d_0/\sin\theta$, and $x = -d_0 \cot\theta$, which gives $dx = (d_0 d\theta)/\sin^2\theta$.

Therefore, the momentum transferred to the electron during the entire passage is

$$I_y = 2ze^2/(d_0 v) \quad , \tag{B.1}$$

and the energy acquired by the electron is

$$E(d_0) = \frac{I_y^2}{2m} = \frac{2z^2e^4}{mv^2}\left(\frac{1}{d_0^2}\right) \quad . \tag{B.2}$$

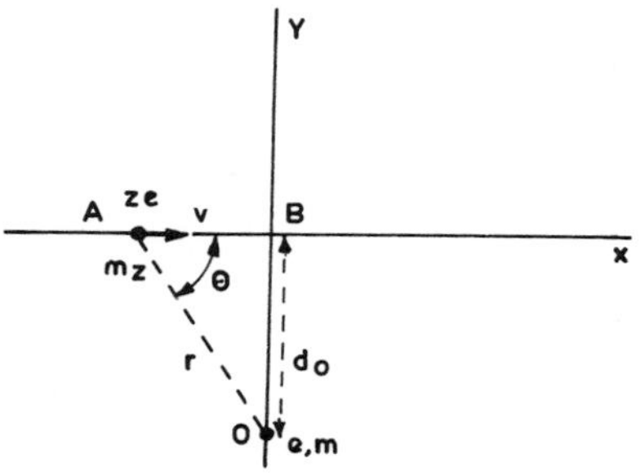

Fig. B.1. A heavy particle of charge ze and mass m_Z passing an electron at distance d_0

A particle passing through matter sees electrons at various distances from its path. If L is the number of target atoms per cm^3, the number of electrons located at impact parameters between d_0 and $d + dd_0$ in a thickness dx of target is $L\ 2\pi d_0\ dd_0\ dx$. Therefore, the energy loss is

$$- \frac{dE}{dx} = 2\pi L \int E(d_0)\ d_0 dd_0 = 4\pi L\ \frac{z^2 e^4}{mv^2} \int_{min}^{max} \frac{1}{d_0^2} d_0 dd_0 = 4\pi L\ \frac{z^2 e^4}{mv^4} \ln \frac{d_{max}}{d_{min}} \quad . \tag{B.3}$$

The integral over dd_0 is not taken from 0 to ∞ because then dE/dx would be infinite.

We have now to select values for d_{min} and d_{max}. The maximum change of kinetic energy possible in an elastic collision is $\frac{1}{2}$ m (change of velocity)^2 $= \frac{1}{2} m(2v)^2 = 2mv^2$. If we insert this value of $E_{max}(d_0)$ in (B.2), we get

$$d_{min} = \frac{ze^2}{mv^2} \quad . \tag{B.4}$$

The force ($\propto 1/r^2$) experienced by the electron is large only when the charged particle remains within a distance of the order of d_0 from the point B of closest approach; thereafter it decreases very rapidly with increasing distance. Thus, the time during which energy transfer takes place is of the order of $d_0/v = \tau$. In quantum theory, $d_0/v \approx \hbar/(E_n - E_0) = 1/\omega$, where E_n and E_0 are atomic energy levels and $\hbar\omega$ the energy difference. Because $\hbar/(E_n - E_0)$ is usually of the same order as the classically calculated periods, we can write

$$\tau(d_{max}) \simeq 1/\omega_j = 1/(2\pi\nu_j) \quad , \tag{B.5}$$

where ω_j is a characteristic atomic frequency. Then

$$d_{max} \simeq v/(2\pi\nu_j) \quad . \tag{B.6}$$

Thus

$$- \frac{dE}{dx} = 4\pi L\ \frac{z^2 e^4}{mv^2} \ln \frac{mv^3}{2\pi\nu_j e^2} \quad .$$

Taking the sum over all frequencies ν_j of electrons, we can write

$$\sum_{j=1}^{Z} \ln \frac{mv^3}{2\pi\nu_j e^2} = Z \ln \frac{mv^3}{2\pi\bar{\nu}_j ze^2} = Z\ l \quad , \tag{B.7}$$

where $\bar{\nu}_j$ is some average value. The quantity l takes average value of 6. Because $dv^4/dx = (8E/m^2)dE/dx = (4v^2/m)dE/dx$, we finally get

$$\frac{dv^4}{dx} = -bZ \quad , \quad b = 16\pi L \frac{z^2e^4}{m^2} 1 \quad . \tag{B.8}$$

This formula was developed by BOHR [1.36] for the passage of heavy particles, like α particles, through matter. However, KRAMERS has used it for the passage of electrons through a target. For fast electrons, it overestimates the energy loss. A more satisfactory formula for the stopping of non-relativistic electrons has been derived by BETHE [1.37,38].

Appendix C
X-Ray Atomic Energy Levels

SIEGBAHN [2.55] first evaluated the atomic energy levels from X-ray absorption edges and emission lines. K edges were used for low Z and L_{III} edges for the rest, as reference levels. Other levels were determined on a relative scale from the emission lines as suggested by IDEI [2.44]. The principles involved in evaluating atomic energy levels are shown in Fig.C.1. Usually, the energy difference between two levels can be obtained in two or more ways. For example, the L_{II} and L_{III} difference can be found from $K\alpha_1 - K\alpha_2$, or from $L\beta_1$ $(L_{II} \to M_{IV}) - L\beta_2$ $(L_{III} \to M_{IV})$. The number of available lines is usually much greater than the number of levels involved. In place of the X-ray absorption edge, photoelectron measurement values can be used, if available [2.68], as they establish the absolute scale. For thorium, as many as ninety-nine equations (including sixteen photoelectric measurements) can be set up to determine only twenty-five unknown levels. A least-square solution of this set gives the required energy levels.

As improved X-ray and photoelectron data have become available, several reviews have appeared [2.5,38,67-73]. The K atomic energy levels are given in Appendix D. E_K is the energy difference between the Fermi level (zero energy) and the K state.

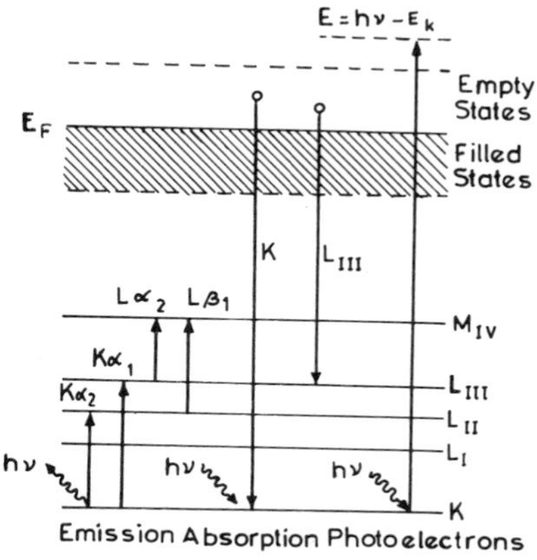

Fig. C.1. Principles involved in evaluating atomic energy levels from measurements in solids. E_F is the Fermi energy

Appendix D
Electron Distribution Among the Levels of Free Atoms

K [eV] Binding Energies	Element		Level													
			1s	2s	2p	3s	3p	3d	4s	4p	4d	4f	5s	5p	5d
13.6	1 H	Hydrogen	1												
24.6	2 He	Helium	2												
55	3 Li	Lithium	K	1											
111	4 Be	Beryllium	K	2											
188	5 B	Boron	K	2	1										
284	6 C	Carbon	K	2	2										
402	7 N	Nitrogen	K	2	3										
532	8 O	Oxygen	K	2	4										
685	9 F	Fluorine	K	2	5										
867	10 Ne	Neon	K	2	6										
1072	11 Na	Sodium	KL			1									
1305	12 Mg	Magnesium	KL			2									
1560	13 Al	Aluminum	KL			2	1								
1839	14 Si	Silicon	KL			2	2								
2146	15 P	Phosphorus	KL			2	3								
2472	16 S	Sulfur	KL			2	4								
2822	17 Cl	Chlorine	KL			2	5								
3203	18 Ar	Argon	KL			2	6								
3607	19 K	Potassium	KL			2	6		1						
4038	20 Ca	Calcium	KL			2	6		2						
4493	21 Sc	Scandium	KL			2	6	1	2						
4966	22 Ti	Titanium	KL			2	6	2	2						
5465	23 V	Vanadium	KL			2	6	3	2						
5989	24 Cr	Chromium	KL			2	6	5	1						
6539	25 Mn	Manganese	KL			2	6	5	2						
7112	26 Fe	Iron	KL			2	6	6	2						
7709	27 Co	Cobalt	KL			2	6	7	2						
8333	28 Ni	Nickel	KL			2	6	8	2						
8979	29 Cu	Copper	KL			2	6	10	1						
9659	30 Zn	Zinc	KL			2	6	10	2						
10367	31 Ga	Gallium	KLM						2	1					
11103	32 Ge	Germanium	KLM						2	2					
11867	33 As	Arsenic	KLM						2	3					
12658	34 Se	Selenium	KLM						2	4					
13474	35 Br	Bromine	KLM						2	5					
14326	36 Kr	Krypton	KLM						2	6					
15200	37 Rb	Rubidium	KLM						2	6			1		
16105	38 Sr	Strontium	KLM						2	6			2		
17038	39 Y	Yttrium	KLM						2	6	1		2		

K [eV] Binding energies	Element			Level												
				1s	4s	4p	4d	4f	5s	5p	5d	5f	6s	6p	6d	7s
17998	40	Zr	Zirconium	KLM	2	6	2		2							
18986	41	Nb	Niobium	KLM	2	6	4		1							
20000	42	Mo	Molybdenum	KLM	2	6	5		1							
21044	43	Tc	Technetium	KLM	2	6	6		1							
22117	44	Ru	Ruthenium	KLM	2	6	7		1							
23220	45	Rh	Rhodium	KLM	2	6	8		1							
24350	46	Pd	Palladium	KLM	2	6	10		1							
25514	47	Ag	Silver	KLM	2	6	10		1							
26711	48	Cd	Cadmium	KLM	2	6	10		2							
27940	49	In	Indium	KLM	2	6	10		2	1						
29200	50	Sn	Tin	KLM	2	6	10		2	2						
30491	51	Sb	Antimony	KLM	2	6	10		2	3						
31814	52	Te	Tellurium	KLM	2	6	10		2	4						
33169	53	I	Iodine	KLM	2	6	10		2	5						
34561	54	Xe	Xenon	KLM	2	6	10		2	6						
35985	55	Cs	Cesium	KLM	2	6	10		2	6			1			
37441	56	Ba	Barium	KLM	2	6	10		2	6			2			
38925	57	La	Lanthanum	KLM	2	6	10		2	6	1		2			
40443	58	Ce	Cerium	KLM	2	6	10	2	2	6			2			
41991	59	Pr	Praseodymium	KLM	2	6	10	3	2	6			2			
43569	60	Nd	Neodymium	KLM	2	6	10	4	2	6			2			
45184	61	Pm	Promethium	KLM	2	6	10	5	2	6			2			
46834	62	Sm	Samarium	KLM	2	6	10	6	2	6			2			
48519	63	Eu	Europium	KLM	2	6	10	7	2	6			2			
50239	64	Gd	Gadolinium	KLM	2	6	10	7	2	6	1		2			
51996	65	Tb	Terbium	KLM	2	6	10	9	2	6			2			
53789	66	Dy	Dysprosium	KLM	2	6	10	10	2	6			2			
55618	67	Ho	Holmium	KLM	2	6	10	11	2	6			2			
57486	68	Er	Erbium	KLM	2	6	10	12	2	6			2			
59390	69	Tm	Thulium	KLM	2	6	10	13	2	6			2			
61332	70	Yb	Ytterbium	KLM	2	6	10	14	2	6			2			
63314	71	Lu	Lutetium	KLMN					2	6	1		2			
65351	72	Hf	Hafnium	KLMN					2	6	2		2			
67416	73	Ta	Tantalum	KLMN					2	6	3		2			
69525	74	W	Tungsten	KLMN					2	6	4		2			
71676	75	Re	Rhenium	KLMN					2	6	5		2			
73871	76	Os	Osmium	KLMN					2	6	6		2			
76111	77	Ir	Iridium	KLMN					2	6	7		2			
78395	78	Pt	Platinum	KLMN					2	6	8		2			
80725	79	Au	Gold	KLMN					2	6	10		1			
83102	80	Hg	Mercury	KLMN					2	6	10		2			
85530	81	Tl	Thallium	KLMN					2	6	10		2	1		
88005	82	Pb	Lead	KLMN					2	6	10		2	2		
90526	83	Bi	Bismuth	KLMN					2	6	10		2	3		
93105	84	Po	Polonium	KLMN					2	6	10		2	4		
95730	85	At	Astatine	KLMN					2	6	10		2	5		
98404	86	Rn	Radon	KLMN					2	6	10		2	6		
101137	87	Fr	Francium	KLMN					2	6	10		2	6		1
103921	88	Ra	Radium	KLMN					2	6	10		2	6		2
106755	89	Ac	Actinium	KLMN					2	6	10		2	6	1	2
109651	90	Th	Thorium	KLMN					2	6	10		2	6	2	2
112601	91	Pa	Protactinium	KLMN					2	6	10	2	2	6	1	2
115606	92	U	Uranium	KLMN					2	6	10	3	2	6	1	2
118678	93	Np	Neptunium	KLMN					2	6	10	4	2	6	1	2

K [eV] Binding energies	Element				Level							
					5s	5p	5d	5f	6s	6p	6d	7s
121818	94	Pu	Plutonium	KLMN	2	6	10	5	2	6	1	2
125027	95	Am	Americium	KLMN	2	6	10	6	2	6	1	2
128220	96	Cm	Curium	KLMN	2	6	10	7	2	6	1	2
131590	97	Bk	Berkelium	KLMN	2	6	10	8	2	6	1	2
135960	98	Cf	Californium	KLMN	2	6	10	10	2	6		2
139490	99	Es	Einsteinium	KLMN	2	6	10	11	2	6		2
143090	100	Fm	Fermium	KLMN	2	6	10	12	2	6		2
146780	101	Md	Mendelevium	KLMN	2	6	10	13	2	6		2
150540	102	No	Nobelium	KLMN	2	6	10	14	2	6		2
154380	103	Lw	Lawrencium	KLMN	2	6	10	14	2	6	1	2

Appendix E
Curves Representing Values of Electron Energies

Curves representing values of electron energies, as a function of atomic number Z (compare with Appendix D for the filling of shells), are given in Fig.E.1. They have been calculated by the Thomas-Fermi method.

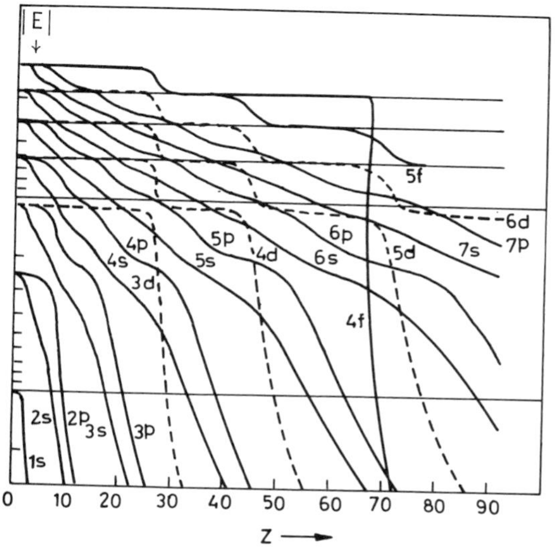

Fig. E.1. Electron energies as a function of Z

Appendix F
Quantization of the Electromagnetic Field

The classical harmonic oscillator of angular frequency $\omega = 2\pi\nu$ is described by a single coordinate q that obeys the equation of motion

$$\ddot{q} + \omega^2 q = 0 \quad . \tag{F.1}$$

If the oscillating particle has mass m, the momentum is $p = m\dot{q}$ and the total energy is given by the Hamiltonian

$$H = \frac{p^2}{2m} + \frac{1}{2} m\omega^2 q^2 \quad . \tag{F.2}$$

In quantum mechanics, we regard q and p as Hermitian matrices. The elements q_{lk}, p_{lk} refer to a transition from the energy state E_k to E_l. The time variation is given by

$$\dot{q} = \frac{1}{i\hbar}[q,H] = \frac{1}{i\hbar}(qH-Hq) = \frac{p}{m} \quad , \tag{F.3}$$

$$\dot{p} = \frac{1}{i\hbar}[p,H] = -m\omega^2 q \quad , \tag{F.4}$$

because $[q_l,q_k] = [p_l,p_k] = 0$ and $[q_l,p_k] = i\hbar\delta_{lk}$. It follows that

$$\ddot{q} = \dot{p}/m = -\omega^2 q \quad . \tag{F.5}$$

Every matrix element of q and p obeys (F.5), that is

$$\ddot{q}_{lk} + \omega^2 q_{lk} = 0 \quad . \tag{F.6}$$

It is satisfied by

$$q_{lk} = a_{lk}\exp(-i\omega_{lk}t) \quad , \quad \omega_{lk} = (E_k - E_l)/\hbar \quad , \tag{F.7}$$

if

$$\left(\omega^2-\omega_{lk}^2\right)q_{lk} = 0 \quad . \tag{F.8}$$

Hence all elements q_{lk} vanish except those for which $\omega_{lk} = \pm\omega$. A simple choice is to have all of the elements q_{lk} vanish except those with the frequencies

$$\omega_{n-1,n} = (E_n-E_{n-1})/\hbar = \omega \quad , \quad \omega_{n,n-1} = (E_{n-1}-E_n)/\hbar = -\omega \quad , \tag{F.9}$$

corresponding to the emission and absorption, respectively, of a photon of energy $\hbar\omega$ by the oscillator.

The momentum-matrix elements are $p_{lk} = m\dot{q}_{lk} = -i\omega_{lk}mq_{lk}$, with the nonvanishing elements

$$p_{n-1,n} = -i\omega mq_{n-1,n} \quad ; \quad p_{n,n-1} = i\omega mq_{n,n-1} \quad . \tag{F.10}$$

Let us number the rows and columns as 0,1,2,..., where zero labels the ground state. Thus, the commutation relation $qp - pq = i\hbar$ gives

$$q_{01}\, q_{10} = \frac{\hbar}{2\omega m} \quad , \quad q_{12}\, q_{21} - q_{01}\, q_{10} = \frac{\hbar}{2\omega m} \quad ,\ldots$$

$$q_{n,n+1}\, q_{n+1,n} - q_{n-1,n}\, q_{n,n-1} = \frac{\hbar}{2\omega m} \quad .$$

Therefore, on addition, we have

$$q_{n,n+1}\, q_{n+1,n} = (n+1)\hbar/(2\omega m) \quad .$$

Without loss of generality, we can write

$$q_{n,n+1} = \left[\frac{(n+1)\hbar}{2\omega m}\right]^{\frac{1}{2}} e^{-i\omega t} \quad , \quad q_{n+1,n} = \left[\frac{(n+1)\hbar}{2\omega m}\right]^{\frac{1}{2}} e^{i\omega t} \quad . \tag{F.11}$$

From the matrices of q and p, we can calculate

$$H = \omega^2 m \begin{pmatrix} q_{01}q_{10} & 0 & 0 & \cdot \\ 0 & q_{01}q_{10}+q_{12}q_{21} & 0 & \cdot \\ 0 & 0 & q_{12}q_{21}+q_{23}q_{32} & \cdot \\ \cdot & \cdot & \cdot & \cdot \end{pmatrix} .$$

Thus, H is diagonal, the nth level being

$$E_n = H_{nn} = \omega^2 m(q_{n-1,n}\, q_{n,n-1} + q_{n,n+1}\, q_{n+1,n})$$
$$= \frac{1}{2}\hbar\omega(n+n+1) = \hbar\omega\left(n+\frac{1}{2}\right) \quad . \tag{F.12}$$

$E_0 = \hbar\omega/2$ is the zero-point energy.

Maxwell's equations in free space are

$$\mathrm{curl}\ \underline{E} + \frac{1}{c}\dot{\underline{H}} = 0 \quad , \quad \mathrm{curl}\ \underline{H} - \frac{1}{c}\dot{\underline{E}} = 0 \quad ,$$
$$\mathrm{div}\ \underline{E} = 0 \quad , \quad \mathrm{div}\ \underline{H} = 0 \quad . \tag{F.13}$$

In terms of potentials A, ϕ, we have (see Chap.1),

$$\underline{E} = -\frac{1}{c}\dot{\underline{A}} - \mathrm{grad}\phi \quad , \quad \underline{H} = \mathrm{curl}\ \underline{A} \quad . \tag{F.14}$$

For waves without charge, $\phi = 0$, $\underline{A}$ satisfies the equations

$$\nabla^2 A - \frac{1}{c^2}\ddot{\underline{A}} = 0 \quad , \quad \mathrm{div}\ \underline{A} = 0 \quad , \tag{F.15}$$

where the last equation defines the Coulomb gauge and implies that the waves are transverse.

We can expand $\underline{A}$ in a Fourier series throughout a cubical region of dimension L ($L^3=V$). Thus,

$$\underline{A} = V^{-\frac{1}{2}} \sum_l \underline{a}_l\left(q_l\, e^{i\underline{k}_l\cdot\underline{r}} + q_l^*\, e^{-i\underline{k}_l\cdot\underline{r}}\right) \quad , \tag{F.16}$$

where $\underline{a}_l$ is the polarization vector. We can express it as

$$\underline{A} = V^{-\frac{1}{2}} \sum_l \underline{a}_l\left[Q_l \cos(\underline{k}_l\cdot\underline{r}) - \frac{1}{\omega_l} P_l \sin(\underline{k}_l\cdot\underline{r})\right] \quad , \tag{F.17}$$

$$Q_l = q_l + q_l^* \quad , \quad P_l = \dot{Q}_l = -i\omega_l(q_l - q_l^*) \quad . \tag{F.18}$$

With the normalizing condition, $\underline{a}_l = \sqrt{(4\pi c^2)}\underline{e}_l$, the Hamiltonian of the whole field can be written as

$$H_f = \frac{1}{8\pi}\int (\underline{E}^2+\underline{H}^2)d\tau = \sum_l H_l = \sum_l \frac{1}{2}\left(P_l^2+\omega_l^2 Q_l^2\right) \quad , \tag{F.19}$$

where we have used (F.14). The Hamiltonian equations of motion give

$$\dot{Q}_1 = \frac{\partial H_1}{\partial P_1} = P_1 \quad , \quad \ddot{Q}_1 = \dot{P}_1 = -\frac{\partial H_1}{\partial Q_1} = -\omega_1^2 Q_1 \quad , \quad \ddot{Q}_1 + \omega_1^2 Q_1 = 0 \quad . \quad (F.20)$$

To quantize H_1 we note that (F.20) resembles (F.6) for a harmonic oscillator and $H_1 = \frac{1}{2}\left(P_1^2 + \omega_1^2 Q_1^2\right)$ resembles the Hamiltonian (F.2) for an $m = 1$ oscillator. Therefore, from (F.11) we can write the nonvanishing matrix elements as

$$(Q_1)_{n,n+1} = \left[\frac{(n+1)\hbar}{2\omega_1}\right]^{\frac{1}{2}} e^{-i\omega_1 t} \quad , \quad (P_1)_{n,n+1} = -i\omega_1\left[\frac{(n+1)\hbar}{2\omega_1}\right]^{\frac{1}{2}} e^{-i\omega_1 t} \quad ,$$

$$(Q_1)_{n+1,n} = \left[\frac{(n+1)\hbar}{2\omega_1}\right]^{\frac{1}{2}} e^{i\omega_1 t} \quad , \quad (P_1)_{n+1,n} = i\omega_1\left[\frac{(n+1)\hbar}{2\omega_1}\right]^{\frac{1}{2}} e^{i\omega_1 t} \quad . \quad (F.21)$$

Because of the quantization Q_1 and P_1 are now self-adjoint operators ($Q_1 = Q_1^\dagger$, $P_1 = P_1^\dagger$) and obey the commutation relations

$$[Q_1, Q_m] = [P_1, P_m] = 0 \quad , \quad [Q_1, P_m] = i\hbar\delta_{1m} \quad . \quad (F.22)$$

Thus, the pure radiation field is equivalent to a system of harmonic oscillators described by an enumerably infinite set of canonically conjugate variables Q_1, P_1. From (F.18)

$$q_1 = \frac{1}{2}\left(Q_1 + \frac{i}{\omega_1} P_1\right) \quad , \quad q_1^\dagger = \frac{1}{2}\left(Q_1 - \frac{i}{\omega_1} P_1\right) \quad . \quad (F.23)$$

In view of (F.21), the nonvanishing elements of q_1, $q_1^\dagger$ are

$$(q_1)_{n,n+1} = \left[\frac{(n+1)\hbar}{2\omega_1}\right]^{\frac{1}{2}} e^{-i\omega_1 t} \quad ; \quad (q_1)_{n+1,n} = \left[\frac{(n+1)\hbar}{2\omega_1}\right]^{\frac{1}{2}} e^{i\omega_1 t} \quad . \quad (F.24)$$

The operator q_1 is the operator for the *emission* by the 1th oscillator, of a photon of angular frequency ω_1, and $q_1^\dagger$ is the operator for the *absorption* of such a photon.

Appendix G
Dipole Sum Rule

Consider the commutation relation $xp - px = i\hbar$. Its matrix element between the normalized states $|a\rangle$ is

$$\langle a|xp-px|a\rangle = i\hbar\langle a|a\rangle = i\hbar \quad . \tag{G.1}$$

If we have a complete set of normalized states $|b\rangle$, then we can apply the closure property, $\sum_b |b\rangle\langle b| = 1$, to write (G.1) as

$$\sum_b (\langle a|x|b\rangle\langle b|p|a\rangle - \langle a|p|b\rangle\langle b|x|a\rangle) = i\hbar \quad . \tag{G.2}$$

We can now use the equation of motion $\dot{A} = [A,H]/i\hbar$, to write

$$\begin{aligned}\langle a|p|b\rangle &= \langle a|m\dot{x}|b\rangle = (m/i\hbar)\langle a|[x,H]|b\rangle \\ &= (m/i\hbar)\langle a|xH-Hx|b\rangle = (m/i\hbar)(E_b-E_a)\langle a|x|b\rangle \quad .\end{aligned} \tag{G.3}$$

From (G.2,3)

$$\sum_b (E_b-E_a)|\langle b|x|a\rangle|^2 = \hbar^2/(2m) \quad . \tag{G.4}$$

This is called the dipole sum rule or Thomas-Reiche-Kuhn sum rule.

Note that for the dipole moment

$$\sum_b |x_{ba}|^2 = \sum_b x_{ab}x_{ba} = (x^2)_{aa} = \int |\psi_a|^2 x^2 \, dx \quad . \tag{G.5}$$

If the state a is isotropic,

$$\sum_b |x_{ba}|^2 = \frac{1}{3}(r^2)_{aa} \quad . \tag{G.6}$$

Appendix H
Calculation of the Photoabsorption Coefficient

The radial equation for a positive-energy electron ($E > 0$) that moves in an attractive Coulomb field is

$$\frac{1}{r^2}\frac{d}{dr}\left(r^2\frac{d^2R_E}{dr}\right)+\left[k^2+\frac{2\eta k}{r}-\frac{l(l+1)}{r^2}\right]R_E(r) = 0 \quad , \tag{H.1}$$

where

$$k^2 = \frac{2mE}{\hbar^2} \quad , \quad \eta = \frac{mZe^2}{\hbar^2 k} \quad .$$

We shall, for convenience, work in the atomic units ($m=1$, $e=1$, $a_0=\hbar^2/me^2=1$, $e^2/a_0=1$). Then $k^2 = 2E$ and $\eta = Z/k$. If we substitute

$$R_E(r) = r^l e^{ikr} f_l(r) \quad , \tag{H.2}$$

we get

$$rf_l'' + [2ikr+2(l+1)]f_l' + [2ik(l+1)-2\eta k]f_l = 0 \quad . \tag{H.3}$$

The confluent hypergeometric equation

$$z\frac{d^2F}{dz^2} + (b-z)\frac{dF}{dz} - aF = 0 \quad , \tag{H.4}$$

that has the solution

$$F(a,b,z) = \sum_{s=0}^{\infty}\frac{\Gamma(a+s)\Gamma(b)z^s}{\Gamma(a)\Gamma(b+s)\Gamma(1+s)} = 1 + \frac{az}{b1!} + \frac{a(a+1)z^2}{b(b+1)2!} + \dots \quad , \tag{H.5}$$

is equivalent to (H.3) if we put

$$f_l(r) = C_k F(l+1+i\eta, 2l+2, -2ikr) \quad . \tag{H.6}$$

The integral representation [3.58]

$$F(a,b,z) = \frac{\Gamma(b)}{\Gamma(a)\Gamma(b-a)} \int_0^1 e^{zt}\, t^{a-1}(1-t)^{b-a-1}\, dt \quad , \tag{H.7}$$

can be used to write the radial function as

$$\begin{aligned} R_E &= C(-i\rho)^l\, e^{-\rho/2} \cdot \frac{1}{2\pi i} \int (x+\rho)^{-i\eta-l-1}\, x^{+i\eta-l-1}\, x^{-x}\, dx \quad , \\ &= C\,\frac{(i\rho)^{-l-1}}{2\pi} \int e^{-\rho u}\left(u+\frac{1}{2}\right)^{-i\eta-l-1}\left(u-\frac{1}{2}\right)^{+i\eta-l-1} du \quad , \end{aligned} \tag{H.8}$$

where $\rho = -2ikr$ and $x = \rho\left(u - \frac{1}{2}\right)$. The final form of the normalized function is

$$\begin{aligned} R_E &= (-1)^{l+1}\,\frac{2Z^{\frac{1}{2}}}{(1-e^{-2\pi\eta})^{\frac{1}{2}}} \prod_{s=1}^{l} (s^2+\eta^2)^{\frac{1}{2}}(2kr)^{-l-1} \\ &\times \frac{1}{2\pi} \int e^{-\rho u}\left(u+\frac{1}{2}\right)^{-i\eta-l-1}\left(u-\frac{1}{2}\right)^{i\eta-l-1} du \quad . \end{aligned} \tag{H.9}$$

In our problem, an s-state electron *enters* a p state (dipole case) of the continuous spectrum. Therefore, the solution should represent at infinity a plane wave and an *ingoing* spherical wave. The wave function (H.2) represents at infinity a plane wave and an outgoing wave. To obtain the ingoing wave from the outgoing wave, we have to replace k by -k in (H.9).

The ground-state wave function of the hydrogenlike atom is

$$\psi_K = 2Z^{3/2}\, e^{-Zr} \text{ at. un.} \quad . \tag{H.10}$$

We are interested in calculating the dipole-matrix element $x_{K\to E}$. The continuum eigenfunction $\psi_E = N_{n,l,m} R_E(r) Y_{l,m}(\theta,\phi)$ depends not only on E but also on l,m. Because ψ_K is spherically symmetric, only those ψ_E contribute that have an angular dependence of the form $\sin\theta\,\cos\phi$ (that is, Y_{00} and Y_{11}). Therefore, in spherical coordinates,

$$\begin{aligned} x_{K\to E} &= \int_0^\infty r^3\, dr\, R_{E,l=1}(r) 2Z^{3/2}\, e^{-Zr} \int_0^\pi \sin\theta\, d\theta \int_0^{2\pi} d\phi \\ &\times \left(\frac{3}{4\pi}\right)^{\frac{1}{2}} \sin\theta\,\cos\phi \left(\frac{1}{4\pi}\right)^{\frac{1}{2}} \sin\theta\,\cos\phi \quad , \\ &= \frac{4Z^2(1+\eta^2)^{\frac{1}{2}}}{3^{\frac{1}{2}}(1-e^{-2\pi\eta})^{\frac{1}{2}}}\,\frac{1}{2\pi} \int_0^\infty (2kr)^{-2}\, r^3\, dr\, e^{-Zr} \oint \left(u+\frac{1}{2}\right)^{-i\eta-2} \\ &\times \left(u-\frac{1}{2}\right)^{i\eta-2} e^{-2ikru}\, du \quad . \end{aligned} \tag{H.11}$$

Interchanging the order of integration with respect to r and u, and disregarding the coefficient in front of the first integral sign, we get

$$J = \frac{1}{4k^2} \oint du\left(u+\frac{1}{2}\right)^{-i\eta-2}\left(u-\frac{1}{2}\right)^{i\eta-2}(Z+2iku)^{-2}$$

$$= \frac{-1}{16k^4} \int du\left(u+\frac{1}{2}\right)^{-i\eta-2}\left(u-\frac{1}{2}\right)^{i\eta-2}\left(u-\frac{1}{2}i\eta\right)^{-2} \quad . \tag{H.12}$$

The branch points are at $u = 1/2$ and $u = -1/2$. Because the integrand goes to zero as $1/u^6$, the contour may be extended to ∞. The contour goes around the pole $u = i\eta/2$ in the negative sense. Therefore, the integration gives

$$J = \frac{2\pi i}{16k^4} \cdot \frac{d}{du}\left[\left(u+\frac{1}{2}\right)^{-i\eta-2}\left(u-\frac{1}{2}\right)^{i\eta-2}\right]_{u=i\eta/2}$$

$$= \frac{(64\eta)2\pi}{16k^4(1+\eta^2)^3}\left(\frac{i\eta-1}{i\eta+1}\right)^{i\eta} \quad , \tag{H.13}$$

where $\eta = Z/k$ in at.un. Use of $[(\eta+i)/\eta-i]^i = \exp(-2\eta \cot^{-1}\eta)$, finally gives

$$|x_{K\to E}|^2 = \frac{2^8}{3Z^4}\left(\frac{\eta^2}{1+\eta^2}\right)^5 \frac{e^{-4\eta \cot^{-1}\eta}}{1-e^{-2\pi\eta}} \text{ at.un.} \quad . \tag{H.14}$$

The dimensions are of area/energy; in at.un. a_0^3/e^2. Because[4] $\nu_K = R_\infty cZ^2 = Z^2$ Rydberg and $\nu = (Z^2+k^2)$ Rydberg, we can express $\eta = Z/k$ as

$$\eta = \left(\frac{\nu_K}{\nu-\nu_K}\right)^{\frac{1}{2}} \quad . \tag{H.15}$$

Because the K level has 2 electrons, the photoabsorption coefficient is

$$\tau_K = 2(2\pi a_0)^3\, n\frac{\nu}{c}\,|x_{K\to E}|^2 = \frac{2^8\pi e^2}{3mc}\, n\, \frac{\nu_K^3}{\nu^4}\, \frac{e^{-4\eta \cot^{-1}\eta}}{1-e^{-2\pi\eta}}$$

$$= 4.1 \times 10^8 \frac{\rho}{AZ^2}\left(\frac{\nu_K}{\nu}\right)^4 \frac{e^{-4\eta \cot^{-1}\eta}}{1-e^{-2\pi\eta}} \quad , \tag{H.16}$$

where ρ is the density, n the number of atoms per cm^3 in state K and A the atomic weight.

---

[4]The quantity $R_\infty hc$ is called the Rydberg (Ry) and is a unit of energy. 2.17972×10^{-18} J; the quantity $R_\infty c = 3.289842 \times 10^{-15}\ s^{-1}$ is a unit of frequency.

Appendix I
Screening Effect, According to Slater

The ionization energy for one inner electron in a hydrogenlike atom is given approximately by

$$E_i = hc\tilde{\nu}_i = R_\infty hc(Z/n)^2 \quad . \tag{I.1}$$

In a many-electron atom, all of the energies result from attractive interactions of electrons with positive-ion cores and repulsive interactions of electrons among themselves. Therefore, for better agreement with experimental results, we can write

$$\tilde{\nu}_i/R_\infty = \left[(Z-\sigma_i)/n_i^*\right]^2 \quad , \tag{I.2}$$

where σ_i is the internal screening constant, and n_i^* is the effective principal quantum number for the i level. The energy of the normal atom can be expressed in Rydbergs as

$$\tilde{\nu}/R_\infty = \sum_i z_i(\tilde{\nu}_i/R_\infty) = \sum_i z_i\left[(Z-\sigma_i)/n_i^*\right]^2 \quad , \tag{I.3}$$

where z_i is the number of electrons in the level i. Similar equation will hold for an ion with energy denoted by $(\tilde{\nu}/R_\infty)_q$.

For an atom with one electron removed from the q level to a final level,

$$\tilde{\nu}_q/R_\infty = (\tilde{\nu}/R_\infty)_q - \tilde{\nu}/R_\infty \quad , \tag{I.4}$$

where $\tilde{\nu}_q/R_\infty$ gives the energy of the q level or of the q absorption edge.

SLATER (1930) gives the following values for n_i^*:

n_i :	1	2	3	4	5	6
n_i^* :	1	2	3	4.7	4.0	4.2 .

Table I.1. The fractions $\Delta\sigma_j$ of the inner screening constants σ_i of electrons of the i level

Electrons of group i		Electrons of internal level j ($n_j \leq n_i$)		
i = 1s	i ≠ 1s	i = s,p		i = d,f
$n_j = n_i$	$n_j = n_i$	$n_j = n_i - 1$	$n_j \leqq n_i - 2$	$n_j \leqq n_i - 1$
$\Delta\sigma_j$ = 0.30	0.35	0.85	1.00	1.00

Table I.2. Calculation of the energy of the tin atom

Group	$Z - \sigma_i$	n_i^*	z_i	$z_i(\tilde{\nu}_i/R_\infty)$
1s	49.70	1	2	4940
2s,p	45.85	2	8	4200
3s,p	38.75	3	8	1336
3d	28.85	3	10	925
4s,p	22.25	3.7	8	289
4d	10.85	3.7	10	86
5s,p	5.65	4.0	4	8
	$\tilde{\nu}/R_\infty = \sum_i z_i(\tilde{\nu}_i/R_\infty) = 11784$			

Further, he assigns a certain fraction $\Delta\sigma_j$ (Table I.1) of the internal screening constant to each electron, such that

$$\sigma_i = \sum_{j=1}^{i-1} z_j\Delta\sigma_j + (z_i-1)\Delta\sigma_i = \left(\sum_{j=1}^{i} z_j\Delta\sigma_j\right) - \Delta\sigma_i \quad . \tag{I.5}$$

Here j is the order number of the electron groups into which SLATER distributes all of the electron shells of the atom,

1s; 2s,p; 3s,p; 3d; 4s,p; 4d; 4f; 5,sp; 5d; ...

Thus, ns and np valence electrons are treated on the same footing, independent of n, except that 1s electrons interact with each other more strongly (Table I.1). The nd electrons are found to lie outside the ns and np ones, or more precisely between ns, np and (n+1)s, (n+1)p.

Consider the tin atom (^{50}Sn). Slater's groups are

$$1s^2 2(s,p)^8 3(s,p)^8\ 3d^{10}\ 4(s,p)^8\ 4d^{10}\ 5(s,p)^4 \quad .$$

The screening constants for these groups are:

$$\sigma_{1s} = 1 \times 0.30 = 0.30 \quad ,$$

$$\sigma_{2s,p} = 7 \times 0.35 + 2 \times 0.85 = 4.15 \quad ,$$

$$\sigma_{3s,p} = 7 \times 0.35 + 8 \times 0.85 + 2 \times 1.00 = 11.25 \quad ,$$

$$\sigma_{3d} = 9 \times 0.35 + 18 \times 1.00 = 21.15 \quad ,$$

$$\sigma_{4s,p} = 7 \times 0.35 + 18 \times 0.85 + 10 \times 1.00 = 27.75 \quad ,$$

$$\sigma_{4d} = 9 \times 0.35 + 36 \times 1.00 = 39.15 \quad ,$$

$$\sigma_{5s,p} = 3 \times 0.35 + 18 \times 0.85 + 28 \times 1.00 = 44.35 \quad .$$

The calculation of $\tilde{\nu}/R_\infty$, (I.3), is shown in Table I.2. For a recent application of Slater's rule, see SNYDER [3.59].

SLATER takes into account only the inner screening effect. To improve agreement with the observed values,

$$E_{nl} = -R_\infty hc\left[(Z-\sigma)/n^*\right]^2 + V_0$$

is usually used. The difference $n - n^*$ is called the *quantum defect* [3.60]; V_0 takes care of the *outer* screening effect *after an absorption process*. As the ejected electron moves outward through the region occupied by the outer electrons, it is repelled by them and consequently gains energy V_0. The outer screening potential V_0 depends on the number and radial distribution of electrons and ranges up to ~20 keV near the nucleus of heavy atoms [3.61].

Appendix J
Electronegativity Scale

The electronegativity scale of PAULING [6.5] is given below:

Li	Be	B											C	N	O	F
1.0	1.5	2.0											2.5	3.0	3.5	4.0
Na	Mg	Al											Si	P	S	Cl
0.9	1.2	1.5											1.8	2.1	2.5	3.0
K	Ca	Sc	Ti	V	Cr	Mn	Fe	Co	Ni	Cu	Zn	Ga	Ge	As	Se	Br
0.8	1.0	1.3	1.5	1.6	1.6	1.5	1.8	1.8	1.8	1.9	1.6	1.6	1.8	2.0	2.4	2.8
Rb	Sr	Y	Zr	Nb	Mo	Te	Ru	Rh	Pd	Ag	Cd	In	Sn	Sb	Te	I
0.8	1.0	1.2	1.4	1.6	1.8	1.9	2.2	2.2	2.2	1.9	1.7	1.7	1.8	1.9	2.1	2.5
Cs	Ba	La - Lu	Hf	Ta	W	Re	Os	Ir	Pt	Au	Hg	Tl	Pb	Bi	Po	At
0.7	0.9	1.1-1.2	1.3	1.5	1.7	1.9	2.2	2.2	2.2	2.4	1.9	1.8	1.8	1.9	2.0	2.2
Fr	Ra	Ac	Th	Pa	U	Np-No										
0.7	0.9	1.1	1.3	1.5	1.7	1.3										

These values are for the common oxidation states (see also [6.199])

Wavelength Tables

Table 1. K absorption edges in the region 2 kXU to 0.5 kXU in pure elements. The Kl (longer wavelength) edge is given (marked *) where the main edge shows clear split structure. The values are based on CAUCHOIS and HULUBEI[a], SANDSTRÖM[b], and BEARDEN[c]. The conversion factor used is E[keV] = 12372.42/λ [XU][c]

Element	XU	[keV]	Element	XU	[keV]
24 Cr	2065.91	(5.9888)	36 Kr	863.73	(14.3244)
25 Mn	1892.50	(6.5376)	37 Rb*	814.39[f]	(15.1923)
26 Fe	1739.85	(7.1112)	38 Sr	768.35[g]	(16.1026)
27 Co*	1604.8(2)[d]	(7.7095)	39 Y	726.15	(17.0388)
28 Ni	1484.99	(8.3317)	40 Zr*	687.57[h]	(17.9944)
29 Cu*	1377.79[e]	(8.9799)	41 Nb	651.63	(18.9869)
30 Zn	1280.7	(9.6607)	42 Mo	618.50	(20.0039)
31 Ga	1193.3	(10.3682)	43 Te	587.84	(21.0473)
32 Ge	1114.27	(11.1036)	44 Ru	559.35	(22.1193)
33 As	1042.8	(11.8646)	45 Rh	532.84	(23.2198)
34 Se	977.71	(12.6545)	46 Pd	508.15	(24.348)
35 Br (gas)	918.5	(13.4702)			
(KBr)	918.09	(13.4763)			

(a) Y. Cauchois, H. Hulubei: Longueurs d'Onde des Emissions X et des Discontinuitês d'Absorption X (Hermann and Co., Paris 1947)
(b) A.E. Sandström: Handbuch der Physik *30*, 78 (1957)
(c) J.A. Bearden: X-Ray Wavelengths NYO-10586, USAEC, Oak Ridge, Ten. 1964
(d) A.K. Dey, B.K. Agarwal: J. Chem. Phys. *59*, 1397 (1973)
(e) B.K. Agarwal, C.B. Bhargava, A.N. Vishnoi, V.P. Seth: J. Phys. Chem. Sol. *37*, 725 (1976)
(f) B.K. Agarwal, R.K. Johri: J. Phys. F *7*, 1607 (1977)
(g) V.B. Singh, B.K. Agarwal: J. Phys. Chem. Sol. *35*, 465 (1974); interpolated value.
(h) B.K. Agarwal, R.K. Johri: Phys. Status Solidi (b) *88*, 309 (1978)

Table 2. L absorption edges in the wavelength region 2 kXU to 0.75 kXU. Measurements made on compounds are marked as (+) (BEARDEN 1964)

Element	L_I [XU]	L_{II} [XU]	L_{III} [XU]
57 La	1973.9^+	2100.9^+	2256^+
58 Ce	1889.5^+	2008.2^+	2162^+
59 Pr	1810.3^+	1921.5^+	2074.8^+
60 Nd	1735.4^+	1840.2^+	1992.6^+
61 Pm	1663.9	1763.9	1915.1
62 Sm	1596.9^+	1691.8^+	1841.9^+
63 Eu	1534.9^+	1623.7^+	1772.4^+
64 Gd	1475.3^+	1561.44[a]	1708.66[b,c]
65 Tb	1419.4^+	1499.2^+	1646.3^+
66 Dy	1366.4^+	1442.18[a]	1588.41[b]
67 Ho	1316.3^+	1388.02[b]	1533.9[d]
68 Er	1268.0^+	1337.09[a]	1480.4^+
69 Tm	1222.5^+	1286.5^+	1430.4^+
70 Yb	1179.4^+	1240.2^+	1384.54[e]
71 Lu	1137.8^+	1196.0^+	1337.7^+
72 Hf	1097.4	1152.4	1294.5^+
73 Ta	1059.1	1111.4	1252.7
74 W	1022.55	1072.3	1213.0
75 Re	987.4	1035.0	1174.9
76 Os	953.8	999.3	1138.4
77 Ir	921.7	965.1	1103.5
78 Pt	891.2	932.2	1070.1
79 Au	861.97	900.72	1037.85
80 Hg	833.6	870.4	1006.8[f]
81 Tl	806.4	841.7	977.16[g]
82 Pb	780.34	813.69	948.78[g]
83 Bi	755.5	787.1	921.49[g]

a B.K. Agarwal, B.R.K. Agarwal: X-ray Spectrom. *7*, 12 (1978)
b B.R.K. Agarwal, L.P. Verma, B.K. Agarwal: Nuovo Cimento Lett. *2*, 581 (1971)
c P. Deshmukh, P. Deshmukh, C. Mande: J. Phys. C *10*, 3421 (1977) give for Gd L_{III} edge - 1709.04 XU
d B.R.K. Agarwal, L.P. Verma, B.K. Agarwal: Nuovo Cimento Lett. *1*, 781 (1971)
e P. Deshmukh, C. Mande: Pramana *2*, 138 (1974)
f B.K. Agarwal, L.P. Verma: J. Phys. C *2*, 104 (1969)
g A.N. Vishnoi, B.K. Agarwal: Nuovo Cimento Lett. *4*, 771 (1970)

Table 3. Some useful reference lines[1] (BEARDEN 1964)

Element	λ [XU]		Element	λ [XU]	
24 Cr	$K\alpha_2$	2288.854	47 Ag	$K\alpha_2$	562.630
	α_1	2284.96		α_1	558.2486
25 Mn	α_2	2101.42	50 Sn	α_2	494.027
	α_1	2097.466		α_1	489.583
26 Fe	α_2	1935.961	74 W	$L\beta_4(L_IM_{II})$	1298.92
	α_1	1932.031		$\beta_3(L_IM_{III})$	1260.07
27 Co	α_2	1789.136		$\gamma_2(L_IN_{II})$	1065.85
	α_1	1785.259		$\gamma_3(L_IN_{III})$	1059.80
				$\eta\ (L_{II}M_I)$	1418.16
28 Ni	α_2	1658.304		$\beta_1(L_{II}M_{IV})$	1279.153
	α_1	1654.475		$\gamma_5(L_{II}N_I)$	1130.00
29 Cu	α_2	1541.198		$\gamma_1(L_{II}N_{IV})$	1096.27
	α_1	1537.370		$l\ (L_{II}M_I)$	1674.7
	$\beta_{1,3}$	1389.334		$\alpha_2(L_{III}M_{IV})$	1484.35
	β_2	1378.23		$\alpha_1(L_{III}M_V)$	1473.33
	β_5	1378.7		$\beta_6(L_{III}N_I)$	1287.22
30 Zn	α_2	1436.019		$\beta_{15}(L_{III}N_{IV})$	1243.73
	α_1	1432.182		$\beta_2(L_{III}N_V)$	1242.02
42 Mo	α_2	712.112			
	α_1	707.831			
	β_3	631.55			
	β_1	630.978			
	β_2	619.70			
	$\beta_5^{II}\ (KM_{IV})$	625.78			
	$\beta_5^{I}\ (KM_V)$	625.62			
	$\beta_4^{II}\ (KN_{IV})$	618.95			
	$\beta_4^{I}\ (KN_V)$	618.73			

[1]National Bureau of Standards Special Publication 398 (issued August 1974) gives kXU to Å ratio as 1.0020772 based on $\lambda(CuK\alpha_1) \equiv 1.537400$ kXU.

References

1.1 C.G. Barkla: Phil. Trans. Roy. Soc. *204*, 467 (1905)
1.2 H. Kulenkampff: Ann. Phys. *87*, 597 (1928)
1.3 B. Dasannacharya: Phys. Rev. *35*, 129 (1930)
1.4 G. Stokes: Proc. Manchester Lit. Phil. Soc. (1898)
1.5 J.J. Thomson: Philos. Mag. *45*, 172 (1898)
1.6 A. Sommerfeld: Phys. Z. *10*, 969 (1909)
1.7 M. Scheer, E. Zeitler: Z. Phys. *140*, 642 (1955)
1.8 W.W. Nicholas: U.S. Bur. Stand. J. Res. *2*, 837 (1929)
1.9 K. Bohm: Ann. Phys. *33*, 315 (1938)
1.10 R. Honerjäger: Ann. Phys. *38*, 33 (1940)
1.11 K. Harworth, P. Kirkpatrick: Phys. Rev. *60*, 163 (1941)
1.12 K. Harworth, P. Kirkpatrick: Phys. Rev. *62*, 334 (1942)
1.13 R. Kerscher, H. Kulenkampff: Z. Phys. *140*, 632 (1955)
1.14 S. Thordarson: Ann. Phys. *35*, 135 (1939)
1.15 G. Sesemann: Ann. Phys. *40*, 66 (1941)
1.16 H. Determann: Ann. Phys. *30*, 481 (1937)
1.17 H.P. Hanson, S.I. Salem: Phys. Rev. *124*, 16 (1961)
1.18 H.A. Kramers: Philos. Mag. *46*, 836 (1923)
1.19 G. Wentzel: Z. Phys. *27*, 257 (1924)
1.20 W. Duane, F.L. Hunt: Phys. Rev. *6*, 166 (1915)
1.21 P. Ohlin: Nature *152*, 392 (1943)
1.22 P. Ohlin: Ark. Fys. *4*, 387 (1952)
1.23 J.A. Bearden, F.T. Johnson, H.M. Watts: Phys. Rev. *81*, 70 (1951)
1.24 G.L. Felt, J.N. Harris, J.W.M. Dumond: Phys. Rev. *92*, 1160 (1953)
1.25 R. Whiddington: Proc. Roy. Soc. London A*86*, 360 (1912)
1.26 H. Kulenkampff: Ann. Phys. *69*, 548 (1922)
1.27 H. Kulenkampff, L. Schmidt: Ann. Phys. *43*, 494 (1943)
1.28 C.J. Ulrey: Phys. Rev. *11*, 401 (1918)
1.29 D.L. Webster: Phys. Rev. *9*, 220 (1917)
1.30 W. Rump: Z. Phys. *43*, 254 (1927)
1.31 A. Sommerfeld: Ann. Phys. *11*, 257 (1931); also see *Atomic Structures and Spectral Lines*, English ed. (Methuen, London 1923)
1.32 P. Kirkpatrick, L. Wiedmann: Phys. Rev. *67*, 321 (1945)
1.33 V.B. Berestetskii, E.M. Lifshitz, L.P. Pitaevskii: *Relativistic Quantum Theory* (Pergamon Press, Oxford 1971)
1.34 W. Heitler: *Quantum Theory of Radiation*, 3rd ed. (Oxford Univ. Press 1954)
1.35 B.K. Agarwal: *Quantum Mechanics and Field Theory* (Asia Publishing House, Bombay, London, New York 1976) p.128
1.36 N. Bohr: Philos. Mag. *25*, 10 (1913)
1.37 H. Bethe: Ann. Phys. *5*, 325 (1930)
1.38 H. Bethe: Z. Phys. *76*, 293 (1932)
2.1 W. Kossel: Dtsch. Phys. Ges. *16*, 899, 953 (1914)
2.2 D.L. Webster: Phys. Rev. *7*, 599 (1916)
2.3 H.G. Moseley: Philos. Mag. *27*, 703 (1914)
2.4 W.L. Bragg: J. Sci. Instrum. *24*, 27 (1947)

2.5 J.A. Bearden: "*X-Ray Wave Lengths*", Tech. Rpt. NYO 10586, U.S. Atomic Energy Commission, Division of Technical Information Extension, Oak Ridge, Tenn. (1964)
2.6 W. Heitler: *The Quantum Theory of Radiation*, 3rd ed. (Oxford 1954) p.179
2.7 B.K. Agarwal: *Quantum Mechanics and Field Theory* (Asia Publishing House, Bombay, London, New York 1976)
2.8 E. Merzbacher, H.W. Lewis: *Handbuch der Physik*, Vol.XXXIV, ed. by S. Flügge (Springer, Berlin, Göttingen, Heidelberg 1958)
2.9 G.N. Ogurtsov: Rev. Mod. Phys. *44*, 1 (1972)
2.10 B. Davis: Phys. Rev. *11*, 433 (1918)
2.11 B.A. Wooten: Phys. Rev. *13*, 71 (1919)
2.12 J. Hill, H. Terrey: Philos. Mag. *23*, 339 (1937)
2.13 S. Rosseland: Philos. Mag. *45*, 65 (1923)
2.14 A. Jönsson: Z. Phys. *41*, 221 (1927)
2.15 D.L. Webster, W. Hansen, F. Duveneck: Phys. Rev. *43*, 839 (1933)
2.16 L. Pockman, D.L. Webster, P. Kirkpatrick, K. Harworth: Phys. Rev. *71*, 330 (1947)
2.17 J.J. McCue: Phys. Rev. *65*, 168 (1944)
2.18 M. Green, V.E. Cosslett: Brit. J. Appl. Phys. (J. Phys. D) *1*, 425 (1968)
2.19 W. Pauli: Z. Phys. *31*, 778 (1925)
2.20 F. Constantinescu, E. Magyari: *Problems in Quantum Mechanics* (Pergamon Press, New York 1971)
2.21 P.S. Bagus: Phys. Rev. *A139*, 619 (1965); also see, T. Åberg: Phys. Rev. *162*, 5 (1967)
2.22 T. Koopmans: Physica *1*, 104 (1934)
2.23 R. Manne, T. Åberg: Chem. Phys. Lett. *7*, 282 (1970)
2.24 H.C. Burger, H.B. Dorgelo: Z. Phys. *23*, 258 (1924)
2.25 L.S. Ornstein, H.C. Burger: Z. Phys. *24*, 41 (1924)
2.26 E.U. Condon, G.H. Shortley: *The Theory of Atomic Spectra* (University Press, Cambridge 1935) p.238
2.27 J.H. Williams: Phys. Rev. *44*, 146 (1933)
2.28 A. Jönsson: Z. Phys. *36*, 426 (1926)
2.29 A. Jönsson: Z. Phys. *46*, 383 (1928)
2.30 J.H. Scofield: Phys. Rev. *179*, 9 (1969)
2.31 H.R. Rosner, C.P. Bhalla: Z. Phys. *231*, 347 (1970)
2.32 R.K. Smither, M.S. Freedman, F.T. Porter: Phys. Lett. *A32*, 405 (1970)
2.33 P.V. Rao, J.M. Palms, R.E. Wood: Phys. Rev. *A3*, 1568 (1971)
2.34 P. Tothill: Brit. J. Appl. Phys. (J. Phys. D) *1*, 1093 (1968)
2.35 M.H. Unsworth, J.R. Greening : Phys. Med. Biol. *15*, 621, 631 (1970)
2.36 N. Bohr, D. Coster: Z. Phys. *12*, 342 (1923)
2.37 G. Hertz: Z. Phys. *3*, 19 (1920)
2.38 A.E. Sandström: *Handbuch der Physik*, Vol.XXX, ed. by S. Flügge (Springer, Berlin, Göttingen, Heidelberg 1957)
2.39 A. Sommerfeld: *Atomic Structure and Spectral Lines*, English ed. (Methuen, New York, London 1928)
2.40 C. Mande, P.S. Damle: Indian J. Pure Appl. Phys. *3*, 142 (1965)
2.41 W.J. Veigele, D.E. Stevenson, E.M. Henry: J. Chem. Phys. *50*, 5404 (1969) and references given therein
2.42 A. Landé: Z. Phys. *19*, 112 (1923)
2.43 S. Goudsmit: Phys. Rev. *31*, 946 (1928)
2.44 S. Idei: Sci. Rep. Tohuku Imperial University I *19*, 559, 641 (1930)
2.45 W. Duane: Phys. Rev. *37*, 1017 (1931)
2.46 P.A. Ross: Phys. Rev. *39*, 536, 748 (1932)
2.47 P.A. Ross: Phys. Rev. *43*, 1036 (1933)
2.48 E. Carlsson: Z. Phys. *80*, 640 (1933)
2.49 H. Hulubei: Compt. Rend. *201*, 1356 (1935)
2.50 Y. Cauchois, H. Hulubei: *Longueurs d'Onde des Emissions X et des Discontinuités d'Absorption X* (Hermann and Co., Paris 1947)

2.51 O. Beckman: Ark. Fysik *9*, 495 (1955)
2.52 S.K. Allison, A. Armstrong: Phys. Rev. *26*, 714 (1925)
2.53 H.T. Meyer: Wiss. Veröff. Siemens-Konzern *7*, 108 (1929)
2.54 A.L. Catz: Phys. Rev. Lett. *24*, 127 (1970)
2.55 M. Siegbahn: *Spektroskopik der Röntgenstrahlen*, 2nd ed. (Springer, Berlin 1931)
2.56 A.E. Lindh: *Handbuch der Experimentalphysik*, Bd. 24, Teil 2 (Leipzig 1930)
2.57a W. Kossel: Z. Phys. *1*, 119 (1920)
2.57b W. Kossel: Z. Phys. *2*, 470 (1920)
2.58 A. Sandström: Z. Phys. *66*, 784 (1930)
2.59 E. Ingelstam: Nova Act. Reg. Svi. Soc. Upsala *10*, no.5 (1937)
2.60 E.J. Willimas: Proc. Roy. Soc. London *A130*, 310 (1930)
2.61 A. Jönsson: Z. Phys. *43*, 845 (1927)
2.62 S.K. Allison: Phys. Rev. *32*, 1 (1928)
2.63 S.K. Allison: Phys. Rev. *34*, 7 (1929)
2.64 C.G. Barkla, C.A. Sadler: Philos. Mag. *16*, 550 (1908)
2.65 B.K. Agarwal: Sci. Culture *17*, 479 (1952)
2.66 A.J.C. Wilson: Phys. Abstr. *55*, 932 (1952)
2.67 J.A. Bearden: Rev. Mod. Phys. *39*, 78 (1967)
2.68 S. Hagström, C. Nordling, K. Siegbahn: *α-, β- and γ-Ray Spectroscopy*, Vol.1, ed. by K. Siegbahn (North-Holland, Amsterdam 1965)
2.69 E. Saurl: *Landoft-Börnstein*, Vol.1, ed. by A. Eucken (Springer, Berlin 1950)
2.70 Y. Cauchois: J. Phys. Radium *13*, 113 (1952)
2.71 Y. Cauchois: J. Phys. Radium *16*, 253 (1955)
2.72 R.D. Hill, E.L. Church, J.W. Mihelich: Rev. Sci. Instrum. *23*, 523 (1952)
2.73 J.C. Slater: Phys. Rev. *98*, 1039 (1955)
2.74 W. Kossel: Dtsch. Phys. Ges. *18*, 339, 396 (1916)
2.75 W. Kossel: Phys. Z. *18*, 240 (1917)
2.76 A.F. Burr, J.K. Carson: J. Phys. B*7*, 451 (1974)
2.77 B.G. Gokhale, U.D. Misra: J. Phys. B*10*, 3599 (1977); *11*, 2077 (1978)
3.1 H.A. Lorentz: *The Theory of Electrons* (G.E. Stechert, New York 1923)
3.2 A. Larsson: Thesis, Upsala (1929)
3.3 W.L. Bragg: Proc. Cambridge Philos. Soc. *17*, 43 (1913)
3.4 A.H. Compton: Philos. Mag. *45*, 1121 (1923)
3.5 R.L. Doan: Philos. Mag. *4*, 100 (1927)
3.6 S.W. Smith: Phys. Rev. *40*, 156 (1932) and references therein
3.7 W. Stenström: Thesis, Lund (1919)
3.8 C.C. Hatley, B. Davis: Phys. Rev. *23*, 290 (1924)
3.9 C.C. Hatley: Phys. Rev. *24*, 486 (1924)
3.10 B. Davis, R. von Nardoff: Phys. Rev. *23*, 291 (1924)
3.11 R. von Nardoff: Phys. Rev. *24*, 143 (1924)
3.12 A. Larsson, M. Siegbahn, I. Waller: Naturwiss. *12*, 1212 (1924)
3.13 B. Davis. C.M. Slack: Phys. Rev. *25*, 881 (1925)
3.14 B. Davis, C.M. Slack: Phys. Rev. *27*, 18 (1926)
3.15 C.M. Slack: Phys. Rev. *27*, 691 (1926)
3.16 R. Ladenberg: Z. Phys. *4*, 451 (1921)
3.17 G.E.M. Jauncey: Philos. Mag. *48*, 81 (1924)
3.18 J.A. Prins: Z. Phys. *47*, 479 (1928)
3.19 H. Kallmann, H. Mark: Naturwiss. *14*, 648 (1926)
3.20 H. Kallmann, H. Mark: Ann. Phys. *82*, 585 (1927)
3.21 W. Kuhn: Z. Phys. *33*, 408 (1925)
3.22a W. Thomas: Naturwiss. *13*, 627 (1925)
3.22b F. Feiche, W. Thomas: Z. Phys. *34*, 510 (1925)
3.23 H.A. Bethe, E.E. Salpeter: *Handbuch der Physik*, Vol.XXXV, ed. by S. Flügge (Springer, Berlin, Göttingen, Heidelberg 1957) p.88
3.24 S.T. Manson, J.W. Cooper: Phys. Rev. *165*, 126 (1968)

3.25 E.J. McGuire: Phys. Rev. *175*, 20 (1968)
3.26 U. Fano, J.W. Cooper: Rev. Mod. Phys. *40*, 441 (1968)
3.27 Y. Sugiura: J. Physique *8*, 113 (1927)
3.28 R. de L. Kronig, H.A. Kramers: Z. Phys. *48*, 174 (1928)
3.29 M. Wolf: Ann. Phys. *10*, 973 (1933)
3.30a H. Hönl: Z. Phys. *84*, 1 (1933)
3.30b H. Hönl: Ann. Phys. *18*, 625 (1933)
3.31 J.A. Wheeler, J.A. Bearden: Phys. Rev. *46*, 755 (1934)
3.32 F. Herman, S. Skillman: *Atomic Structure Calculation* (Prentice-Hall Inc., N.J. 1963)
3.33 E.N. Lassettre, S.A. Francis: J. Chem. Phys. *40*, 1208 (1964)
3.34 L.G. Parratt, C.F. Hempstead: Phys. Rev. *94*, 1593 (1954)
3.35 V. Weisskopf, E. Wigner: Z. Phys. *63*, 54 (1930)
3.36 F.K. Richtmyer, S.W. Barnes, E. Ramberg: Phys. Rev. *46*, 843 (1934)
3.37 P. Bouguer: *Essai d'Optique sur la Gradation de la Lumière* (Hause, France 1729)
3.38 J.H. Lambert: *Photometria suie de Mensura et Gradibus Luminis* (Colorum et Umbrae, Augsburg 1760)
3.39 A. Beer: Ann. Phys. *86*, 78 (1852)
3.40 K. Lonsdale: *International Tables for X-Ray Crystallography*, ed. by K. Lonsdale (Kynoch, England 1962) Sect.3.2
3.41 M. Stobbe: Ann. Phys. *7*, 661 (1930)
3.42 J.C. Slater: Phys. Rev. *36*, 57 (1930)
3.43 R.T. Berger: Radiat. Phys. *15*, 1 (1961)
3.44 J.W. Allison: Aust. J. Phys. *14*, 443 (1961)
3.45 J.H. Hubbell: Radiat. Res. *70*, 58 (1977)
3.46 E. Jönsson: Thesis, Upsala (1928)
3.47 H. Rindfleisch: Ann. Phys. *28*, 409 (1937)
3.48 S. Laubert: Ann. Phys. *40*, 553 (1941)
3.49 H. Tellez-Plasencia: J. Phys. Radium. *10*, 14 (1949)
3.50 B. Walter: Fortschritte a.d. Geb. der Röntgen. *35*, 929, 1308 (1927)
3.51 B.K. Agarwal: Curr. Sci. *23*, 357 (1954)
3.52 R. Böklen, S. Geiling: Z. Metallkunde *40*, 157 (1949)
3.53 R. Forster: Helv. Phys. Acta *1*, 18 (1927)
3.54 R. Forster: Naturwiss. *15*, 969 (1927)
3.55 J. Thibaud: Phys. Rev. *35*, 1452 (1930)
3.56 F. Jentzsch, H. Steps: Z. Phys. *91*, 151 (1934)
3.57 H. Kiessig: Ann. Phys. *10*, 715, 769 (1931)
3.58 T.D. Sanders: *Modern Physical Theory* (Addison-Wesley, U.S.A. 1970) p.542
3.59 L.C. Snyder: J. Chem. Phys. *55*, 95 (1971)
3.60 E.J. McGuire: Phys. Rev. *161*, 51 (1967)
3.61 A.R.P. Rau, U. Fano: Phys. Rev. *167*, 7 (1968)
3.62 Z.G. Pinsker: *Dynamical Scattering of X-Rays in Crystals* (Springer, Berlin, Heidelberg, New York 1978) pp.82,359
4.1 F. Sauter: Ann. Phys. *11*, 454 (1931); also see H. Hall: Rev. Mod. Phys. *8*, 358 (1936)
4.2 G. Schur: Ann. Phys. *4*, 433 (1930)
4.3 J. Cooper, R.N. Zare: J. Chem. Phys. *48*, 942 (1968)
4.4 T. Koopmans: Physica *1*, 104 (1934)
4.5 R.J.W. Henry, L. Lipsky: Phys. Rev. *153*, 51 (1967)
4.6 L.J. Aarons, M.F. Guest, I.H. Hillier: J. Chem. Soc. Faraday Trans. *68*, 1866 (1972)
4.7 H. Basch: J. Electron. Spectrosc. *5*, 463 (1975)
4.8 C.S. Fadley: In *Electron Emission Spectroscopy*, ed. by DeKeyser, Fiermans, Van der Kalen, Vennik (Reidel, Dordrecht 1973)
4.9 M.A. Brisk, A.D. Baker: J. Electron. Spectrosc. *7*, 197 (1975)
4.10 H. Robinson, W.F. Rawlinson: Philos. Mag. *28*, 277 (1914); Philos. Mag. *50*, 241 (1925);

H.R. Robinson, A.M. Cassie: Proc. Roy. Soc. *113*, 282 (1927);
H.R. Robinson, C.L. Young: Proc. Roy. Soc. *128*, 92 (1930)
4.11 B.M. Hagström, C.S. Fadley: In *X-Ray Spectroscopy*, ed. by L.V. Azaroff (McGraw-Hill, New York 1974)
4.12 J.A. Bearden, A.F. Burr: Rev. Mod. Phys. *39*, 125 (1967)
4.13 K. Siegbahn, C. Nordling, A. Fahlman, R. Nordberg, K. Hamrin, J. Hedman, G. Johansson, T. Bergmark, S.-E. Karlsson, I. Lindgren, B. Lindberg: Nova Acta Reg. Soc. Sci. Upaslien, Ser.IV, Vol.20 (1967)
4.14 W. Lotz: J. Opt. Soc. Am. *60*, 206 (1970)
4.15 C. Nordling: Arkiv Fys. *15*, 397 (1959)
4.16 W. Bothe: Z. Phys. *26*, 59 (1924)
4.17 E.C. Watson: Phys. Rev. *30*, 479 (1927); E.C. Watson, J.A. van den Akker: Proc. Roy. Soc. London A*126*, 138 (1929)
4.18 L. Meitner: Z. Phys. *9*, 131 (1922)
4.19 H. Robinson: Proc. Roy. Soc. *104*, 455 (1923)
4.20 P. Auger: C.R. Acad. Sci. *178*, 929 (1924); *180*, 65 (1925); *182*, 773, 1215 (1926); J. Phys. Rad. *6*, 205 (1925); Ann. Phys. *6*, 183 (1926)
4.21 E.H.S. Burhop, W.N. Asaad: In *Advances in Atomic and Molecular Physics*, Vol.8, ed. by D.R. Bates (Academic Press, New York 1972)
4.22 W. Bambynek, B. Crasemann, R.W. Fink, H.U. Freund, H. Mark, C.D. Swift, R.E. Price, P.V. Rao: Rev. Mod. Phys. *44*, 716 (1972)
4.23 E.H.S. Burhop: *The Auger Effect* (University Press, Cambridge 1952)
4.24 B.K. Agarwal: *Quantum Mechanics and Field Theory* (Asia Publishing House, New York, London, Bombay 1976)
4.25 H.A. Kramers: Z. Phys. *39*, 828 (1926)
4.26 D. Coster, R.L. Kronig: Physica *2*, 13 (1935)
4.27 H.S.W. Massey, E.H.S. Burhop: Proc. Camb. Phil. Soc. *32*, 461 (1936)
4.28 J. Cooper: Phys. Rev. *65*, 155 (1944)
4.29 Y. Cauchois: J. Phys. Radium *5*, 1 (1944)
4.30 L. Asplund: Phys. Scr. *16*, 268 (1977)
4.31 F.K. Richtmyer, S.W. Barnes, E.G. Ramberg: Phys. Rev. *46*, 843 (1934)
4.32 J. Cooper: Phys. Rev. *61*, 234 (1942)
4.33 A. Bril: Physica *13*, 481 (1947)
4.34 E.G. Ramberg, F.K. Richtmyer: Phys. Rev. *51*, 913 (1937)
4.35 L. Pincherle: Mem. Accad. Lincei *20*, 29 (1934); Physica *2*, 596 (1935)
4.36 D. Coster, A. Bril: Physica *9*, 84 (1942)
4.37 E.J. McGuire: Phys. Rev. A*3*, 1801 (1971)
4.38 C.D. Ellis: Proc. Roy. Soc. London A*139*, 336 (1933)
4.39 L.M. Steffen, O. Huber, F. Humbel: Helv. Phys. Acta *22*, 167 (1949)
4.40 G.L. Locher: Phys. Rev. *40*, 484 (1932)
4.41 L.H. Martin, J.C. Bower, T.H. Laby: Proc. Roy. Soc. A*148*, 40 (1935)
4.42 L.H. Martin, F.H. Eggleston: Proc. Roy. Soc. A*158*, 46 (1937)
4.43 J.C. Bower: Proc. Roy. Soc. A*157*, 662 (1936)
4.44 J.J. Lander: Phys. Rev. *91*, 1382 (1953)
4.45 J.F. McGilp, P. Weightman, E.J. McGuire: J. Phys. C*10*, 3445 (1977)
4.46 M. Siegbahn, W. Stenström: Physikal. Z. *17*, 318 (1916)
4.47 G. Wentzel: Ann. Phys. *66*, 437 (1921); Z. Phys. *31*, 445 (1925)
4.48 M.J. Druyvesteyn: Z. Phys. *43*, 707 (1928)
4.49 L.G. Parratt: Phys. Rev. *50*, 1 (1936)
4.50 R.M. Langer: Phys. Rev. *37*, 457 (1931)
4.51 H.C. Wolfe: Phys. Rev. *43*, 221 (1933)
4.52 E.H. Kennard, E. Ramberg: Phys. Rev. *46*, 1040 (1934)
4.53 D.J. Candlin: Proc. Phys. Soc. *68*A, 322 (1955)
4.54 B. Nordfors: Ark. Fys. *10*, 279 (1956)
4.55 Z. Horák: Proc. Phys. Soc. *77*, 980 (1961)
4.56 H. Hartman, L. Papula, W. Strehl: Theor. Chim. Acta Berlin *20*, 243 (1971)
4.57 V.F. Demekhin, V.P. Sachenko: Bull. Acad. Sci. USSR, Phys. Ser. *31* (6), 913 (1967)

4.58 R.E. LaVilla: Phys. Rev. A*4*, 476 (1971)
4.59 C. Bonnelle, C. Senemaud: C.R. Acad. Sci. Paris *268*, 65 (1969)
4.60 T. Åberg, G. Graeffe, J. Utrianen, M. Linkoaho: Phys. C*3*, 1112 (1970)
4.61 M. Sawada, K. Tsutsumi, T. Shiraiwa, T. Ishimura, M. Obashi: Ann. Rep. Sci. Works, Osaka Univ. *7*, 1 (1959)
4.62 R.D. Deslattes: Phys. Rev. *133*, 4, 399 (1964)
4.63 T.A. Carlson, M.O. Krause: Phys. Rev. *137*, A1655 (1965); *140*, A1057 (1965)
4.64 V.P. Sachenko, V.F. Demekhin: i Teor. Fiz. *49*, 765 (1965) [Sov. Phys.-JETP *22*, 532 (1966)]
4.65 T. Åberg: Phys. Rev. *156*, 35 (1967)
4.66 A.E. Sandström: *Handbuch der Physik*, Vol.XXX, ed. by S. Flügge (Springer Berlin, Göttingen, Heidelberg 1957) p.78
4.67 F.K. Richtmyer: Philos. Mag. *6*, 64 (1928); J. Franklin Inst. *208*, 325 (1929)
4.68 F. Bloch: Phys. Rev. *48*, 187 (1935)
4.69 S. Idei: Sci. Rep. Tohoku Symp. Univ. I, A*13*, 383 (1930)
4.70 H. Hulubei: C.R. Acad. Sci. *201*, 1356 (1935); *205*, 440 (1937)
4.71 F.R. Hirsh: Phys. Rev. *48*, 722 (1935); *50*, 191 (1936)
4.72 L.G. Parratt: Phys. Rev. *50*, 598 (1936); *54*, 99 (1938)
4.73 F.K. Richtmyer, E.G. Ramberg: Phys. Rev. *51*, 925 (1937)
4.74 L. Pincherle: Phys. Rev. *61*, 225 (1942)
4.75 Y. Cauchois: J. Phys. Radium *1*, 1 (1944)
4.76 M.O. Krause, F. Wuilleumier, C.W. Nestor: Phys. Rev. A*6*, 871 (1972)
4.77 H. Beuthe: Z. Phys. *60*, 603 (1930)
4.78 O.R. Ford: Phys. Rev. *41*, 577 (1932)
4.79 J. Valasek: Phys. Rev. *53*, 274 (1938)
4.80 M.A. Blokhin: JETP *9*, 1515 (1939)
4.81 M. Sawada, K. Tsutsumi, T. Shiraiwa, M. Obashi: J. Phys. Soc. Jpn. *10*, 647 (1955)
4.82 H. Hulubei: C.R. Acad. Sci. *224* 770 (1947)
4.83 H. Hulubei, Y. Cauchois, J. Manescu: C.R. Acad. Sci. *226*, 764 (1948)
4.84 T. Åberg, J. Utriainen: Phys. Rev. Lett. *22*, 1346 (1969); J. Physique *32*, 295 (1971)
4.85 J. Utriainen, T. Åberg: J. Phys. C*4*, 1105 (1971)
4.86 F. Bloch, P.A. Ross: Phys. Rev. *47*, 884 (1935)
4.87 T. Åberg: Phys. Rev. A*4*, 1735 (1971)
4.88 G.A. Rooke: Phys. Lett. *3*, 234 (1963)
4.89 D.S. Urch: J. Phys. C*3*, 1275 (1970)
4.90 J.P. Briand, P. Chevallier, M. Tavernier, J.P. Rozet: Phys. Rev. Lett. *27*, 777 (1971)
4.91 T. Sagawa: J. Physique *32*, 186 (1971)
4.92 P. Richard, W. Hodge, C.F. Moore: Phys. Rev. Lett. *29*, 393 (1972)
4.93 W.F. Hanson, E.T. Arakawa: Z. Phys. *251*, 271 (1972)
4.94 F.R. Hirsh: Rev. Mod. Phys. *14*, 45 (1942)
4.95 G.B. Deodhar: Proc. Nat. Acad. Sci. India A*32*, 320 (1962)
4.96 S.J. Edwards: Contemp. Phys. *11*, 195 (1970)
4.97 T. Åberg: *Proc. of the International Symposium on X-ray Spectra and Electronic Structure of Matter*, Vol.I, ed. by A. Faessler, G. Wiech (München 1973) p.1
4.98 E.H.S. Burhop: Proc. Roy. Soc. London A*198*, 272 (1935)
4.99 L. Pincherle: Nuovo Cimento *12*, 81, 122 (1935)
4.100 H.S.W. Massey, E.H.S. Burhop: Proc. Roy. Soc. London A*153*, 661 (1936)
4.101 H.J. Leisi, J.H. Brunner, C.F. Perdrisat, P. Scherrer: Helv. Phys. Acta *34*, 161 (1961)
4.102 B.G. Gokhale: Ann. Phys. (Paris) *7*, 852 (1952)
4.103 W. Laskar: J. Phys. Radium *16*, 644 (1955)
4.104 V.O. Kostroun, M.H. Chen, B. Crasemann: Phys. Rev. A*3*, 533 (1971)
4.105 J.H. Scofield: Phys. Rev. *179*, 9 (1969)

4.106 P.A. Ross: Phys. Rev. *28*, 425 (1926)
4.107 A.H. Compton: Philos. Mag. *8*, 961 (1929)
4.108 R.J. Stephenson: Phys. Rev. *51*, 637 (1937)
4.109 W.J. Campbell, J.V. Gilfrich: Anal. Chem. *42*, 248R (1970)
4.110 C.D. Ellis, B.A. Skinner: Proc. Roy. Soc. A*105*, 185 (1924)
4.111 V.L. Fitch, J. Rainwater: Phys. Rev. *92*, 789 (1953)
4.112 D. Kessler, H.L. Anderson, M.S. Dixit, H.J. Evans, R.J. McKee, C.K. Hargrove, R.D. Barton, E.P. Hincks, J.D. McAndrew: Phys. Rev. Lett. *18*, 1179 (1967)
4.113 E.H.S. Burhop: Contemp. Phys. *11*, 335 (1970)
4.114 D.A. Shirley: In *Photoemission in Solids I*, Topics in Applied Physics, Vol.26, ed. by M. Cardona, L. Ley (Springer, Berlin, Heidelberg, New York 1978)
5.1 A.H. Compton, C.F. Hagenow: J. Opt. Soc. Am. Rev. Sci. Instrum., p.487 (1924)
5.2 R.W. James: *The Optical Principles of the Diffraction of X-Rays* (Bell 1967)
5.3 A.H. Compton: Phys. Rev. *21*, 207A (1923); *22*, 409 (1923)
5.4 C.T.R. Wilson: Proc. Roy. Soc. London *A104*, 1 (1923)
5.5 A.A. Bless: Phys. Rev. *30*, 871 (1927)
5.6 R. Hofstadter, J.A. McIntyre: Phys. Rev. *78*, 24 (1950)
5.7 O. Klein, Y. Nishina: Z. Phys. *52*, 853 (1929)
5.8 B.K. Agarwal: *Quantum Mechanics and Field Theory* (Asia Publishing House, Bombay, London, New York 1976)
5.9 R.D. Evans: *Handbuch der Physik*, Vol.XXXIV, ed. by S. Flügge (Springer, Berlin, Göttingen, Heidelberg 1958)
5.10 C.V. Raman: Indian J. Phys. *2*, 387 (1928)
5.11 A. Smekal: Naturwiss. *11*, 873 (1923)
5.12 J.W.M. Dumond: Rev. Mod. Phys. *5*, 11 (1933)
5.13 A. Sommerfeld: Phys. Rev. *50*, 38 (1936)
5.14 K. Das Gupta: Phys. Rev. Lett. *3*, 38 (1959); also see Phys. Rev. *128*, 2181 (1962); Phys. Rev. Lett. *13*, 338 (1964)
5.15 A. Faessler, P. Mühle: Phys. Rev. Lett. *17*, 4 (1966)
5.16 T. Suzuki: J. Phys. Soc. Jpn. *22*, 1139 (1967)
5.17 G. Priftis: Phys. Lett. *49A*, 281 (1974)
5.18 D. Pines: *Elementary Excitations in Solids* (Benjamin, New York 1963) pp.204-207
5.19 G. Priftis, A. Theodossiòn, K. Alexopoulos: Phys. Lett. *27A*, 577 (1968)
5.20 M. Suzuki: Proc. 24th Annual Meeting of Phys. Soc. Jpn. *4*, 120 (1969); A. Tanokura, N. Hirota, T. Suzuki: J. Phys. Soc. Jpn. *27*, 515 (1969)
5.21 G.G. Cohen, N.G. Alexandropoulos: Solid State Commun. *10*, 95 (1972)
6.1 F. Hullinger, E. Mooser: "The Bond Description of Semiconductors: Polycompounds", in *Progress in Solid State Chemistry*, ed. by H. Reiss (Pergamon, Oxford 1965)
6.2 J.C. Phillips: Rev. Mod. Phys. *42*, 317 (1970)
6.3 J.C. Phillips: *Bonds and Bands in Semiconductors* (Academic Press, New York 1973)
6.4 C.A. Coulson, L.R. Redei, D. Stocker: Proc. Roy. Soc. *270*, 357 (1962)
6.5 L. Pauling: *The Nature of the Chemical Bond* (Cornell Univ. Press, Ithaca, N.Y. 1960)
6.6 A.L. Allred, E.G. Rochow: J. Inorg. Nucl. Chem. *5*, 264 (1958)
6.7 J.A. Van Vechten: Phys. Rev. *182*, 891 (1969); *187*, 1007 (1969); *B7*, 1479 (1973)
6.8 J.C. Phillips: Phys. Rev. Lett. *20*, 550 (1968)
6.9 J.P. Suchet: J. Phys. Chem. Sol. *21*, 156 (1961)
6.10 R.T. Sanders: *Inorganic Chemistry* (van Nostrand Reinhold, New York 1967)
6.11 N.F. Mott, H. Jones: *Theory of the Properties of Metals and Alloys* (Dover, New York 1958)

6.12 D.R. Penn: Phys. Rev. *128*, 2093 (1962)
6.13 H.R. Phillips, M. Ehrenreich: Phys. Rev. *129*, 1550 (1963)
6.14 B.F. Levine: J. Chem. Phys. *59*, 1463 (1973); Phys. Rev. *B7*, 2591, 2600 (1973); *B10*, 1655 (1974)
6.15 B. Szigeti: Proc. Roy. Soc. London *A204*, 51 (1950)
6.16 E. Mooser, W.B. Pearson: Nature *190*, 406 (1961); *192*, 335 (1961)
6.17 K. Siegbahn, C. Nordling, A. Fahlman, R. Nordberg, K. Hamrin, J. Hedman, G. Johansson, T. Bergmark, S.-E. Karlsson, I. Lindgren, B. Lindberg: *ESCA Atomic Molecular and Solid State Structure Studied by Means of Electron Spectroscopy* (Almqvist and Wiksell, Stockholm 1967)
6.18 K. Siegbahn: Phil. Trans. Roy. Soc. *A268*, 33 (1970)
6.19 W.L. Jolly: J. Amer. Chem. Soc. *92*, 3360 (1970)
6.20 M.E. Schwartz, J.D. Switalski, R.E. Stronski: In *Electron Spectroscopy*, ed. by D.A. Shirley (North-Holland, Amsterdam 1972)
6.21 J.P. Suchet: *Chemical Physics of Semiconductors* (van Nostrand, London 1965)
6.22 Y.K. Syrkin, M.E. Dyatkina: *Structure of Molecules* (Butterworths, London 1950)
6.23 S.S. Batsanov, I.A. Ovsyannikova: In *Chemical Bonds in Semiconductors and Thermodynamics*, ed. by N.N. Sirota (Consultants Bureau, New York 1968) p.65
6.24 A.E. Lindh, O. Lundquist: Ark. Mat. Astro Och Fysik *18*, Nos.14, 34, 35 (1924)
6.25 B.B. Ray: Philos. Mag. *49*, 168 (1925)
6.26 E. Bäcklin: Z. Phys. *33*, 547 (1925); *38*, 215 (1926)
6.27 G.B. Deodhar: Proc. Roy. Soc. London *A131*, 647 (1931)
6.28 F.A. Gianturco, C.A. Coulson: Mol. Phys. *14*, 223 (1968)
6.29 H. Schrenk: Thesis, University of Munich (1969)
6.30 A. Fahlman, K. Hamrin, J. Hedman, R. Nordberg, C. Nordling, K. Siegbahn: Nature *210*, 4 (1966)
6.31 A. Faessler, M. Goehring: Naturwiss. *39*, 1969 (1952); Z. Phys. *142*, 558 (1952)
6.32 E. Gilberg, B. Kern: Z. Phys. *174*, 372 (1963)
6.33 A. Faessler: "X-Ray Emission Spectra and Chemical Bonds", in Proc. of the Xth Colloquium Spectroscopium Internationale (Univ. of Maryland 1963) p.307
6.34 R. Jenkins: *An Introduction to X-Ray Spectrometry* (Heyden, London 1976)
6.35 B.G. Gokhale, R.B. Chesler, F. Boehm: Phys. Rev. Lett. *18*, 957 (1967)
6.36 E.V. Petrovich et al.: Sov. Phys. JETP *34*, 935 (1972)
6.37 M.A. Coulthard: J. Phys. B*7*, 440 (1974)
6.38 A.S. Koster, H. Mendel: J. Phys. Chem. Sol. *31*, 2511, 2523 (1970)
6.39 S. Hagström, S.E. Karlsson: Ark. Fys. *26*, 451 (1964)
6.40 W. Fischer: J. Chem. Phys. *42*, 3814 (1965)
6.41 A. Fahlman, K. Hamrin, R. Nordberg, C. Nordling, K. Siegbahn: Phys. Rev. Lett. *14*, 127 (1965)
6.42 Landolt-Börnstein: *Zahlenwerte und Funktionen*, Vol.I (Springer, Berlin 1955) p.4
6.42a D.W. Fischer: Adv. X-ray Anal. *13*, 173 (1969)
6.43 J.H.O. Varley: Nature *178*, 939 (1956)
6.44 A.T. Shubaev: Bull. Acad. Sci. USSR, Phys. Ser. [English transl.] *24*, 434 (1960)
6.45 W.J. Veigele, D.E. Stevenson, E.M. Henry: J. Chem. Phys. *50*, 5404 (1969)
6.46 J.C. Slater: *Quantum Theory of Atomic Structure*, Vols.1,2 (McGraw-Hill, New York 1960)
6.47 V.I. Nefedov: J. Struct. Chem. USSR [English transl.] *7*, 518 (1966)
6.48 A.T. Shubaev: Bull. Acad. Sci. USSR, Phys. Ser. [English transl.] *25*, 998 (1961)
6.49 K. Alder, G. Baur, U. Raff: Helv. Phys. Acta *45*, 765 (1972)
6.50 I. Lindgren: *Proc. Intern. Symp. Röntgenspektren und Chemische Bindung* (Verlag B.G. Teubner, Leipzig 1965)

6.51 F. Herman, S. Skillman: *Atomic Structure Calculations* (Prentice-Hall, N.J. 1963)
6.52 C. Froese: Cam. J. Phys. *41*, 1895 (1963)
6.53 C.J. Vesely, D.W. Langer: Phys. Rev. *4B*, 451 (1971)
6.54 S.M. Karalnik: Bull. Acad. Sci. USSR, Phys. Ser. [English transl.] *20*, 739 (1956); *21*, 1432 (1957)
6.55 R.L. Barinskii, V.I. Nefedov: *Röntgenspektroskopische Bestimmung der Atomladungen in Molekülen* (Akademische Verlagsgesellschaft Geest and Portig K.-G., Leipzig 1969)
6.56 G. Leonhardt, A. Meisel: J. Chem. Phys. *52*, 6189 (1970)
6.57 E. Clementi: IBM J. Res. Dev. Suppl. *9*, 2 (1965)
6.58 A.T. Shubaev: Bull. Acad. Sci. USSR, Phys. Ser. [English transl.] *27*, 6667 (1963)
6.59 O.I. Sumbaev: Soviet Phys. JETP [English transl.] *30*, 927 (1970)
6.60 J. Tilgner, I. Topol, G. Leonhardt, A. Meisel: J. Phys. Chem. Sol. *30*, 27 (1975)
6.61 K. Siegbahn, C. Nordling, G. Johansson, J. Hedman, P.F. Hedên, K. Hamrin, U. Gelius, T. Bergmark, L.O. Werme, R. Manne, Y. Bear: *ESCA Applied to Free Molecules* (North-Holland, Amsterdam 1969)
6.62 A. Goldman, J. Tejeda, N. Shevchik, M. Cardona: Phys. Rev. B*10*, 4388 (1974)
6.63 J.A. Bearden, H. Friedman: Phys. Rev. *58*, 387 (1940)
6.64 W.W. Beeman, H. Friedman: Phys. Rev. *56*, 392 (1939)
6.65 J. Farineau: Ann. Phys. *10*, 20 (1938)
6.66 H.W.B. Skinner: Phil. Trans. Roy. Soc. A*239*, 95 (1940)
6.67 R.H. Kingston: Phys. Rev. *84*, 944 (1951)
6.68 H.W.B. Skinner, T.G. Bullen, J.E. Johnston: Philos. Mag. *45*, 1070 (1954)
6.69 L.G. Parratt: Phys. Rev. *50*, 1 (1939)
6.70 L.G. Parratt: Rev. Mod. Phys. *31*, 616 (1959)
6.71 C. Mande: Thesis, Paris (1960)
6.72 C.G. Dodd, G.L. Glen: J. Appl. Phys. *39*, 5377 (1968)
6.73 C.H.M. O'Brien, H.W.B. Skinner: Proc. Roy. Soc. London *A176*, 229 (1940)
6.74 D.S. Urch: J. Phys. C*3*, 1275 (1970)
6.75 C.J. Ballhausen, H.B. Gray: *Molecular Orbital Theory* (Benjamin, New York 1964)
6.76 A. Sandström: Z. Phys. *65*, 632 (1930)
6.77 C. Kurylenko: Thesis, Paris (1939)
6.78 T.V. Krishman, A.N. Nigam: Proc. Indian Acad. Sci. *65*, 45 (1967)
6.79 L.G. Parratt, C.F. Hempstead, E.L. Jossem: Phys. Rev. *105*, 1228 (1957)
6.80 B. Nordfors: Ark. Fys. *18*, 37 (1960)
6.81 T. Magnusson: Thesis, Uppsala (1938)
6.82 B.K. Agarwal: Z. Phys. *142*, 161 (1955)
6.83 A.E. Sandström: In *Handbuch der Physik*, ed. by S. Flügge, Vol.XXX (Springer, Berlin, Göttingen, Heidelberg 1957)
6.84 W.F. Peed, L.E. Burkhart, R.A. Staniforth, L.G. Fauble: Phys. Rev. *105*, 588 (1957)
6.85 G.D. Matthews: Master's Essay, Johns Hopkins Univ. (1964)
6.86 O. Beckman, B. Axelsson, P. Bergvall: Ark. Fys. *15*, 567 (1959)
6.86a W. Kossel: Z. Phys. *1*, 119 (1920); *2*, 470 (1920)
6.87 L.G. Parratt: Phys. Rev. *56*, 295 (1939)
6.88 K. Schnopper: Phys. Rev. *131*, 2558 (1963)
6.89 T. Watanabe: Phys. Rev. *A139*, 1747 (1965)
6.90 P.S. Bagus: Phys. Rev. *A139*, 619 (1965)
6.91 G.R. Mitchell: Dev. Appl. Spectrosc. *4*, 109 (1965)
6.92 E.E. Vainshtein, K.I. Narbutt: Izvest. Akad. Nauk. SSSR, OkhN *1*, 71 (1945)
6.93 E.G. Nadzhakov, R.L. Barinskii: Soviet Phys. Doklady [English transl.] *4*, 1319 (1960)
6.94 Y. Sugiura: J. Phys. Rad. *8*, 113 (1927)

6.95 R.L. Barinskii, E.G. Nadzhakov: Bull. Acad. Sci. USSR, Phys. Ser. [English transl.] *24*, 419 (1960)
6.96 R.L. Barinskii: Bull. Acad. Sci. USSR, Phys. Ser. [English transl.] *25*, 958 (1961)
6.97 B.K. Agarwal, B.R.K. Agarwal: X-Ray Spectrom. *7*, 12 (1978); J. Phys. C *20*, 4223 (1978)
6.98 Y. Cauchois, N.F. Mott: Philos. Mag. *40*, 1260 (1949)
6.99 J.A. Bearden, T.M. Snyder: Phys. Rev. *59*, 162 (1941)
6.100 D. Coster, A. Bril: Physica *10*, 391 (1943)
6.101 D. Coster, H. DeLang: Physica *15*, 351 (1949)
6.102 Y. Cauchois, C. Bonnelle: C.R. Acad. Sci. Paris *245*, 1230 (1957)
6.103 C. Bonnelle: C.R. Acad. Sci. Paris *254*, 2313 (1962); Ann. Phys. Paris *1*, 439 (1966)
6.104 M.A. Blokhin, V.F. Demekhin, I.G. Shveitser: Bull. Acad. Sci. USSR *28*, 742 (1964)
6.105 M.F. Sorokina, C. Memnonov: Izv. Akad. Nauk. SSSR *31*, 1023 (1968)
6.106 S. Kawata, K. Maeda: J. Phys. Soc. Jpn. *32*, 778 (1972)
6.107 S. Kawata: J. Phys. F. *5*, 324 (1975)
6.108 B.K. Agarwal, L.P. Verma: J. Phys. C*1*, 208 (1968)
6.109 L.P. Verma, B.K. Agarwal: J. Phys. C*1*, 1658 (1968)
6.110 G.L. Glen, C.G. Dodd: J. Appl. Phys. *39*, 5372 (1968)
6.111 W. Seka, H.P. Hanson: J. Chem. Phys. *50*, 344 (1969)
6.112 A.K. Dey, B.K. Agarwal: J. Chem. Phys. *59*, 1397 (1973)
6.113 P. Sakellaridis: C.R. Acad. Sci. Paris *236*, 1014 (1953)
6.114 B.R.K. Agarwal, L.P. Verma, B.K. Agarwal: Nuovo Cimento Lett. *1*, 581, 781 (1971)
6.115 P. Deshmukh, C. Mande: Pramana *2*, 138 (1974)
6.116 B.D. Padalia, S.N. Gupta, V.P. Vijayavargiya, B.C. Tripathi: J. Phys. F*4*, 938 (1974)
6.117 P. Sakellaridis: Personal communication (1971)
6.118 M. Sawada, K. Tsutsumi, T. Shiraiwa, T. Ishimura, M. Obashi: Rep. Sci. Wks. Osaka Univ. *7*, 1 (1959)
6.119 N.F. Mott, K.W.H. Stevens: Philos. Mag. *2*, 1364 (1957)
6.120 W.W. Beeman, J.A. Bearden: Phys. Rev. *61*, 455 (1942)
6.121 E.E. Vainshtein: DAN *69*, 771 (1949)
6.122 V.G. Bhide, N.V. Bhat: J. Chem. Phys. *48*, 3103 (1968)
6.123 V.B. Singh, B.K. Agarwal: J. Phys. C*7*, 831 (1974)
6.124 B.K. Agarwal, R.K. Johri: J. Phys. F*7*, 1607 (1977)
6.125 H.C. Yeh, L.V. Azaroff: J. Appl. Phys. *38*, 4034 (1967)
6.126 J.O. Dimmock: Solid State Phys. *26*, 103 (1971)
6.127 Y.P. Irkhin: Fiz. Metal. Metalloved *11*, 10 (1961) [English transl.] Phys. Metals Metallog. USSR *11*, 9 (1961)
6.128 G.A. Burdick: Phys. Rev. *129*, 138 (1963)
6.129 L.V. Azaroff: J. Appl. Phys. *38*, 2809 (1967)
6.130 L.V. Azaroff, D.M. Pease: In *X-Ray Spectroscopy*, ed. by L.V. Azaroff (McGraw-Hill, New York 1974) p.284
6.131 V.G. Bhide, N.V. Bhat: J. Chem. Phys. *50*, 42 (1969)
6.132 B.K. Agarwal, V.B. Singh: Nuovo Cimento Lett. *4*, 765 (1972)
6.133 V.G. Bhide, S.K. Kaicker: J. Phys. Chem. Sol. *35*, 695 (1974)
6.134 V.O. Kostroun, R.W. Fairchild, C.A. Kukkoner, J.W. Wilkins: Phys. Rev. *B13*, 3268 (1976)
6.135 A. Kotani, Y. Toyozawa: J. Phys. Soc. Jpn. *35*, 1073, 1082 (1973)
6.136 J. Bergengren: Z. Phys. *3*, 247 (1920)
6.137 A.E. Lindh: Z. Phys. *6*, 303 (1921)
6.138 Y. Cauchois: *Les Spectres de Rayons X et la Structure Electronique de la Matière* (Gauthier-Villars, Paris 1948)
6.139 A. Meisel: Phys. Status Solidi *10*, 365 (1965)
6.140 D.J. Nagel, W.L. Baun: In *X-Ray Spectroscopy*, ed. by A. Azaroff (McGraw-Hill, New York 1974)

6.141 B.K. Agarwal, C.B. Bhargava, A.N. Vishnoi, V.P. Seth: J. Phys. Chem. Sol. *37*, 725 (1976)
6.142 G. Wiech: Z. Phys. *207*, 428 (1967; *216*, 472 (1968)
6.143 B.K. Agarwal, L.P. Verma: J. Phys. C*3*, 535 (1970)
6.144 V. Kunzl: Coll. Trav. Chim. Tchecoslovaquie *4*, 213 (1932)
6.145 G. Boehm, A. Faessler, G. Ruttmayer: Naturforsch. *9b*, 509 (1954)
6.146 C. Mande, A.R. Chetal: *Röntgenspektren und Chemische Bindung - Vorträge des Internationalen Symposiums* (Institut für Physikalische Chemie der Karl-Marx Univ., Leipzig 1966)
6.147 A. Miller: J. Phys. Chem. Sol. *29*, 633 (1968)
6.148 V.B. Sapre, C. Mande: J. Phys. *C5*, 793 (1972)
6.149 J. Prasad, V. Krishna, H.L. Nigam: J. Phys. Chem. Sol. *38*, 1149 (1977)
6.150 I.A. Ovsyannikova, S.S. Batsanov, L.I. Nanonova, L.R. Batsanova, E.A. Nekrasova: Bull. Acad. Sci. USSR, Phys. Ser. *31*, 936 (1967)
6.151 A.K. Dey, B.K. Agarwal: Nuovo Cimento Lett. *1*, 803 (1971)
6.152 A.N. Vishnoi, B.K. Agarwal: Phys. Lett. *29A*, 105 (1969)
6.153 S.V. Adhyapak, S.M. Kanetkar, A.S. Nigavekar: Nuovo Cimento *35B*, 179 (1974)
6.154 D.K. Kulkarni, C. Mande: Acta Cryst. *B27*, 1044 (1971)
6.155 V.K. Kondawar, C. Mande: X-ray Spectrom. *5*, 2 (1976)
6.156 R.K. Johri, B.K. Agarwal: J. Phys. F*8*, 555 (1978)
6.157 R.M. Levy, J.R. Van Wazer: J. Am. Chem. Soc. Abstract Papers No.150, 26B (1965)
6.158 L.V. Azaroff: Rev. Mod. Phys. *35*, 1012 (1963)
6.159 R. de L. Kronig: Z. Phys. *75*, 468 (1932)
6.160 R.L. Kronig: Z. Phys. *70*, 317 (1931); *75*, 191 (1932)
6.160a B.K. Agarwal: *Quantum Mechanics and Field Theory* (Asia Publishing House, Bombay, London, New York 1976)
6.161 H. Petersen: Z. Phys. *80*, 258 (1933)
6.162 D.R. Hartree, R.L. Kronig, H. Petersen: Physica *1*, 895 (1934)
6.163 E.M. Corson: Phys. Rev. *70*, 645 (1946)
6.164 J.N. Singh: Physica *28*, 131 (1962)
6.165 W.M. Weber: Phys. Rev. *B11*, 2744 (1975)
6.166 T. Hayasi: Sci. Rept. Tohoku Univ. First Ser. *33*, 123 (1949)
6.167 A.I. Kostarev: Zh. Eksp. Teor. Fiz. *11*, 60 (1941); *19*, 413 (1949); *21*, 917 (1951)
6.168 T. Shiraiwa, T. Ishimura, M. Sawada: Phys. Soc. Jpn. *12*, 788 (1957)
6.169 J.N. Singh: Phys. Rev *123*, 1724 (1961)
6.170 W.F. Nelson, I. Siegel, R.W. Wagner: Phys. Rev. *127*, 2025 (1962)
6.171 A.N. Vishnoi, B.K. Agarwal: Proc. Phys. Soc. London *89*, 799 (1966)
6.172 A.I. Kozlenkov: Bull. Acad. Sci. USSR, Phys. Ser. [English transl.] *25*, 968 (1961)
6.173 H. Petersen: Z. Phys. *98*, 569 (1936)
6.174 T. Shiraiwa, T. Ishimura, M. Sawada: Phys. Soc. Jpn. *13*, 847 (1958)
6.175 N.F. Mott, H.S.W. Massey: *Theory of Atomic Collisions* (Clarendon Press, Oxford 1933)
6.176 V.V. Schmidt: Bull. Acad. Sci. USSR, Phys. Ser. [English transl.] *25*, 988 (1961)
6.177 J.A. Jope: J. Phys. C*2*, 1817 (1969)
6.178 E.A. Stern: Phys. Rev. *B10*, 3027 (1974)
6.179 F.W. Lytle, D.E. Sayers, E.A. Stern: Phys. Rev. *B11*, 4825 (1975)
6.180 E.A. Stern, D.E. Sayers, F.W. Lytle: Phys. Rev. *B11*, 4836 (1975)
6.181 P.A. Lee, J.B. Pendry: Phys. Rev. *B11*, 2795 (1975)
6.182 R.M. Levy: J. Chem. Phys. *43*, 1846 (1965)
6.183 V.B. Singh, B.K. Agarwal: J. Phys. Chem. Sol. *35*, 465 (1974)
6.184 F.W. Lytle: Adv. X-ray Anal. *9*, 398 (1966)
6.185 J.C. Slater: In *Handbuch der Physik*, Vol.XIX, ed. by S. Flügge (Springer, Berlin, Göttingen, Heidelberg 1956)
6.186 B.K. Agarwal, R.K. Johri: J. Phys. C*10*, 3213 (1977)

6.187 P. Chivate, P.S. Damle, N.V. Joshi, C. Mande: J. Phys. *C1*, 1171 (1968)
6.188 L.I. Schiff: *Quantum Mechanics* (McGraw-Hill, New York 1955) p.79
6.189 P. Deshmukh, P.C. Deshmukh, C. Mande: Pramana *6*, 305 (1976)
6.190 J.W.M. DuMond, V.L. Bollman: Phys. Rev. *51*, 400 (1937)
6.191 P. Ohlin: Thesis, Uppsala (1941); Ark. Fys. *4*, 387 (1952)
6.192 A. Nilsson: Ark. Fys. *6*, 513 (1953)
6.193 P. Johansson: Ark. Fys. *18*, 329 (1960)
6.194 J.J. Spijkerman, J.A. Bearden: Phys. Rev. *A134*, 871 (1964)
6.195 B.R.A. Nijboer: Physica *12*, 461 (1946)
6.196 K. Ulmer: Proc. Intern. Symp. (Kiev 1968) *2*, 79 (1969)
6.197 G. Böhm, K. Ulmer: J. Phys. Paris *Suppl.10*, C4-241 (1971)
6.198 S. Bergwoll, R.K. Tyagi: Ark. Fys. *29*, 439 (1965)
6.199 W. Gordy, W.J.O. Thomas: J. Chem. Phys. *24*, 439 (1956)
6.200 L. Fonda, R.G. Newton: Ann. Phys. (N.Y.) *9*, 416 (1960)
7.1 H.W.B. Skinner: Phil. Trans. Roy. Soc London A*239*, 95 (1940)
7.2 D.H. Tomboulian: In *Handbuch der Physik*, Vol.XXX, ed. by S. Flügge (Springer, Berlin, Göttingen, Heidelberg 1957)
7.3 J.A.R. Samson: *Techniques of Vacuum Ultraviolet Spectroscopy* (John Wiley New York 1967)
7.4 D.J. Fabian: Crit. Rev. Solid State Sci. USA *2*, 255 (1971)
7.5 D.J. Fabian, L.M. Watson, C.A.W. Marshall: Rep. Prog. Phys. *34*, 601 (1971)
7.6 F.C. Brown: Solid State Phys. *29*, 1 (1974)
7.7 V. Schumann: Akad. Wiss. Wien *102*, (2A), 625 (1893)
7.8 M. Plato: Z. Naturforsch. *19a*, 1324 (1964)
7.9 W.R.S. Garton, M.S.W. Webb, P.C. Wildy: J. Sci. Instr. *34*, 496 (1957)
7.10 P.G. Wilkinson: J. Opt. Soc. Am. *45*, 1044 (1955)
7.11 N.N. Axelrod: J. Opt. Soc. Am. *53*, 297 (1963)
7.12 F.C. Fehsenfeld, K.M. Evenson, H.P. Broida: Rev. Sci. Instr. *36*, 294 (1965)
7.13 D.H. Tomboulian, P.L. Hartman: Phys. Rev. *102*, 1423 (1956)
7.14 R.P. Godwin: *Springer Tracts in Modern Physics*, Vol.51 (Springer, Berlin Heidelberg, New York 1969)
7.15 K. Feser, J. Müller, G. Wiech, A. Faessler: J. Physique *32*, C4-333 (1971
7.16 W. Hayes: Contemp. Phys. *13*, 441 (1972)
7.17 K. Codling: Rep. Prog. Phys. *36*, 541 (1973)
7.18 J.D. Jackson: *Classical Electrodynamics* (John Wiley, New York, London 1962)
7.19 G.C. Sprague, D.H. Tomboulian, D.E. Bedo: J. Opt. Soc. Am. *45*, 756 (1955
7.20 A.L. Morse, G.L. Weissler: Sci. Light *15*, 22 (1966)
7.21 G.R. Harrison, R.C. Lord, J.R. Loofbourow: *Practical Spectroscopy* (Prentice Hall, N.J. 1948)
7.22 D.L. MacAdam: J. Opt. Soc. Am. *23*, 178 (1933)
7.23 R.A. Sawyer: *Experimental Spectroscopy* (Dover, New York 1963)
7.24 G.A. Sawyer, A.J. Bearden, I. Henins, F.C. Jahoda, F.L. Ribe: Phys. Rev. *131*, 1891 (1963)
7.25 E.R. Piore, G.G. Harvey, E.M. Gyorgy, R.H. Kingston: Rev. Sci. Instr. *23*, 8 (1952)
7.26 J.L. Rogers, F.C. Chalkin: Proc. Phys. Soc. London B *67*, 348 (1954)
7.27 D.L. Ederer, D.H. Tomboulian: Appl. Opt. *3*, 1073 (1964)
7.28 A.J. Tuzzolino: Rev. Sci. Instr. *35*, 1332 (1964); Phys. Rev. *A134*, 205 (1964)
7.29 B.K. Agarwal: *Quantum Mechanics and Field Theory* (Asia Publishing House, Bombay, London, New York 1976)
7.30 D.E. Bedo, D.H. Tomboulian: Phys. Rev. *109*, 35 (1958)
7.31 J.A. Catterall, J.A. Trotter: Philos. Mag. *4*, 1164 (1959)
7.32 R.S. Crisp, S.E. Williams: Philos. Mag. *5*, 525 (1960)
7.33 F.S. Ham: Phys. Rev. *128*, 82, 2524 (1962)
7.34 D.A. Goodings: Proc. Phys. Soc. London *86*, 75 (1965)

7.35 G.A. Ausman, A.J. Glick: Phys. Rev. *183*, 687 (1969)
7.36 A.G. Mathewson, H.P. Myers: Philos. Mag. *25*, 853 (1972)
7.37 G.D. Mahan: Solid State Phys. *29*, 75 (1974)
7.38 N.V. Smith: Phys. Rev. *183*, 634 (1969)
7.39 G.D. Mahan: Phys. Rev. *163*, 612 (1967)
7.40 Y. Mizuno, K. Ishikawa: J. Phys. Soc. Jpn. *25*, 627 (1968)
7.41 P. Nozieres, C.T. DeDominicis: Phys. Rev. *178*, 1097 (1969)
7.42 B. Roulet, J. Gavoret, P. Nozieres: Phys. Rev. *178*, 1072 (1969)
7.43 W.L. Baun, D.W. Fischer: Adv. X-ray Anal. *8*, 371 (1965)
7.44 H. Neddermeyer: Z. Phys. *271*, 329 (1974)
7.45 K. Laüger: In *Soft X-ray Band Spectra and Electronic Structure of Metals and Materials*, Part I, ed. by D.J. Fabian (Academic Press, New York 1969)
7.46 A. Appleton: Contemp. Phys. *6*, 50 (1964)
7.47 G.A. Rooke: In *X-ray Spectroscopy*, ed. by L.V. Azaroff (McGraw-Hill, New York 1974)
7.48 J.J. Lander: Phys. Rev. *91*, 1382 (1953)
7.49 Y. Cauchois: Philos. Mag. *44*, 173 (1953)
7.50 H.W.B. Skinner, T.G. Bullen, J.E. Johnston: Philos. Mag. *45*, 1070 (1954)
7.51 E.M. Gyorgy, G.G. Harvey: Phys. Rev. *87*, 861 (1952); *93*, 365 (1954)
7.52 D.J. Fabian (ed.): *Soft X-ray Band Spectra* (Academic Press, London, New York 1968) and references herein
7.53 D.H. Tomboulian, D.E. Bedo: Rev. Sci. Instr. *26*, 747 (1955)
7.54 M.P. Givens, W.P. Siegmund: Phys. Rev. *85*, 313 (1952)
7.55 J.W. Cooper: Phys. Rev. *128*, 681 (1962)
7.56 J. Friedel: Philos. Mag. *43*, 153, 1115 (1952); also see Comments Solid State Phys. *2*, 21 (1969)
7.57 P.W. Anderson: Phys. Rev. Lett. *18*, 1049 (1967)
7.58 M. Combescot, P. Nozieres: J. Physique *32*, 913 (1971)
7.59 H.W.B. Skinner, J.E. Johnston: Proc. Roy. Soc. London *A161*, 420 (1937)
7.60 D.H. Tomboulian, D.E. Bedo, W.M. Neupert: J. Phys. Chem. Sol. *3*, 282 (1957)
7.61 B.K. Agarwal, M.P. Givens: Phys. Rev. *107*, 62 (1957); *108*, 658 (1957); J. Phys. Chem. Sol. *6*, 178 (1958)
7.62 B. Sonntag: DESY - F41/1 Rep. (Hamburg 1969); also see B. Sonntag, R. Haensel, C. Kunz: Solid State Commun. *7*, 597 (1969)
7.63 F.C. Brown, C. Gähwiller, A.B. Kunz: Solid State Commun. *9*, 487 (1971)
7.64 W. Gudat, C. Kunz: Phys. Status Solidi B *52*, 433 (1972)
7.65 N.F. Mott, K.W.H. Stevens: Philos. Mag. *2*, 1364 (1957)
7.66 B. Sonntag, F.C. Brown: Phys. Rev. *B10*, 2300 (1974)
7.67 A. Kotani, Y. Toyozawa: In *Synchrotron Radiation*, Topics in Current Physics, Vol.10, ed. by C. Kunz (Springer, Berlin, Heidelberg, New York 1979)
8.1 A. Hadding: Z. Phys. *3*, 369 (1920)
8.2 B.K. Agarwal: *Elements of Statistical Mechanics* (Pothishala, Allahabad 1970) p.177
8.3 W.L. Bragg: J. Sci. Instrum. *24*, 27 (1947)
8.4 J.A. Bearden: Phys. Rev. B*137*, 455 (1965)
8.5 E.P. Bertin: *Principles and Practice of X-ray Spectrometric Analysis* (Plenum, New York 1975)
8.6 R.W.G. Wyckoff: *The Structure of Crystals* (Chemical Catalog Co., New York 1931)
8.7 M. Siegbahn: *Spektroskopie der Röntgenstrahlung* (Springer, Berlin 1931)
8.8 B.K. Agarwal, A.N. Vishnoi: Naturwiss. *49*, 178 (1962)
8.9 S.K. Allison: Phys. Rev. *38*, 203 (1931)
8.10 Y. Cauchois: J. Phys. Rad. Paris *3*, 320 (1932); *4*, 61 (1933); Compt. Rend. *194*, 362, 1479 (1932)
8.11 G.L. Clark: *Applied X-rays* (McGraw-Hill, New York 1955) p.544

8.12 J.W.M. DuMond: Rev. Sci. Instr. *18*, 626 (1947)
8.13 H. Johann: Z. Phys. *69*, 185 (1931)
8.14 T. Johansson: Z. Phys. *82*, 507 (1933)
8.15 M. Schwarzschild: Phys. Rev. *32*, 162 (1928)
8.16 R.C. Spencer: Phys. Rev. *38*, 618 (1931)
8.17 A. von Laue: Z. Phys. *72*, 472 (1931)
8.18 L.P. Smith: Phys. Rev. *46*, 343 (1934)
8.19 A.H. Compton, S.K. Allison: *X-rays in Theory and Experiment* (van Nostrand, New York 1935)
8.20 J.S. Thomsen: In *X-ray Spectroscopy*, ed. by L.V. Azaroff (McGraw-Hill, New York 1974)
8.21 B. Davis, W. Stempel: Phys. Rev. *17*, 608 (1921); *19*, 504 (1922)
8.22 S.K. Allison: Phys. Rev. *41*, 1 (1932)
8.23 L.G. Parratt: Phys. Rev. *41*, 561 (1932)
8.24 A.H. Compton, R.L. Doan: Proc. Nat. Acad. Sci. *11*, 598 (1926)
8.25 J. Thibaud: J. Phys. Paris *8*, 13 (1927)
8.26 C. Eckart: Phys. Rev. *44*, 12 (1933)
8.27 J.R. Cuthill: In *X-ray Spectroscopy*, ed. by L.V. Azaroff (McGraw-Hill, New York 1974)
8.28 D.A. Gedcke: X-ray Spectrom. *1*, 129 (1972)
8.29 W. Rump: Z. Phys. *43*, 254 (1927)
8.30 M. Yoshimatsu, S. Kozaki: In *X-ray Optics*, Topics in Applied Physics, Vol.22, ed. by H.J. Queisser (Springer, Berlin, Heidelberg, New York 1977)
8.31 H.S. Uhler, C.D. Cooksey: Phys. Rev. *10*, 645 (1917)
8.32 H.S. Uhler: Phys. Rev. *11*, 1 (1918)
8.33 C.D. Cooksey, D. Cooksey: Phys. Rev. *36*, 85 (1930)
8.34 Z.G. Pinsker: *Dynamical Scattering of X-rays in Crystals* (Springer, Berlin, Heidelberg, New York 1978) p.302

Author Index

Subject Index